W0079850

Springer
Proceedings in Physics 2

Springer Proceedings in Physics is a new series dedicated to the publication of conference proceedings. Each volume is produced on the basis of camera-ready manuscripts prepared by conference contributors. In this way, publication can be achieved very soon after the conference and costs are kept low; the quality of visual presentation is, nevertheless, very high. We believe that such a series is preferable to the method of publishing conference proceedings in journals, where the typesetting requires time and considerable expense, and results in a longer publication period. Springer Proceedings in Physics can be considered as a journal in every other way: it should be cited in publications of research papers as *Springer Proc. Phys.*, followed by the respective volume number, page and year.

EXAFS and
Near Edge Structure III

Proceedings of an International Conference,
Stanford, CA, July 16–20, 1984

Editors: K. O. Hodgson, B. Hedman,
and J. E. Penner-Hahn

With 392 Figures

Springer-Verlag Berlin
Heidelberg GmbH 1984

Dr. Keith O. Hodgson
Dr. Britt Hedman
Dr. James E. Penner-Hahn
Department of Chemistry, Stanford University,
Stanford, CA 94305, USA

Originally published by

ISBN 978-3-642-46524-6   ISBN 978-3-642-46522-2 (eBook)
DOI 10.1007/978-3-642-46522-2

This work is subject to copyright. All rights are reserved, whether the whole or part of the material is concerned, specifically those of translation, reprinting, reuse of illustrations, broadcasting, reproduction by photocopying machine or similar means, and storage in data banks. Under § 54 of the German Copyright Law where copies are made for other than private use, a fee is payable to "Verwertungsgesellschaft Wort", Munich.

© Springer-Verlag Berlin Heidelberg 1984

Originally published by Springer-Verlag Berlin Heidelberg New York Tokyo in 1984

Softcover reprint of the hardcover 1st edition 1984

The use of registered names, trademarks, etc. in this publication does not imply, even in the absence of a specific statement, that such names are exempt from the relevant protective laws and regulations and therefore free for general use.

2153/3130-543210

Preface

This volume contains the Proceedings of the Third International EXAFS Conference, hosted by Stanford University and the Stanford Synchrotron Radiation Laboratory on July 16-20, 1984. The meeting, co-chaired by Professors Arthur Bienenstock and Keith Hodgson, was attended by over 200 scientists representing a wide range of scientific disciplines. The format of the meeting consisted of 51 invited presentations and four days of poster sessions. This Proceedings is a compilation of 139 contributions from both invited speakers and authors of contributed posters.

The last ten years has seen the rapid maturation of x-ray absorption spectroscopy as a scientific discipline. The vitality of the field is reflected in the diversity of applications found in the Proceedings. Recent work continues to probe the limits of x-ray spectroscopy, with proven techniques being extended to, for example, very low or high energy studies, to very dilute systems, and to studies of surface structure. In fact, the title of the conference does not at all reflect the breadth of the science discussed at this meeting. The number of fields in which x-ray absorption spectroscopy is finding applications has increased dramatically even in the two years since the previous International Conference held in Frascati*. The prospects for continued growth and innovation will be even further enhanced if a new generation 6 GeV storage ring is constructed in the next five years.

Many of the contributions to this Proceedings could have been classified under several headings. The order chosen is that of the Editors. A brief discussion of the different chapters and their contents follows. Chapter I deals with fundamental aspects of EXAFS and XANES spectroscopy. EXAFS theory papers are followed by theoretical and experimental work on XANES. Chapter II deals with EXAFS data analysis, including such topics as choice of E_0 and accuracy of bond length and coordination number determination. Chapter III is on biological systems and is organized by metal type. Several papers dealing with time-resolved EXAFS are at the end of Chapter III. Chapter IV contains contributions on catalytic systems and small metal clusters. Surface structure is discussed in Chapter V and includes both SEXAFS and x-ray scattering theory and applications. Amorphous systems are covered in Chapter

---

A. Bianconi, L. Incoccia, S. Stipcich (eds.): *EXAFS and Near Edge Structure*, Springer Ser. Chem. Phys., Vol. 27 (Springer, Berlin, Heidelberg 1983)

VI and include amorphous semiconductors followed by glassy materials (grouped as silicates, oxides, and metals). Chapter VII deals with geology, geochemistry and high pressure studies, and is organized with sections on, among others, nuclear waste storage, minerals, coal and steel, and high pressure structures. In Chapter VIII, a variety of other applications are found. These include conducting polymers, complexes in aqueous solutions, and solid solutions and binary alloys. The last chapter encompasses a variety of related techniques and instrumentation, including dispersive EXAFS, reflection EXAFS, x-ray optical luminescence, crystal glitches, and laboratory EXAFS spectrometers.

The Editors thank the attendees for their enthusiastic participation and for their prompt preparation of manuscripts for these Proceedings. Both the Conference and Proceedings would not have been possible without the dedicated help of Katherine Cantwell, Carol Mitchell, and Shirley Robinson of the SSRL Staff. Financial assistance was graciously provided by the National Science Foundation, the National Institutes of Health, the U.S. Department of Energy, and by the XEROX, General Electric, EXXON Research and Engineering, and Chevron Corporations. We look forward to the next International Conference to be held in France in the summer of 1986.

Stanford, California
September, 1984

Keith O. Hodgson
Britt Hedman
James E. Penner-Hahn

Contents

Part IV   Catalytic Systems and Small Metal Clusters

Part V   Surface Structure

Part VI Amorphous Materials and Glasses

Part VII   Geology and Geochemistry, and High Pressure

Part VIII   Other Applications

Part IX   Related Techniques and Instrumentation

Part I Fundamental Aspects of EXAFS and XANES

Single Scattering Theory of X-Ray Absorption

W.L. Schaich

Physics Department and Materials Research Institute, Indiana University
Bloomington, IN 47405, USA

This paper reviews the formulae and justifications for a single scattering
approach to the calculation of the final state in x-ray absorption. It is
argued that the results can be useful all the way down to an absorption edge.
Corrections due to multiple scattering are considered and it is shown how
they may be either suppressed by Lorentzian energy averages if they involve
long scattering paths or included via renormalization factors if they arise
from shadowing effects.

1. Introduction

EXAFS has become a powerful tool because of its relative ease of interpretation. This follows in turn from the simplicity of the formulae describing
it. One writes for the contribution to the x-ray absorption coefficient due
to the dipole excitation of a deep core level

$$\mu_c = 16 \frac{\pi^3}{3} \frac{e^2}{\hbar c} n_c \hbar \omega N_0 [\ell M^2_{\ell,\ell-1} \chi_{\ell-1} + (\ell+1) M^2_{\ell,\ell+1} \chi_{\ell+1}] \tag{1}$$

where we have averaged over the x-ray polarization and summed over the azimuthal quantum number of the core level. The quantity ℓ is the orbital
angular momentum of the core level, n_c is the density of absorbing atoms, ω
is the x-ray frequency, and $N_0 = mk/\pi^2\hbar^2$ is the free electron density of
states, with k determined by the final state kinetic energy $E = \hbar^2 k^2/2m$.
The M's are atomic radial integrals and the χ's incorporate the influence
of the neighboring atoms. In EXAFS one focusses on the deviation of the
χ's from unity using the now standard formula [1,2]

$$\chi_\ell = 1 + (-)^\ell \sum_j N_j \mathrm{Im}\{e^{2i\delta'_\ell} \frac{e^{2ikR_j}}{kR_j^2} f_j(\pi)\} \tag{2}$$

where Im denotes imaginary part and the sum on j is over shells of neighbors,
each containing N_j identical atoms at a distance R_j from the absorber. For
simplicity the effects of disorder of either thermal or other origin have
been omitted. The δ'_ℓ are the scattering phase shifts of the absorbing atom
(possibly different from those of the unexcited neighbors) and $f_j(\pi)$ is the
backscattering amplitude of a single neighboring atom in the jth shell. One
has

$$f(\pi) = \frac{1}{k} \sum_\ell (2\ell+1)(-)^\ell e^{i\delta_\ell} \sin\delta_\ell \tag{3}$$

All these scattering properties are to be evaluated at the energy E. The
physics behind the above equations is that within a single particle (muffin

2

tin potential) model the effect of the absorbing atom is fully included while that of the neighboring atoms enters only via an approximate single backscattering event.

It has long been believed that such a transparent description as the above could not work near threshold. The main argument for this assertion was the claim that many body effects and/or multiple scattering processes are dominant there. However, until recently there have been few studies that examined the validity of these claims.

This situation has been remedied in the last few years. The extensive calculations by MULLER and collaborators [3-5] of x-ray absorption in crystalline transition metals have shown that a single particle approach can reproduce experimental data quite well. Further use of these results was made to demonstrate that a single scattering approximation can (sometimes) quickly and accurately duplicate the elaborate calculations [6].

The formulae used in the single scattering approximation are (1) and a modified version of (2). This modification properly treats the spatial variation of the outgoing and (single) backscattered wavelets. Mathematically it corresponds to not replacing in various expansions $i^\ell h_\ell^+(\rho)$ by $e^{i\rho}/\rho$, where $h_\ell^+(\rho)$ is a spherical Bessel function [7] and $\rho = kR$. This improvement has been used before - see [8] and references in [6] - but one did not realize that it alone could give useful results for the complete spectrum. Since derivations of the formulae have been published [9,10], we only quote the results:

$$\chi_\ell = 1 + \sum_j N_j \text{Im}\{e^{2i\delta_\ell'} \sum_{\ell'} (2\ell'+1)e^{i\delta_{\ell'}} \sin\delta_{\ell'} H(\ell,\ell';R_j)\} \qquad (4)$$

where we omit a j index from the $\delta_{\ell'}$ and

$$H(\ell,\ell';R) = \sum_{\bar\ell} (2\bar\ell+1) \left[\begin{pmatrix}\ell\ell'\bar\ell\\000\end{pmatrix}h_{\bar\ell}^+(\rho)\right]^2 \qquad (5)$$

with the first factor in the square bracket being a 3j symbol [7]. This symbol enters as the result of several sums over azimuthal quantum numbers.

Far above the edge (when $\rho >> \ell^2$), (5) becomes $(-)^{\ell+\ell'}e^{2i\rho}/\rho^2$ and (4) becomes (2). We note further that an extension of (1) to include spin orbit coupling in the initial state and an analogous formula for high energy electron loss spectroscopy have been derived [9].

2. Justifications

In [6] a comparison of three separate calculations of the K-edge absorption spectrum of crystalline copper is shown. These three calculations are each done with the same atom locations, the same scattering potential, and the same broadening function. All use (1) but make different approximations to the required χ_1. The first is based on a band structure scheme [3-5], which represents within a single particle muffin tin model an exact solution; the second comes from (4,5); and the third is the conventional EXAFS expression (2). The first two agree well over the first 100 eV above the edge while the last two become the same beyond 200 eV above the edge. The latter observation is expected, as noted below (5), while the former is possibly a surprise. We found similar behavior for other edges and other transition metals [11].

Here we briefly discuss reasons for the often (but not universal) agreement between an exact multiple scattering calculation of χ_ℓ and the much more tractable single scattering expressions (4,5). The formal difference between the two theories lies in higher order scattering processes for the final state electron. If all scattering by this electron is weak, which does happen when its kinetic energy becomes large, then one can justifiably neglect the extra terms. This is the main reason the standard EXAFS (2) works so well. However, the extra scattering is not always weak, as for instance in the shadowing effect [8] and certainly near threshold. We show in the next section how (5) may be corrected for the shadowing effect and here concentrate on how multiple scattering effects may be suppressed even though the scattering may be strong.

The key lies in the need to energy average any single particle calculation in order to approximately account for various decay processes and for experimental resolution. It is easy to show [9] that a Lorentzian energy average of a spectrum is equivalent to forcing all the intersite propagators to decay with scattering path length. Both long multiple scattering paths and distant single scattering events should produce no structure that survives the energy average. With sufficient broadening then the remnant structure in a spectrum should be modulated from the atomic result only by the effect of a single backscattering from the nearest neighbors. Our calculations [11] indicate that one is close (within an order of magnitude) to the required broadening for this simple situation when the experimental value of broadening is used. We found that slight increases in the broadening above its experimental value removed the discrepancies (if any were present) between the band structure and single scattering results.

The above argument may be rephrased as follows. Unlike multiple scattering calculations, the single scattering approach using near neighbors retains only those processes whose structure will survive a severe energy broadening. In the absence of sufficient energy broadening it will generally produce a poor spectrum, but in many practical cases its omissions are washed out. By not computing structure that won't survive the broadening the single scattering approach gains much in speed and transparent interpretation over multiple scattering calculations. We urge those who perform multiple scattering calculations [12-15] to confirm this prediction of progressive agreement with single scattering results as the broadening is increased.

3. Multiple Scattering Corrections

In a recent letter [16] BUNKER and STERN have argued that except for very close to an absorption edge and/or for very short bond lengths that the shadowing effect of atoms in one shell by those in another is the major multiple scattering effect missing from the single scattering approach. Fortunately one may include this effect by a simple application of renormalized forward scattering theory [17] without changing the basic form of (4,5).

Begin with the general expression for χ_ℓ [9],

$$\chi_\ell = 1 + \text{Im}\{\frac{e^{2i\delta'_\ell}}{2\ell+1} \sum_m [\sum_{L'\underset{\sim}{R}} C_{L\underset{\sim}{0},L'\underset{\sim}{R}} \, t^+_{\ell'}(\underset{\sim}{R}) \, C_{L'\underset{\sim}{R},L\underset{\sim}{0}} + \ldots]\} \tag{6}$$

where $t^+_\ell = e^{i\delta_\ell}\sin\delta_\ell$ and the propagator

$$C_{L\underset{\sim}{R},L'\underset{\sim}{R}'} = 4\pi \sum_{\bar{\ell}} i^{\bar{\ell}} \, h^+_{\bar{\ell}}(kr) Y^*_{\bar{L}}(\hat{r}) \langle Y_L | Y_{\bar{L}} Y_{L'}\rangle \, (1-\delta_{\underset{\sim}{R},\underset{\sim}{R}'}) \tag{7}$$

with $\underline{r} = \underline{R}-\underline{R}'$. In (6,7) L denotes the pair of angular momentum quantum numbers ℓ and m. The Y_L are spherical harmonics and $\langle Y_L | Y_{\Gamma} Y_{L'} \rangle$ is the overlap integral between three of them. We have shown explicitly in (6) only the contribution that determines the single scattering approximation. Omitted are all the multiple scattering terms that take the final state electron from the combined angular momentum- (excited) atom location, LO, through all possible paths in angular momentum-atom site space that end at LO.

If the atoms of one shell of radius R_1 shadow those in another shell of radius R, the relevant multiple scattering terms are those in which the electron in either going to or returning from the outer shell atom may either forward scatter once or not at all from the inner shell atom. Rewriting the single scattering approximation of χ_ℓ as $1 + \sum_R \delta \chi_\ell (R)$, then the shadowing effects change $\delta \chi_\ell(R)$ to

$$\delta \chi_\ell^{(s)} (R,R_1) = \text{Im}\{\frac{e^{2i\delta_\ell'}}{2\ell+1} \sum_m [\sum_{L'} D_{L\underline{O},L'\underline{R}} \; t_\ell^+(\underline{R})D_{L'\underline{R},L\underline{O}}]\}$$ (8)

where the propagators in (6) have been replaced by

$$D_{L\underline{R},L'\underline{R}'} = C_{L\underline{R},L'\underline{R}'} + \sum_{\bar{L}} C_{L\underline{R},\bar{L}\underline{R}_1} \; t_{\bar{\ell}}^+(R_1)C_{\bar{L}\underline{R}_1,L'\underline{R}'}$$ (9)

with $\underline{R}$, $\underline{R}'$, and $\underline{R}_1$ colinear.

An approximate evaluation of (9) may be quickly made by using (far above threshold) the replacement $i^\ell h_\ell^+(\rho) = e^{i\rho}/\rho$. Then

$$C_{L\underline{R},L'\underline{R}'} \rightarrow 4\pi \frac{e^{ikr}}{kr} \; Y_L^*(\hat{r})Y_{L'}(\hat{r})$$ (10)

and for D between the excited atom and the outer shell

$$D_{L\underline{R},L'\underline{R}'} \rightarrow C_{L\underline{R},L'\underline{R}'}(1 + \frac{R}{R_1 R_2} f(0))$$ (11)

where $R_2 = R-R_1$ and $f(0)$ is the forward scattering amplitude of an inner shell atom. The shadowing modification described by (11) would be easy to include. Indeed with the approximate h_ℓ^+'s, one may readily write down relatively simple expressions for other multiple scattering paths [18,19].

Rather than repeat these here, we note that considerable simplification of (9) is possible even with the exact h_ℓ^+'s. We find that both $D_{L\underline{O},L'\underline{R}}$ and $D_{L'\underline{R},L\underline{O}}$ may be expanded as in (7) with only the change $h_{\bar{\ell}}^+(\rho) \rightarrow H_{\bar{\ell};\ell\ell'}(\tilde{\rho};\rho_1\rho_2)$ where $\tilde{\rho}_1 = kR_1$, etc. and

$$H_{\bar{\ell};\ell\ell'}(\rho;\rho_1\rho_2) = h_{\bar{\ell}}^+(\rho)$$

$$+ (-i)^{\bar{\ell}} \sum_{\ell_1\ell_2\ell_3} i^{\ell_1} h_{\ell_1}^+(\rho_1) i^{\ell_2} h_{\ell_2}^+(\rho_2) t_{\ell_3}^+ \; a(\ell_1\ell_2\ell_3;\ell\ell'\bar{\ell}) \quad \text{with} \quad (12)$$

$$a(\ell_1\ell_2\ell_3;\ell\ell'\bar{\ell}) = (-)^{\ell_1+\ell_2+\ell_3}(2\ell_1+1)(2\ell_2+1)(2\ell_3+1)\{ \begin{matrix} \ell & \ell'\bar{\ell} \\ \ell_2\ell_1\ell_3 \end{matrix} \}$$

$$\times \begin{pmatrix} \ell_1\ell_3\ell \\ 0\;0\;0 \end{pmatrix}\begin{pmatrix} \ell_3\ell_2\ell' \\ 0\;0\;0 \end{pmatrix}\begin{pmatrix} \ell_2\ell_1\bar{\ell} \\ 0\;0\;0 \end{pmatrix}/\begin{pmatrix} \ell\ell'\bar{\ell} \\ 0\;0\;0 \end{pmatrix} \; .$$ (13)

5

The quantity inside {} is the 6j symbol [7]. As a consequence of (12) the only difference between $\delta\chi_\ell(R)$ and $\delta\chi_\ell^{(s)}(R,R_1)$ is the replacement in (5) of $h_\ell^+(\rho) \rightarrow H_{\ell;\ell\ell'}(\rho;\rho_1\rho_2)$. One may check that when every $i^\ell h_\ell^+(\rho) \rightarrow e^{i\rho}/\rho$, that

$$H_{\ell;\ell\ell'}(\rho;\rho_1\rho_2) \rightarrow h_\ell^+(\rho) \left(1 + \frac{R}{R_1 R_2} f(0)\right) \tag{14}$$

as expected from (11).

One knows that the modification (11) can significantly improve single scattering theory well above threshold; e.g., for E~100-200 eV in Cu [8]. It would be interesting to see what happens closer to threshold, say in NiO, where both data and calculations exist which indicate that (4,5) is a poor description.

Acknowledgment

This work was supported in part by the NSF, grant DMR81-15705.

References

1. P.A. Lee, P.H. Citrin, P. Eisenberger, and B.M. Kincaid: Rev. Mod. Phys. 53, 769 (1981).
2. T.M. Hayes and J.B. Joyce: Solid State Phys. 37, 173 (1982).
3. J.E. Müller, O. Jepsen, O.K. Andersen, and J.W. Wilkins: Phys. Rev. Lett. 40, 720 (1978).
4. J.E. Müller, O. Jepsen, and J.W. Wilkins: Solid State Commun. 42, 365 (1982).
5. J.E. Müller and J.W. Wilkins: Phys. Rev. B29, 4331 (1984).
6. J.E. Müller and W.L. Schaich: Phys. Rev. B27, 6489 (1983).
7. A. Messiah: Quantum Mechanics (North-Holland, Amsterdam 1966).
8. P.A. Lee and J.B. Pendry: Phys. Rev. B11, 2795 (1975).
9. W.L. Schaich: Phys. Rev. B29, 6513 (1984).
10. S.J. Gurham, N. Binsted, and I. Ross: J. Phys. C17, 143 (1984).
11. J.E. Müller and W.L. Schaich: unpublished.
12. C.R. Natoli, D.K. Misemer, S. Doniach, and F.W. Kutzler: Phys. Rev. A22, 1104 (1980).
13. F.W. Kutzler, D.K. Misemer, S. Doniach, and K.O. Hodgson: J. Chem. Phys. 73, 3274 (1980).
14. P.J. Durham, J.B. Pendry, and C.H. Hodges: Comput. Phys. Commun. 25, 193 (1982).
15. T. Fujikawa, T. Matsuura, and H. Kuroda: J. Phys. Soc. Japan 52, 905 (1983).
16. G. Bunker and E.A. Stern: Phys. Rev. Lett. 52, 1990 (1984).
17. J.B. Pendry: Low Energy Electron Diffraction (Academic, New York 1974).
18. B.-K. Teo: J. Am. Chem. Soc. 103, 3990 (1981).
19. J.J. Boland, S.E. Crane, and J.D. Baldeschwieler: J. Chem. Phys. 77, 142 (1982).
20. D. Normal, J. Stöhr, R. Jaeger, P.J. Durham, and J.B. Pendry: Phys. Rev. Lett. 51, 2052 (1983).

Band Structure Approach to the X-Ray Spectra of Metals

J.E. Müller

Institut für Festkörperforschung der Kernforschungsanlage Jülich
D-5170 Jülich, Fed. Rep. of Germany

1. Introduction

The non-trivial part in the computation of the X-ray absorption spectrum (XAS)

$$\mu_i(E) = \text{const} \sum_f <\psi_i|\vec{r}\cdot\hat{\varepsilon}|\psi_f>^2 \, \delta(E-E_f+E_i) \quad , \tag{1}$$

due to excitation of a core level $i=(n\ell J)$, is the evaluation of the final states ψ_f. From a mathematical point of view, the scattering and the band structure formalisms are two different ways of calculating the same quantity (1). In the scattering formalism one takes ψ_f to be a solution for the excited atom potential and improves it systematically by including the single backscattering from neighboring atoms [1,2], the spatial variation of the scattered wavelets [3], and multiple scattering events [4], as needed. In the alternative approach presented here ψ_f is a Bloch state resulting from a band structure calculation. Unlike the scattering approach, which is of general applicability, the band structure approach requires a periodic potential. The appealing feature of the scheme is that it includes multiple scattering to infinite order. Figure 1 shows that the band structure result reproduces all features of the single-particle spectrum.

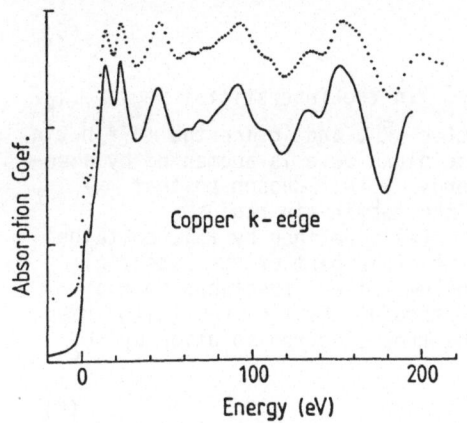

Copper k-edge

Fig. 1. Calculated K-edge absorption coefficient of copper (solid line) compared with the experimental results of Lengeler (dotted line)

From a physical point of view the two schemes lead to different interpretations of the spectra. For instance, in band theory one explains the observed structures in terms of hybridization and densities of states, while in scattering theory one speaks of matrix elements and nearest neighbour distances. The aim of this article is to reconcile these two points of view.

We also discuss how hybridization effects can be so strong at 200 eV above threshold, where the electrons are usually thought of as plane waves. Finally, we make the connection with bremsstrahlung isochromat spectroscopy (BIS).

2. Band Structure Problem

Consider the eigenvalue problem

$$(-\nabla^2 + V - E_{\vec{k}_j}) \psi_{\vec{k}_j} = 0 \tag{2}$$

for a periodic muffin-tin potential $V=V(\vec{r})$. In the conventional APW and KKR band structure methods one solves (2) explicitly by finding the roots of $\det(H-E)=0$ for each reduced vector $\vec{k}$ and band j. A more efficient approach is provided by the linear APW method [5], which yields all bands for each $\vec{k}$ at once. This is achieved by expanding $\psi_{\vec{k}_j}$ in energy-independent basis functions as

$$\psi_{\vec{k}_j} = \sum_{\vec{K}} A_{\vec{K}_j} \chi_{\vec{K}} \tag{3}$$

and converting (2) in a secular problem. Here is $\vec{K}=\vec{k}+\vec{G}$, i.e. the basis set is labelled by the reciprocal lattice vectors $\vec{G}$. A convenient choice of basis functions is given by the energy-independent augmented plane wave

$$\chi_{\vec{K}} = \begin{cases} 4\pi \sum\limits_{\ell m} i^\ell Y_{\ell m}^*(\hat{K})Y_{\ell m}(\hat{r}) \, j_\ell(Kr) = e^{i\vec{K}\cdot\vec{r}} & \text{for } r \geq S \ , \\ \\ 4\pi \sum\limits_{\ell m} i^\ell Y_{\ell m}^*(\hat{K})Y_{\ell m}(\hat{r}) \, \phi_\ell(K,r) & \text{for } r \leq S \ , \end{cases} \tag{4}$$

where S is the muffin-tin radius and $\phi_\ell(K,r)$ is a solution of the radial Schrödinger equation with energy $\epsilon_\ell(K)$ determined by the logarithmic derivative boundary condition

$$\frac{S\phi_\ell'(K,S)}{\phi_\ell(K,S)} = \frac{x \, j_\ell'(x)}{j_\ell(x)} \bigg|_{x=KS} = D_\ell(K^2) \tag{5}$$

and normalization fixed by $\phi_\ell(K,S)=j_\ell(KS)$. In the interstitial region $\chi_{\vec{K}}$ is defined to be the plane wave of wave vector $\vec{K}$, and inside the muffin-tin sphere each of the partial components of the plane wave is augmented by a solution of the Schrödinger equation with energy $\epsilon_\ell(K)$ chosen so that $\chi_{\vec{K}}$ and its first derivative are continuous at the muffin-tin radius.

The set of crystal potential functions $\epsilon_\ell(K)$, defined by (5), contains all the information about the spherically symmetric part of the potential which is relevant to the band structure problem. These functions bear a simple numerical relation with the energy-dependent phase shifts $\delta_\ell(E)$ used in scattering theory. There, one shifts the free-electron solution by δ_ℓ outside the atomic sphere, namely

$$\phi_\ell(k^2,r) = \cos \delta_\ell \, j_\ell(kr) - \sin \delta_\ell \, n_\ell(kr) \tag{6}$$

in order to have continuous and differentiable partial waves, while in (4) we shift the energy inside the muffin-tin sphere by $[\epsilon_\ell(K)-K^2]$ and retain the unshifted sinusoidal form of the plane waves in the interstitial region. The plane waves provide a very convenient way of fulfilling the periodic boundary condition. Notice that $\chi_{\vec{K}}$, although it satisfies Bloch's theorem with Bloch vector $\vec{k}$, is not a solution for the solid because its energy

$\in_\ell(K)$ inside the muffin-tin sphere is different for each ℓ and is also different from its energy K^2 in the interstitial region (except for the trivial case of free electrons).

From (1) one obtains the following secular equation [6]

$$\sum_{\vec{K}'} \left\{ (K^2-E)\ \delta_{\vec{K}\vec{K}'} + \sum_\ell w_\ell(\vec{K},\vec{K}')\ [\Gamma_\ell(K,K')-E\Delta_\ell(K,K')] \right\}\ A_{\vec{K}'} = 0 \ . \tag{7}$$

The dependence on the crystal structure is given by

$$w_\ell(\vec{K},\vec{K}') = \frac{4\pi S}{\Omega}\ (2\ell+1)\ P_\ell(\hat{K}\cdot\hat{K}')\ j_\ell(KS)j_\ell(K'S)\ \frac{D_\ell(K^2)-D_\ell(K'^2)}{K^2 - K'^2} \tag{8}$$

for each ℓ , while the dependence on the potential is determined by the pseudopotential function $\Gamma_\ell(K,K')=m_\ell(K,K')\in_\ell(K)-K^2$ and by the overlap function $\Delta_\ell(K,K')=m_\ell(K,K')-1$. Here

$$m_\ell(K,K') = \frac{K^2 - K'^2}{\in_\ell(K^2) - \in_\ell(K'^2)} \tag{9}$$

is a non-diagonal effective mass.

3. X-ray Spectrum

Within the muffin-tin approximation for the potential the X-ray spectrum (1) can be expressed as [6]

$$\mu_i(\in) = \text{const} \left[\frac{\ell}{2\ell-1}\ r^2_{i,\ell-1}(E)\ N_{\ell-1}(E) + \frac{\ell+1}{2\ell+1}\ r^2_{i,\ell+1}(E)\ N_{\ell+1}(E) \right] \tag{10}$$

with

$$r^2_{i,\ell}(E) = \frac{<\psi_i\ \vec{r}\ \phi_\ell(E)>^2}{<\phi^2_\ell(E)>} \quad , \tag{11}$$

$$N_\ell(E) = 2 \sum_{k_j} \delta(E-E_{\vec{k}_j})\ \sum_{j\ m} |<\ell m|\psi_{\vec{k}_j}>|^2 \quad . \tag{12}$$

Using the normalization (6) for the final state wave function, one can write

$$r^2_{i,\ell}(E)\ N_\ell(E) = M^2_{i,\ell}(E)\ \chi_\ell(E) \tag{13}$$

with

$$M^2_{i,\ell}(E) = (2\ell+1)N^{fe}(E)\ <\psi_i\ \vec{r}\ \phi_\ell(E)>^2 \tag{14}$$

$$\chi_\ell(E) = \frac{N_\ell(E)}{(2\ell+1)N^{fe}(E)<\phi^2_\ell(E)>} \quad , \tag{15}$$

N^{fe} being the free-electron density of states.

Equation (13) provides the connection between the band structure and the scattering formalisms. In the latter, the excited electron wave function is normalized to scattering states (6) which assume an asymptotic free-electron density of states. This normalization permits to factorize the spectrum in a solid state term χ_ℓ , which describes the scattering on the surrouding atoms, and an atomic matrix element $M^2_{i,\ell}$, which characterizes the effect of the central potential. In the band structure formalism the spectrum is expressed in terms of partial densities of states $N_\ell(E)$, which includes *both* atomic and solid state effects. For instance, the $L_{2,3}$ white lines of the transition metals are interpreted as a density of states effect in band structure calculations but as a matrix element effect in the scattering approach. The energy dependence of the band structure matrix elements (11) is rather trivial, determined mainly by the normalization of the wave function in the primitive cell, i.e. $r^2_{c,\ell}(E) \sim \langle\phi^2_\ell(E)\rangle^{-1}$.

4. Physical Interpretation

The energy bands are classified according to the number of nodes and orbital symmetry of the wave functions. While for the occupied states they retain much of their atomic character (e.g. in the 3d band of Cu), this correspondence disappears for the higher ones. Higher bands arise from additional nodes in the wave function in the region between the core and the cell boundary, which produce features in the density of states roughly periodic in $k=\sqrt{E}$. The amplitude of features due to specific bands are proportional to the strength with which they hybridize. Hybridization (i.e. the mixing of different angular momentum components) occurs when the functions $\epsilon_\ell(K)$ have different energy dependence for different ℓ-values. Notice that if $\epsilon_\ell(K)= \in(K)$ for all values of ℓ one would obtain the free electron band structure rescaled by the function $\in(K)$ and the angular momentum content of a particular eigenstate with energy E would be that of a plane wave of energy $K^2(E)$. It is only when the various $\epsilon_\ell(K)$ are different from one another that the free electron distribution of angular momenta is altered and hybridization can take place. Hybridization of a given eigenstate results from the fact that its various angular momentum components tend to be displaced in energy by different amounts.

The connection with the scattering approach can be best seen from the standard EXAFS expression given by

$$\chi_1(k^2) = - \sum_j N_j \text{ Im} \left\{ \frac{e^{2i(kR_j+\delta_1)}}{kR^2_j} \ f(\pi) \right\} \quad , \tag{16}$$

$$f(\pi) = \frac{1}{k} \sum_\ell (-)^\ell \ (2\ell+1) \ \sin \delta_\ell \tag{17}$$

for the K-edge. The position $k_n \sim n\pi/R_j + const$ of the peaks of the spectrum is related to the number of nodes n of the wave function between neighboring atoms, and the magnitude of the peaks is given by the backscattering amplitude $f(\pi)$. Notice that successive partial terms in (17) tend to cancel each other, so that $f(\pi)$ is large when the corresponding δ_ℓ's have different energy dependence.

5. Free-electron Limit

Even for the largest energies above threshold of interest in X-ray absorption, the kinetic energy of the excited electron is small compared with the potential energy close to the nuclei. With increasing energy, however, the spectrum becomes dominated by the large ℓ-components of the wave function which, because of the large centrifugal repulsion, are confined outside of the core region and exhibit free-electron-like behavior. This can be seen by the following simple argument [7]. The number of states function for a single muffin-tin potential embedded in the electron gas is given by

$$n(E) = \frac{\Omega}{3\pi^2} \ E^{3/2} + \sum_\ell \frac{2(2\ell+1)}{\pi} \ \delta_\ell(E) \tag{18}$$

and the density of states by $N(E)=dn/dE$. One may view (18) as the first two terms of an expansion of the crystal number of states function, with successive terms introducing the effects of multiple scattering of the electrons by the lattice. The evolution towards free-electron behavior can be seen in the second term (Friedel sum). At high energies, one can write [8]

$$\delta_\ell \sim -k \int_0^\infty dr\ r^2\ j_\ell^2(kr)\ V(r) \quad , \tag{19}$$

$$\sum_\ell \frac{2(2\ell+1)}{\pi}\ \delta_\ell = -\frac{\Omega}{2\pi^2}\ E_0\ E^{1/2} \tag{20}$$

where

$$E_0 = \frac{4\pi}{\Omega} \int_0^S dr\ r^2\ V(r) \tag{21}$$

and $\sum_\ell (2\ell+1)\ j_\ell^2 = 1$ has been used. In this limit (18) becomes the first two terms in a Taylor expansion of $(\Omega/3\pi^2)(E-E_0)^{3/2}$. However, a comparison between the Friedel sum and its limit (20) for Cu (Fig. 2) indicates that one must look at $\gtrsim 100$ Ry above threshold to find free-electron-like behavior.

6. Connection Between BIS and XAS

The BIS spectrum is also given by (1) with a high energy Bloch state $\psi_{\vec{k}_i}$ as initial state. Unlike the XAS case, the BIS matrix element $B=\langle\psi_{\vec{k}_i}|\vec{r}\cdot\hat{\epsilon}|\psi_{\vec{k}_j}\rangle$ weights preferentially the outer region of the atom, where

$\psi_{\vec{k}_j} \sim e^{i(\vec{k}+\vec{G}_j)\cdot\vec{r}}$, and probes all ℓ-components of the final state $\psi_{\vec{k}_j}$ simultaneously. Therefore $B \sim \langle e^{i(\vec{G}_i-\vec{G}_j)\cdot\vec{r}}\ \vec{r}\cdot\hat{\epsilon}\rangle$ has a simple behavior, decreasing smoothly with increasing final state energy and the BIS spectrum can be written as $B^2N(E)$. Under these conditions the connection with XAS is given by (15), i.e.

$$\mu_i(E) = B^2N^{fe}(E)\ \sum_\ell (2\ell+1)\ \langle\phi_\ell^2(E)\rangle\ \chi_\ell(\epsilon) \quad . \tag{22}$$

The BIS spectrum is thus proportional to a superposition of all possible XAS spectra and it cannot be simply interpreted in terms of interatomic distances. However, it must be emphasized that BIS offers an alternative way of probing unoccupied states without core excitation. Detailed compar-

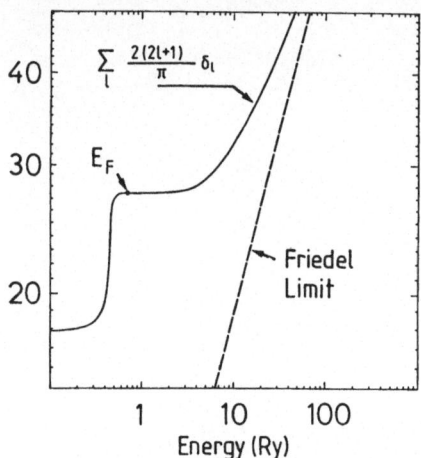

Fig. 2. Friedel sum contribution (solid line) to the number of states per atom $n(E)$ and its free-electron limit (dashed line)

isons between BIS and XAS spectra should provide important information about
the effect of the core hole on the excitation spectra.

References

1. C.A. Ashley and S. Doniach, Phys. Rev. B 11, 1279 (1975)
2. P.A. Lee and J.B. Pendry, Phys. Rev. B 11, 2795 (1975)
3. J.E. Müller and W.L. Schaich, Phys. Rev. B 27, 6489 (1983)
4. P.J. Durham, J.B. Pendry, and C.H. Hodges, Solid State Commun. 38,
       159 (1981)
5. O.K. Andersen, Phys. Rev. B 12, 3060 (1975)
6. J.E. Müller and J.W. Wilkins, Phys. Rev. B 29, 4331 (1984)
7. R.C. Albers, A.K. McMahan, and J.E. Müller, submitted to Phys. Rev. B 15
8. See, for instance, L.I. Schiff: Quantum Mechanics (McGraw Hill,
       New York 1955), second edition, p. 167

Self-Consistent Density Functional Methods in Photoabsorption

A. Zangwill

Dept. of Physics, Polytechnic Institute of New York
Brooklyn, NY 11201, USA

I. Introduction

The realization that important geometrical and electronic structure information is contained in the near-edge photoabsorption spectrum of molecules and
solids has increased the pressure on theorists to provide accurate cross section calculations for atoms in a variety of local environments. The quantity
of most direct experimental interest for these studies is the absorption
coefficient, $\mu(\omega)$. A typical calculation [1] proceeds by evaluation of Golden
Rule matrix elements. In the dipole approximation one finds,

$$\mu(\omega) = \frac{4\pi^2}{\Omega}\alpha h\omega \sum_{i,j} |<i|\hat{\epsilon}\cdot\vec{r}|f>|^2 \delta(h\omega - E_f + E_i) \qquad (1)$$

where Ω is the unit cell volume, α is the fine structure constant, and $\hat{\epsilon}$ is
the polarization vector of the incident radiation. The transition operator
reflects the external electric potential associated with long wavelength
($\lambda >> a_B$) light of unit field strength. If $|i>$ and $|f>$ denote exact many-
electron states of the system of interest, the Golden Rule expression provides
an exact route to the absorption coefficient. However, in practical calculations, the initial and final state wave functions are obtained as the eigen-
states of a one-electron potential which takes account of Coulomb interactions
in an average self-consistent manner. The Hartree-Fock and local density
approximations are popular examples of this *independent particle approximation*.
In many cases, calculations performed in this way are in good agreement with
experiment. Here, I will be concerned with the remedy necessary when agreement
is not so good.

Discrepancies between absolute cross section measurements and calculations based on the independent particle approximation are well known in atomic
physics literature. For excitation energies in the sub-kilovolt range it has
been established that dielectric effects must be considered properly in order
to recover agreement with experiment [2]. The basic idea is that an external
field polarizes the electronic charge cloud of an atom. Let us characterize
this polarization by the frequency dependent deviation of the charge density
from its ground state value, $\delta n(\vec{r}|\omega)$. By Poisson's law, $\delta n(\vec{r}|\omega)$ acts as the
source for a non-uniform *internal* field which can screen or anti-screen the
external field. When the net charge distortion and electric field within the
atom are mutually consistent (in a manner to be quantified below), one has obtained an average account of the residual Coulomb interactions between electrons embodied in a *complex, frequency-dependent, effective* excitation field
which simply replaces the external field dipole operator in all Golden Rule
matrix elements for absorption.

If one begins with a Hartree-Fock description of electronic structure,
the self-consistent field theory sketched above is simply the familiar random

13

phase approximation (with exchange). If instead one begins with a local density approximation (LDA) to the electronic structure, the analogous prescription has been termed a *time-dependent local density approximation* (TDLDA) [3].
For low energy (hν < 200 eV) atomic photoabsorption problems the latter approach is computationally straightforward, remarkably accurate, and lends
itself to ready physical interpretation of polarization-type many-body phenomena. Since the TDLDA and a number of illustrative examples have been reviewed recently [4] I will focus here, after a brief resumé, on recent developments of this method most relevant to potential XANES applications.

II. Time-Dependent Local Density Approximation

A useful measure of the influence of an external perturbation on a finite
electronic system is the dipole polarizability,

$$\alpha(\omega) = e \int d\vec{r} \; z \delta n(\vec{r}|\omega). \tag{2}$$

For frequencies above the first ionization $\alpha(\omega)$ is a *complex* function. This
simply reflects the fact that the charge density generally oscillates out of
phase with the external field. Furthermore, the system is lossy and absorbs
energy according to

$$\mu(\omega) = \frac{4\pi\omega}{\Omega c} \; \text{Im} \; \alpha(\omega). \tag{3}$$

Hence, the absorption coefficient is obtained directly from the perturbed
charge density, $\delta n(\vec{r}|\omega)$. The self-consistency condition on this quantity can
be stated compactly within linear response theory as

$$\delta n(\vec{r}|\omega) = \int d\vec{r}' \; \chi_0(\vec{r},\vec{r}'|\omega) \; \{U_{ext}(\vec{r}') + U_{ind}(\vec{r}'|\omega)\} \tag{4}$$

$$U_{ind}(\vec{r}|\omega) = \int d\vec{r}' \frac{\delta n(\vec{r}'|\omega)}{|\vec{r}-\vec{r}'|} + \frac{\partial V}{\partial n} xc[n] \; \delta n(\vec{r}|\omega) \tag{5}$$

Here, $V_{xc}[n(\vec{r})]$ is the piece of the LDA potential function which accounts
for the quantum mechanics of exchange and correlation. $\chi_0(\vec{r},\vec{r}'|\omega)$ is the
Fourier transform of the LDA density-density response function. It is obtained
from the LDA potential function and implicitly contains the eigenfunction sums
evident in the Golden Rule. Indeed, the TDLDA method is, in fact, nothing more
than self-consistent, time-dependent, first-order perturbation theory built
on an LDA orbital basis.

Upon simultaneous solution of (4) and (5), the absorption coefficient
can be found in two ways. First, as noted, $\delta n(\vec{r}|\omega)$ leads immediately to $\mu(\omega)$
from (2) and (3). Equivalently, the effective field, $U_{eff}(\vec{r}|\omega) = U_{ext}(\vec{r}) + U_{ind}(\vec{r}|\omega)$, can be inserted directly into the Golden Rule. The latter procedure,
useful for calculation of partial *photoionization* cross sections, emphasizes
the fact that the concept of excitations between one-electron eigenfunctions
can be retained as long as the atom is regarded properly as a dielectric
medium. In this light, an important question must be asked: when are dielectric effects expected to be important? A recent calculation [5] will serve.

The figure 1 illustrates the absorption spectrum of a small metal
particle as *simulated* by a sphere of jellium containing 92 electrons. The

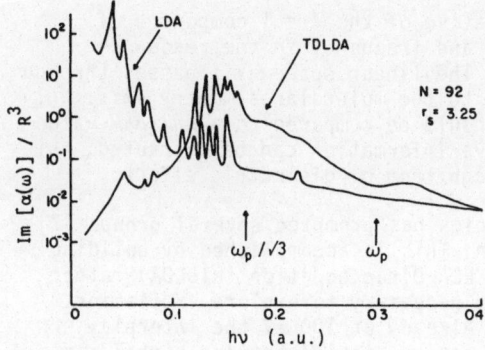

Figure 1

calculations were performed both in an independent particle model (LDA) and the TDLDA. The latter clearly exhibits a surface plasmon peak as well as a nascent bulk plasmon peak. The fine structure is a quantum size effect. By contrast, the LDA curve merely decreases (on average) from a very large value near threshold. The connection between these two is readily understood within the general theory of self-consistent collective oscillations. Therein, one finds that a large concentration of spectral oscillator strength confined within a narrow energy range can be *transferred* to higher energy by the coherent polarization of one-particle transitions. If we regard the electron-hole pair formed by near threshold absorption as an elementray dipole we should expect strong dielectric effects when the self dipole-dipole interaction is large. In this example, strong one-electron transitions occur near threshold to continuum states trapped behind a substantial angular momentum barrier. In just the same way, the strongest dielectric effects in atoms are found when independent particle oscillator strength is concentrated in large-overlap transitions to real or virtual bound states near the vacuum threshold.

III. Recent Results

Recent progress in self-consistent local density calculations of optical absorption has focused on molecules at low energies and relativistic effects in atoms at high energies. I will emphasize the latter although some very nice results [6] have been obtained in small molecules. For example, no independent particle calculation is capable of reproducing the rich autoionization structure which dominates molecular VUV photoabsorption spectra. This phenomenon arises from the dipole-dipole interaction between degenerate excitation channels. In acetylene, near 14 eV, a $2\sigma_u \to 1\pi_g$ transition competes with continuum photoemission from the $1\pi_u$ valence level. Rather than illustrate the accurately calculated Fano resonance observed experimentally, Fig. 2 shows the effective

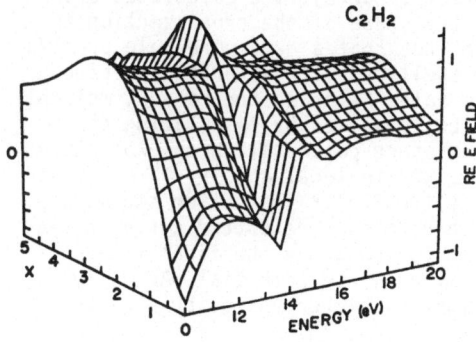

Figure 2

electric field [actually the radial derivative of the $\ell = 1$ component of $U_{eff}(\vec{r}|\omega)$] plotted as a function of space and frequency in the resonance region obtained from a TDLDA calculation. The linear space axis passes through the molecular center and is perpendicular to the molecular symmetry axis. Only the real part of the field is shown and should be compared to a *uniform* value of $E_{ext} = 1$. Although detailed quantitative information can be extracted, this representation clearly demonstrates the magnitude of dielectric effects.

The success of the TDLDA at low energies has prompted several groups [7] to explore its relativistic generalization. This is accomplished by building the self-consistent field procedure on an LDA Dirac equation (RTDLDA) rather than a Schrodinger equation and is clearly necessary to explore excitation from deep inner shells at x-ray energies. Already at 100 eV the interplay between spin-orbit and dielectric effects can be significant for high Z materials. As an illustration, Fig. 3 shows the measured photoabsorption spectrum (solid line) for metallic uranium near the 5d shallow core threshold [8]. In an independent particle model, the absorption is dominated by *discrete* transitions from the spin-orbit split $5d_{5/2}$ & $5d_{3/2}$ levels to bound 5f levels. The depicted atomic RTDLDA calculation pushes this oscillator strength above the vacuum threshold into two broad peaks whose areas are *not* in statistical ratio. It can be shown [9] that the majority of the discrepancy between this theory and experiment is due to the neglect of open-shell multiplet structure.

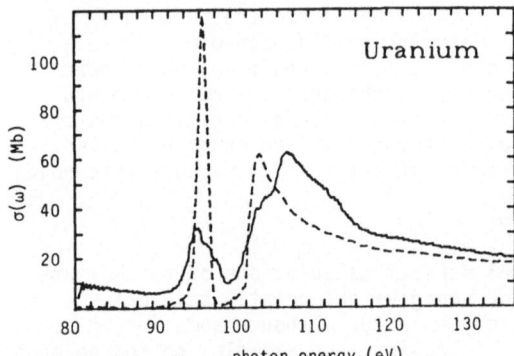

Figure 3

Excitation from the deep $M_{4,5}$ core levels of xenon was recently investigated with the RTDLDA formalism [10]. A serious disagreement between absolute absorption measurements [11] and a relativistic LDA calculation indicated that dielectric effects continue to be important at near-kilovolt energies. However, as Fig. 4 shows, the RTDLDA calculation (dashed curve) also yields a poor representation of the data (solid curve). The calculated absorption threshold is in the wrong place and the shape of the cross section is far too broad. The cause is the well-known fact that a deep core hole provides an attractive scattering center for outer shell and virtual orbitals. Electrons tend to relax inward toward the core hole. As an *ad hoc* remedy, multiple scattering from a core hole potential can be *simulated* by calculating the final photoabsorption states in a self-consistent potential which explicitly contains a hole in the relevant core level. Within local density theory, the most natural choice for such a potential is the Slater transition state, long known to yield quite accurate ionization thresholds. The dotted curve in Fig. 4 illustrates the results of an RTDLDA calculation where the final states were derived from a relativistic LDA transition state potential. Actually, the calculated curve has been rigidly shifted to higher energy by 4 eV to

16

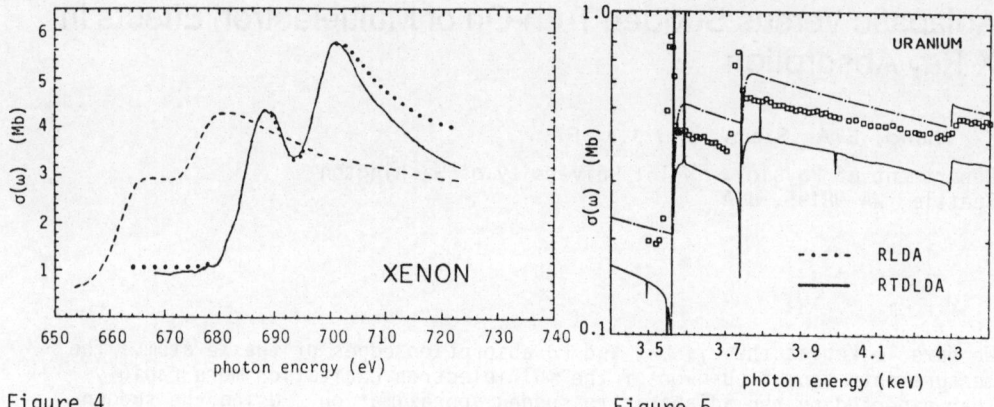

Figure 4   Figure 5

facilitate comparison with experiment. Nevertheless, an important point should be clear: significant rearrangement of near edge photoabsorption fine structure can result from dielectric screening, core hole relaxation, or some subtle combination of both *even in the atomic limit*.

At sufficiently high energy, dielectric effects must become negligible and an extensive series of RTDLDA calculations [12] which span the periodic table show this indeed to be the case, except very near threshold. Nonetheless, some mysteries remain. Figure 5 shows recent high resolution absorption data near the $M_{4,5}$ edges of uranium [13] along with a relativistic LDA and RTDLDA calculation. Although the observed sharp autoionization resonances are picked up in the dielectric theory the absolute magnitude of the cross section has not been accurately predicted. Indeed, the uniform diminution of the RTDLDA result relative to the independent particle curve is curious in itself, although a similar effect was found [14] for 3d absorption in copper, albeit at very much lower energy. Clearly, more work remains to be done at high energies, including for example, the introduction of higher multipoles.

1. See, for example, J.E. Muller and J.W. Wilkins, Phys. Rev. B 29, 4331 (1984)
2. M. Ya. Amusia, Adv. At. Mol. Phys. 17, 1 (1981); G. Wendin, in New Trends in Atomic Physics, edited by G. Grynburg and R. Stora (North-Holland, Amsterdam, 1983)
3. A. Zangwill and P. Soven, Phys. Rev. Lett. 45, 204 (1980)
4. A. Zangwill, in Atomic Physics 8, edited by I. Lindgren, A. Rosen and S. Svanberg (Plenum, New York, 1983)
5. M. Puska, M. Manninen and R.M. Nieminen (unpublished); W. Ekhardt, Phys. Rev. Lett. 52, 1925 (1984)
6. Z.H. Levine and P. Soven, Phys. Rev. A 29, 625 (1984)
7. F.A. Parpia and W.R. Johnson, Phys. Rev. A 29, 3173 (1984); D.A. Liberman and A. Zangwill, Comp. Phys. Commun. 32, 75 (1984)
8. M. Cukier, P. Dhez, B. Gauthe, P. Jaegle, Cl. Wehenkel and F. Combet-Farnoux, J. Physique 39, L315 (1978)
9. A. Zangwill and D.A. Liberman (unpublished)
10. A. Zangwill and D.A. Liberman, J. Phys. B 17, L253 (1984)
11. O. Yagci and J.E. Wilson, J. Phys. C 16, 383 (1983)
12. G. Doolen (unpublished)
13. N. Kerr Del Grande and A.J. Oliver, in Low Energy X-ray Diagnostics, edited by D.T. Attwood and B.L. Henke (AIP, New York, 1981)
14. A. Zangwill and P. Soven, Phys. Rev. B 24, 4121 (1981)

Adiabatic Versus Sudden Turn-On of Multielectron Effects in X-Ray Absorption

Ke Zhang, E.A. Stern, and J.J. Rehr

Department of Physics, FM-15, University of Washington
Seattle, WA 98195, USA

We have looked at the L_1, L_2, and L_3 absorption edges of the Xe atom. The measurements show a turn-on of the multielectron excitation more rapidly than expected by the adiabatic to sudden approximation. Using the sudden approximation we made a ΔSCF calculation for both one- and two-electron excitations of the L-edges. The result shows that the calculation at least qualitatively explains the multielectron excitations past the L-edges.

1. Experimental Evidence

An accurate x-ray absorption experiment at the L-edges of xenon atoms was performed at SSRL. The results for the L_1-edge are shown in Fig. 1. The experiment shows the following: (a) L_1, L_2, and L_3 edges all have similar fine structure added onto a smooth background. Three bumps sit above the edge at 13, 24, and 80 eV; (b) Both the main peak and the bumps have about equal onset widths of about 5 eV; (c) After each bump the slope of the curve changes. The L_1-edge shows a different behavior for the slope change compared with L_2 and L_3 edges at the 80 eV onset. The bumps correspond to double electron excitation and are located at the Z+1 atom's (Cs) absorption threshold for the same excitation. The experiment is not consistent with the adiabatic to sudden model for the onset of double excitations. In that model the escaping photoelectron only turns on the core hole potential rapidly enough to excite a given shell electron when its energy is much greater than the binding energy of that electron. For the 80 eV onset the adiabatic to sudden approximation would predict an 80 eV width to the onset, clearly in disagreement with the measurements.

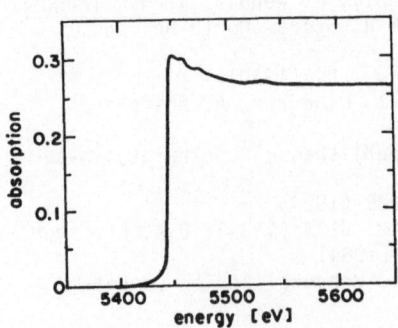

Fig. 1. Measured L_1 absorption edge of Xe vapor. The vertical scale is arbitrary

2. Simple Calculation

The x-ray absorption was calculated by the ΔSCF method in the dipole approximation. The one-electron states were calculated self-consistently by the Herman-Skillman method, which uses local exchange but does not include spin-

orbit coupling or other relativistic corrections. The N-electron wave
function was a Slater determinant of the one-electron states. Since the final
state contained an L shell hole, the self-consistent one-electron potential
of the final state is different from that of the initial one, leading to
different one-electron states in the two cases. Only the relaxation of the
outer electrons, namely the 4d, 5p, and 5s, were taken into account. The
changes in the rest of the electrons, the inner core ones, were neglected as
a calculation indicated was justified. For comparison we made a non-self-
consistent calculation in which the final states were left the same as the
initial states.

3. Results and Discussion

The calculated results shown in Fig. 2 neglect lifetime effects. The calculated two-electron absorption step near the edge is found to be 20% of the
one-electron excitation. The step at 80 eV corresponds to exciting both an
L shell electron and a 4d electron. Its calculated step agrees reasonably
well with the measurement. The contribution to the absorption where all (one
or two) excited electrons are bound is added as a hatched rectangle to give
the correct area over their energy range. The actual shape will be greatly
affected by lifetime broadening. The comparison between experiment and calculation,neglecting lifetime broadening, appears to be poor near the edge.
However, the lifetime broadening is 5 eV and will substantially fill in the
dip before the onset of the first multielectron excitation and lower the peak
at the edge, giving better agreement between theory and experiment. We plan
to include the effects of lifetime broadening in the future to be able to
assess more definitely the accuracy of the ΔSCF method.

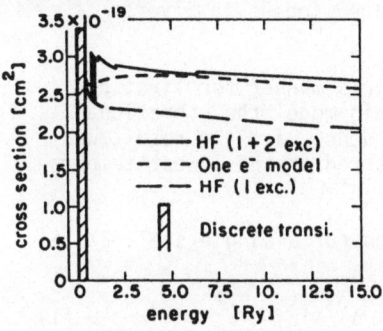

Fig. 2. Calculated L_1 absorption
edge of Xe vapor using the ΔSCF
method. The hatched rectangles
are due to transitions in which
all excited electrons are in
bound states. No lifetime
broadening is included.

4. Conclusions

Our experimental results are inconsistent with an adiabatic to sudden turn-on
of the multielectron excitations, expecially for the higher energy ones. A
simple ΔSCF calculation including only the averaged potential self-consistently
is qualitatively correct, but lifetime broadening needs to be added for a more
accurate comparison. The ΔSCF calculation assumes a sudden turn-on of the
core hole potential.

   This work was supported by the National Science Foundation, grants DMR80-
22221 (Zhang and Stern) and DMR82-07357 (Rehr). We thank the SSRL for its
assistance. SSRL is supported by the DOE Office of Basic Energy Sciences,
and the NIH, Biotechnology Resources Program, Division of Research Resources.

Many-Body Effects on EXAFS Amplitudes

K. Kim and E.A. Stern

Department of Physics, FM-15, University of Washington
Seattle, WA 98195, USA

1. Experiment

Extended x-ray absorption fine structure was measured at K edges of both
elements in InAs and GaAs. The measurements were made at SSRL (Stanford
Synchrotron Radiation Laboratory) during dedicated runs. The samples were
measured at three temperatures (80 K, 180 K, and room temperature) to determine the temperature dependency of the disorders.

2. Analysis

The analysis of the data used the same method as used by STERN et al. [1].
Linear pre-edge background is removed before the data are normalized by the
edge step. The energy origin is chosen at the half-step point, and smooth
background is removed by the cubic spline fit method. The Fourier filtering
method is used to extract single-shell EXAFS data, which are expressed in
phases and amplitudes. Care has been taken to reduce the finite window
size effect by using the same size coordinate space windows in inverse
transformation.

The single-shell EXAFS data were obtained in this manner for first and
second shells at both edges. By using the Einstein model the vibrational
disorder was estimated from the temperature dependence, and the result was
used to eliminate the disorder from the data measured at the lowest temperature, which is then used in the subsequent analysis.

The analysis is based on the following equation for a single-shell EXAFS
amplitude:

$$\|\chi(k)\| = S_0^2(k) \frac{N_i}{R_i^2} F_i(k) \exp(-2k^2\sigma_i^2) \exp[-2(R_i - \Delta)/\lambda] , \qquad (1)$$

where the phase term has been dropped. As the coordination number and radial
distance are known, it is possible to correct for these quantities, and a
new term E can be defined as

$$E(k) = \|\chi(k)\| \frac{R_i^2}{N_i} \exp(2k^2\sigma_i^2) = S_0^2(k) F_i(k) \exp[-2(R_i - \Delta)/\lambda] . \qquad (2)$$

Then the ratio between first shell and second shell can be taken to get

$$R_A(k) = \frac{E_{A,1}(k)}{E_{A,2}(k)} = \frac{F_{B,1}(k)}{F_{A,2}(k)} \exp[2(R_2 - R_1)/\lambda] , \qquad (3)$$

where the $S_0^2(k)$ term cancels out and A and B represent the center atom and
the first-neighbor atom, respectively. Finally, two ratios at different

edges can be multiplied to give

$$\frac{1}{\lambda_A} + \frac{1}{\lambda_B} = \frac{1}{2(R_2 - R_1)} \log\left(R_A R_B \frac{F_{A,2}}{F_{A,1}} \frac{F_{B,2}}{F_{B,1}}\right) \ . \tag{4}$$

The mean free path in EXAFS comes from two sources, the core hole lifetime and the photoelectron lifetime. It is given by

$$\frac{1}{\lambda_A} = \frac{1}{\lambda_A} \ (\text{core hole}) + \frac{1}{\lambda} \ (\text{photoelectron}) \ , \tag{5}$$

and λ (core hole) is given by

$$\lambda = v\tau = \frac{\hbar k}{m} \frac{\hbar}{\Gamma} \ , \tag{6}$$

where k is the photoelectron wave number and Γ is the width of the core hole state. The values for Γ were taken from the work by KESKI-RAHKONEN et al. [2]. The ratios F_1/F_2 were estimated by using the method by MÜLLER et al. [3] which accounts for the spherical nature of the photoelectron as required at low k.

After the mean free paths have been obtained by this method, a second method was used to get the mean free path at the As edge. Assuming that λ_{As} in GaAs is correct, we compared the second-shell data at As edges from GaAs and InAs. Both center atom and backscatter are equal, and the ratio is given by

$$\frac{E}{E'} = \frac{\exp[-2(R_2 - \Delta)/\lambda]}{\exp[-2(R' - \Delta')/\lambda']} \ , \tag{7}$$

where primed quantities correspond to InAs and unprimed ones to GaAs. Using the relation $\Delta \simeq R_1$, λ' can be determined from (7).

3. Conclusion and Discussion

The mean free path λ_{As} in InAs obtained in the two ways are shown in Fig. 1. The bottom curve is the one obtained by the second method, and the top one is by the first method. The error bar is mainly from the uncertainties in the disorder term.

The discrepancy is too big to be explained by any experimental errors, and we conclude that the reduction of amplitudes for higher shells in EXAFS cannot be described by a simple mean free path term.

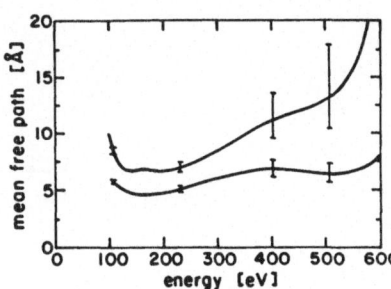

Fig. 1. Mean free path in InAs at As edge

There are many possible sources of invalid approximations, but the most probable one is that the inelastic scattering of the photoelectron is anisotropic in these covalently bounded compounds, and not isotropic as usually assumed.

This research was supported by the National Science Foundation, grant no. DMR80-22221. We thank the staff of SSRL, at which these experiments were conducted. SSRL is supported by the DOE Office of Basic Energy Sciences, and by the NIH, Biotechnology Resources Program, Division of Research Resources.

References

1. E.A. Stern, B.A. Bunker, and S.M. Heald: Phys. Rev. B <u>21</u>, 5521 (1982)
2. O. Keski-Rahkonen and M.O. Krause: Atomic Data and Nuclear Data Tables <u>14</u>, 139 (1974)
3. J.E. Müller and W.L. Schaich: Phys. Rev. B <u>27</u>, 6489 (1983)

XANES and EXAFS Measurements on Pd_2Si: One-Electron or Many Body Effects?

M. De Crescenzi
Dipartimento di Fisica, Università dell'Aquila, I-67100 L'Aquila, Italy
E. Colavita
Dipartimento di Fisica, Università della Calabria, I-87036 Cosenza, Italy
U. Del Pennino, P. Sassaroli, and S. Valeri
Dipartimento di Fisica, Università di Modena, Modena, Italy
C. Rinaldi, L. Sorba, and S. Nannarone
Dipartimento di Fisica, Università di Roma, Roma, Italy

X-ray absorption spectroscopy has been used to investigate near L-edge structures of Pd in pure Pd and bulk Pd_2Si and PdSi silicides (1).

The experiment was performed at the X-ray beam line of the Frascati synchrotron radiation facility. The joint analysis of EXAFS and XANES of the spectra (Fig.1 and 2) reveal the occurrence of physical processes in between the one electron and the many body description.

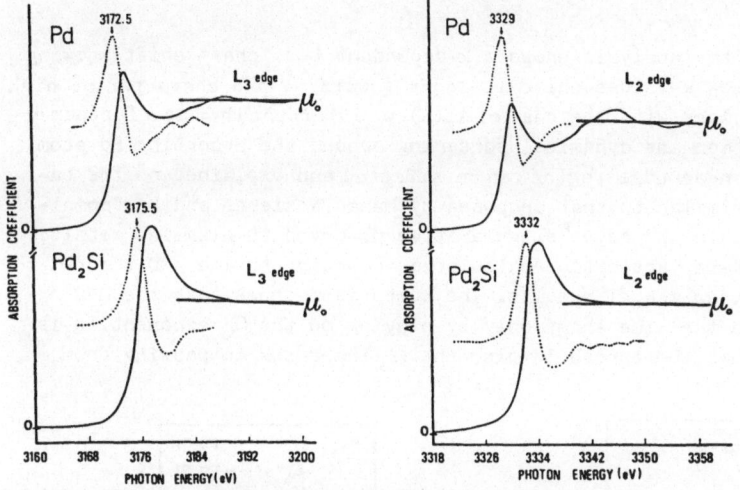

Fig.1 -L_3 and L_2 XANES of Pd in Pd metal and Pd_2Si.

The main result is an absorption increase of the $L_{2,3}$ white line decreasing the Pd concentration,which cannot be explained in terms of p-d partial density of states (2).

We suggest that this enhancement can be ascribed to the excitonic nature of the $L_{2,3}$ which requires a localization and dynamical screening of the core hole. This picture is in agreement with the observed L_3/L_2 branching ratio evolution towards the statistic atomic ratio,asymmetric lineshape and energy shifts (~3 eV). The picture we suggest stems from the following evidences in the EXAFS analysis:a) the F(R) is characterized by the presence of only the first Si neighbour shell and this is assumed as an indication that the photoelectron is confined inside the first coordination shell. b) The $L_{2,3}$ phase shift (Fig.3) follows a behaviour not foreseen by the optical (p-d) selection

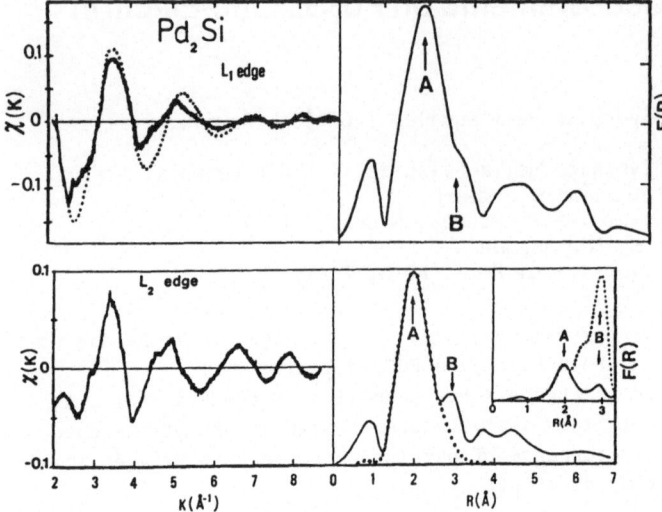

Fig.2 -EXAFS modulation χ(k) for Pd in Pd_2Si and Fourier transform F(R) around L_1 and L_2 edges. The inset shows the EXAFS calculations with the inclusion of Pd-Si and Pd-Pd coordination shells.

Note that the two χ(k) have the same phase shift sign.

rules.In particular the analysis shows a k-dependent $L_{2,3}$ phase shift showing a p-s character at low k-values while it tends towards a p-d character at high k-values (3). In analogy with the case of Al(4) we interpret this as a further e- vidence of localization and dynamical screening around the absorbing Pd atom. We suggest that the near edge region can be affected and explained on the ba- sis of a mechanism similar to that proposed by Mahan, Nozieres and De Domini- cis(MND) (4) to explain the edges' singularities observed in alkaline metals.

Fig.4 shows the X-ray absorption calculated according to the MND theory compared with the experimental results. The comparison shows that the MND theory can reproduce the line shape only by playing on the α_o exponent.It is important to note that the threshold exponent α_o increases on passing from

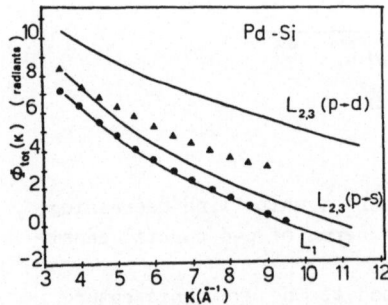

Fig.3 -Pd_2Si experimental phase shift for L_1 edge (•••) and L_2 edge (▲▲▲) compared with Teo and Lee calculations.

Fig.4 -a) and b) L_3 and L_2 XANES spec- tra of Pd_2Si and PdSi compared with (c) with the MND model (4) for two values of α_o.

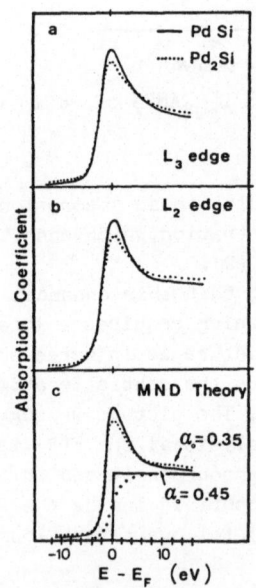

Pd_2Si to PdSi,thus indicating a predominance of s wave scattering at E_F,increasing the Si concentration. This supports once again the localization of the deep $L_{2,3}$ core hole.

Similar effects (enhancement,asymmetry and energy shifts) have been recently observed by Mason in the $L_{2,3}$ XANES of Pd clusters where the Pd absorbing atom is likely to be in a more atomic and unscreened situation (6).

Although the above description of physical processes occurring at the L-edges of Pd in Pd-Si compounds is reasonable, detailed theoretical calculations (2) could allow the passage from a qualitative to a quantitative model and clarify the role of a deep core hole in the XANES lineshape of these systems. Moreover,possible information on the homogeneity of the d-metal-silicon compound at the interface can be obtained by monitoring the chemical shift and the asymmetry of deep-core lineshapes.

Furthermore the conclusion of our work should suggest a great care when bulk compound phase shift are used in the interpretation of SEXAFS data in the case of structural analysis of non-stoichiometric surface compounds(6).

References

(1) M.De Crescenzi,E.Colavita,U.Del Pennino,P.Sassaroli,S.Valeri,C.Rinaldi,
    L.Sorba,S.Nannarone to be published
(2) O.Bisi,C.Calandra,J.Phys.C14,5479(1981)
    J.E.Müller,J.W.Wilkins,Phys.Rev.B29,4331(1984)
(3) C.Noguera,D.Spanjaard,J.Phys.F11,1133(1981)
(4) G.D.Mahan,Phys.Rev.B11,4814(1975)
    P.Nozieres,C.T.De Dominicis,Phys.Rev.178,1697(1969)
(5) M.G.Mason,Phys.Rev.B27,748(1984)
(6) J.Stöhr,R.Jaeger,J.Vac.Sci.Technol.21,619(1982)

Effect of Intermediate Range Order on the Lineshape of the Ge K Edge

M.G. Proietti, S. Mobilio, and A. Gargano
PULS, Laboratori Nazionali di Frascati, I-00044 Frascati, Italy
L. Incoccia
Istituto di Struttura della Materia del C.N.R., I-Frascati, Italy
F. Evangelisti
Dipartimento di Fisica, Università "La Sapienza", I-00185 Roma, Italy

1 Introduction

EXAFS investigations on a-Ge [1-3] deposited at different temperatures pointed to the nucleation of ordered grains in the amorphous matrix, whose size increases with temperature. This process was revealed by important modifications of the EXAFS spectrum, $\chi(k)$, and consequently of its Fourier transform. Besides these modifications in the EXAFS part of the spectrum, significant changes in the shape of the near edge structure (XANES) occur which should be correlated to the structural evolution of the system. It is the purpose of this paper to show that this is indeed the case and that a careful analysis of the near edge structure (associated with a calculation of the absorption cross section) can elucidate the relation between fluctuations of bond distances and edge shape.

2 Experimental

The samples were obtained by thermal evaporation onto Si(111) crystal substrates. The evaporation temperature, T_S, was varied in the range 130-350°C. The smoothness of the Ge films and the absence of macroscopic inhomogeneities were checked by a Scanning Electron Microscope. The optical thickness of the films ranged from $\mu x = 0.4$ to $\mu x = 0.7$. The crystalline sample was a Ge single crystal polished down to an optical thickness of $\mu x = 1.8$. The X ray spectra at the Ge K edge were taken at the Synchrotron Radiation Facility in Frascati, at LN temperature. The radiation was monochromatized by a Si(220) crystal, and the resolution was $\Delta E \sim 2$ eV. After background removal, the near edge spectra were normalized to the absorption far above the edge in order to eliminate their thickness dependence.

3 Results and discussion

In Fig.1 we report the experimental Ge K edges of several samples, normalized as described above, as well as their first derivatives. The main feature common to all spectra is the strong "white line" peak labelled A, due to the high density of empty p-like states. At $\Delta E \sim 15$ eV from the main peak a second structure B is present which exhibits significant variations as a function of deposition temperature: it is barely visible for the sample grown at T = 130°C, while it shows up more and more clearly at increasing temperature. In crystalline Ge peak B is well defined and prominent. The largest variation of this structure occurs in the temperature range 200 - 240°C, where the system undergoes the amorphous-to-crystal transition [1].

The question arises, then, about what is the correlation between the local order and this peak. One could argue that other processes (not structural in nature) could be at the origin of this structure. For in-

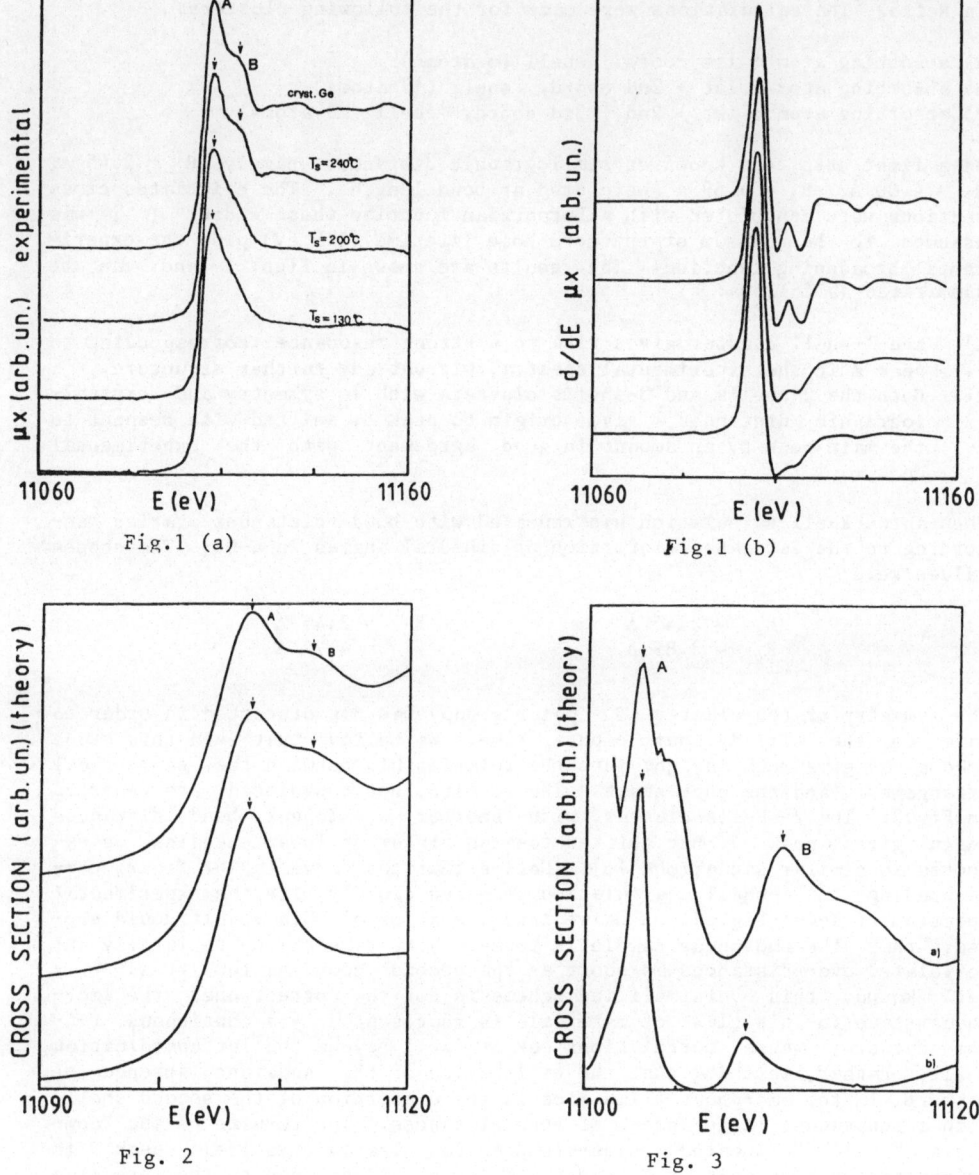

Fig.1 (a)

Fig.1 (b)

Fig. 2

Fig. 3

Fig. 3 Theoretical cross section obtained for a two shells cluster with a) 2nd shell "compressed" and b) "elongated". The spectra are not convoluted with a lorentzian function.

stance, it has been shown [4] that a similar structure occurring in some transition-metal chlorides can be explained in terms of multielectron excitations, as due to shake up and shake off processes. We believe though, that this is not the case for Ge: in fact, were a multielectron process at the origin of peak B, there is no way to explain why it should appear in crystalline and not in amorphous Ge. In order to ascertain the effect which originates that structure, we performed a calculation of the absorption cross section following the multiple scattering approach already used in Ref.5. The calculations were made for the following clusters:

a) absorbing atom + 1st coord. shell (5 atoms)
b) absorbing atom + 1st + 2nd coord. shell (17 atoms)
c) absorbing atom + 1st + 2nd + 3rd coord. shell (29 atoms)

As a first step, the known crystallographic distances, namely $R_1 = 2.45$ Å, $R_2 = 4.00$ Å, $R_3 = 4.69$ Å, were used as bond lengths. The calculated cross sections were convoluted with a Lorentzian function whose width, Γ, was assumed to be the sum of the core hole lifetime (1.9 eV) plus the experimental broadening function. The results are shown in Figs. 2 and can be summarized as follows:

i)   the 1-shell cluster gives rise to a strong resonance, corresponding to peak A in the experimental spectra, without any further structure.
ii) Both the 2-shells and 3-shells clusters with T_d symmetry and crystallographic distances, give origin to peak B, shifted with respect to the main peak by an amount in good agreement with the experimental one.

Then a two-shell calculation was repeated with bond distances varied according to the estimated distortion of dihedral angles in a-Ge. The chosen values were:

$$R_1 = 2.45 \text{ Å} \qquad\qquad R_1 = 2.45 \text{ Å}$$
$$R_2 = 3.89 \text{ Å} \qquad\qquad R_2 = 4.22 \text{ Å}$$

The symmetry of the cluster (T_d point group) was not distorted in order to save on the already long computer time. We believe that even this crude scheme can give much insight into the relationship between the geometrical arrangement and the edge shape. The results, not convoluted, are reported in Fig.3. The 2-shells cluster, with shorter or longer bond distances again gives peak B, but shifted towards higher or lower energies, as expected in similar situations [6]. Notice that the curve (a) of Fig.2, corresponding to 1-shell calculation compares closely with the experimental spectrum of a-Ge (Fig.1). A naive interpretation of this result could suggest that the amorphous sample is formed by tetrahedral units largely uncorrelated over distances as short as the second coordination shell. As is well known, this oversimplified scheme is not the correct one: the amorphous state in this class of materials is represented by a continuous random network, where correlations exist well beyond the 1st coordination shell. Rather, we think that the explanation of the apparent absence of peak B in the amorphous films lies in the distortion of the second shell, with a consequent large spread of bond distances. The results on the "compressed" and "elongated" clusters bear out clearly this idea: peak B is originated by a transition to empty scattering states, due to the presence of the second (and third) coordination shell. When these shells are distorted, the total absorption cross-section can be visualized as the superposition of several contributions corresponding to different distorted clusters, each of them contributing a B-like structure at a different energy. If the spread of bond distances is large enough, the resulting peak will be smeared out.

References

1 F.Evangelisti, M.G.Proietti, A.Balzarotti, F.Comin, L.Incoccia and
   S.Mobilio: Solid State Commun. $\underline{37}$, 413 (1981)
2 E.A.Stern, C.E.Bouldin, B. von Roedern and J.Azoulay: Phys. Rev. B$\underline{27}$,
   6557 (1983)
3 M.A.Paesler. D.E.Sayers, R.Tsu and J.Gonzalez-Hernandez: Phys. Rev.
   B$\underline{28}$, 4550 (1983)
4 E.A.Stern: Phys. Rev. Lett. $\underline{49}$, 1353 (1982)
5 C.R.Natoli, D.K.Misemer, S.Doniach and F.W.Kutzler: Phys. Rev. A$\underline{22}$,
   1104 (1980); F.W.Kutzler, C.R.Natoli, D.K.Misemer, S.Doniach and
   K.O.Hodgson: J. Chem. Phys. $\underline{73}$, 3274 (1980)
6 C.R.Natoli: "EXAFS and Near Edge Structure", Eds. A.Bianconi,
   L.Incoccia and S.Stipcich, Springer Verlag, 1983, p.43

Debye-Waller-Factor Modification in EXAFS with "Focussed" Multiple Scattering

N. Alberding and E.D. Crozier

Department of Physics, Simon Fraser University
Burnaby, B.C., Canada V5A 1S6

1. Analysis of Effects of Thermal Motion

The Debye-Waller factor of the single-scattering EXAFS formalism arises from the average of the static EXAFS spectrum over a Gaussian distribution of interatomic distances. This factor is

$$\exp(-2\sigma^2 k^2) = (1/\sqrt{2\pi\sigma^2}) \int_{-\infty}^{\infty} \exp[-(r-\bar{r})^2/2\sigma^2]e^{2ikr}\,dr \qquad (1)$$

where r is the interatomic distance, σ is its standard deviation from the mean distance $\bar{r}$ and k is the photoelectron wavevector. This single Debye-Waller factor no longer applies to thermal motion of a nearly colinear three-atom system such as shown in Fig. 1.

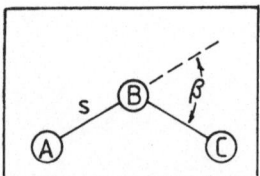

Fig. 1. Geometry for the case of "focussed" multiple scattering. The bond length to the bridging atom is s and β is the forward scattering angle.

For such a three-atom system, the effect of photoelectron backscattering from atom C on the EXAFS spectrum of atom A is composed of contributions from three pathways: A-C-A, A-B-C-A and A-B-C-B-A. The analysis of thermal motion in this case must include the following considerations:

1. Each photoelectron pathway has a different σ [1].
2. The magnitude of contributions of pathways ABCA and ABCBA depend on the forward scattering amplitude of atom B, $F_B(\beta,k)$, which strongly depends on the angle β.
3. A normal mode involves both bending and stretching. Because the photoelectron path distances are non-linear in bending coordinate, Gaussian distributions of the harmonic normal coordinates become asymmetric distributions of pathlengths.

BOLAND and BALDESCHWIELER [1] have addressed point 1 where through a normal-mode analysis they derive a Debye-Waller factor for each pathway. We have discussed point 2 in a previous paper [2]. Including all three considerations in the case of a harmonic system, the Debye-Waller factor for each pathway should be replaced by

$$(1/2\pi\sigma_1\sigma_2)\int_{-\infty}^{\infty}\int_{-\infty}^{\infty} \exp[-q_1^2/2\sigma_1^2-q_2^2/2\sigma_2^2]F_B^n(\beta,k)\exp[ikr_n(q_1,q_2)]dq_1dq_2 \qquad (2)$$

where q is the deviation of a normal coordinate from its equilibrium position, n is the order of scattering from atom B (n=0,1,2) and $r(q_1,q_2)$ is the pathlength expressed as a function of q. Generalization of (2) for more complicated systems with additional normal modes is straightforward.

In practice (2) may be integrated numerically. The derivation of approximate analytical expressions must not neglect the strong non-linearity of $F_B(\beta,k)$. In the case of oxygen multiple scattering [3], we obtain analytic expressions consistent with numerical integrations if, for $\beta < 50°$, we write

$$F_B(\beta,k) = [f_0 + f_1 \exp(-\gamma k^2 \beta^2/4)] \exp[i(\phi_0 + \phi_2\beta^2)] \tag{3}$$

where f_0, f_1, ϕ_0, ϕ_2 and γ may be k-dependent and are adjusted to best fit tabulated values of TEO [3]. Furthermore, we express r_0 and r_1 as quadratic in β (r_3 is independent of β). Thus changing variables from q_1 and q_2 to β and s gives Gaussian integrals which may be evaluated in closed form using the relation

$$\int_{-\infty}^{\infty} \exp[a\beta^2 + b\beta + i(u\beta^2 + v\beta)]d\beta =$$

$$(1/\sqrt{2}|a^2+u^2|^{\frac{1}{2}}\pi) \exp\left\{ \frac{a(b^2-v^2) + 2uvb}{4(a^2+u^2)} \right.$$

$$\left. +i\left[\frac{2abv - u(b^2-v^2)}{4(a^2+u^2)} - 0.5\ \tan^{-1}(u/a) \right]\right\}. \tag{4}$$

2. Examples

As an illustration we consider a hypothetical bridged Fe-O-Fe complex where the stretching and bending force constants are k_s = 3.0 mdynes/Å and k_b = .1 mdynes/Å. For a given bond angle and temperature the σ for each mode may be calculated as in [1]. A comparison of simulated spectra where integration of multiple scattering effects (1, 2 and 3) are included with those where only effect 1 is considered, reveals the importance of the appropriate treatment. For the linearly bridged geometry (β=0) Fig. 2a shows effects 2

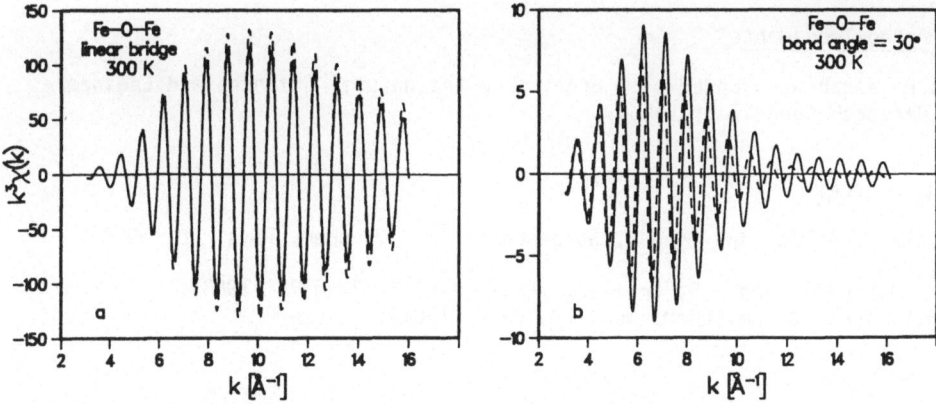

Fig. 2. Simulated Fe-edge EXAFS spectra where integration of all multiple scattering effects is included (solid curve) compared to neglecting variation of forward scattering amplitude and asymmetry of pathlengths over the range of thermal motion. Simulations are for 300 K for (a) the linearly bridged case β=0° and (b) the bent case β=30°.

and 3 cause a reduction in the amplitude because, at 300 K, thermal vibra-
tions may cause deviations of as much as 10° from the peak forward scattering configuration. Fig. 2b shows that for a bent system (β=30°) the combination of thermal vibration and asymmetry in pathlength-probability distribution results in a phase-shift and an increase of amplitude of the spectrum.
The amplitude depends critically on the interference of the three components. Fig. 3 shows that, as temperature increases, the amplitude of the
total spectrum may grow as the asymmetry and average forward-scattering
amplitude increase with greater vibrational amplitude.

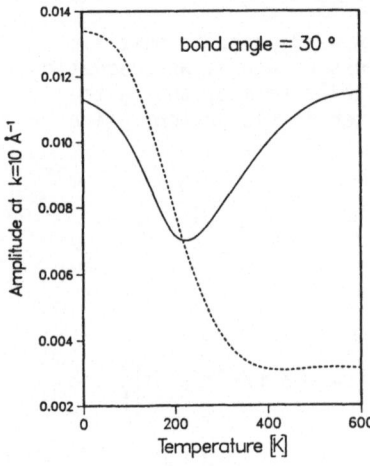

Fig. 3. Effect of integration of
multiple scattering effects
(solid) with their neglect for a
range of temperatures (dashed).
The crossover of the two curves
is due to the changing interference of the three pathways.

3. Conclusions

For the multiple atom geometry where "focussed" multiple scattering is
important, thermal motion can result in large variations of scattering
amplitude. Improperly averaging over the thermal distribution in such a
case would result in significant errors in determining such parameters as R,
N or σ^2.

4. Acknowledgements

This research was funded by a grant from the Natural Sciences and Engineering Research Council of Canada.

5. References

1. John J. Boland and John D. Baldeschwieler: J. Chem. Phys. 80, 3005
   (1984).
2. N. Alberding and E.D. Crozier: Phys. Rev. B 27, 3374 (1983).
3. B.K. Teo: J. Am. Chem. Soc. 14, 3992 (1981).

Theory of X-Ray Absorption Edge Spectra[+]

S. Doniach and M. Berding

Department of Applied Physics, Stanford University, Stanford, CA 94305, USA

T. Smith and K.O. Hodgson

Department of Chemistry, Stanford University, Stanford, CA 94305, USA

X-ray absorption edge spectra constitute a very tantalizing set of data since they are usually the easiest part of the x-ray absorption spectrum to measure in a complex chemical system. In this talk we focus on the measurement of absorption spectra for heavy atoms (principally copper) in low-symmetry environments. Our motivation is to understand how metallo-proteins "do their thing." Before getting to specific model systems, we discuss some of the theoretical problems in general.

1. Comparison of Theoretical Approaches

Two principal approaches have been tried: one-electron theory computed in a self-consistent molecular electron potential by the $X\alpha$ or Hedin-Lundqvist prescription (SCF-$X\alpha$), and the many-body Hartree-Fock plus configuration interaction approach, which has been applied to x-ray absorption edges by BAIR and GODDARD [1].

In the one-electron SCF approach, the one-electron Schrödinger equation is solved using a $\rho^{1/3}$ potential in the presence of the core hole created by absorption of the x-ray photon. The resulting electron configuration is allowed to relax by iteration of the Schrödinger equation for an electron in the potential determined by the electron density obtained from the previous step in the calculation. For states very close to threshold (bound levels), we have found that the convergence of this iteration can be very unstable in certain classes of molecular clusters, but can be stabilized by introducing a Fermi distribution at rather high temperature. After the convergence has started, the temperature may be reduced successively (Peter Feibelman, personal communication) until self consistency is obtained at zero temperature.

The above procedure defines an artifically metastable electronic state in which the core hole is frozen into the electronic configuration. Since the electron distribution has been allowed to relax, the resulting computation implicitly includes "shakedown" contributions to the lowest energy core-hole state of the molecular cluster.

The question then arises as to the importance of "shake up" satellites of this lowest-energy core-hole distribution, which are known to be very important in the XPS core-hole spectrum [2]. It has been known for quite a while that, in the case of plasmon satellites, the strength of the satellite goes to zero at threshold and increases quite slowly, possibly over a range of several hundred eV to its full strength in XPS when the escaping photoelectron has a large kinetic energy [3]. Recently E. STERN [4] has suggested that similar effects could account for the low intensity of shakeup satellites in x-ray absorption edge spectra. An example of such a satellite may be seen in

---

[+]Research supported by NSF Grant PCM 82-08115-A02

ref. [5], Fig. (5), where a one-electron SCF-Xα calculation was unable to find any dipole oscillator strength in the region of 8989 eV, but there appears to be a significant absorption bump in the experiment. We presumed that this was a satellite including more than one core hole in the final state electron distribution.

Bair and Goddard have proposed that a number of shakeup satellites occur in x-ray absorption edge spectra of copper compounds. Their calculations are based on the use of an "improved virtual orbital" method in which a set of localized Slater orbitals are used in a variational principle. Some of the resulting eigenvalues can have positive energy even though the trial wave functions used are localized in space. Thus, although this calculation is capable of treating the electron-electron interactions in much more generality than the Xα approach, it suffers from the defect that the continuum states are modelled by localized orbitals so that the fact that the photoelectron is actually leaking out of the molecule is not taken into account. It is hard to quantify what kind of errors are introduced by this assumption.

2. Use of the One-Electron Method for Interpretation of Copper K-Edge Spectra for Low Symmetry Model Compounds

Shown in Fig. 1 are edge spectra for a couple of model compounds; for further spectra see SMITH et al. [6]. As may be seen, there is a general chemical feature which is readily observed using the polarization dependence of the absorption spectra: for the E-vector polarized normal to the square planar complex, i.e., along the z-direction, two well defined absorption peaks are seen which are rather narrow in energy. On the other hand, when the polarization vector is in the x-y plane, a broadened peak is seen which is shifted up in energy relative to the lowest z-polarized peak. A simple chemical explanation is that because the axial ligands in these compounds are quite far (>2.5 Å) from the copper atom and do not have bonding character, the orbitals into which the photoelectron is excited in the z-direction are non-bonding orbitals, which are quasi-atomic in character. On the other hand, in the x-y plane, the ligand orbitals are strongly involved in bonding so that the unoccupied orbitals to which the photoelectron is excited are antibonding orbitals and hence shifted up in energy relative to the z-polarized orbitals.

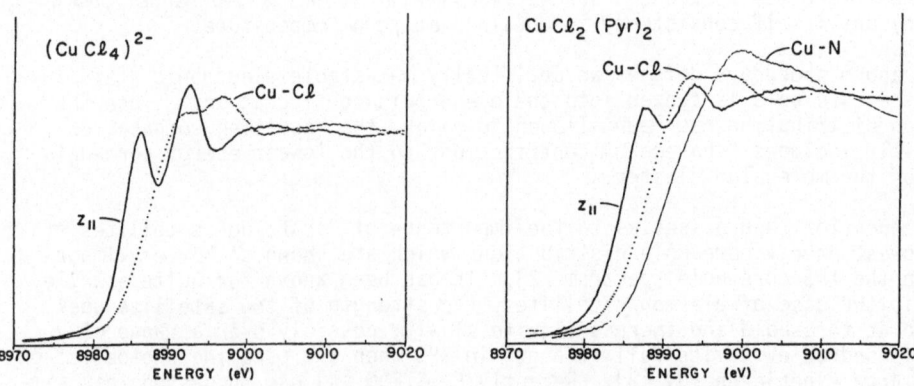

Fig. 1: K-edge absorption spectra as a function of E-polarization for $(CuCl_4)$ -bis creatinium (left panel) and $Cu-Cl_2$-di-pyridine (right panel). The instrumental resolution is lower (∿ 4 eV) in the right panel than in the left panel (∿ 1 eV).

This explanation is consistent with the observation that in the more covalent compound, copper tetraimidazole, the x-y absorption maximum is pushed to higher energies than in the more ionic compound CuCl₄. The mixed ligand compound in the x-y polarization has an energy-dependent absorption edge which varies between that for pure CuCl₄ when the E-vector is pointed towards the chlorine ligand and that for the pure Cu-N complex when the E-vector is pointed towards the nitrogen ligand.

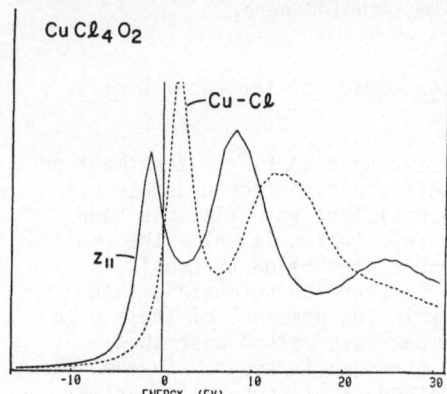 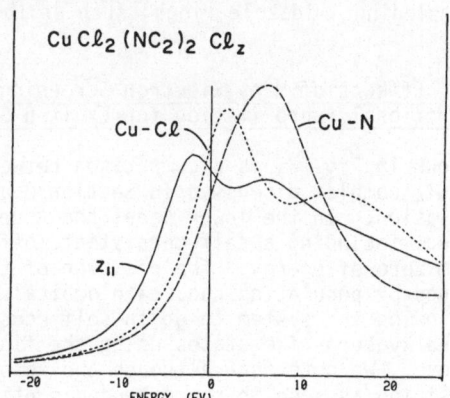

Fig. 2: Absorption cross sections as a function of E-polarization calculated for model clusters using the 'extended continuum' approach. Note the effect of the axial O-ligand producing an extra bump at +23 eV in the (CuCl₄)²⁻ cluster.

One-electron Xα calculations for model clusters simulating the above compound are shown in Fig. 2. In order to obtain the spectra shown, the outer sphere dividing the muffin-tin region from the asymptotic part of the wavefunction is removed (a Watson sphere charge is maintained to neutralize the charge of the cluster, however). By this means, the constant muffin-tin potential, which is below the zero of energy, is artificially extended to infinity, so that the states which would be bound in the self-consistent molecular cluster potential are now above the muffin-tin continuum edge. This "extended continuum" is an artifact which we have introduced to allow the appearance of bound states to be estimated in terms of a continuum T-matrix calculation. The effects of a true-bound-state calculation will be discussed below. As may be seen, the calculations give a reasonable qualitative account of the spectra seen experimentally but are quantitatively inadequate, both as to the relative absorption cross section in the different polarization directions and also as to peak positions. The calculations are able to account for the fairly well defined absorption peaks in the z direction, compared with an up-shifted resonance in the x-y direction. Note that the calculations have been done with quite small clusters which do not include second neighbors (in the case of CuCl₄²⁻). The splitting of the antibonding peak in the x-y plane (Fig. 1) is probably a second neighbor effect.

We believe that the quantitative inadequacy of the method has two sources:

a) the muffin-tin approximation is clearly rather poor for low symmetry compounds of the square planar type being discussed here. Earlier work [5] on tetrahedrally coordinated transition metal compounds gave rather better results than are found in this case;

b) energy dependence of the effective molecular potential seen by the photo-electron in the threshold region may be very important. This is well known in the case of band-structure calculations as, for instance, that for crystal-line silicon. Although groundstate properties are well calculated by the local density functional method, as used here, excited state energies are rather poorly calculated. In particular, the band gap of silicon is off by a factor of two [7], so it is not too surprising that a local density func-tional potential does not do a very good job for highly polarizable complexes (including imidazole rings) such as those being studied here.

3. Effects of Many-Electron Screening and Inadequacy of the Local Density Functional Approximation for Excited States

Shown in Fig. 3, is a comparison between the extended continuum treatment of $CuCl_4$ complex discussed in Section 2 and a self-consistent groundstate cal-culation. In the lower panel the true self-consistent potential has been used, including a self-consistent muffin-tin zero, which is below the contin-uum zero of energy. The position of the bound states below threshold is found by populating candidate orbitals in the relaxed ion potential, then allowing the system to go to self consistency in the presence of these occu-pied Rydberg-like states using the finite temperature method described above. We note that although the absorption strength is roughly in the same position as seen in the extended continuum method (i.e., non-self-consis-tent) the amplitudes are too small by something like a factor of two. This shows that the effect of the bound highly excited Rydberg electron on the relaxation of the other electrons in the cluster is having a sizeable effect on the one-electron wave function for the excited Rydberg state. This may be a reasonable treatment of the problem just below threshold, but immedi-ately above threshold the continuum calculation is performed using a poten-tial which is fully relaxed in the presence of the core hole but without the escaping photoelectron. Since, just above threshold, the escaping photo-electron moves quite slowly, the lack of screening due to this electron pre-sumably has biased the calculation of the threshold continuum cross section by quite a large amount.

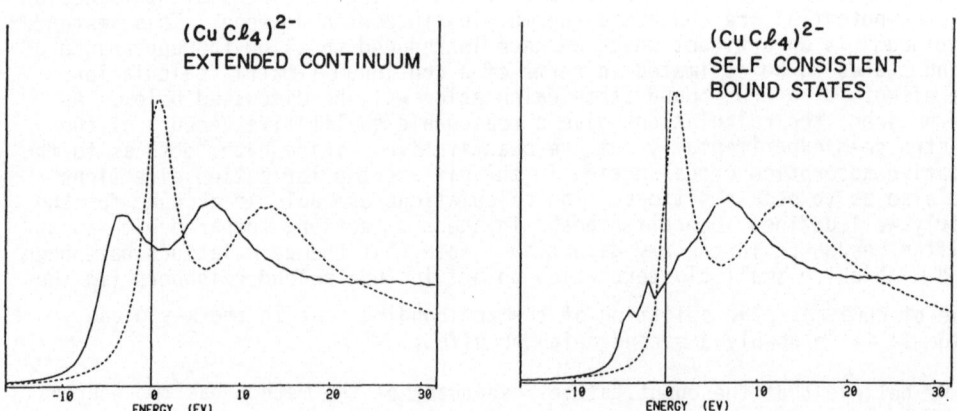

Fig. 3: Comparison of edge spectra computed using the 'extended continuum' approach with those obtained using a fully self-consistent bound-state cal-culation. The axial oxygen is not present in these calculations.

4. Conclusions

Although the one-electron Xα method appears to account for the principal spectral features seen in these low symmetry model compounds (i.e., shakeup satellites do not seem to be important for the interpretation of the spectra) the quantitative features of the calculated spectra are rather inadequate. Current work by C.R. Natoli to extend the muffin-tin method for molecular clusters in order to take account of the full nonspherical corrections to the molecular potential may help improve the accuracy of these calculations. However, estimates of the effects of screening due to the very slowly outgoing photoelectron just above threshold suggest that a highly energy-dependent molecular potential may be needed for these very polarizable complexes.

References

1. R.A. Bair and W.A. Goddard: Phys. Rev. B 22, 2767 (1980).
2. D.C. Frost, A. Ishitani, and C.A. McDowell: Molec. Phys. 24, 861 (1972)
3. J.J. Chang and D.C. Langreth: Phys. Rev. B 5, 3512 (1972) and B 8, 4638 (1973)
4. E.A. Stern: Phys. Rev. Lett. 49, 1353 (1982)
5. F.W. Kutzler, C.R. Natoli, D.K. Misemer, S. Doniach, and K.O. Hodgson: J. Chem. Phys. 73, 3274 (1980)
6. T.A. Smith, J. E. Penner-Hahn, K.O. Hodgson, M. Berding, and S. Doniach: article in this volume.
7. W. Hanke and L.J. Sham: Phys. Rev. B 21, 4656 (1980).

Distance Dependence of Continuum and Bound State of Excitonic Resonances in X-Ray Absorption Near Edge Structure (XANES)

C.R. Natoli

Laboratori Nazionali di Frascati - I.N.F.M., C.P. 13
I-00044 Frascati, Italy

The rising interest in XANES is due to the experimental evidence that information on coordination geometry and bonding angles, not given by EXAFS, can indeed be extracted from XANES [1,2]. In this paper I would like to point out that certain features of XANES, being sensitive to interatomic distances, can also be exploited to determine first coordination bond lengths to a few **percent** accuracy (5% or less) [3]. This aspect of XANES is particularly useful whenever EXAFS modulations are too weak to be measured, as in the case of low Z backscatterers. For the sake of concreteness, I shall limit myself to K-edge absorption spectra of simple molecules containing C-C, C-N, C-O type of bonds and of two series of Mn and V oxides (see Tables I and II). These cases are general enough to serve as an illustration of other cases.

In ref. [4], I suggested that the following relation

$$(E_r - \bar{V}) R^2 = C_r = const \qquad (1)$$

should correlate the energy position $E_r = \bar{E}_r - \bar{V}$ of a continuum resonance (measured from an average interstitial potential $\bar{V}$ in the material under consideration) and the bond length R between the absorbing atom and its nearest neighbor(s), responsible for the resonance through a "caging" effect.

Eq. (1) was obtained by imposing the condition Det M = 0 at the resonant energy, where M is the multiple scattering matrix [4]. Here I observe that the same relation can be obtained in m.s. theory on the basis of a simple scaling argument in case the maximum in the spectrum is not a resonant maximum, but is still due to neighbors at distance R from the photoabsorber.

In general, application of Eq. (1) is impaired by the fact that $\bar{V}$ is not an experimentally accessible quantity and may be different in the systems under comparison. To circumvent this difficulty, I suggest here the use of a relation similar to Eq. (1) for bound or excitonic resonances, i.e.

$$(E_b - \bar{V}) R^2 = C_b \qquad (2)$$

Subtraction of Eq (1) and Eq (2) then yields

$$(E_r - E_b) R^2 = C_r - C_b \qquad (3)$$

a relation containing only measurable quantities and of immediate application. By the term "bound or excitonic resonances" I mean those spectral pre-edge features due to transition to truly bound states in molecules (e.g., the π^* transition in diatomic molecules [5]) or to antibonding states of 3d character in metal oxides [6]. Both have in common the feature that they fall in an energy region where atomic resonances of the constituent atoms occur and, so to speak, are driven by them.

However, use of Eq. (2) needs some discussion, since the quantity C_b would in general depend itself on R, due to the **non-negligible energy dependence of the atomic phase shifts in** the resonant region [4]. A simple model of an excitonic resonance between two like atoms at distance R apart, driven by an atomic resonance in s state, located at energy E_a and with half width Γ, will illustrate the point. In such a case the condition for the two atom cluster resonance is

$$\cot g\,\delta(E) = \frac{\cos(\kappa R)}{\kappa R} \qquad (4)$$

where $\kappa = [E-\bar{V}]^{1/2} = \bar{E}^{1/2}$. Putting $\rho_a = \kappa_a R = [E_a-\bar{V}]^{1/2} R$, $\varepsilon = \bar{E}_a/\Gamma$, $\cot g\,\delta(E) \simeq (\bar{E}-\bar{E}_a)/\Gamma$ and, to within 10% accuracy, $\arccos(x) \simeq \pi/2(1-x)^{1/2}$, Eq. (4) can be solved analytically to give the wanted relation between the cluster resonance E_b and the bond length R. The result is

$$(E_b-\bar{V})^{1/2}R = \kappa_b R\pi/2 \left| \varepsilon + [2/\pi + 2\varepsilon/\rho_a + \varepsilon^2]^{1/2} \right| (1+\kappa\pi\varepsilon/\rho_a)^{-1} \qquad (5)$$

It is interesting to consider the two limiting cases, $\Gamma \to 0$ ($\varepsilon \to \infty$) and $\Gamma \to \infty$ ($\varepsilon \to 0$). For $\Gamma \to 0$, limit of infinitely sharp atomic resonance, $\kappa_b R = \rho_a = \kappa_a R$, implying that, whatever the bond length, the molecular resonance E_b is pinned down at the atomic value $\bar{E}_a$ and C_b in Eq (2) is $\bar{E}_a R^2$.

For $\Gamma \to \infty$, limit of very weak dependence of atomic phase shifts on energy $\kappa_b R = (\pi/2)^{1/2}$, recovering eq. (1), since in this case the resonance is a continuum resonance. The true state of affairs for a bound-excitonic resonance lies in between. For typical values of $E_a \lesssim 5$ eV, $1.2 \lesssim \rho_a \lesssim 2.5$, $\Gamma \sim 1.0$ eV, one finds $\pi\varepsilon/\rho_a \sim 1$, so that one is in the range of small deviation ($\sim 10\%$) of the r.h.s. of Eq. (3) from constancy, considering that usally $C_r \sim 5 C_b$ and $\Delta R/R \sim 30\%$. Although oversimplified, the model contains all of the relevant features of real systems and one can expect the above estimates to apply to them. One can then assume constancy of the r.h.s. of Eq. (3) and read off the consequences. When applied to free molecules, Eq. (3) states that the energy difference between the σ^* and the π^* resonance depends only on the bond length, provided initial state differences, like more or less covalency or polarity of the bond, have the same effect on the two excited states. For chemisorbed molecules one is led to the same conclusions, provided the two resonances are affected by the metal substrate shielding in the same way. Under these assumptions, the constant should be transferable from one phase to another.

If applied to metal oxides, Eq. (3) implies that the energy separation between the pre-edge excitonic feature and the first strong absorption maximum after the rising edge is a function of the bond length. Since the inflection point of the rising edge tends to follow this maximum, care should be exercised not to attribute its energy shift to an initial state effect, i.e., different valence state of the metal ion. Any initial state effect should therefore be assessed after correcting for this "bond length effect". Analysis of the edge shifts in the Mn and V oxide series along these lines reveals that initial state effects due to the different chemical states of the metal ion amount to less than 1 eV as opposed to 5-8 eV shifts caused by bond length variation. This result is in keeping with the conclusions of ligand field theory, whereby back-bonding effects tend to neutralize charge transfer from the metal to the ligands so that the net charge on the metal ion remains roughly constant along the series.

Table I

Molecule	Ref	1s Hole State	$E(\pi^*)$	I.P.	$E(\sigma^*)$	$\Delta_{\pi\sigma}$ (eV)	$R(\overset{o}{A})$	$\Delta_{\pi\sigma}R^2$ (ev$\overset{o}{A}^2$)
C_2H_2	7	C	285.6	291.1	310	24.4	1.204	35.4
C_2H_2	7	C	284.7	290.8	306	21.3	1.337	38.0
HCN	9	C	286.4	293.4	307.9	21.5	1.156	28.7
		N	399.7	406.8	420.8	21.1		28.2
C_2N_2	9	C	286.3	294.5	306.3	20.0	1.16	26.9
		N	398.9	407.4	419.3	20.4		27.4
N_2	8	N	400.9	409.9	418.9	18	1.095	21.6
CO	8	C	287.3	296.2	303.9	16.5	1.128	21.0
		O	534.1	542.6	550.9	16.8		21.4
H_2CO	8	C	286.0	294.5	300.9	14.9	1.21	21.8
		O	530.8	539.4	544.0	13.2		19.3
CH_3COH	8	C	286.3	294.0	301.0	14.7	1.22	21.9
		O	531.1	538.0	544.5	13.4		19.9
$(CH_3)_2CO$	8	C	286.8	293.8	301.0	14.2	1.22	21.1
		O	531.3	537.9	545.0	13.7		20.4

Table I summarizes the experimental data for the energy position (eV) of the π^* and σ^* resonances, the bond length (Å) and the ionization potential of several molecules in their gas phase. The last column gives the product $\Delta_{\pi\sigma}R^2$ (evA^2). In keeping with theory one finds that this quantity depends on the atom forming the pair and it is roughly constant within the same pair. (According to the Z + 1 rule, at Carbon and Nitrogen K-edge CO and N_2 have the same final state, so they are listed together).

Table II

Oxide	Ref	E_A(3d) (eV)	E_C(4p) (eV)	$\overline{R}$(Å)	$\Delta_{AC}R^2$ (evÅ^2)	Coordination
MnO	6	0.3	16.0	2.22	77.4	6
Mn_2O_3	6	0.5	20.0	2.01	78.8	6
MnO_2	6	0.7	22.0	1.90	76.9	6
$KMnO_4$	6	1.3	28.0	1.6	68.3	4
VO	10	3.0	20.4	2.05	73.1	6
V_2O_3	10	3.4	23.5	2.01	81.1	6
V_4O_7	10	4.1	25.1	1.98	82.3	6
V_2O_4	10	4.5	26.2	1.93	80.8	6
V_2O_5	10	5.6	30.1	1.83	82.0	5
NH_4VO_3	10	4.8	26.6	1.73	65.2	4
$CrVO_4$	10	4.8	24.2	1.76	60.1	4
Vanadinite	10	4.5	25.1	1.75	63.1	4
VP_c	10	3.9	27.3	1.58	58.4	1
VTPP	10	4.1	25.7	1.62	56.7	1

Similarly Table II summarizes, for the series of Mn and V oxides, the energy position E_A(3d) of the transition to an antibonding state of 3d character on the metal ion (feature A in ref [6]), the position E_C(4p) of the first strong maxiumum after the rising edge (feature C in ref [6]), the length of the metal to ligands bond $\overline{R}$ (or the average value if they are not all equal) and the product $\Delta_{AC}\overline{R}^2$ (evÅ^2).

Acknowledgments

Illuminating discussions with Dr. F. Sette on the subject of molecular resonances are gratefully acknowledged. Thanks are due to him and Dr. J. Wong for making preprints of their papers available to me prior to publication.

References

1. A. Bianconi, M. Dell'Ariccia, P.J. Durham and J. P. Pendry, Phys. Rev. B26, 6502 (1982).
2. G. Bunker and E. Stern, Phys. Rev. Lett. 52, 1990 (1984).
3. M. Dell'Ariccia, A. Gargano, C.R. Natoli, A. Bianconi, subm. for publ. to Phys. Rev. A.
4. C. R. Natoli, Proceedings of the 1st International Conference on EXAFS and XANES, Frascati, Italy 1982, Springer Series on Chem. Phys. Vol. 27 (1983) p. 43.
5. D. Dill and J.L. Dehmer, J. Chem. Phys. 65, 5327 (1976); F. Sette, these Proceedings.

6. M. Belli, A. Scafati, A. Bianconi, S. Mobilio, L. Palla-
    dino, A. Reale and E. Burattini, Solid State Comun. <u>35</u>,
    355 (1980).
7. A. P. Hitchcock and C.E. Brion, J. Electron Spectrosc. <u>10</u>,
    317(1977); ibid. <u>22</u>, 283(1981).
8. A. P. Hitchcock and C.E. Brion, J. Electron Spectrosc. <u>18</u>,
    1(1980); ibid, <u>19</u>, 231 (1980).
9. A. P. Hitchcock and C.E. Brion, Chem. Phys. <u>37</u>, 319(1979).
10. J. Wong, F.W. Lytle, R.P. Messmer and D.H. Maylotte, to be
    published in Phys. Rev. B.

Molecular Bond Length Determination from σ Resonances in Near Edge Spectra

A.P. Hitchcock
Institute for Materials Research, McMaster University
Hamilton, Canada L8S 4M1
F. Sette
NSLS, Brookhaven, Upton, NY 11973, USA
J. Stöhr
Corporate Research Science Laboratories, Exxon Research and Engineering Co.,
Annandale, NJ 08801, USA

1. Introduction

In recent years the K-shell excitation spectra of numerous gas phase molecules containing B, C, N, O and F have been studied by electron energy loss
or photoabsorption spectroscopy. A common feature of these spectra is one
or more relatively broad peaks in the region of the K-shell ionization threshold. Within a molecular orbital description these features may be attribu-
ted to $1s \rightarrow \sigma^*$ transitions. Alternatively, they may be described as σ shape
resonances associated with scattering of the excited core electron by adjac-
ent atoms. The scattering process and in particular, the energy at which
the excited electron is trapped by the intramolecular potential, is expected
to be a function of the local structure of the core excited atom. We have
examined literature molecular K-shell spectra [1] to investigate the relationship between molecular structure and σ resonances.

2. Results and Discussion

If one defines δ as the position of a σ resonance relative to the K-shell
ionization threshold, then one finds in almost all cases that δ varies linearly with the internuclear distance between the pairs of atoms giving rise
to the scattering resonance [2]. This linear relationship (Fig. 1) holds
remarkably well within classes of molecules characterized by the same Z_T,
the sum of atomic numbers of the pair of atoms involved (for example Z_T =
14 for B-F, C-O and N-N bonds).

Natoli [3] has used multiple scattering theory to predict the relationship $[k_r \cdot R = \text{constant}]$ between the bond length (R) and the photoelectron
wave vector (k_r) at the σ resonance. This implies a relationship

$$(\delta - V_0) \cdot R^2 = \text{constant} \tag{1}$$

where the inner potential (V_0) may be taken as the multiple scattering muffin tin potential in the interstitial region. In order to use this formalism to explain the observed linear correlation of δ and R, V_0 must vary
systematically with Z_T and R. Analysis of the empirical results within
this framework allows prediction of the average intramolecular muffin tin
potential in the presence of the core hole.

These K-shell σ resonances appear to be remarkably localized. This is
reflected in the near independence of their location on the presence of
other σ resonances. A typical example is the carbon 1s spectrum of solid
CH_3CN, recorded by synchrotron electron yield. When this is compared to
the C 1s gas phase electron energy loss spectra of C_2H_6 and HCN [4] (Fig. 2)
sigma resonances associated with scattering along the C-C and C≡N bonds in

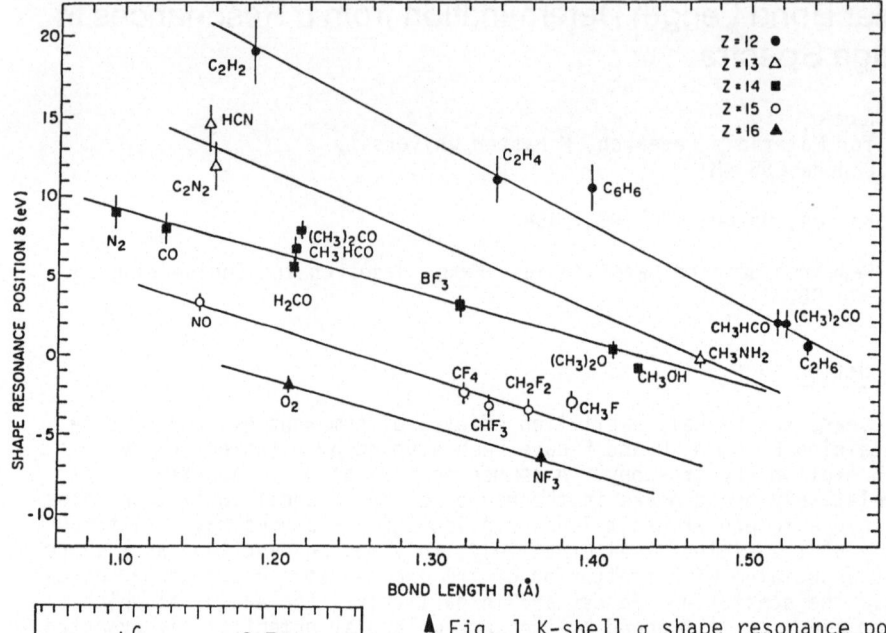

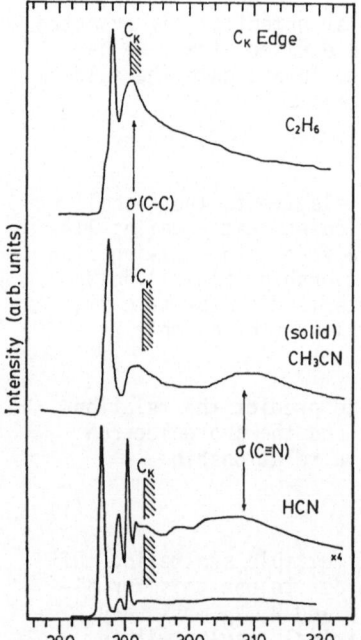

▲ Fig. 1 K-shell σ shape resonance position ($\delta = E_r - IP$) versus bond length. The lines are linear least squares fits to points in the same class of Z_T, the sum of atomic numbers of the bonded atoms

Fig. 2 The carbon 1s synchrotron electron yield spectrum of solid CH_3CN compared to the C 1s electron energy loss spectra of gas phase C_2H_6 and HCN [4]. The hatched vertical lines indicate the locations of the carbon 1s ionization potentials

CH_3CN are observed at the same location as the σ(C-C) resonance in C_2H_6 and the σ(C≡N) resonance in HCN. Since these resonances arise from localized intramolecular scattering, they appear in the spectra of both condensed phase and free molecules. With appropriate corrections the empirical linear relations derived from σ resonances in gaseous molecules can be used to quantitatively determine intramolecular distances in molecular solids and chemisorbed molecules. This, along with the directional nature of the resonance excitation, makes σ resonance features an excellent, quantitative probe of the orientation and bond lengths of surface adsorbed molecules [5].

The empirical σ(R) relationships derived from these observations may be used to aid spectral assignment or to determine bond lengths from the core spectra of free or surface adsorbed molecules. Analysis of the carbon 1s electron energy loss spectrum of gas phase perfluoro-2-butene (Fig. 3) [6] is a good example of this approach. In fluorine containing molecules the σ(C-F) resonances are very intense and located at negative δ (see Fig. 1). Thus we assign features 5 and 8 (Fig. 3) to σ(C-F) resonances involving excitation of the vinyl (-CF=) and methyl (-CF$_3$) carbon 1s electrons; feature 6 to the σ(C-C) resonance and feature 10 to the σ(C=C) resonance, both involving excitation of the (-CF=) vinyl carbon 1s electrons. The interatomic distances derived from the resonance positions and the least squares correlation lines (Fig. 1) for Z_T = 12 and 15 are 1.29 (C-F), 1.32 (C=C) and 1.54 Å (C-C). Although the structure of perfluoro-2-butene has not been reported previously, these values are in good agreement with bond lengths in similar molecules.

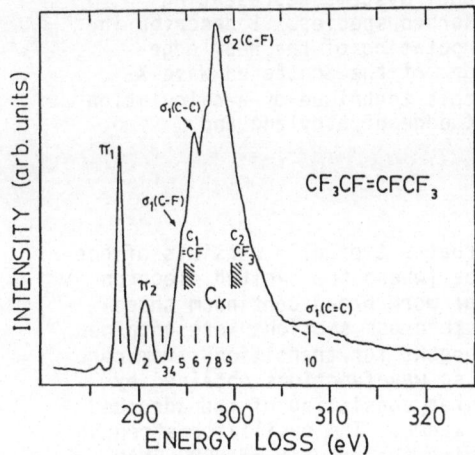

Fig. 3 The carbon 1s electron energy loss spectrum of gaseous perfluoro-2-butene (2.5 keV impact energy and 2° scattering angle). Hatched lines indicate XPS ionization potentials [7]

3. Conclusions

We have shown that there is a general correlation between σ-shape resonance position and bond length in a wide variety of molecules. The empirical δ(R) correlation [2] allows determinations of intramolecular distances accurate to ±0.05 Å in simple gas phase and chemisorbed molecules [5].

1. A.P. Hitchcock, J. Electron. Spect. 25, 245 (1982) (gas phase core-excitation bibliography).
2. F. Sette, J. Stöhr and A.P. Hitchcock, J. Chem. Phys. in press.
3. R. Natoli, EXAFS and Near Edge Structure, Springer Series in Chem. Phys. 27, 43 (1983).
4. A.P. Hitchcock and C.E. Brion, J. Electron. Spect. 10, 317 (1977); Chem. Phys. 37, 319 (1979).
5. F. Sette and J. Stöhr, this volume.
6. A.P. Hitchcock, S. Beaulieu, T. Steel, J. Stöhr and F. Sette, J. Chem. Phys. 80, 3927 (1984).
7. D.W. Davis, M.S. Banna and D.A. Shirley, J. Chem. Phys. 60, 237 (1974).

Calculation of the Near Edge X-Ray Absorption Structure of Adsorbates by the Scattered Wave Xα Method

J.A. Horsley

Corporate Research Science Laboratories, Exxon Research and Engineering Co.
Clinton Township, Annandale, NJ 08801, USA

Introduction

Near edge x-ray absorption spectroscopy is now being used to obtain information on the geometry and chemical bonding of adsorbed species [1,2]. In cases where the adsorbate geometry is not known with certainty it is useful to be able to compare the experimental near edge spectrum with a calculated spectrum for model systems representing various possible configurations of the adsorbed species. I describe in this paper a general technique for the computation of the near edge structure of adsorbates within the framework of the Scattered Wave Xα method and illustrate the application of this technique by a calculation of the near edge structure at the carbon K edge of ethylene and ethylidyne (CCH_3) on Pt(111).

Computational Method

The near edge structure of adsorbates typically consists of one or more fairly sharp bound state resonances (where the excited electron remains on the molecule) followed by one or more broad continuum shape resonances. Bound state and continuum state cross sections were obtained in separate calculations. Oscillator strengths for transitions from core levels to bound states were calculated using wavefunctions obtained by carrying out SCF-Xα calculations on a cluster consisting of the adsorbed molecule and the nearest neighbor surface atoms. The oscillator strength was calculated in the acceleration form using the program of NOODLEMAN [3]. The full core hole potential was used, with one electron promoted to the appropriate antibonding orbital. The calculated oscillator strength was converted to an absorption cross section, following DEHMER and DILL [4], using the relationship

$$\sigma(E) = (\pi e^2 h/mc)\ df/dE = (\pi e^2 h/mc)f\ dn/dE$$

where f is the oscillator strength and n is the principal quantum number. The height of the resonance is given by $\sigma(E)$ and the width by $(dn/dE)^{-1}$. As the latter quantity is difficult to estimate for valence levels I take a value of 2 eV for the resonance width, this being a typical value of the bound state resonance width at the carbon K edge, and adjust the height accordingly. The bound state resonance is then represented by a Lorentzian of height $\sigma(E)$ and full width at half maximum of 2 eV.

The position of the bound state relative to the ionization threshold was obtained from transition state calculations on the bound excited state and the fully ionized cluster. Absorption cross sections for the continuum resonances were calculated in the acceleration form using the program of DAVENPORT [5], modified to allow the use of a core orbital as the initial state. These continuum resonances appear to be localized on the adsorbate molecule as any shifts in the position of the resonance on chemisorption can be accounted for by changes in bond length [2]. These resonances were therefore calculated using the potential of the adsorbed molecule only, with the appropriate molecular geometry. The transition state potential, with half an electron removed from the core orbital, was used.

Near Edge Structure of Ethylene and Ethylidyne on Pt(111)

Near edge x-ray absorption spectra of ethylene and ethylidyne on Pt(111) have recently been obtained by KOESTNER et al [6]. The polarization dependence of the σ shape resonances shows that the ethylene lies down parallel to the surface plane while the ethylidyne (CCH$_3$) stands up perpendicular to the surface plane. For ethylene, two possible modes of bonding must be considered. One mode consists of the ethylene molecule π bonded to a single Pt atom, the hybridization remaining approximately sp^2. In the other configuration, known as the di-σ configuration the ethylene is bonded to two adjacent Pt atoms, and the hybridization should become approximately sp^3. The latter configuration is thought to be the correct one for ethylene on Pt(111) as surface vibrational spectroscopy [7] shows that the vibrational frequencies on the chemisorbed ethylene are very close to the di-σ bonded analogue C$_2$H$_4$Br$_2$. The observed near edge structure of ethylene on Pt(111) apparently contradicts this assignment however because the bound state resonance completely disappears for polarized light incident perpendicular to the carbon-carbon bond (with electric field vector along the bond). This would be expected for sp^2 hybridization as the π and π^* orbitals remain perpendicular to the bond, but not for sp^3 hybridization where the π and σ orbitals should strongly mix. I have calculated the near edge structure at the carbon K edge for a Pt$_2$C$_2$H$_4$ cluster in a geometry appropriate to sp^3 hybridized carbon. The results for polarized x-ray incidence perpendicular and parallel to the carbon-carbon bond are shown in Fig. 1. The agreement with experiment [6] is excellent, and in particular, the bound state resonance almost completely disappears for x-ray incidence perpendicular to the carbon-carbon bond. This shows that the observed near edge structure is indeed consistent with the di-σ configuration proposed previously, although it would appear that the ethylene π^* orbitals retain their identity in this configuration and do not mix with the σ orbitals. Molecular orbital contour plots of the

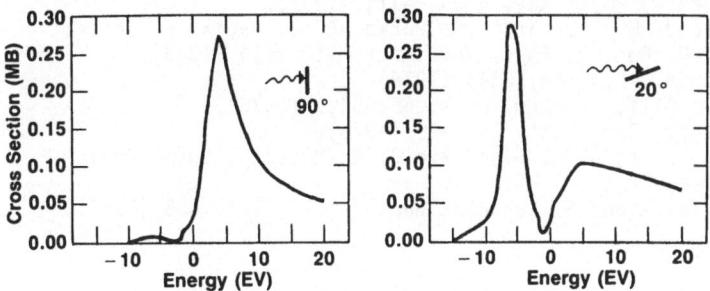

Fig. 1 Calculated Near Edge Spectra at the C K Edge for Pt$_2$C$_2$H$_4$ Cluster for Incident Angles 90° and 20° (with respect to the C-C bond)

empty orbitals of the $Pt_2C_2H_4$ cluster confirm this. Evidently the simple description of the bonding in the di-σ configuration in terms of sp^3 hybridization is not appropriate.

The near edge structure of ethylidyne on Pt(111) shows significant bound state resonances for both normal and grazing x-ray incidence (relative to the surface plane). The nature of these resonances has been determined by SCF-Xα calculations on a cluster consisting of the ethylidyne species bonded symmetrically to three Pt atoms, the ethylidyne being perpendicular to the plane of the Pt triangle. The empty states were found to be antibonding combinations of the Pt d orbitals and the p orbitals on the adjacent carbon atom. Transitions to states involving the carbon p orbitals parallel to the surface give rise to the resonance observed for normal x-ray incidence and transitions to states involving carbon p orbitals normal to the surface give rise to the resonance observed at grazing incidence. This assignment is confirmed by the near edge structure calculated using the Pt_3CCH_3 cluster, which is shown in Fig. 2. It can be seen that there is a significant bound state resonance for both angles of incidence. The agreement with experiment [6] is again good except for the Rydberg region between the bound and continuum resonances. The computational technique has therefore been able to reproduce the main features of the near edge structure of two adsorbates with very different kinds of bonding to the surface. It should prove particularly useful as an aid to interpreting the near edge spectra of complex adsorbed species, where assignment of the resonances may present serious problems.

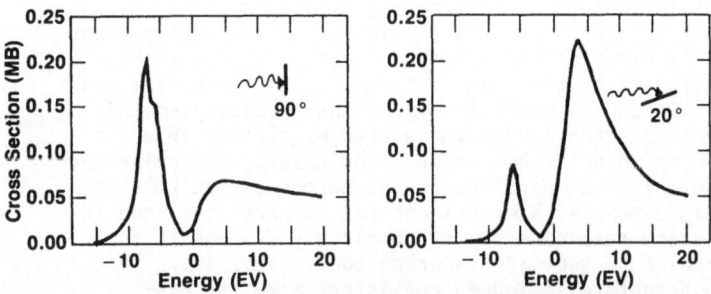

Fig. 2 Calculated Near Edge Spectra at the C K Edge for Pt_3CCH_3 Cluster for Incident Angles 90° and 20° (with respect to the Pt_3 plane)

References

1. J. Stohr and R. Jaeger, Phys. Rev. B 26, 4111 (1982).
2. J. Stohr, J. L. Gland, W. Eberhardt, D. Outka, R. J. Madix, F. Sette, R. J. Koestner and U. Dobler, Phys. Rev. Lett. 51, 2414 (1983).
3. L. Noodleman, J. Chem. Phys. 64, 2343 (1976).
4. J. L. Dehmer and D. Dill, J. Chem. Phys. 65, 5327 (1976).
5. J. W. Davenport, Phys. Rev. Lett. 36, 945 (1976).
6. R. J. Koestner, J. Stohr, J. L. Gland and J. A. Horsley, Chem. Phys. Lett. 105, 332 (1984).
7. H. Steininger, H. Ibach and S. Lehwald, Surface Sci. 117, 685 (1982).

Experimental Study of the L_3, L_2 and L_1 White Line Intensities in Rare-Earth Compounds

Joe Wong, Scott H. Lamson, and K.J. Rao

GE Corporate Research and Development, P.O. Box 8
Schenectady, NY 12301, USA

1. Introduction

This investigation was motivated by a recent observation that the intensity of the white lines at the L_3 and L_2 absorption edges of Nd^{3+} is markedly increased in going from pure NdF_3 crystal, which is highly ionic in character, to a NdF_3-doped BeF_2 glass [1]. The normal electronic configuration of the rare-earth (RE) ions in the trivalent state may be written as $(Xe)4f^n$, with n = 0, 1,...14 for La^{3+}, Ce^{3+}...Lu^{3+} respectively. The white lines in the L_3 and L_2 XANES spectra arise from an L-core → bound state transition. For the rare-earth ions, these transitions correspond to 2p → 5d ($^2P_{3/2} \rightarrow {}^2D_{5/2}$) and 2p → 5d ($^2P_{1/2} \rightarrow {}^2D_{3/2}$) respectively. Since the d final states in all the RE^{3+} ions are empty, one would expect a strong resonance and with uniform intensity across the 4f series. The present experimental study provides such a test and consists of a systematic measurment of the white line intensities at the L_3, L_3 and L_1 absorption edges in the complete RE oxide and fluoride series. Effects of ligand type and structure variation are also studied for the case of Nd halides (NdX_3, X = Fe, Cl, Br, and I) and chalcogenides (Nd_2Y_3, Y = S, Se and Te).

2. Experimental

L-edge spectra of the RE compounds were recorded with the C2 spectrometer at CHESS and CESR operating at an electron energy of 5.2 GeV and injection current of ~15 mA. The x-ray beam from CESR was monochromatized with a channel-cut Si(220) crystal and detuned to 50% to minimized harmonic contents at the energy of each L-edge. All spectra were recorded at room temperature in the transmission mode. Spectral specimens were prepared by mixing fine powders (~400 mesh) of the materials with Duco cement and casting the mull into films between two microscope slides. All RE compounds were purchased commercially as reagent grade and verified with an in-house x-ray diffraction analysis.

3. Results and Discussion

A typical experimental scan of the $L_{3,2}$ edge spectra of RE compounds is shown in Fig. 1 for the case Tb_2O_3. For both the RE oxide and fluoride series, the peak intensities of the white lines at the L_3 and L_2 edges bear an approximate ratio of 2 as shown in Table 1.

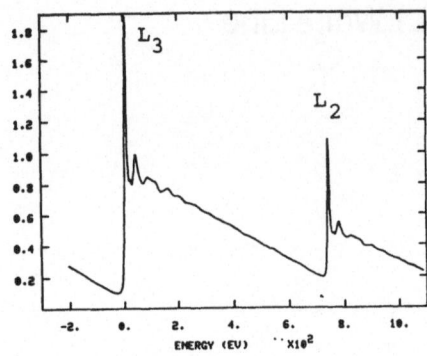

Fig. 1 Experimental Scan of the L_3 and L_2 spectra of Tb in Tb_2O_3. The zero of energy is taken at the L_3 edge of Tb Metal at 7514 eV.

Fig. 2 Normalized Intensities of L_3, L_2 and L_1 White Lines in RE_2O_3.

Fig. 3 Normalized Intensities of L_3, L_2 and L_1 White Lines in REF_3.

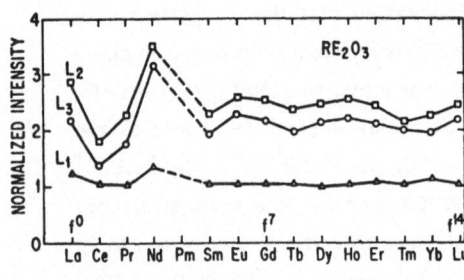

Fig.2

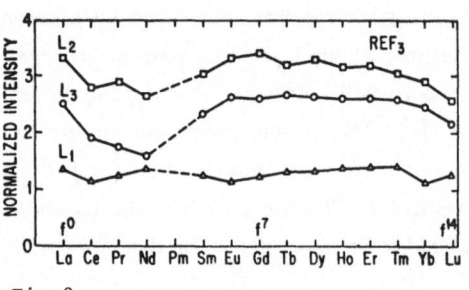

Fig.3

Table 1. Ratio of White Line Intensities at L_3 and L_2 Edges

	La	Ce	Pr	Nd	Pm	Sm	Eu	Gd	Tb	Dy	Ho	Er	Tm	Yb	Lu
Oxide	1.8	2.3	2.1	1.8	–	2.0	1.9	1.9	2.0	1.9	2.1	2.0	1.9	2.4	2.1
Fluoride	1.9	1.6	1.6	2.0	–	1.7	1.8	1.9	2.0	1.8	1.9	1.9	1.8	1.9	–

The true peak intensity can be depressed by thickness effects [2], which cause a deviation of this ratio from 2.0.

The XANES spectra at each L-edge were normalized with respect to its respective pre-edge and post-edge (EXAFS) backgrounds according to a procedure described elsewhere[3]. The normalized peak intensity is plotted for the oxide and fluoride series in Fig. 2 and 3. For L_1, the transition corresponding to 2p → 6p ($^2S_{1/2}$ → $^2P_{3/2}$), the normalized intensity is fairly constant and structureless across the 4f series. For the L_3 and L_2 transitions, however, the intensity variation is not constant. In both the oxide and fluoride series, the normalized intensity of the L_2 white line is always higher than that of the L_3 white line. In the fluorides, the normalized intensities of both the L_2 and L_3 white lines are higher than the corresponding ones in the oxide, indicating a ligand electronegativity and/or coordination number effect. In the RE fluorides, each RE ion is coordinated by 9 F^- ions, whereas in the RE oxides each RE ion is coordinated by 7 $O^=$ ions.

Further ligand effect may be seen in the isostructural series of Nd halides and chalcogenides. The normalized intensities of the white lines at the three L-edges

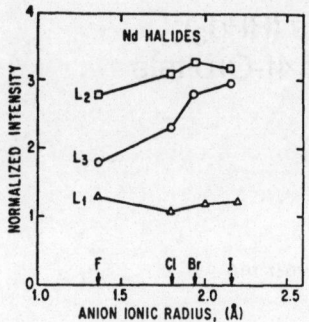

Fig. 4 Normalized Intensities of L_3, L_2 and L_1 White Lines in Nd Halides.

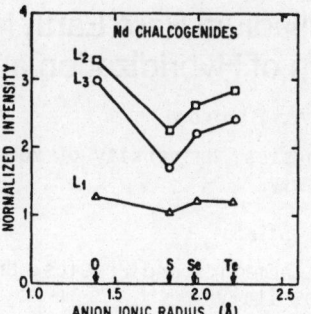

Fig. 5 Normalized Intensities of L_3, L_2 and L_1 White Lines in Nd Chalcogenides.

are plotted in Fig. 4 and 5 as a function of anion radius size. Again the intensity variation is not constant. The normalized L_3 and L_2 white line intensities in the halides are again higher than those of the chalcogenide compounds.

4. Concluding Remarks

From the present systematic study the following conclusions may be drawn. The intensity of the white lines at the L_3 and L_2 edges is a function of (a) the filling of the 4f levels across the rare-earth series, (b) ligand type and/or (c) coordination number. These factors contribute to the final density of states of the d-level which in turn alter the matrix elements associated with the $2p \rightarrow 5d$ transitions in the L-edge spectra. These effects will be analyzed more quantitatively in the future.

5. Acknowledgment

We are grateful for experimental opportunities at CHESS which is supported by NSF.

6. References

1. K.J. Rao, J. Wong and M.J. Weber: J. Chem. Phys. 28, 6228 (1983)
2. E.A. Stern and K. Kim: Phys. Rev. B23, 3781 (1981)
3. J. Wong, F.W. Lytle, R.P. Messmer and D.H. Maylotte: Phys. Rev. B (1984) in press.

XANES of Palladium Rare Earth Intermetallics (RPd$_3$): Determination of Hybridization and Mixing of 4f-Orbitals

A. Marcelli and A. Bianconi

Department of Physics, University of Rome, 'La Sapienza'
I-00185 Roma, Italy

I. Davoli and S. Stizza

Department of Mathematics and Physics, University of Camerino
I-62035 Camerino, Italy

INTRODUCTION

The single site promotional model of heavy rare earth valence fluctuating compounds has been supported by XPS and XANES experiments [1,2]. Light rare earth compounds exibith anomalous electronic properties due to hybridization of the 4f-orbitals of the rare earth with the conduction band.

In valence fluctuating materials the white line in the L_{III}-XANES of rare earth atoms due to the one-electron atomic-like transition $2p \leftrightarrow 5, \epsilon d$ transition appear to be splitted in two final state configurations $2p^5 4f^{n-1} c^{m+1} \overline{5, \epsilon d^1}$ (where c is the conduction band) and $2p^5 4f^n c^m \overline{5, \epsilon d^1}$ which are derived from the degenerate $4f^{n-1} c^m \leftrightarrow 4f^n c^{m-1}$ ground state configurations.

We have recently shown [3] that the L_{III}-XANES of LaPd$_3$, a system considered to have $4f^0$ configuration in the ground state, shows a weak low energy satellite at ~ 3 eV lower energy than the main line due to 4f-hybridization.

Here we show that the joint analysis of the palladium and the rare earth L_{III}-XANES of RPd$_3$ (R=La,Ce,Pr,Nd,Sm) intermetallic compounds gives the possibility to discriminate experimentally between the intra-atomic hybridization of $4f \leftrightarrow (5d6s)$ orbitals and the interatomic mixing of rare earth 4f-orbitals with the palladium 4d orbitals.

EXPERIMENTAL

The experiment was performed by measuring the x-ray trasmission through metal films at the Frascati Synchrotron Radiation facility using the storage ring ADONE.

We have performed experiments on the same sample using both Si(220) and Si(111) monochromators. The metal powder was obtained by filing single crystal under inert N$_2$ atmosphere. The powder was pound on agata mortar, always under inert atmosphere, to obtain much thinner grains. This procedure was found to be determinant to obtain homogeneous and pin-holes free layers.

RESULTS AND DISCUSSION

Figure 1 shows the measured absorption spectra at the L_{III} edge of intermetallics compounds RPd$_3$ (R=La,Ce,Pr,Nd,Sm) and CeCu$_2$Si$_2$. The vertical scales have been adjusted so that the maximum of the absorption coefficent is the same for all compounds. The photon energy scale is given by (hν-E$_0$), where E$_0$ is the energy of the white line maximum. A detailed analysis of the absorption spectrum of LaPd$_3$ and CePd$_3$ shows the presence of a double structure due to the $4f^{n+1}$ and $4f^n$ final states. In the mixed valence compound a third structure is present at higher energy due to a $4f^{n-1}$ final state ($4f^0$ in cerium).

The intensity of $4f^{n+1}$ configuration decreases going from La to Pr and Nd where it is nearly negligible within the experimental errors. Moreover the relative intensity decreases with the same trend observed in XPS 3d core lines [4] but with much lower intensity (in lanthanum the $4f^{n+1}/4f^n$ relative intensity is about 70% in XPS, to be compared with $\sim$12% in XANES).

In these compounds the hybridization decrases from La to Sm due to the contraction and shift of 4f-orbitals. The shake-down intensity ($4f^{n+1}$ configurations in all systems) decrease like the ratio $V/\Delta E$ where V is the total hybridization energy of 4f-electrons and the conduction states and ΔE is the "main line-satellite" energy distance [5]. We estimate from our date, a value of V of about 0.4 eV both from lanthanum and cerium compounds.

In fig.2 we report the palladium L_{III} absorption edge in the same RPd$_3$ compounds, with the normalized absorption coefficient alligned at the maximum of the white line. The great analogy of these spectra confirm the interpretation of XANES by the one-electron theory in the framework of the multiple scattering because the photoelectron excited at the palladium atoms along the series (RPd$_3$) is scattered in the same geometrical cluster and with pratically the same inter-atomic distance (the lattice constants change about 3%).

In the Pd L_{III}-XANES spectrum of LaPd$_3$ there is a unique structure at about 3 eV above the white line maximum which by comparison with other spectra cannot be assigned to elastic one-electron excitations. From our study of the rare earth L_{III}-XANES we know that the ground state of LaPd$_3$ can be described by two virtual degenerate configuartions $4f^0c^n$-$4f^1c^{n-1}$ where c is the valence band of the metal.

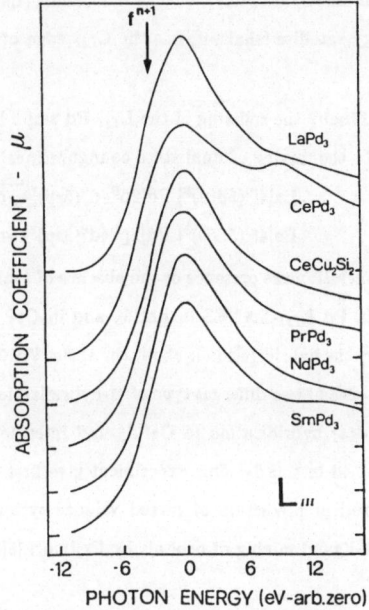

FIG.1 L_{III}-XANES of rare earths R in intermetallic compounds RPd$_3$. The CePd$_3$ spectrum shows the $4f^0$ peak at the high energy side of the "white line" typical of a mixed valent system.

The valence band of LaPd$_3$ is formed by (5d6s) lanthanum orbitals and by palladium 4d (5sp) orbitals forming a bonding Pd 4d valence band. If the two configurations are due to intra-atomic hybridization i.e. $4f^0(5d6s)^3 \leftrightarrow 4f^1(5d6s)^2$ at the L_{III} edge of Pd we do not expect to observe a splitting of the $2p \rightarrow 4,\epsilon d$ atomic like transition on the Pd site. This is demonstrated by the experimental observation that no splitting of the Pd $2p \rightarrow 4,\epsilon d$ transition appear in the L_{III} edge of CePd$_3$. In this mixed valence compound it is well established that there are fluctuations due to $4f^0c^{n+1} \leftrightarrow 4f^1c^n$ configurations. Therefore the electron should fluctuate between 4f and (5d6s) cerium derived conduction band and in this way there is not change in the local Pd valence charge.

The splitting of the palladium $2p \rightarrow 4, \epsilon d$ transition in $LaPd_3$ should be assigned to two final state configurations involving the palladium 4d valence band $La(4f^0)Pd(4d^9) \leftrightarrow La(4f^1)Pd(4d^8)$. The mixing of these configurations should arise from the mixing of La 4f and Pd 4d-orbitals in the valence band. So this mixing of interatomic orbitals gives in the XANES low energy satellite (shake-down) at the L_{III} edge of lanthanum (La $4f^1$) and a high energy satellite (shake-up) at the L_{III} edge of palladium (Pd $4d^8$).

Finally the splitting of the L_{III} Pd white line of $LaPd_3$ is due to the mixing of final state configurations:

$$La[4f^0(5d6s)^3] \; Pd[2p^5 4d^9 (5sp)^1 \overline{4, \epsilon d^1}]$$
$$La[4f^1(5d6s)^3] \; Pd[2p^5 4d^8 (5sp)^1 \overline{4, \epsilon d^1}]$$

In conclusion the presence or the absence of shake-up $4d^8$ satellite in Pd L_{III}-XANES in $LaPd_3$ and in $CePd_3$ respectively, where the hybridization is about the same ($V \simeq 0.4$ eV) show the existence of two different type of 4f-hybridization: intra-atomic 4f-(5d6s) hybridization in $CePd_3$ and interatomic mixing La 4f-Pd 4d in $LaPd_3$. Our experiment give first evidence of the theoretical prediction of mixed valence systems due to rare earth-ligand mixing of orbitals by Fujimori [6].

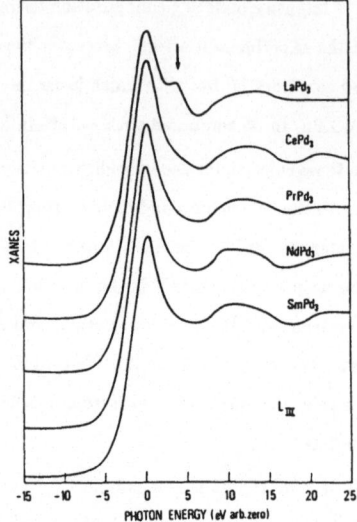

FIG.2 L_{III}-XANES of palladium in the intermetalli compounds RPd_3. The arrow shows the final stat configuration $La(4f^1)Pd(4d^8)$.

REFERENCES

1) N.Marthensson,B.Reihl,R.A.Pollak,F.Holtzberg,G.Kaindl and D.E.Eastman Phys.Rev.B 26,(1982)648

2) A.Bianconi,S.Modesti,M.Campagna,K.Fisher and S.Stizza J.Phys.C 14,(1981)4737

3) A.Bianconi,A.Marcelli,I.Davoli,S.Stizza and M.Campagna Solid State Commun. 49,5,(1984)409

4) F.U.Hillebrecht and J.C.Fuggle Phys.Rev.B 25,6,(1982)3550

5) S.-J.Oh and S.Doniach Phys.Rev. B 26,4,(1982)2085

6) A.Fujimori Phys.Rev.B 28,8(1983)4220

Shake-Down Phenomena and Polarized XANES Spectra of Cu(II) Complexes

Nobuhiro Kosugi, Toshihiko Yokoyama, and Haruo Kuroda

Department of Chemistry and Research Center for Spectrochemistry
Faculty of Science, The University of Tokyo, Hongo, Tokyo 113, Japan

1. Introduction

The shoulder structure about half-way up the Cu K-edge in the x-ray absorption spectra is well known to appear characteristically in Cu (II) compounds. Both the theoretical treatments based on the crystal field theory [1] and on the simple molecular orbital theory [2] failed to explain the origin of such a shoulder, and it was tentatively assigned to the vibronically allowed 1s-4s transition [3,4]. Recently, BAIR and GODDARD [5] ascribed, theoretically from an *ab initio* calculation on a model system $CuCl_2$, the shoulder structure to a group of dipole-allowed transitions involving 1s-4p excitations simultaneous with the *ligand-to-metal charge transfer* process. This newly proposed transition, called *shake-down*, is widely received as a candidate for the shoulder structure, but has not yet been experimentally confirmed.

We observed polarized Cu K-edge spectra of planar Cu (II) complexes by use of synchrotron radiation with the x-ray polarization (electric field vector **E**) perpendicular and parallel to the molecular plane. HAHN et al. [6] examined the intensity variation of the Cu $1s$-$3d_{x^2-y^2}$ pre-edge absorption peak of a planar $CuCl_4^{2-}$ complex on rotating about the normal axis with **E** kept parallel to the molecular plane, but examined nothing about polarization dependence of the shoulder and higher-energy structures. Our approach can separate structures associated with the out-of-plane $1s$-π^* transitions from those with the in-plane $1s$-σ^* transitions, and may provide an evidence for shake-down phenomena. The Cu (II) complexes selected in the present work are $(\text{creatininium})_2CuCl_4$ (nearly D_{4h} $CuCl_4^{2-}$) and $CuCl_2 \cdot 2H_2O$ (D_{2h} $CuCl_2(H_2O)_2$).

2. Experimental

The molecular plane of the square planar $CuCl_4^{2-}$ ion in the single crystal of $(\text{creat})_2CuCl_4$ is almost perpendicular to the (100) face and parallel to the b axis as shown in Fig. 1a. The molecular plane of $CuCl_2(H_2O)_2$ is perpendicular to the (101) face and to the b axis as shown in Fig. 2a. We measured the x-ray absorption spectra for the polarizations parallel and perpendicular to the b axis by rotating the sample crystal around the axis normal to the (100) or (101) face which had been set normal to the x-ray beam, by use of the EXAFS facilities at Beam Line 10B of Photon Factory in National Laboratory for High Energy Physics (KEK-PF). A Pb plate with an orifice of the diameter of about 1 mm was placed in front of the crystal to make the beam size smaller than the crystal face in order to prevent the leakage of x-ray. The channel-cut Si(311) monochromator was operated with the energy step of 0.09 eV in the pre-edge and XANES regions and with the step of 0.5 eV in the EXAFS region.

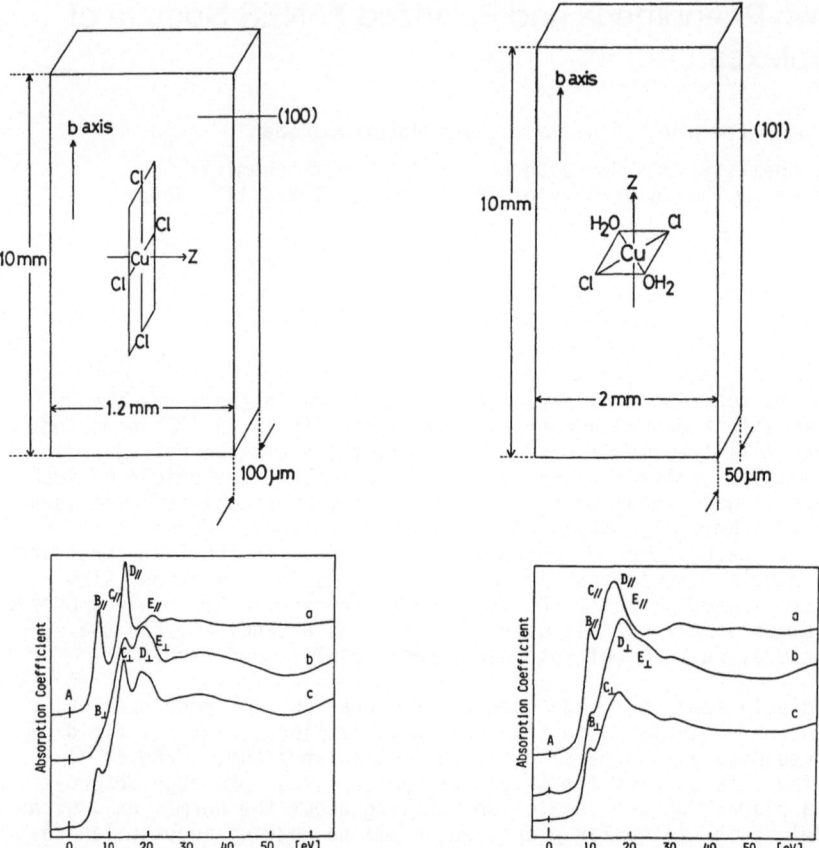

Fig.1a,b. (a) The dimension and morphology of the single crystal of
(creat)$_2$CuCℓ_4. (b) XANES spectra of
(creat)$_2$CuCℓ_4 (a) // z (out-of-plane)
polarization (b) ⊥z (in-plane) polarization (c) powder spectrum

Fig.2a,b. (a) The dimension and morphology of the single crystal of
CuCℓ_2·2H$_2$O. (b) XANES spectra of
CuCℓ_2^2·2H$_2$O (a) // z (out-of-plane)
polarization (b) ⊥z (in-plane) polarization (c) powder spectrum

3. Results and Discussion

The Cu K-edge spectra of the single crystals for the x-ray polarizations
parallel (// z) and perpendicular (⊥z) to the z axis normal to the molecular planes are shown in Figs. 1 and 2 together with the spectra of the
powder samples. The structures in the // z spectra should be attributed to
the excitations with the transition moments perpendicular to the molecular
plane and those in the ⊥z spectra to the ones with the transition moments
within the molecular plane. Distinct difference can be seen between the // z
and ⊥z spectra, and the powder spectra nearly correspond to the superposition
of the // z and ⊥z spectra. The weak pre-edge peaks A arise mainly from 1s-
3d$_{x^2-y^2}$ transitions [6]. For the (creat)$_2$CuCℓ_4 samples, the structure B,
which appears as a shoulder in the powder spectrum, appears as a sharp and
strong peak in the // z spectrum, but no corresponding peak can be found in

the $\perp z$ spectrum. From this polarization character, we can conclude that the structure B is associated with an out-of-plane transition and cannot be due to the vibronically allowed 1s-4s transition. In the spectra of $CuC\ell_2 \cdot 2H_2O$, the shoulder structure B is associated partially with an in-plane transition. The most prominent structure must be associated with the dipole-allowed 1s-4p transition; it is most likely that the peaks $D_{//}$ and $D_{\perp}$ are associated with 1s-4pπ and 1s-4pσ transitions, respectively. The structures E are likely to be due to 1s-5p transition. In any case, the structure B and C cannot be interpreted within the single-electron excitation picture.

The results of the theoretical calculations [5,7] show that every 1s-np excitation splits into two: the one for the *normal* transition and the other for the process involving *shake-down*. We can interpret the structures observed in the polarized XANES spectra of $(creat)_2CuC\ell_4$ and $CuC\ell_2 \cdot 2H_2O$ as shown in Table 1. The results of the present polarized XANES measurements provide an evidence for shake-down phenomena.

Table 1. Energies [eV] and assignments of the polarized XANES

Assignment	$(creat)_2CuC\ell_4$		$CuC\ell_2 \cdot 2H_2O$	
	$// z$ (π)	$\perp z$ (σ)	$// z$ (π)	$\perp z$ (σ)
A : 1s-3d	0.0	0.0	0.0	0.0
B : 1s-4p + shake-down	7.06	(sh.)	9.71	10.21
C : 1s-5p + shake-down	} 13.63	13.80	~13.8	(sh.)
D : 1s-4p		18.27	15.20	17.23
E : 1s-5p	21.47	~21.5	(sh.)	~22.5

References

1. F. A. Cotton and C. J. Ballhausen: J. Chem. Phys. 25 , 617 (1956)
2. W. Seka and H. P. Hanson: J. Chem. Phys. 50 , 344 (1969)
3. F. A. Cotton and H. P. Hanson: J. Chem. Phys. 25 , 619 (1956)
4. S. I. Chan, V. W. Hu, and R. C. Gamble: J. Mol. Struct. 45 , 239 (1978)
5. R. A. Bair and W. A. Goddard III: Phys. Rev. B22 , 2767 (1980)
6. J. E. Hahn, R. A. Scott, K. O. Hodgson, S. Doniach, S. R. Desjardins, and E. I. Solomon: Chem. Phys. Letters 88 , 595 (1982)
7. N. Kosugi, T. Yokoyama, K. Asakura, and H. Kuroda, submitted to Chem. Phys.

Polarized Cu K-Edge Studies

Teresa A. Smith, James E. Penner-Hahn, and Keith O. Hodgson
Department of Chemistry, Stanford University, Stanford, CA 94305, USA
Martha A. Berding and Sebastian Doniach
Department of Applied Physics, Stanford University, Stanford, CA 94305, USA

1. Introduction

The X-ray edge and near-edge region, typically 0-50 eV above the threshold
energy, is often rich in structure which reflects the local environment of the
absorbing atom. Edge and near-edge features are poorly resolved in the K-edge
spectra of most amorphous samples. Polarized studies result in the enhancement of the edge and near-edge feature intensities, and the symmetry properties of the final states responsible for these features may be determined from
their angular properties. We have utilized the high flux and plane-polarized
nature of synchrotron radiation to study the polarized K-edge spectra of
oriented single crystals of a number of square-planar or flattened tetrahedral
(T_d) Cu(II) and linear Cu(I) model compounds. The intensity, shape and location of observed polarized edge features are examined as probes of the electronic and geometric environment of the metal center.

2. Experimental

Spectra were measured at SSRL using a Si(220) double crystal monochromator
with the storage ring operating in dedicated mode (3.0 GeV electron energy,
55-70 mA current) or on a wiggler beam line in parasitic mode (1.8 GeV,
12-15 mA). Powder data were measured as transmission using N_2-filled ionization chambers. Single crystal data were measured as excitation fluorescence
detected by an array of NaI(Tl) scintillation detectors. Single crystals were
oriented on a Syntex $P2_1$ diffractometer, and alignment in the SSRL experimental station was achieved on a two-circle lucite goniometer. The spectra
presented, which represent the average of 2-5 data scans, are pre-edge background-subtracted and normalized to give an edge jump of 1.0.

3. Results and Discussion

The Cu(II) complexes consist of CuN_4, $CuCl_4$ and trans mixed-ligand complexes
(CuN_2Cl_2), where N = imidazole or pyridine ligands. A series of complexes
with varying axial atom distances and equitorial Cu-ligand bond lengths was
studied. The flattened T_d complexes are characterized by the flattening
angle, θ, which is a measure of the extent of distortion from a purely T_d to
a square-planar geometry ($\theta=109.5°$ and $180°$, respectively). The Cu(I) complexes have linear CuN_2 coordination environment (N = pyrazole). The molecular orientations studied are defined by the orientation of the direction of
polarization ($\vec{e}$) of the incoming radiation with respect to the molecular axes.
The resulting polarized spectra are referred to as $\vec{z}||\vec{e}$ or $\vec{z}\perp\vec{e}$ for the Cu(II)
complexes and Cu-N$||\vec{e}$ or Cu-N$\perp\vec{e}$ for the Cu(I) complexes. A pictorial summary
of the molecular orientations is shown in Fig. 1.

Typical unresolved fine structure observed in the Cu K-edge spectra of some
amorphous samples may be seen in Fig. 2a, which compares the K-edge spectra of

58

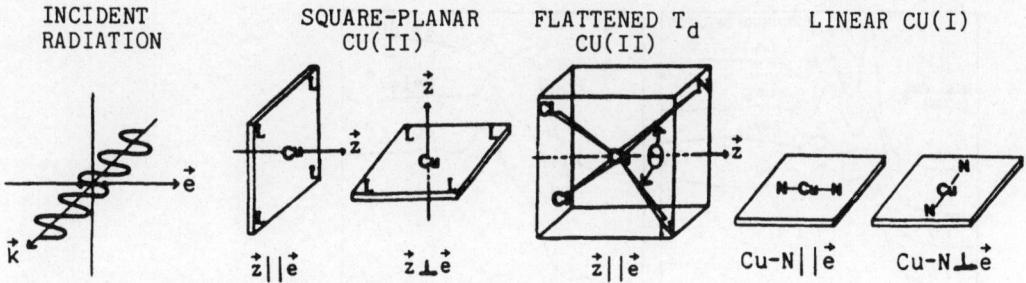

INCIDENT RADIATION   SQUARE-PLANAR CU(II)   FLATTENED T_d CU(II)   LINEAR CU(I)

$\vec{z}||\vec{e}$ $\vec{z}\perp\vec{e}$ $\vec{z}||\vec{e}$ Cu-N$||\vec{e}$ Cu-N$\perp\vec{e}$

Fig. 1.   Summary of molecular orientations.

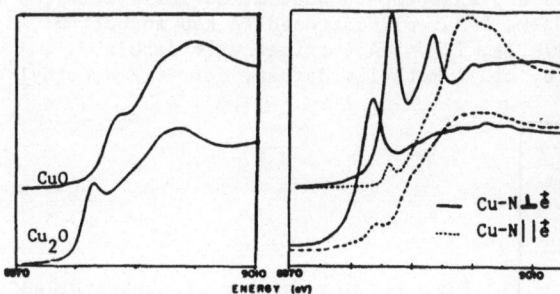

— Cu-N$\perp\vec{e}$
···· Cu-N$||\vec{e}$

Fig. 2. Comparison of Cu(I) and Cu(II) spectra. a) Powder spectra of Cu(I) and Cu(II) oxides. b) Polarized single crystal spectra. Upper: square-planar Cu(II) [Cu(TRI)$_4$ 2ClO$_4$], Lower: linear Cu(I) [Cu(TMP)$_2$BF$_4$]. TRI=1,4,5-trimethylimidazole TMP=1,3,5-trimethylpyrazole

Cu(I) and Cu(II) oxide. Our polarized studies of square-planar Cu(II) and linear Cu(I) reveal that the unresolved low-energy features in powder spectra appear as sharp, 'white line' type features which are highly polarized in the direction perpendicular to the Cu-ligand bonds (Fig. 2b). The highly directional and intense nature of these features indicate transitions to bound or localized continuum final states of p_z or p_π character for the square-planar and linear complexes, respectively. Our polarized theoretical calculations of square-planar CuCl$_4$ using a symmetry-adapted multiple scattered wave SCF X-α method successfully reproduce the marked dichroism exhibited by the square-planar complexes, and there is qualitative agreement with the general features observed in both the $\vec{z}||\vec{e}$ and the $\vec{z}\perp\vec{e}$ spectra [1].

All square-planar Cu(II) spectra in Fig. 3a show two prominent features in the $\vec{z}||\vec{e}$ orientation at ~8986 and 8993 eV, while the linear Cu(I) spectrum (Fig. 2b) shows only a single sharp feature at ~8983 eV in the Cu-N$\perp\vec{e}$ orientation. Changes in equitorial ligands and axial atom distance in the square-planar compounds have little effect on the positions of the two $\vec{z}||\vec{e}$ features (<2eV shifts); however, the relative intensities of the two features are sensitive to these structural differences. The general features observed in the square planar spectra are present but less pronounced in the analogously polarized spectra of flattened T_d complexes. The $\vec{z}||\vec{e}$ spectra of a series of three mixed-ligand complexes with geometries ranging from nearly T_d (θ=115°) to square planar are shown in Fig. 3b. Examination of these spectra reveals the following trends in response to increasing θ: a) The intensity of the lower-energy 8986 eV peak increases in intensity and b) The 1s → 3d pre-edge transition at ~8979 eV changes from a z-polarized transition, indicating d-p$_z$ mixing in the nearly T_d geometry, to an xy-polarized, quadrupole-coupled transition in the square planar geometry [2].

In contrast to the $\vec{z}||\vec{e}$ spectra, the $\vec{z}\perp\vec{e}$ spectra are characterized by relatively featureless edges with a single broad or split principal maximum lo-

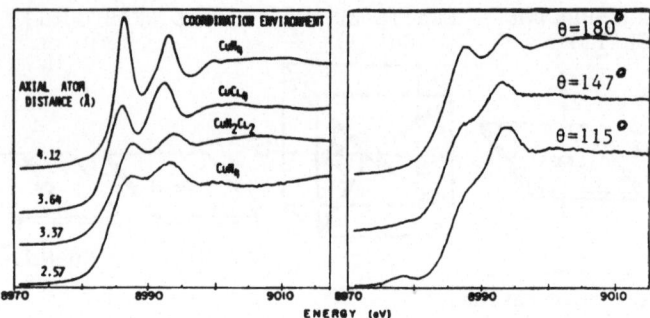

Fig. 3. $\vec{z}\|\vec{e}$ Cu(II) spectra a) square-planar top to bottom: Cu(TRI)$_4$ 2ClO$_4$, (creatinium)$_2$-CuCl$_4$, Cu(mp)$_2$Cl$_2$, CuIm$_4$ 2NO$_3$. b) Flattened Td top to bottom: Cu(mp)$_2$Cl$_2$, Cu(mi)$_2$Cl$_2$, and Cu(dmi)$_2$Cl$_2$. TRI = 1,4,5-trimethylimidazole, mp = 2-methylpyridine, IM = imidazole, mi = 1-methylimidazole, dmi = 1,2-dimethyl-imidazole

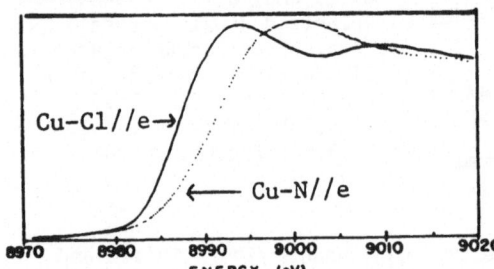

Fig. 4. $\vec{z}\perp\vec{e}$ spectra of square-planar CuIm$_2$Cl$_2$. Comparison of Cu-Cl$\|\vec{e}$ and Cu-N$\|\vec{e}$.

cated between ~8990 and 9000 eV. The position and shape of the $\vec{z}\perp\vec{e}$ maxima are sensitive to the nature and distance of the ligands in the direction of polarization. Polarized studies of mixed-ligand complexes permit the investigation of the effect of Cu-ligand bond distance in the absence of other electronic and structural changes. Figure 4 shows the Cu-N and Cu-Cl$\|\vec{e}$ spectra for the mixed-ligand complex, CuIm$_2$Cl$_2$, which exhibit an increase in the principal maximum location of ~7eV when polarized along the shorter Cu-N bond direction. Our theoretical calculations of $\vec{z}\perp\vec{e}$ spectra for a mixed-ligand model are successful in reproducing the experimentally observed shift in response to the bond distance change of 0.36 A [1]. A shift of this magnitude is consistent with the $1/r^2$ dependency of continuum shape resonance positions predicted by NATOLI [3]. The position of the principal maximum in the Cu-N$\|\vec{e}$ spectrum of linear Cu(I) is observed at approximately the same energy as the comparable Cu-N$\|\vec{e}$ peak for Cu(II) (Fig. 2b), although the Cu-N distance in the Cu(I) complex is ~0.13 A shorter. Based on bond distance effects alone, a 2-3 eV shift to higher energy would be predicted for the Cu(I) complex. The absence of a shift of this magnitude indicates that additional electronic differences between the Cu(I) and Cu(II) complexes are responsible for a 2-3 eV shift in the opposite direction. A comparable shift to lower energy is observed in the position of the first Cu-N$\perp\vec{e}$ peak located at 8983 and 8986 eV for Cu(I) and Cu(II), respectively.[4]

References

1. S. Doniach, M. Berding, T. Smith, and K. O. Hodgson: this volume.
2. J. E. Hahn, R. A. Scott, K. O. Hodgson, S. Doniach, S. R. Desjardins, and E. I. Solomon: Chem. Phys. Lett., 88(6), 595-598 (1982).
3. C. R. Natoli: Springer Ser. Chem. Phys., 27, 43-56 (1983).
4. This work is supported by NSF Grant #PCM-82-08115.

Near Edge Structure of Highly Ionized Manganese in Mn-doped Chevrel Compounds

F.C. Brown, W.M. Miller, D.M. Ginsberg, and K. Stolt

Department of Physics and Materials Research Laboratory,
University of Illinois, Urbana, IL 61801, USA

This paper was withdrawn after the proceedings had been assembled in its final form. In order to keep to a rapid publication schedule the succeeding pages have not been renumbered; pages 62 and 63 are therefore missing

Molybdenum $L_{II,III}$ Edge Studies

Britt Hedman, James E. Penner-Hahn, and Keith O. Hodgson

Department of Chemistry, Stanford University, Stanford, CA 94305, USA

1. Introduction

X-ray absorption edge structure has proven quite useful in the study of
first-row transition metals; Some of these applications are discussed elsewhere in this book. There have been fewer studies of edge structure for
second-row transition metals, due in part to the fact that the K absorption
edges for these metals occur at higher energy (>17keV). At higher energy,
the monochromator energy bandpass and the intrinsic (core hole lifetime)
broadening of the edge features both increase. These effects combine to
give an effective resolution of, for example, 10 eV or worse at the Mo K
edge. Low resolution thus hinders detailed studies of the edge structure
of many important metalloproteins. These resolution degrading effects are
less severe at lower energies, thus suggesting that edge structure which is
unresolvable at the Mo K edge may be resolved at the L edges. An additional advantage of L edge spectroscopy is the ability to study both **s** and **p**
symmetry initial states, thereby enhancing the electronic information which
can be obtained.

The Mo L edges occur at an energy which is experimentally difficult to
reach (2500-2900 eV). The traditional approach to this energy range has
been via UHV systems, thus making most biological systems impossible to
study. The alternative experimental approach, using hard x-ray beam lines,
is not straightforward, since absorption in the beam lines (Be and Kapton
windows etc.), in the scintillation detectors (Be windows), as well as in
the beam paths (He and air) greatly reduces the available x-ray flux at
these low energies. However, LYTLE et al. [1] have recently shown that an
experimental setup with a fluorescence ionization chamber detector [2] has
sufficient sensitivity to permit measurement of edge spectra at energies as
low as at the S K edge (2476 eV). We have recently utilized this detector
system in a preliminary study of a series of molybdenum compounds having
varying oxidation states and coordination environments.

Although the detailed origin of different edge features is not well
understood, this technique is still quite useful as an empirical probe of
the geometrical and electronic structure of a metal site. Since x-ray edge
structure is sensitive to the precise geometrical arrangement of ligands
around an absorbing site, it is potentially extremely useful for addressing
structural biological questions. In order to utilize more fully the information contained in x-ray absorption edge spectra, it is necessary to
understand in detail the relationships between molecular structure and edge
structure. The experiments described herein, while lacking immediate biological applications, are directed towards this goal of understanding
better the physical origin of x-ray edge structure.

2. Experimental

Data were collected at the Stanford Synchrotron Radiation Laboratory (SSRL) during dedicated and parasitic conditions, using unfocussed beam lines I-5 and IV-2 (wiggler), respectively. Harmonic rejection was accomplished by detuning the Si(111) double crystal monochromator to give a decrease in I_0 of 40 to 80 %. A He beam path minimized the air absorption. The fluorescence detector was filled with N_2. Samples were powdered and mounted on Mylar tape either directly, or mixed in Duco^R cement. The spectra shown below, which represent data from one to three scans, were pre-edge background subtracted and normalized to give an edge jump (post-edge minus pre-edge) of 1.0. The data were smoothed by convolution with a Gaussian of 0.23 eV fwhm. The energy resolution was estimated to be better than 0.8 eV.

3. Results and Discussion

A variety of molybdenum compounds having oxygen, nitrogen and sulfur ligation and representing a variety of coordination geometries and oxidation states were investigated. Preliminary results for octahedral, distorted octahedral and tetrahedral coordination will be discussed. As shown in Fig. 1, a consistent increase in the energy of the absorption edge is found as the oxidation state of Mo (in octahedral geometry) is increased, suggesting the potential utility of Mo L edges for studying oxidation changes. The $L_{II,III}$ edges, representing transitions from $2p_{1/2}$ and $2p_{3/2}$ initial states, have a pronounced "white-line" feature for all compounds examined. This is assigned to the allowed $2p \to 4d$ transition. For high oxidation states (+VI, +V), the white-line is split, reflecting the ligand field splitting of the d orbitals (see Fig. 2). The splitting is smaller for tetrahedral

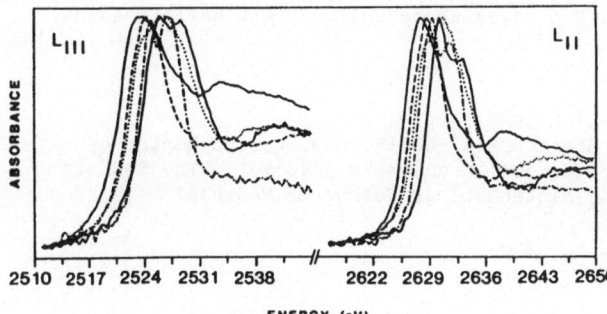

Fig. 1. Dependence of Mo L edge energy on oxidation state of the metal. Edges are (from left to right): Mo(0) metal, $Mo(II)_2$-(acetate)$_4$, $Mo(IV)O_2$, $Mo(V)O(OCH_2CH_2S)TPZB$, and $Mo(VI)O_3$. Similar effects are observed for both L_{II} and L_{III} edges. (The spectra are plotted to have the same main peak height)

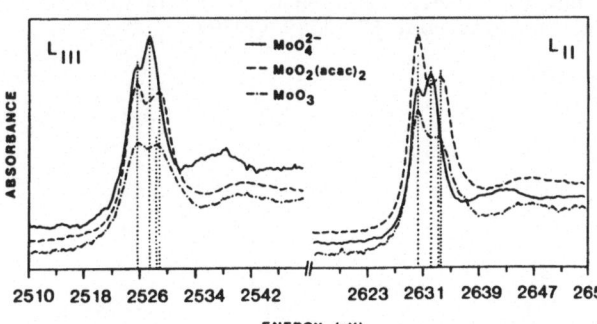

Fig. 2. Detail of the white line for L_{II} and L_{III} edges of three Mo(VI) compounds. The splitting is larger for the distorted octahedral, and smaller for the tetrahedral coordination

(1.8 eV) than for distorted octahedral (2.8 eV) coordination, consistent with predictions from ligand-field theory. Both the splitting and the relative intensities of the white-line features can be rationalized qualitatively using a simple ligand-field approach (Fig. 3). For MoO_3, the more intense component of the split $2p \rightarrow 4d$ transition is at lower energy, while for MoO_4^{2-} it is at higher energy. This may reflect the number of available orbitals (e:2, t_2:3), since t_2 will be higher in energy for the tetrahedral MoO_4^{2-}, while for the distorted octahedral MoO_3 the t_{2g} orbitals will be at lower energy.

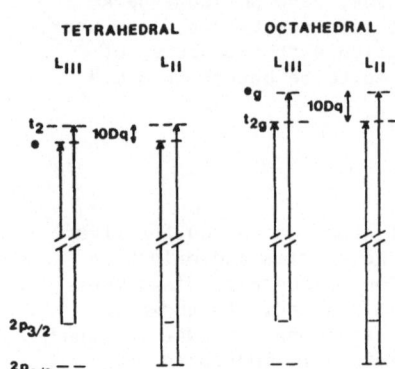

Fig. 3. Ligand-field interpretation of Mo(VI) edges. Spin-orbit splitting of **d** levels and ligand-field splitting of **p** levels are undetectable at this resolution. This model can rationalize the relative energies and intensities of the three Mo L edges in Fig. 2

The line width for the $2p \rightarrow 4d$ lines is consistently larger for the L_{III} edge than for the L_{II} edge. This may be due to splitting of the $2p_{3/2}$ ground level. The relative intensities of the white-lines differ consistently between the L_{III} and L_{II} edges. In addition, for a given compound the ratio of the intensity of the first to the second peak in the split line increases for L_{II} relative to L_{III}. The origin of this effect is not understood. We are presently extending this work to a wider variety of molybdenum compounds and to the S K edges where sulfur ligation is present.

4. Acknowledgements

This work was supported by NSF Grant PCM-82-08115. B.H. was supported by Swedish Natural Science Research Council Grant K-PD 3890-101. The SSRL is supported by the DOE and by NIH, Biotechnology Resources Program.

5. References

1. F.W. Lytle, R.B. Greegor, D.R. Sandstrom, E.C. Marques, J. Wong, C.L. Spiro, G.P. Huffman and F.E. Huggins: Nucl. Inst. Meth. In press (1984).
2. E.A Stern and S.M. Heald: Rev. Sci. Instrum. 50, 1579 (1979).

The Role of Two-Electron Threshold Resonances in Si Near Edge Structure

M.H. Hecht and F.J. Grunthaner

Jet Propulsion Laboratory, California Institute of Technology, USA

P. Pianetta, H.Y. Cho, M.L. Shek, and P. Mahowald

Stanford Synchrotron Radiation Laboratory, Stanford, CA 94305, USA

We have observed a giant two-hole resonance near the 1s threshold of Si in SiO_2 which accounts for much of the white line structure in the KLL Auger yield. The resonance appears as a peak in the energy distribution curve with linear dispersion, at an energy near the KLL line and thus not corresponding to a one-hole photoemission line. This peak is identified with a final state consisting of two 2p holes, a localized excited state, and a continuum electron, and corresponds to direct 2p photoemission with a 2p shakeup hole. The process responsible for this peak can either be viewed as resonant shakeup due to autoionization of the 2p electron [1] or as an Auger Resonant Raman transition [2]. A corresponding line is observed in elemental Si, but it is strongly suppressed due, presumably, to the shorter excitonic lifetime.

We describe photoemission measurements performed on the JUMBO double crystal monochromator at the Stanford Synchrotron Radiation Laboratory using InSb crystals for monochromatization. The measurements were of silicon wafers covered with thin, chemically etched thermal oxides. The results reported here pertain only to thick (300Å) oxides and clean silicon (100) samples.

While measuring Near Edge Structure in SiO_2 with a KLL final state, an unusual white line was observed, as shown in figure 1. The intensity

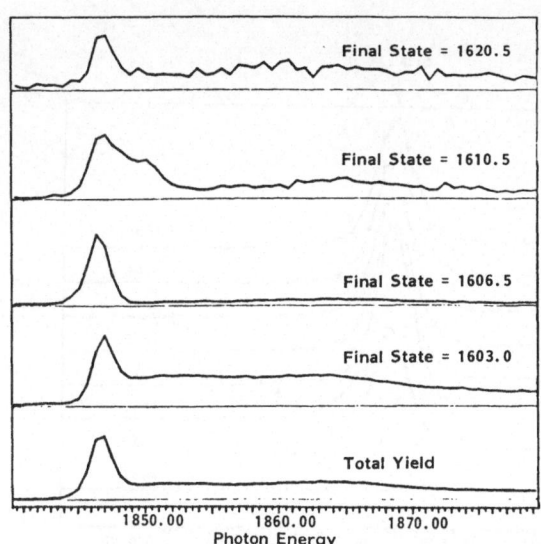

Figure 1: SEXAFS of SiO_2 with KLL final state

Final State = 1620.5

Final State = 1610.5

Final State = 1606.5

Final State = 1603.0

Total Yield

1850.00   1860.00   1870.00

Photon Energy

of the line relative to the absorption above the edge varied over an order of magnitude with final state energy. The lineshape of the line also varied significantly.

In order to understand this phenomenon, we performed photoemission measurements of the KLL region at photon energies near the K edge (figure 2) What appears to be a dispersive photoemission line passes through the KLL window near the threshold, resonating sharply. Since the apparent binding energy of this line (235 eV relative to E_f) does not correspond to any single electron photoexcitation in either Si or oxygen, a final state with two excited electrons (as in a shakeup final state) is suggested. In particular, the proximity in energy (near threshold) to the KLL final state suggests a final state with two 2p shell vacancies. Such a process can be described (figure 3) as

$$1s^2 2p^6 \ ---> \ 1s2p^6 np \ ---> \ 1s^2 2p^4 np + e^- \quad .$$

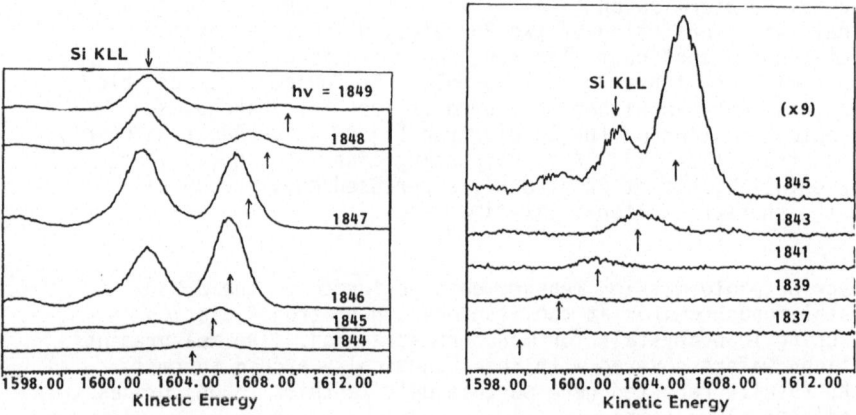

Figure 2: Photoemission from SiO_2 near Si K edge

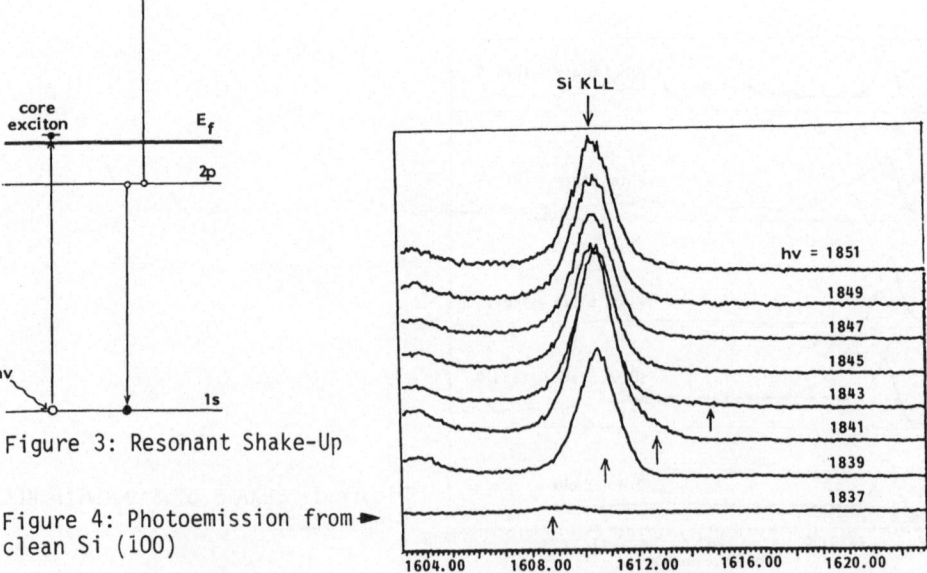

Figure 3: Resonant Shake-Up

Figure 4: Photoemission from clean Si (100)

68

This autoionization of the 2p electron differs from a KLL Auger decay only in that the process is sufficiently rapid that the initial excited electron stays in a fixed energy excitonic state, while energy is conserved by the outgoing 2p electron via post-collisional interaction. The relative importance of the KLL vs. the autoionization decay is determined by the lifetime of the excitonic state.

The arrows in figure 2 indicate the position of the peak under the assumption of linear dispersion. The measured peak position appears to be shifted to higher kinetic energy below resonance, and to lower kinetic energy above resonance. We attribute this shift simply to the extreme narrowness of the resonance compared to the relatively broad photon energy linewidth (>1eV). The shifted line reflects the convolution of a sharply peaked cross section with a broad intensity distribution.

The corresponding spectra for elemental Si are shown in figure 4. The principal KLL line is shifted by 7.5 eV with respect to the SiO_2 line. The corresponding resonance line is indicated with the arrows. It can be seen that the Auger line dominates the autoionization line in this case, presumably due to the shorter relative lifetime of the excitonic state.

In the context of near edge structure analysis, we note that such resonance phenomena are probably fairly common and can not be overlooked as a contribution to Auger yield spectra and, to a lesser extent, total yield spectra. (It is also noted that this structure should be detectable in Auger Electron Spectroscopy measurements of SiO_2.) Photoemission measurements in the vicinity of the threshold are thus a critical aspect of an XANES measurement. We also note in passing that a similar resonance has been suggested to explain valence band structure in Si near the 2p threshold [3]. The existence of a resonance in this case is not as clear as in the present example [4]. This work suggests that it may be profitable to seek such resonances in SiO_2 rather than in elemental silicon.

We are grateful to A. Zangwill for useful discussions about the resonance process. The research described in this paper was performed by the Jet Propulsion Laboratory, California Institute of Technology, under contract with the National Aeronautics and Space Administration. The work was performed at (and in collaboration with the staff of) the Stanford Synchrotron Radiation Laboratory, which is supported by the DOE, Office of Basic Energy Sciences, and NSF, Division of Materials Research.

1. A. Zangwill & P. Soven, Phys. Rev. B24, 4121 (1981).
2. G.S. Brown, M.H. Chen, B. Crasemann, G.E. Ice, Phys. Rev. Lett. 45, 1937 (1980).
3. K.L.I. Kobayashi, H. Daimon, and Y. Murata, Phys. Rev. Lett. 50,1701, (1983).
4. R. Reidel, M. Turowski, G. Margaritondo, Phys. Rev. Lett. 52, 1568, (1984).

Polarization Dependent Near-Edge X-Ray Absorption Fine Structure of Single Crystal Graphite

R.A. Rosenberg, P.J. Love, and Victor Rehn

Michelson Laboratories, NWC, China Lake, CA 93555, USA

C(K)-edge photoabsorption spectra from graphite as a function of incidence angle, α, are shown in Fig. 1. Dramatic changes in the spectral structure are observed as α is varied. The most noticeable change is the disappearance of the intense, sharp peak (peak A) at 285.5 eV when α is decreased to zero degrees. This and the other changes are due to different final states being excited as the crystal is rotated. When the E-vector lies completely in the basal plane ($\alpha=0$), only states of σ symmetry may be excited, while states of π symmetry become dipole-allowed as the E-vector is rotated out of the plane. Comparison of the 0 degree spectrum (σ only) to the other spectra ($\sigma + \pi$) allows one to deduce the symmetry of the final states whose decay results in the peaks seen in Fig. 1. The symmetry of the states deduced in this manner is indicated at the bottom of Fig. 1.

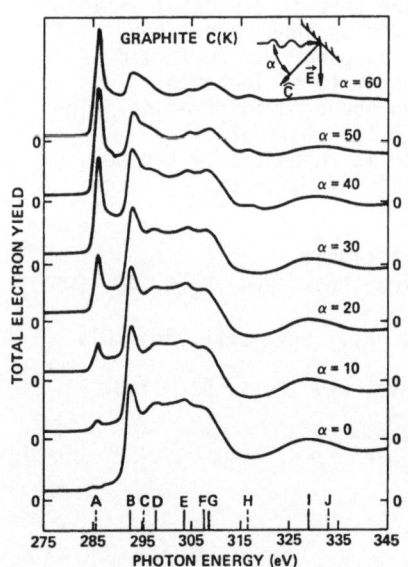

Fig. 1: C(K) edge absorption spectra of graphite as a function of incidence angle, α. At the bottom of the figure, peak positions and symmetries are indicated.

dashed lines= π states
solid lines= σ states.

The angular variation of the peak intensity should be propotional to $\sin^2(\alpha)$ for a 1s to π transition and to $\cos^2(\alpha)$ for a 1s to σ transition.[1,2] In Fig. 2 the relative intensity of peak A is plotted as a function of $\sin^2(\alpha)$. The linear relationship proves unequivocally that peak A results from excitation to a π final state.

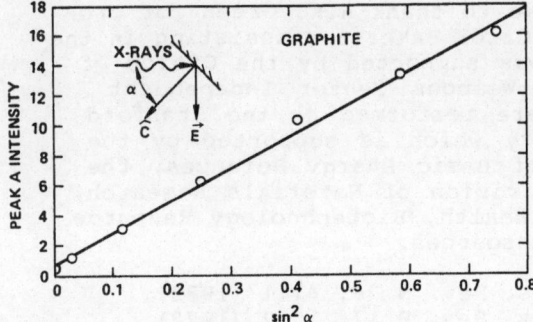

Fig. 2: Relative intensity of Peak A (Fig. 1) as a function of $\sin^2(\alpha)$.

In Fig. 3 is shown the band structure of graphite, as calculated by Willis, Fitton, and Painter.[3] The 285.5 eV peak is assigned to excitation to the π_0 state which crosses Q at around 2 eV, in agreement with previous C(K) excitation studies of graphite.[4,5] On the right side of Fig. 3, the peak positions and symmetries deduced from Fig. 1 have been transposed to facilitate comparison with the band structure. It is thus possible to assign many of the peaks to bands of a particular symmetry. For instance, peak B results from excitation to the σ_1 and σ_2 bands, while peak C results from excitation to the π_1 band. Many of the higher lying peaks seem to correlate with features in the band structure; however, single and multiple scattering events undoubtedly also contribute to this structure. To fully comprehend the spectra of Fig. 1 will require a partial density-of-states band structure calculation coupled with a multiple scattering calculation.

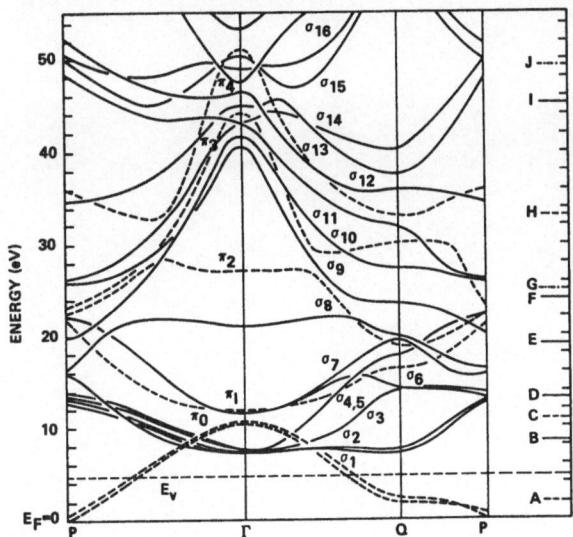

Fig. 3: Calculated band structure of graphite.[3] On the right, the peak positions and symmetries from Fig. 1 have been transposed.

ACKNOWLEDGEMENTS: We would like to thank A.K. Green for pro-
viding the graphite sample and Carol Baker for assisting in the
data acquisition. This work was supported by the Office of
Naval Research and the Naval Weapons Center Independent
Research Fund. Experiments were performed at the Stanford
Synchrotron Radiation Laboratory which is supported by the
Department of Energy, Office of Basic Energy Sciences, the
National Science Foundation, division of Materials Research,
and the National Institute of Health, Biotechnology Resource
Program, Division of Research Resources.

1. J. Stöhr and R. Jaeger, Phys. Rev. B 26, 4111 (1982).
2. S. Wallace and D. Dill, Phys. Rev. B 17, 1692 (1978).
3. R.F. Willis, B. Fitton, and G.S. Painter, Phys. Rev. B 9,
    1926 (1983).
4. D. Denley, P. Perfetti, R.S. Williams, D.A. Shirley, and
    J. Stöhr, Phys. Rev. B 21, 2267 (1980).
5. B.M. Kincaid, A.E. Meixner, and P.M. Platzman, Phys. Rev.
    Lett, 40, 1296 (1978).

Part II **EXAFS Data Analysis**

EXAFS Studies of Anharmonic Solids

J.M. Tranquada

Brookhaven National Laboratory, Upton, NY 11973, USA

EXAFS spectroscopy is a sensitive probe of atomic pair distribution functions (PDF's) and a useful tool for studying the anharmonic, as well as the harmonic, part of interatomic potentials. The thermal average of the interference phase factor $\exp(i2kr)$ appearing in the standard EXAFS parameterization for an atom with a single shell of neighbors is just the Fourier transform of the PDF $P(r)$; the general form of the transform is conveniently expressed in terms of the cumulant expansion [1,2]:

$$\langle e^{i2kr} \rangle = \exp\left[\sum_{n=1}^{\infty} \frac{(i2k)^n}{n!} \sigma^{(n)} \right] .$$

The nth cumulant $\sigma^{(n)}$ can be described in terms of moments of the distribution; $\sigma^{(1)} = R$, the mean interatomic distance, and $\sigma^{(2)} = \sigma^2$, the mean square relative displacement. For most crystalline solids at low temperature, the harmonic approximation is valid and all cumulants beyond the second are insignificant; the cumulant expansion then reduces to the standard Gaussian Debye-Waller factor. As the temperature increases, the PDF begins to deviate from a Gaussian and some of the higher cumulants become important.

Of course, the r-dependent factor $\exp(-2r/\lambda)/r$ is also included in the thermal average of the EXAFS interference function and modifies the results. If one ignores the k-dependence of the mean free path λ, then the lowest order contributions are easily evaluated: besides the amplitude factor $\exp(-2R/\lambda)/R$, there is a correction term which enters the single shell phase so that $2kR$ becomes $2kR - 4\sigma^2(1+R/\lambda)/R$.

If measurements are made on the same sample under two different conditions, changes in the cumulants can be extracted from the data by Fourier filtering to obtain single shell EXAFS (phase and amplitude) and then applying the ratio method [2]. The variations in the EXAFS should be due just to changes in the PDF; it follows, using the general cumulant expansion, that the logarithm of the ratio of single shell amplitudes should correspond to a polynomial containing just even powers of k, while the difference in single shell phases is described by a series of odd-powered terms. The different k-dependences of the cumulant terms make it possible (in principle) to separate them.

NaBr and RbCl, alkali halides with the rocksalt structure, are solids which exhibit significant anharmonicity even at room temperature. (With Debye temperatures of 224 and 165 K [3], respectively, this fact should not be surprising.) Since they are simple, well studied materials, they provide good test cases for demonstrating the use of EXAFS to study anharmonic potentials.

EXAFS measurements were made on the Br K-edge in NaBr and the Rb K-edge in RbCl at 71 and 295 K [4]. After Fourier filtering the EXAFS, the first shell contributions for each material at the two temperatures were analyzed using the ratio method. The amplitude ratio and phase difference data for RbCl (circles and solid lines) are shown in Figs. 1 and 2, respectively, together with the polynomial fits (dashed lines). Both fitted curves were constrained to pass through y-intercepts of zero. The numerical results of the analysis are presented in Table 1.

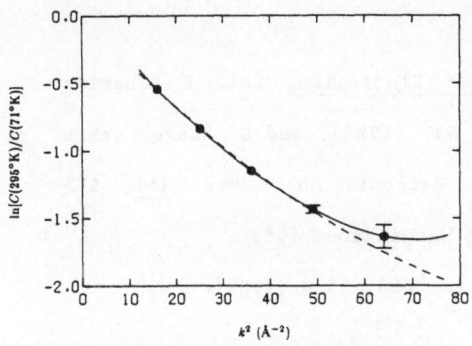

Fig. 1. Logarithm of the ratio of single shell amplitudes at 295 and 71 K for the first shell in RbCl

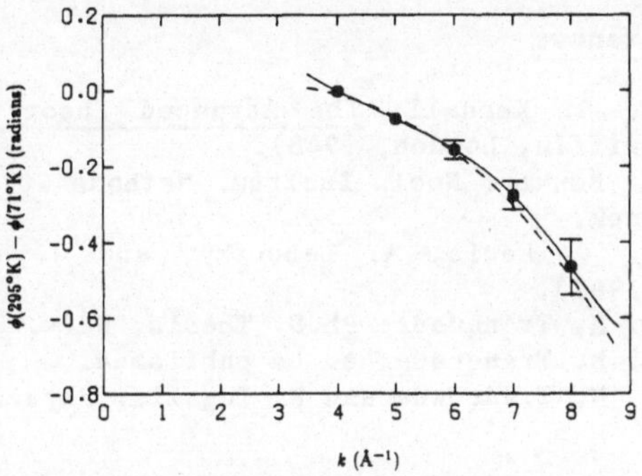

Fig. 2. Difference in single shell phases at 295 and 71 K for the first shell in RbCl

Table 1. Changes in cumulants for RbCl between 71 and 295 K

		$\Delta\sigma(1)$ [Å]	$\Delta\sigma^2$ [10^{-2} Å^2]	$\Delta\sigma(3)$ [10^{-3} Å^2]	$\Delta\sigma(4)$ [10^{-4} Å^4]
NaBr	expt.	0.037±0.010	1.77±0.16	1.7±0.2	2.2±0.8
	theory	0.029	1.78	1.4	2.5
RbCl	expt.	0.024±0.010	1.77±0.13	0.9±0.2	1.8±0.6
	theory	0.034	2.13	1.9	3.8

Because of the simple structure of the rocksalt lattice (simple cubic with two types of ions distributed symmetrically on the lattice) it is possible, using a nearest-neighbor central force model, to calculate expressions for the cumulants in terms of the interatomic potential [5]. Using only the interatomic distance and bulk modulus at T=0, P=0 as input, one obtains the theoretical results shown in Table 1. The experimental and theoretical results are in very good agreement for NaBr, with somewhat worse agreement for RbCl. Part of the discrepancy for RbCl is due to the correlation between the fitted values of $\Delta\sigma(1)$ and $\Delta\sigma(3)$, and of $\Delta\sigma^2$ and $\Delta\sigma(4)$.

EXAFS has also been used to study the anomalous motion of Cu atoms in CuBr [6]. Providing information complementary to that obtained from x-ray diffraction, the EXAFS results showed that (at least up to room temperature) the behavior of the Cu ions is best described in terms of anharmonic motion rather than static displacements.

In conclusion, when one accounts for anharmonic contributions, accurate determinations of interatomic distances and mean square relative displacements can be made. The anharmonic information obtained can be quite useful for describing the shape of interatomic potentials. When properly applied, EXAFS is a sensitive technique for studying anharmonicity in crystalline solids.

Acknowledgment

This work was supported in part by the NSF under Grants No. DMR-78-24995 and DMR-77-27489 (in cooperation with DOE).

References

1. M. G. Kendall: The Advanced Theory of Statistics, Vol. 1 (Charles Griffin, London, 1948).
2. G. Bunker: Nucl. Instrum. Methods 207, 437 (1983), and G. Bunker: this book.
3. J. T. Lewis, A. Lehoczky, and C. V. Briscoe, Phys. Rev. 161, 877 (1967).
4. J. M. Tranquada: Ph.D. Thesis, Univ. of Washington, 1983.
5. J. M. Tranquada: to be published.
6. J. M. Tranquada and R. Ingalls: Phys. Rev. B28, 3520 (1983).

The Selection of E_0 for EXAFS Spectra of Disordered Systems

David V. Baxter

W.M. Keck Laboratory of Engineering Materials,
California Institute of Technology, Pasadena, CA 91125, USA

1. INTRODUCTION

The selection of an appropriate value of the threshold energy, E_0 has long been recognized as an important and somewhat difficult problem which must be confronted when analyzing EXAFS data. Numerous methods for selecting E_0 have been proposed over the last decade (see [1-4] for example). In this paper we emphasize that the assumptions made in justifying most of these methods are invalid in cases where the material under consideration has an asymmetric distribution function. In particular, the practice of treating E_0 as a free parameter has been found to be inappropriate for such systems.

2. THEORETICAL BACKGROUND

Theoretical expressions used in EXAFS analysis are given in terms of the photo-electron wave vector, k, which is related to the experimentally measured photon energy through the relation:

$$k = \sqrt{\frac{2m}{\hbar}} \left(E - E_0 \right) . \tag{1}$$

In (1) the quantity E_0 is the so-called threshold energy and is the subject of this paper. Experimentally, E_0 is not a well defined quantity and indeed, it is not even strictly independent of k [5]. Obviously, assuming an incorrect value for E_0 in defining the experimental wave vector axis, k', will result in the following relation between the experimental k' and the true wave vector, k (using ΔE_0 to denote the error in the assumed E_0):

$$k' \approx k \left(1 + \frac{1}{2} \frac{\Delta E_0}{E - E_0} \right) . \tag{2}$$

This shows that an incorrect value for E_0 will introduce a nonlinear contribution to the phase of $\chi(k)$, beyond that introduced by the scattering phase shift $\alpha(k)$. Most methods proposed for finding E_0 assume that, once the phase shift $\alpha(k)$ has been properly accounted for, any nonlinearity in the phase of a term in $\chi(k)$ must come from an incorrect value of the threshold energy. Methods as apparently different as those of Stearns [3], and Lee and Beni [1], both rely on this common assumption, as do other methods where the application of this assumption is more obvious.

Eisenberger and Brown have pointed out that if the material under study has an asymmetric distribution function, then the usual expression for the EXAFS function must be replaced by an expression such as the following [6]:

$$\chi(k) = \sum_j B_j(k) \frac{|f(k,\pi)|}{k} \sin(2kR_j + \alpha_j(k) + \Sigma_j(k)) . \qquad (3)$$

In (3), B(k) accounts for the effects of structural disorder and many body corrections (such as the passive electron overlap integral and the electron mean free path), and $\Sigma(k)$ is non-zero, and generally nonlinear, in cases where an asymmetric distribution function is considered. Obviously, any attempt to find E_o by linearizing the phase of $\chi(k)$ will experience difficulty unless $\Sigma(k)$ is already known. For nontrivial asymmetric cases this means that the structural problem must have already been solved before E_o can be determined through the use of such a procedure!

3. E_o AS A FREE PARAMETER

In (3) the functions B(k) and $\Sigma(k)$ are not independent and consequently there is a strong correlation between the amplitude and the phase of the terms in the expression for $\chi(k)$ when an asymmetric distribution is consi-dered. To investigate the effect that this correlation has on the practice of selecting E_o on the basis of a best-fit criterion, as has been done by many authors, we considered an EXAFS spectrum measured on an amorphous $Ga_{20}La_{80}$ sample. After preliminary analysis following conventional lines [7], the Fourier filtered EXAFS data was analyzed using the functional form

$$\rho(R) = A (R - R_j)^2 e^{(R - R_j)/\sigma} , \qquad (4)$$

to model the distribution of La atoms around the absorbing Ga atom. In the fit (performed on the interval 4.0 to 10.0 A^{-1}) the k dependence of the many body contribution to B(k) was explicitly taken into account, and the four parameters N, σ, R and E_o were allowed to vary. By fixing the values of σ and E_o and allowing a nonlinear optimization program to find the corre-sponding optimal values of N and R the correlation between the various parameters could easily be determined. The expected strong correlation between N and σ was seen, and was found to be insensitive to the value chosen for E_o. The solid line in Fig. 1 shows the optimal value of N as a function of σ with the error bars showing how much this correlation curve

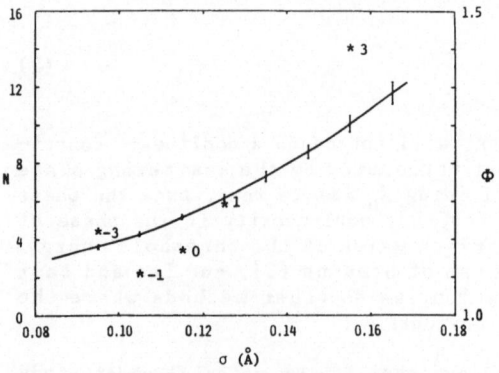

Fig. 1. Correlation between σ and N for various E_o. *E shows the optimal position in the Φ vs σ curve for $\Delta E_o = E$

changes as the threshold is taken at ± 3 eV away from $\Delta E_o = 0$ (defined as the best E_o value found in a study on a crystalline model system Ga_2La).

Although the correlation between N and σ did not change significantly with the assumed threshold energy, the position of the optimal fit along the correlation curve did change significantly. This is indicated by the asterisks in Fig. 1 which give the position of the optimal value of the fit functional, Φ, as a function of σ for each of the indicated values of ΔE_o ($\Delta E_o = -3, -1, 0, 1, 3$). From the figure, it is seen that for $\Delta E_o = -3$eV the optimal fit is obtained at $\sigma = 0.095$, which implies N=3.5, whereas for $\Delta E_o = +3$eV the optimal position has N=10. Furthermore it is seen that the best overall fit is found for $\Delta E_o = -1$eV, at which point the coordination number is an unphysical N=4.0. The actual coordination number for this system has been measured using X-ray diffraction and isomorphous substitution, and was found to be N=7.8, which would roughly correspond to $\Delta E_o = 1.5$ eV [7]. Clearly, selecting ΔE_o through a best-fit criterion would be inappropriate for this spectrum.

The practice of choosing E_o by treating it as another free parameter to be optimized in the curve-fitting part of the analysis was originally justified on the basis of the fact the it could not produce an artificially good fit at an incorrect value of R since these two parameters influence the EXAFS function in different regions of the k axis. We have just shown above that this reasoning is irrelevant in the case of an asymmetric distribution, since in such cases E_o may be strongly correlated to parameters other than R.

4. CONCLUSION

We have shown that many traditional methods of selecting E_o, based strictly on an analysis of $X(k)$, rely on assumptions which fail to be valid for systems with asymmetric distributions. For such systems E_o should therefore be chosen using some method which does not depend solely on $X(k)$, such as that suggested by Boland et al. [4].

ACKNOWLEDGEMENTS

This work was conducted under the U. S. Department of Energy project agreement No. DE-AT03-81ER10870 under contract No. DE-AM03-76SF00767. The author gratefully acknowledges the financial support of NSERC (Canada) and the Standard Oil Company of Ohio.

REFERENCES

1). P. A. Lee, and G. Beni, Phys. Rev. B15 2862 (1977).
2). P. A. Lee, P. H. Citrin, P. Eisenberger, and B. M. Kincaid, Rev. Mod. Phys. 53 769 (1981).
3). M. B. Stearns, Phys. Rev. B25 2382 (1982).
4). J. J. Boland, F. G. Halaka, and J. D. Baldeschwieler, Phys. Rev. B28 2921 (1983).
5). E. A. Stern, D. E. Sayers, and F. W. Lytle, Phys. Rev. B11 4836 (1975).
6). P. Eisenberger, and G. Brown, Sol. Stat. Comm. 29 481 (1979).
7). D. V. Baxter, A. Williams, and W. L. Johnson, J. Non-cryst. Sol. 61&62 409 (1983).

Possibility of Bond-Length Determination in EXAFS Without the Use of Model Compounds or Calculated Phases

Folim G. Halaka and John D. Baldeschwieler

Division of Chemistry and Chemical Engineering, California Institute
California Institute of Technology, Pasadena, CA 91125, USA

John J. Boland

IBM Research (Eastview), Yorktown Heights, NY 10598, USA

1. Introduction:

The Fourier transform (FT) of EXAFS data in photoelectron wavenumber (k) space gives a form of radial distribution function (RDF) in which the absorbing atom is located at the origin. The position of each peak in the RDF does not coincide with the true interatomic distance of the corresponding shell due to the presence of the phase function $\theta_i(k)$ in the sine argument. The correct distance is usually obtained by using a known (model) compound with the same absorber-scatterer pair. The phase function is assumed to be transferable from the model compound to the unknown.[1] Calculated phase shift functions are also used for this purpose[2]. Furthermore, each of the proposed methods for determining the distance assume an a priori knowledge of the scattering atom.

A method or determining bond distances in EXAFS which does not rely on model compounds or theoretical phase shifts is presented. There is no assumption concerning the nature of the scattering atom. In addition, during the distance determination process, the observed phase intercept is a measure of the type of scattering atom involved[3].

2. Basis For The Method:

The physical basis for this scheme is the absence of a linear term in the phase function $O_i(k)$. For the central atom phase shift, this fact is expressed by Levinson's Theorem[4] which states:

$$\delta_L(0) - \delta_L(\,) = n_L \pi \qquad (1)$$

where n_L is the number of bound states of angular momentum L and $\delta_L(k)$ is the Lth partial wave phase shift. Levinson's Theorem is an expression of the fact that each partial wave phase shift is bounded, and thus, every atom has a finite scattering power. Since the scatterer phase may be expressed as a sum of partial waves, the total phase shift $\theta_i(k)$ must also be bounded. It can be shown that the asymptotic behavior of both the central atom phase shift and the phase of the scattering amplitude approach zero as $1/k$ as k tends to infinity[5]. The presence of a linear term would cause the phase to diverge as k approaches infinity. Although the phase function can, in a finite k range, be parameterized by an equation containing a linear term as discussed by Lee et al,[2] it must, however, be emphasized that the parameterization does not represent the true functional form of the phase.

3. A Trial Function:

To exploit the absence of linear terms in the phase, $\Theta_i(k)$ we may write the total phase function $F_i(k)$ as:

$$F_i(k) = 2kr_i + \Theta_i(k) \tag{2}$$

where r_i is the true interatomic distance for the shell i. We now define a function $g_i(k)$ such that:

$$g_i(k) = dF_i(k)/dk - F_i(k)/k \tag{3}$$

Since $O_i(k)$ is non-linear there is no loss of phase information in constructing $g_i(k)$. The linear term in Eq. (3), however, is completely absent from $g_i(k)$. Thus $g_i(k)$ is essentially a differential equation in the phase $\Theta_i(k)$

$$g_i(k) = d\Theta_i(k)/dk - \Theta_i(k)/k \tag{4}$$

A functional form for $\Theta_i(k)$ which satisfies the criterion that it behaves as $1/k$ at large k is

$$\Theta_i(k) = A + B \exp(-Ck) \tag{5}$$

The function $g_i(k)$ (Eq. 4) then takes the form:

$$g_i(k) = -A/k - (BC + B/k) \exp(-Ck) \tag{6}$$

A, B and C are now adjustable parameters in the fit of Eq. (6) to the experimentally determined function $g_i(k)$. When these parameters are determined, the phase function may be calculated from Eq. (6) and the distance from Eq.(2).

4. The E_o Problem:

It is important to note that an error in E_o cannot produce an error in k scale that varies linearly with k. An error in E_o, ΔE_o, will change a given k value into k'

$$k' = [k^2 - (2\Delta E_o/7.62)]^{1/2}$$

$$k' - k + \Delta E_o/7.62k$$

Clearly, a change in E_o will not produce a linear change in k. The 2kr term will remain the only linear in the total phase.

5. Results:

To illustrate the ability of this scheme to determine accurate bond distances, a series of known compounds were studied. The total phase was determined using a scheme due to Lee et al.[5] and the function $g_1(k)$ was constructed. Table I shows the parameters of the fit of $g_1(k)$ to the functional form shown in Eq. (6) along with the first-shell distances for several compounds.

6. Scattering Atom Identification:

The phase intercept has been shown to be a quantitative means of determining the nature of the scattering atom.[4] In this

Table I. The parameters obtained in the fit of Eq.(6) to the function $g_1(k)$ which is calculated from the total phase of the first shell peak for the compounds shown.

Compound	A	B	C	A+B		r(calc)	r(cryst)
Cu foil	-80.15	69.90	0.0106			2.560	2.556
Ni foil	-78.33	71.65	0.0105			2.482	2.492
Co(acac)$_3$	-76.93	72.60	0.0149	-4.33		1.880	1.888
[Co(NH$_3$)$_6$]Cl$_3$	-75.81	74.17	0.0132	-4.78	$(-\pi)$	1.940	1.963
[Co(en)$_3$]Cl$_3$	-75.81	74.20	0.0148	-4.75	$(-\pi)$	1.99	1.99
K$_3$[Co(CN)$_6$]	-77.60	72.40	0.0145	-5.20		1.87	1.89

case, the intercept is given by the sum of the A and B coefficients. A series of cobalt complexes in which the Co is surrounded in the first shell by oxygen, nitrogen and carbon was studied to demonstrate the ability to identify the scattering atom. Table I shows the sum of the coefficients A and B for the series studied. It is therefore possible to simultaneously deterermine the bond distance and the type of neighboring atom involved. The relative values of phase intercepts agree with the values previously reported.[3]

7. Conclusions:

This method may be used to determine bond distances from EXAFS data to within 2% of the accepted crystallographic distance for single shell systems without resorting to model compounds or calculations. No assumptions are made concerning the nature of the scattering atoms; indeed, it is possible to identify such atoms from the observed phase intercept.

References:

1. P.H. Citrin, P. Eisenberger, and B.M. Kincaid, Phys. Rev. B28, 2921 (1983).
2. P.A. Lee, B.K. Teo and A.L. Simons, J. Amer. Chem. Soc. 99, 3856 (1977).
3. F.G. Halaka, J.J. Boland and J.D. Baldeschwieler, J. Amer. Chem. Soc. in press (1983).
4. M.L. Goldberger and K.M. Watson, Collision Theory, Wiley and Sons, New York, 1964, p. 284.
5. See, for example, Messiah Quantum Mechanics, John Wiley & Sons (1961).

Coordination Number Determinations of Gold Complexes by EXAFS Spectroscopy

M.K. Eidsness and R.C. Elder

Department of Chemistry, University of Cincinnati
Cincinnati, OH 45221, USA

1. Introduction

Gold coordination number with sulfur and phosphorus ligands concerns us in an EXAFS study of metabolites containing gold. These form as products in drug therapy against rheumatoid arthritis. In a transferability test of amplitude and phase shift functions from the known, two-coordinate structure, $[Au(PPh_2CH_3)_2]^+$, to the known, four-coordinate structure, $[Au(PPh_2CH_3)_4]^+$, the calculated coordination is two rather than the expected four. Also the results of fitting EXAFS data from gold incorporated into metallothionein suggest a two-coordinate gold site whereas the zinc and cadmium atoms (which gold replaces) are thought to be four coordinate. In light of the difficulties in calculating gold coordination numbers, further tests of empirical curve fitting for several Au-S and Au-P type structures were carried out. Additionally, measurements of EXAFS spectra at -185°C were made to study the effect of reduced thermal motion on the calculated coordination numbers. The EXAFS analysis method in this study follows those developed by Hodgson and coworkers [1]. All experiments monitored the Au L_{III} X-ray absorption edge. Data were collected at the Stanford Synchrotron Radiation Laboratory.

2. Empirical Curve Fitting Results

The difficulty fitting EXAFS data from gold-phosphorus compounds could be anticipated on the basis of the Fourier transforms of these data. The peak position moves to higher values of R with increasing bond distance as expected; however, peak height does not increase with an increase in coordination number. Peak shape becomes skewed instead, with a broad rise on the lower R value side of the peak. Curve fitting results for several, single shell gold-sulfur and gold-phosphorus complexes (based on $[Au(S_2O_3)_2]^{3-}$ and $[Au(PPh_2CH_3)_2]^+$ as models) are given in Table 1. For the gold-sulfur cases agreement is good: bond distances agree within 0.01Å and coordination numbers are in error by, at most, 10%. Note, however, that all gold-sulfur bond distances tested are nearly the same, independent of changes in coordination number or oxidation state for gold. Such good agreement does not hold in the gold-phosphorus cases. Clearly, some calculated coordination numbers are quite wrong. Thus, the four-coordinate structure with methyldiphenylphosphine calculates as two coordinate. Generally, as the bond distance approaches that in the model compound, the error in coordination number decreases. In Fig. 1 the coordination numbers (gold by phosphorus) calculated by EXAFS divided by the crystal value are plotted versus the Au-P bond distance for seven structures. The regularity is striking. It appears that much better values of gold coordination numbers by phosphorus can be achieved by an empirical fit followed by rescaling the derived coordination number as indicated in Fig. 1. The logic of Fig. 1 is obvious: higher coordinate Au-P bonds are longer and weaker than the two-coordinate model and thus, a rescaling of coordination number is necessary with increasing bond distance. This situation may prevail in other cases where metal to ligating-atom distances span a considerable range for a given neighbor atom. Certainly, caution seems warranted in using unknown structures with very different bond lengths from

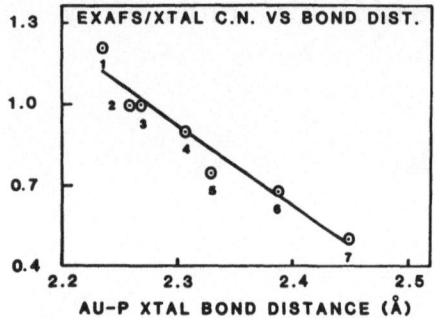

Fig. 1 Ratio of coordination numbers calculated by EXAFS to crystal values plotted versus gold-phosphorus crystal bond distances: 1 = chlorotriphenylphosphinegold(I); 2 = tetraacetylthioglucosetriethylphosphinegold(I); 3 = $[AuSCH_2CH_2P(Et)_2]_2$; 4 = $[Au(dpp)Br]_2$; 5 = chlorobis(triphenylphosphine)gold(I); 6 = $[Au(dpe)_2]$; 7 = $[Au(PPh_2CH_3)_4]$ (See Table 1 for abbreviations).

TABLE 1
Bond distances and coordination numbers for Au-S and Au-P Structures

Structure	xtal dist(Å)	n	Room Temp Data R(Å)	n	Low Temp Data R(Å)	n
$[Au(PPh_2CH_3)_4]^+$	2.449	4	2.44	2.0	2.44	2.8
$[Au(dpe)_2]^+$	2.389	4	2.40	2.7	2.39	3.3
$[Au(dpp)Br]_2$	2.309	2	2.31	1.8	2.31	2.0
$[Au(dte)_2]^-$	2.289	4	2.29	3.9	2.29	4.1
$[Au(dtt)_2]^-$	2.310	4	2.30	4.1	2.30	4.4
$[Au(etu)_2]^+$	2.279	2	2.28	1.8	2.28	2.1

Models: $[Au(S_2O_3)_2]^{3-}$ $Au\text{-}S_{xtal}$ = 2.276 Å
$[Au(PPh_2CH_3)_2]^+$ $Au\text{-}P_{xtal}$ = 2.316 Å

dpe   dpp   dte   dtt   etu

those found for the model structure. We have tried varying the disorder parameter in curve fitting as well. This generally does not change the resultant bond length but does affect coordination numbers. Unfortunately, disorder term optimization does not consistently improve results for coordination number [2].

3. Low Temperature Measurements

The postulate that a difference in disorder occurs between the model and fit structures for gold-phosphorus compounds can be tested by low-temperature data collection. Although the EXAFS disorder term is composed of two parts, one the result of static disorder and the other from thermal vibration, our Au-P test is free from static disorder in the tetrakis (diphenylmethylphosphine) gold(I) case, where the bond lengths are equivalent by crystal symmetry. Data were collected at -185°C using a gravity flow liquid nitrogen apparatus [2]. Low-temperature results are summarized in Table 1. Model compounds had EXAFS data measured at -185°C as well. The bond distances and coordination numbers used for the model compounds are simply the room temperature values. In both the Au-S and Au-P cases, the bond distance from EXAFS and that from the crystal structure determination agree as well as at room temperature. Thus, the changes in bond length with temperature for the model compounds (which are not evaluated here at all) are tracked by the changes which occur in bond lengths for the unknown. The same is not true for coordination number. For sulfur-containing complexes all values increase by about 10% as does that for the two-coordinate gold-phosphorus complex. The four-coordinate gold-phosphorus compounds, which gave the worst room temperature results show significant improvement at -185°C going from a coordination number of 2.0 to 2.8 in one case and from 2.7 to 3.3 in the other. Thus, data collection does improve these calculations for those cases which yield poor room temperature results. Whether even lower temperature would improve matters more remains to be seen.

4. Au(I) Complexes with Metallothionein

Having established that gold coordination number is quite accurate for a variety of sulfur ligated systems, we turn to a complex of some biological interest. Native,-horse-kidney metallothionein (MT) contains zinc and cadmium in a sulfur rich environment with nearly one-third of the amino acids of the protein being cysteine. Incubation with a limited amount of sodium gold(I)thiomalate leads to partial replacement of zinc and essentially no change in cadmium content. In this case the gold is bound to 2.2 sulfur atoms at a distance of 2.29 Å. If an excess of the thiomalate complex is used, then essentially all zinc and cadmium is replaced yielding a coordination number of 1.7 and an Au–S distance of 2.30 Å [3]. Whether there is a significant difference in these two cases is difficult to tell. What is surprising is that the gold coordination number remains about two. The zinc and cadmium atoms which are replaced have been shown to be four coordinate [4,5]. The gold which causes the elimination of zinc and cadmium presumably occupies the same sites and yet only two of the four available sulfur atoms bind gold. This finding is in agreement with the general tendency of gold(I)-sulfur complexes toward two coordination.

Acknowledgements

We thank C. Frank Shaw, III and James Laib for the metallothionein samples, the National Science Foundation (NSF PCM 80-23743) and Smith Kline and French Laboratories for support and Stanford Synchrotron Radiation Laboratories which are operated by the Department of Energy for access to X-rays. MKE thanks the University of Cincinnati for a Distinguished Dissertation Fellowship.

References

1. S.P. Cramer, K.O. Hodgson: Progr. Inor. Chem. 25, 1 (1979).
2. M.K. Eidsness, Ph.D. thesis, University of Cincinnati, 1984.
3. James Laib, C. Frank Shaw III, Kim Melnick, David H. Petering, M. K. Eidsness, R.C. Elder, Jushine S. Garvey: Biochem. in press.
4. C.D. Garner, S.S. Hasnain, I. Bremner, J. Bordas: J. Inorg. Biochem. 16, 253 (1982).
5. Peter J. Sadler, Arne Bakka, Peter J. Beynan: FEBS Lett. 94, 315 (1978).

EXAFS of a-Ni_xAs_{1-x} and the Determination of Charge Transfer

G.P. Tebby and E.A. Marseglia
University of Cambridge, Dept. of Physics, Cambridge, United Kingdom
A.J. Bourdillon
University of Cambridge, Dept. of Metallurgy, Cambridge, United Kingdom
R.T. Phillips
University of Exeter, Dept. of Physics, Exeter, United Kingdom

1. Introduction

While experimental observations on certain ranges of chemically similar
materials have shown that phase shifts are sometimes "chemically transferable"
(i.e. similar for one particular element in a range of compounds)[1] there is
a need for further understanding for precise measurements using differential
Fourier transform techniques [2], least squares fitting with calculated backscattering factors [3] or other similar accurate analyses. This need is
particularly urgent in partly ionic systems, where the charge transfer is unknown, and can in principle be measured, together with the structure by EXAFS.

We describe preliminary results of the calculation of atomic phase shifts
of arbitrary charge, and their application to the fitting of EXAFS data of
a-Ni_xAs_{1-x} ($0 \leq x \leq 1$), and EXELFS (extended electron energy loss fine structure)
data of sapphire (Al_2O_3).

2. Systems studied

Two systems were studied. Amorphous sputtered films of Ni_xAs_{1-x} ($0 \leq x \leq 1$),
covering the metal-insulator transition. Both the nickel and arsenic edges
of a range of samples have been determined at 1-He temperatures by use of the
SERC electron synchrotron at Daresbury Laboratory. In this system, a precise
interpretation of the changes in local structure depends on an understanding
of the charge transfer, albeit fairly small.

Secondly, we have been concerned with the structure of aluminium oxide
films. Some work has been published elsewhere [4], but sapphire, which has an
ionicity of 63% on the Pauling scale, is a useful model in investigating the
dependence of scattering factors on the ionicity of the elements.

The electron energy-loss fine structure (EXELFS) from the oxygen edge was
obtained from an ion-milled foil cooled to 1-N_2 temperature. With minor
differences, the fine structure obtained is the same as that in EXAFS, but with
a matrix operator for momentum transfer, q.r., replacing the dipole operator
e.r. in EXAFS (Bourdillon and Tebby, to be published). The effect of multiple
inelastic scattering was corrected by deconvolution, with the low loss profile.

Scattering factors and charge states

Atomic phase shifts for angular momentum l = 0 to 29 were calculated in the
usual way [5] by use of a version of MUFPOT (Pendry unpublished) with potentials modified to treat ionic species, and with k-dependent Hara exchange [3].
Muffin tin radii equal to half the first shell radius were used, and complex
backscattering factors fed into EXAFSFT [3] which performed a least squares
fit of Gaussian shells to the experimental data.

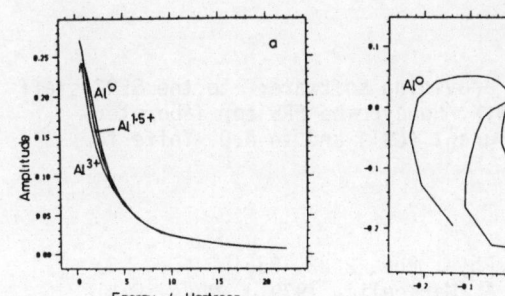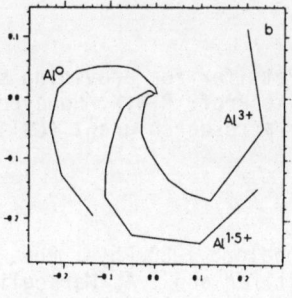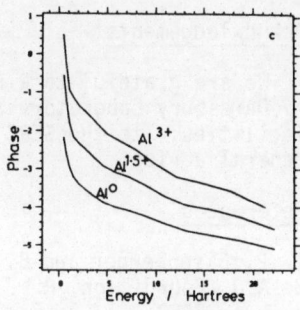

Fig.1. Calculated backscattering factors for aluminium in alpha-alumina for ionic charges of ∅, 1.5 and 3, represented as (a) amplitudes, (b) polar plots, (c) phases

Altering the charge on an atom alters the phase by an amount only weakly dependent on wavevector k (figure 1); in contrast to the change of phase-shift due to a small change in shell radius, which is proportional to k. After a series of calculations, those conditions are found for which the mean square of residuals is minimized. The method appears particularly effective when the first shell about an excited atom contains atoms of one atomic species while the second contains those of another. Figure 2 shows the fit obtained to fine structure from the K edge of Ni in amorphous $Ni_{0.53}As_{0.47}$. The residuals were minimized for a charge transfer of between 0.2 and 0.4 electrons. Preliminary results on fine structure from the K edge of oxygen in α-alumina indicate a charge transfer of about 60 per cent.

This technique offers a rather direct way of measuring charge transfer.

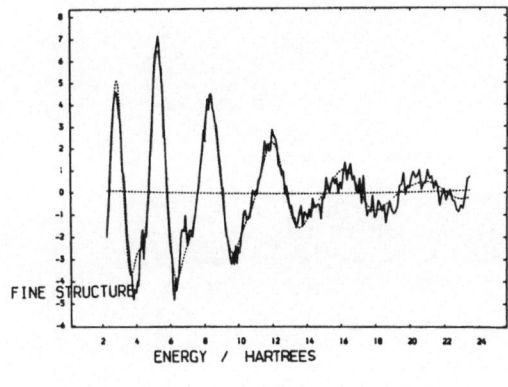

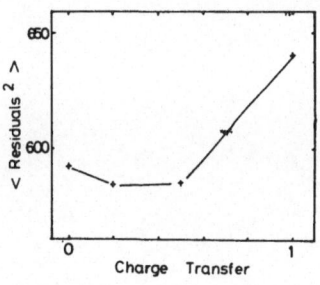

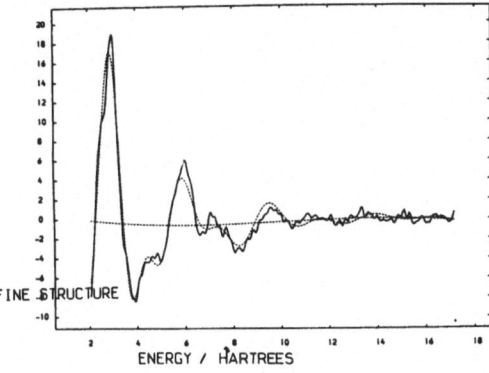

Fig.2. Best fit of a-NiAs data, using backscattering factors calculated with an ionic charge on Nickel of +0.2

Fig.3. A plot of the average squared residuals for best fits of a-NiAs data, as a function of transfer, exhibits a minimum at Ni(∅.2+)As(∅.2-)

Fig.4. Preliminary fit of alpha-alumina indicates a charge transfer of 1.2 electrons

Acknowledgements

We are grateful to R.F. Pettifer for providing software; to the SERC staff of Daresbury Laboratory and to Prof. R.W.K. Honeycombe FRS for laboratory facilities; to the SERC for a research grant (GWT) and to A.D. Yoffe for general advice.

References

1. P. Eisenberger and B. Lengeler: 1980 Phys. Rev. B 22, 3351
2. A.J. Bourdillon, R.F. Pettifer and E.A. Marseglia: 1979 J. Phys. C. 12, 3889
3. R.F. Pettifer and A.D. Cox: 1983 Springer Series in Chemical Physics, Vol. 27, 66
4. A.J. Bourdillon, S.M. El-Mashri and A.J. Forty: 1984 Phil. Mag. A. 341
5. P.A. Lee and J.B. Pendry: 1975 Phys. Rev. B 11, 2795.

The Phase and Amplitude Corrected Fourier Transform for the Detection of Small Signals

J.B.A.D. van Zon, D.C. Koningsberger, and R. Prins

Laboratory of Inorganic Chemistry and Catalysis, Eindhoven University of Technology, P.O. Box 513, NL-5600 MB Eindhoven, The Netherlands

D.E. Sayers

Department of Physics, North Carolina State University
Raleigh, NC 27650, USA

In many cases one is interested in small EXAFS signals which are present along with large signals. When the small signal has a frequency close to the dominant frequency of the EXAFS spectrum it may be fully obscured in the Fourier transform. Detection may be even more difficult when the radial structure function (RSF) of the dominant signal shows a complicated structure due to the k-dependence of the phase and amplitude functions of the dominant absorber scatterer pair. This k-dependence especially appears for high Z elements. When a suitable reference compound for the dominant signal is available, the RSF may be simplified by using a Fourier transform which eliminates the complex structure of the phase shift function and the backscattering amplitude. This can be done by transforming $\chi(k).exp(-i\phi_j(k))/f_j(k)$ instead of $\chi(k)$ (1-4), where $\phi_j(k)$ and $f_j(k)$ are the phase shift function and the backscattering amplitude for the jth shell of the reference compound, respectively. Such a phase and amplitude corrected Fourier transform reduces the complicated peak in the RSF to a single, symmetrical and localized peak which peaks at the correct distance and exhibits a very simple, symmetrical imaginary part. Small signals in the neighbourhood of the main peak will now appear as distortions from the symmetrical peak.

As an example, the corrected Fourier transform is applied to a very highly dispersed rhodium (0.5 wt%) on alumina catalyst. The EXAFS signal of this catalyst shows besides the very large Rh-Rh contributions, small contributions of a Rh-O distance at 2.7 A which is very close to the Rh-Rh distance of 2.68 A. Fig.1a shows a normal k^1-weighed Fourier transform from $k=3.3$ A^{-1}.

Clearly visible is the large sidelobe at the left side of the main peak due to the typical behaviour of the rhodium phase shift function and backscattering amplitude. This sidelobe prevents an easy interpretation of the radial structure function after applying a Fourier transform, which was

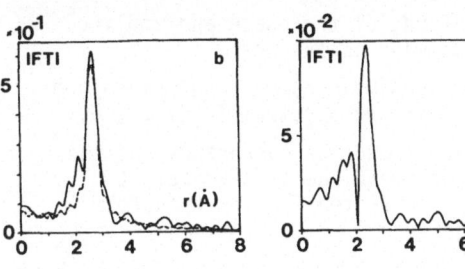

Figure 1

Magnitude of Fourier transform of
a) experimental data (k^1-weighed, $k_{min}=3.3$ A^{-1}, $k_{max}=12.7$ A^{-1})
b) experimental data (solid line) and calculated Rh-Rh EXAFS
(dotted line) (k^1-weighed, Rh-Rh phase and amplitude corr.).

corrected for the Rh-Rh phase shift and backscattering amplitude, obtained
from rhodium metal foil as reference compound. The dotted line indicates
the calculated contribution of the Rh-Rh signal. Since the k-dependence of
the Rh-Rh phase shift and amplitude has been removed, the difference at the
left side of the main peak is not due to a sidelobe but must be ascribed to
a real signal.

The corrected Fourier transform may also be used in the identification of
scatterers with approximately the same backscattering amplitude but with a
difference in phase shift functions. Applying a FT which is corrected for
the right phase shift function will result in a symmetrical imaginary part
which peaks positively with the magnitude of the structure function. Small
deviations from symmetry might indicate a difference in inner potential
between the sample and the used reference compound (5) or a possible
asymmetry in the pair correlation function (6), while large deviations may
point to a mismatch of the phase shift functions. Again, the above mentioned
catalyst may serve as an example. In the catalyst the small signal can be
caused by either chlorine, emanating from the used precursor which was
used for the preparation of the catalyst, or oxygen from the support. After
subtraction of the Rh-Rh content from the spectrum, the RSF of the residual
spectrum was examined with the Rh-O and a Rh-Cl phase shift function. In
this case, no additional amplitude correction was used since it magnifies
the noise to a too large extent, distorting the residual spectrum. The Rh-O
corrected FT gives a positively peaking imaginary part (Fig.2a) while a
Rh-Cl phase shift results in a negatively peaking imaginary part (Fig.2b),
thus indicating the presence of oxygen instead of chlorine as scatterer.

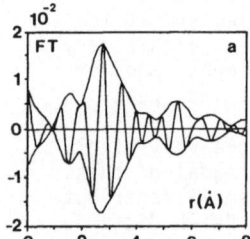 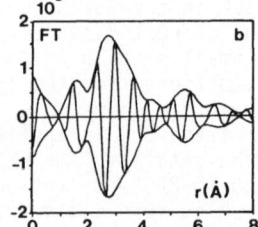

Figure 2

Fourier transform of residual spectrum (k^1-weighed, k_{min}=4.2 Å^{-1},
k_{max}=9.2 Å^{-1})
a) Rh-O phase corr.
b) Rh-Cl phase corr.

The use of the imaginary part and the magnitude of the phase and amplitude
corrected Fourier transform yields the following advantages:

a) it modifies complicated peak structures to simple symmetrical peaks and
   thus allows a more direct interpretation of the spectrum (e.g. Cl, O).

b) once the N constituents of a spectrum are known, the correctness of the
parameters may be verified by analyzing each constituent separately to
verify that the imaginary part of the phase corrected Fourier transform
peaks positively at the maximum of its magnitude.

c) the oscillating behaviour of the imaginary (and real) part of the transform may cause severe interference effects when several components are
contributing in a given region of the spectrum. These interferences may
give rise to peaks which do not correspond to real distances in the material.

By studying imaginary parts under different phase (and amplitude) corrections it is possible to unravel complicated transforms and anticipate these interference effects.

d) the imaginary part can be used for the identification of elements with a similar backscattering amplitude.

1) P.A. Lee and G. Beni, Phys. Rev.B, 15 (1977) 2862.
2) J.B. Pendry, in "EXAFS for Inorganic Systems", p.5, 1981, SERC, Daresbury Laboratory, Engeland.
3) J. Goulon, P. Friant, J.L. Poncet, R. Guiland, J. Fischer, L. Ricard, in "EXAFS and Near Edge Structure", p.100, 1983, ed. A. Bianconi, L. Incoccia and S. Stipcich Springer Verlag, Berlin, Germany.
4) F.W. Lytle, R.B. Greegor, E.C. Marques, D.R. Sandstrom, G.H. Via, J.H. Sinfelt, "Proc. of Advances in Catalytic Chemistry II", p.   , 1982, Salt Lake City, USA.
5) E.A. Stern, D.E. Sayers and F.W. Lytle, Phys. Rev. B, 11 (1975) 4836.
6) E.D. Crozier, J. A. Seary, Can.J.Phys., 58 (1980) 1391.

K Edges EXAFS in Crystalline and Amorphous Germanium

V.D. Chafekar and S.C. Sen

Department of Physics, Indian Institute of Technology, Kanpur 208016, India

1. Introduction

The oscillations of the X-ray absorption coefficient on the high energy side of the absorption edge (EXAFS) provide quantitative information regarding the local structure around the absorbing atom. There has been a revival of interest in this phenomenon due to the improvements in theory and experimentation. A great deal of work is being done on improvements in the data analysis procedures [1,2], a prime factor responsible for determining the accuracy with which EXAFS can give quantitative information of local structure in materials.

Although crystalline and amorphous germanium (C-Ge and a-Ge) have been studied before by other techniques such as X-ray diffraction [3,4] and also by EXAFS [5,6], it is worthwhile, in view of the progress made in theoretical aspects and data analysis procedures to study this system with EXAFS.

Different methods for analysing the EXAFS data have been developed, namely the Fourier transform filtering (FTF) and curve fitting techniques. A quantitative information of phase and amplitude functions is required in these methods to extract the local structure parameters such as coordination numbers, interatomic distance and Debye-Waller factor. In FTF process a smooth filtering window is used to separate out contribution to EXAFS by a particular coordination sphere and the corresponding phase and amplitude functions are subsequently evaluated. In the present work we illustrate the effect of width of the filtering window on the phase and amplitude functions. Also, different analysis procedures for near-neighbour distance determination, in particular, have been examined.

2. Experimental

Crystalline germanium used in the experiment was of ultrapure quality. Amorphous germanium was prepared by evaporation technique with Al-substrate at room temperature. EXAFS data has been collected with a conventional X-ray source coupled with flat crystal spectrometer [7]. The measurements were made in preset count mode (10^5 counts). The angular range for K-edge of germanium was 29.3° to 33.8° in 2θ corresponding to approximately 1000 [eV] above the edge and 200 [eV] below the edge.

3. Results and Discussion

The extraction and normalization of EXAFS $\chi(k)$ has been done by the standard procedure [7]. The incident photon energy, at which the kinetic energy of photoelectron is zero, was determined by taking average of the energy of the point at which slope of the absorption edge starts decreasing and that of the preceeding point (denoted by $E_0 = 0$). The results of the further analysis for C-Ge data are graphically presented in Fig.1. Part (a) of the Fig. 1 shows absorption as a function of incident X-ray photon energy E. The dashed curve is a victoreen curve 0.349 λ^3 to 0.01245 λ^4 fitted to the data on the low-energy side of the edge. In part (c) of the fig. 1, magnitude of Fourier transform (FT) of $k^3 \chi(k)$ is depicted. Following the procedure of STEARNS [2], value of $E_0 = E_c$ is determined at which the position of the first peak is independent of weighting factor.

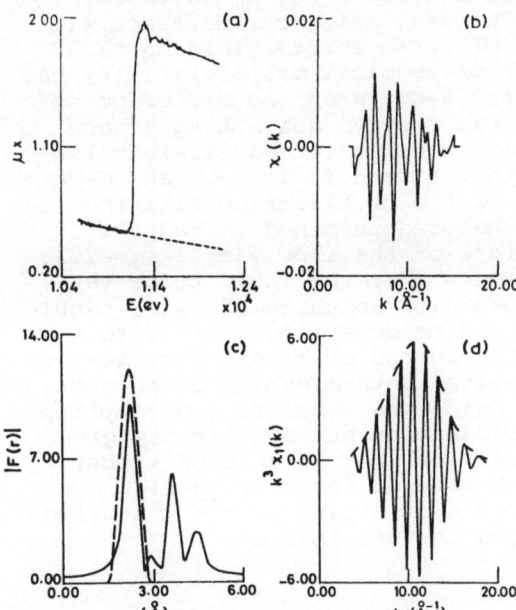

Fig.1. a) Absorption as a function of energy, b) normalised EXAFS, c) magnitude of FT, d) first shell EXAFS from (c).

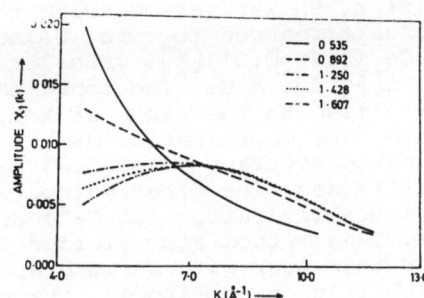

Fig. 2. Amplitude functions for different widths of the filtering window.

The value of E_c was found to be -26 [eV] in close agreement with the one obtained by STEARNS [2]. The amplitude function for the first shell EXAFS separated by the smooth filtering window is shown in part (d) of the Fig.1 by a dashed curve. The effect of width of this filtering window on the envelope of first shell EXAFS is illustrated by plotting amplitude functions for five different window widths in Fig.2. It seems that the envelope of EXAFS is very sensitive to the width of the window function. A too narrow window has completely washed away the details in the amplitude function. With increasing width all amplitude functions start with a lower value and peak around 7 [Å^{-1}], a trend that can be attributed to the backscattering amplitude of germanium. Thus for sufficiently large window it is only low k information which is mostly affected. Non-uniqueness

of this function naturally causes concern and   great caution
should be exercised   . In our analysis
we have used an optimum width (1.25[Å]) for which the details in
the amplitude function are retained. The effect of width of the
window on the phase function could be ascertained by determining
the near neighbour distance by comparison of experimental and
theoretical phase shifts [8]. In a bid to determine near neighbour distance by direct comparison of phase shifts it was found
that it is possible only when the phase function is extracted
with $E_O = E_C$. When near neighbour distances were determined with
values of E_O other than E_C, no physical significance could be
attached to the parameter ΔE_O, the change in the threshold. The
distances also varied by as much as 0.03[Å]. On the other hand
analysis with $E_O = E_C$ yields the bond lengths to an accuracy of
$\pm$0.01[Å]. Also value of ΔE_O was consistent with value of E_C.

A similar analysis was carried out for a–Ge. E_C value for
a–Ge turned out to be -55[eV]. The near neighbour distance was
determined by direct comparison of phase shifts yielding r_1 =
2.49[Å] and E_O = 59[eV]. Concept of chemical transferability has
also been used to determine r_1 for a–Ge. When the difference of
phase function of a–Ge and C–Ge was plotted against k, a nonlinear curve was obtained. A least squares fitted straight line
had intercept -2.31 indicating that threshold in a–Ge and C–Ge
differ. E_O was varied using the method of bisection till the intercept reduced to zero. Using the predetermined value of r_1 for
C–Ge (2.46$\pm$0.01)[Å], value of slope of the line yields r_1 = 2.49$\pm$
0.01[Å] for a–Ge. The coordination number was found to be very
sensitive to the range of k values over which the data is tabulated. The coordination number could be determined only with a
limited accuracy (4.49$\pm$0.8) by the method of transferability of
amplitudes. The other method involved determination of an overall scaling factor for C–Ge and using the same for a–Ge amplitude. This method also yielded coordination number slightly greater than four with as much accuracy as that in case of transferability of amplitudes. The higher value for coordination numbers for a–Ge suggests that the mean free path of the photoelectron is larger in a–Ge as compared to that in C–Ge.

References

1. E.A. Stern and K. Kim; Phys. Rev. B23, 3781 (1981)
2. M.B. Stearns; Phys. Rev. B25, 2382 (1982)
3. N.J. Shevchik, W. Paul; J. Non-Cryst. Solids, 8-10, 381
    (1972)
4. R.J. Temkin, W. Paul and G.A. Connell; N. Adv. Phys. 22,
    581 (1973)
5. D.E. Sayers, E.A. Stern and F.W. Lytte; Phys. Rev. Lett.
    27, 1204 (1971)
6. E.D. Crozier and A.J. Seary; Can. J. Phys. 59, 876 (1981)
7. E.A. Stern, D.E. Sayers and F.W. Lytle; Phys. Rev. B11,
    4836 (1975)
8. B.K. Teo and P.A. Lee; J. Am. Chem. Soc. 101, 2815 (1979)

Part III **Biological Systems**

EXAFS and XANES Studies of Fe and Zn in Horse Spleen Ferritin

E.C. Theil   Department of Biochemistry, North Carolina State University
Raleigh, NC 27695-7622, USA
D.E. Sayers, A.N. Mansour, and F.J. Rennick   Department of Physics,
North Carolina State University, Raleigh, NC 27695-7622, USA
C. Thompson   Department of Chemistry, University of New Hampshire,
Durham, NH 03824, USA

Ferritin, a hollow, spherical protein encasing a polymeric core
of hydrous ferric oxide and ferric oxyphosphate (8:1), stores
iron in a soluble, bioavailable form for plants, animals, and
some bacteria. The core has some properties in common with
geochemical materials, e.g. ferrihydrate and oceanic,
volcanogenic ferric oxides, but in ferritin, core formation
(nucleation) and utilization appear to be influenced by the
protein and by its interactions with iron and other metals such
as zinc. Metal environments of solutions of a Zn(II), Fe(III)-
protein complex and a model Fe(III)ATP complex (4:1) were
examined by XANES and EXAFS to determine the effect of zinc and
phosphorous on the iron environment.

I. Introduction

Ferritin stores ferric iron in a soluble, bioavailable form at concentrations equivalent to ca. 10^{13} times that of the hydrated ion. The iron-
protein complex, widespread among plants and animals, has three components:
the outer layer of protein, the inner core of polymeric hydrous ferric
oxide, and the iron-protein interface [1].

The outer layer of protein is composed of 24 subunits or protomers,
which form a hollow sphere about 23 Å thick containing a double layer of α-
helical bundles. Fourteen channels, eight hydrophilic and six hydrophobic,
pierce the protein shell providing the opportunity for exchange between the
interior of the molecule and the exterior cytoplasm [2]. Cell-specific and
posttranslational variations in the protein can affect the iron storage
function [1]. The inner iron core varies in size from 0 to 4500 iron
atoms, with an average composition of $(FeOOH)_8 \cdot FeO \cdot OPO_3H_2$ and a unique
three-dimensional structure which resembles closely the mineral
ferrihydrate [3]; the local environments examined by XANES and EXAFS
analysis for ferritin, iron-dextran complexes, and ferrihydrate are also
very similar [4, 5]. Since ferrihydrate has no phosphate, the similarity
of the crystal structure of ferritin and ferrihydrate suggest that the
phosphate-containing regions of the ferritin core must not contribute
significantly to the x-ray diffraction pattern, and probably represent
regions of disorder or crystal imperfections [1]. Ultra-high resolution
electron microscopy of the iron cores suggests that imperfections or
disorder affect 20-30% of the iron atoms [6]. Reconstitution of ferritin
from apoferritin and Fe(II) in the presence of air, with and without
phosphate, indicated that phosphate enhanced the formation of iron cores
with a smaller superparamagnetic domain [7], but only a little else is
known about the interaction of phosphate with iron in the ferritin core
[8]. The iron-protein interface, previously observed indirectly by EPR
spectroscopy [9], consists of an average of 12 metal-binding sites per

molecule (0.5/protomer) that accommodates iron as Fe(II) or Fe(III) as well as other metals, e.g. Tb and V; the ligands are mostly O and include at least two carboxylate groups [9 and references therein]. EXAFS analysis of ferritin with small amounts of iron (0.5 Fe(III)/protomer) permitted the detection of an iron environment distinct from the iron core and similar to another iron complex with carboxylate ligands, Fe(III)oxalate (Fig. 1), and suggested that iron bound to the protein was being observed, possibly as an initiation complex [10]. Small differences between the oxalate and protein complex appeared to exist in the nearest, low Z shells, suggesting either heterogeneity of protein ligands (N and O) or greater distortion of the octahedral geometry of the first shell in the Fe(III)apoferritin compared to Fe(III)oxalate [10]. Zinc influences iron core formation in vivo and in vitro and binds to the protein, as judged by the retention of ^{65}Zn [2 and references therein]. However, zinc behaves anomalously in competition studies with the VO^{2+}-apoferritin complex probed by EPR spectroscopy [9].

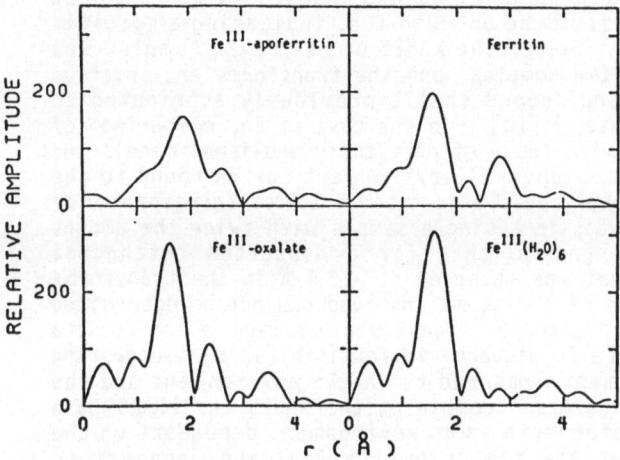

Fig. 1. The magnitude of the Fourier transform $k^3\chi$ vs. r (Å) over a k space of 4-9 Å with evidence of a low Z shell at 2.52-2.64 Å for Fe(III)-apoferritin (10 Fe/molecule) and Fe(III)oxalate in contrast to ferritin (2000 Fe/molecule) and monoatomic Fe(III)$(H_2O)_6$ (data are from [10])

To further understand the interactions of zinc with apoferritin and iron-phosphate interactions that may be important in core formation, XANES and EXAFS were examined for an Fe, Zn-apoferritin complex, Fe(III)ATP (a soluble complex of 4 Fe/ATP), Fe(III)$PO_4 \cdot 2$ H_2O (powder) and ferritin.

II. Experiment

Apoferritin was prepared from horse spleen ferritin (Miles) as previously described [9] using thioglycolic acid in 0.05 M Hepes·Na at pH 7; the iron content was 0.38 Fe/molecule (0.016 Fe/protomer). $ZnSO_4$ was added (0.4 or 0.8/protomer) to apoferritin with Fe(II) (0.4/protomer) oxidized [9] to Fe(III) either before or after the addition of Zn. Final metal concentrations were 2 mM or 4 mM. $ZnSO_4$ in water (pH 5.8) or 0.05 M Hepes·Na, pH 7.0, were used as models. Fe(III)$PO_4 \cdot 2$ H_2O was obtained from Alfa Chemical, and Fe(III)ATP (adenosine triphosphate) was prepared as the 4:1 complex in Hepes·Na, pH 7.0. Ferritin was reconstituted to a core size of 480 Fe atoms using $FeSO_4$; potassium phosphate was added to the protein 15 min before iron at ratios of 0, 0.08, and 0.8/Fe. All measurements were made at the Stanford Synchrotron Radiation Laboratory during dedicated (line I-5) or parasitic (line IV-2) beam time (average beam properties were 3.0 GeV, 80 mA and 1.8 GeV, 15 mA, respectively), using a Si<111> or Si<220> channel-cut crystal and a fluorescence detector [11] with a sample holder of plexiglass and windows of KEL-F, and filters prepared from MnO [12] for Fe and copper foil for Zn. The data are derived from measurements made on duplicate samples.

III. Results and Discussion

A. Fe, Zn-Apoferritin

In an attempt to resolve the dilemma of an effect of zinc on iron core formation in vivo [13] and in vitro [14] with little or no competitive binding with other metals for apoferritin [9], the environments of Zn and Fe in apoferritin were examined by EXAFS and XANES. The XANES for Zn (Fig. 2) and Fe were distinct from the unbound ion, indicating a specific interaction with the protein. For Fe the XANES of the Fe-Zn complex was similar to the Fe(III)apoferritin complex, and the transforms and inverses showed similar low Z first and second shells previously attributed to oxygen and carbon of carboxylate(s) [10]. In the case of Zn, comparison of the XANES and EXAFS structure with those of previously published models and proteins [15-17] suggests an octahedral environment for Zn bound to the Fe(III)apoferritin complex when added at a ratio 1:1:2 of Zn:Fe:protomer (10 Fe, 10 Zn/molecule) (Fig. 2a); in a single sample with twice the amount of added Zn, the XANES changed to that for a distorted octahedral environment (Fig. 2b). A signal was observed at ∼ 3.5 Å in the transforms of the Zn EXAFS, but the nature of the atoms involved can not be determined with certainty at this time. A stronger signal was observed in the Fe data at 3.3 Å, the location of the Fe-Fe distance in ferritin [4, 5], suggesting more order in the Fe environment compared to the Zn environment and the possibility of small metal clusters. Combining the XANES and EXAFS data suggests that Zn binds to apoferritin in an environment dependent on the metal-protein ratio. Zn, at the stoichiometry analyzed, appears to slightly alter the local environment of Fe(III) bound to the protein, but since it is not clear whether or not the Zn is close to Fe, the effects of Zn on hydrous ferric oxide core formation previously observed [13, 14] may be indirect.

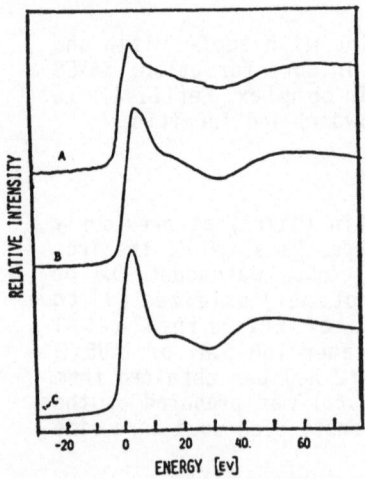

Fig. 2. XANES of (A) Zn, Fe(III)apoferritin (10 Zn, 10 Fe/molecule); (B) Zn, Fe(III)apo-ferritin (20 Zn, 10 Fe/molecule); and (C) $ZnSO_4$; all in 0.05 M Hepes·Na, pH 7.0

B. Fe-Phosphate Interactions

Investigations of iron-phosphate interactions, which in general will aid in understanding the iron core of ferritin (and incidentally, chemical properties of seawater in regions of volcanogenic activity where the ferric oxides released bind phosphate [18]), were made by comparing XANES and EXAFS for solutions of native ferritin, ferritin reconstituted with 0, 0.08, and 0.8 phosphate/Fe, and an Fe(III)ATP complex (phosphate/Fe = 0.75) and $Fe(III)PO_4 \cdot 2 H_2O$. XANES and EXAFS indicated iron in an octahedral first shell environment of great similarity for all the samples. However, higher order interactions were distinctive for ferritin and $Fe(III)PO_4$ with Fe(III)ATP intermediate between the two (Fig. 3). Similar trends were observed with increasing phosphate in the XANES of the reconstituted ferritin samples, but the data quality did not permit further analysis. A strong second shell signal was observed in the transform of the Fe(III)ATP complex (Fig. 3), the position of P in Mn(II)ATP [19] and $Fe(III)PO_4$ and Fe for ferritin (Fig. 3). The data indicate the presence of both Fe and P in the second shell, and thus the presence in the soluble complex of ferric-oxy clusters interacting with triphosphate ester. Further analysis of the Fe(III)ATP complex and comparison to reconstituted cores of ferritin with varying phosphate content will clarify the role of phosphate in core nucleation and core size. In addition, since Fe(III)ATP complexes occur naturally and may play a role in intracellular iron transport, detailed analysis of the structure will be significant for studies of iron metabolism.

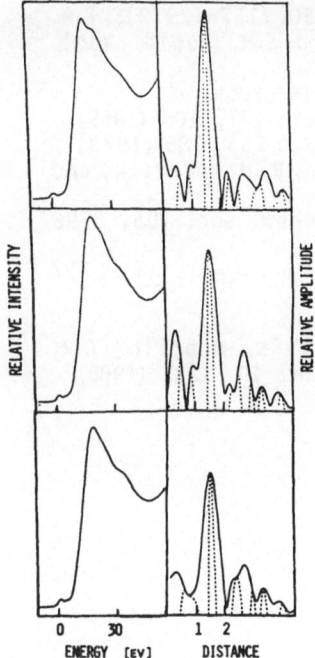

Fig. 3. XANES (left) and Fourier transforms of $k^3 x$ vs. r (Å), not corrected for phase shifts (right) of (top) $Fe(III)PO_4 \cdot 2 H_2O$; (middle) Fe(III)ATP (4:1); (bottom) ferritin

Acknowledgments

This research was supported in part by National Institutes of Health Grant AM20251 and is a contribution from the Schools of Agriculture and Life Sciences and Physical and Mathematical Sciences at North Carolina State

University. The work reported herein was partially carried out at the Stanford Synchrotron Radiation Laboratory, which is supported by the Department of Energy Office of Basic Energy Services and the National Institutes of Health Biotechnology Research Program, Division of Research Resources.

References

1   E.C. Theil: in Advances in Inorganic Biochemistry, edited by E.C. Theil, G.L. Eichhorn, and L.G. Marzilli (Elsevier, New York, 1983), Vol. V, p. 1
2   D.W. Rice, G.C. Ford, J.L. White, J.M.A. Smith, and P.M. Harrison: ibid, p. 39
3   K.M. Towe: J. Biol. Chem. 256, 9377 (1981)
4   E.C. Theil, D.E. Sayers, and M.A. Brown: J. Biol. Chem. 254, 8132 (1979)
5   S.M. Heald, E.A. Stern, B.A. Bunker, E.M. Holt, and S.L. Holt: J. Am. Chem. Soc. 101, 67 (1979)
6   W.H. Massover and J.M. Crowley: PNAS 70, 3847 (1973)
7   J.M. Williams, D.P. Danson, and C. Janot: Phys. Med. Biol. 23, 835 (1978)
8   A. Treffry and P.M. Harrison: Biochem. J. 171, 313 (1978)
9   N.D. Chasteen and E.C. Theil: J. Biol. Chem. 257: 7672 (1982)
10 D.E. Sayers, E.C. Theil, and F.J. Rennick: J. Biol. Chem. 258, 14076 (1983)
11 E.A. Stern and S.M. Heald: Rev. Sci. Instrum. 50, 157 (1979); E.A. Stern, W.T. Elam, B.A. Bunker, K.Q. Lu, and S.M. Heald: Nucl. Instrum. Methods 195, 345 (1982)
12 J. Wong and H. Lieberman: Nucl. Instrum. Methods (in press)
13 C.B. Coleman and G.B. Matrone: Biochim. Biophys. Acta 177, 106 (1969)
14 I.G. Macara, T.G. Hoy, and P.M. Harrison: Biochem. J. 135, 785 (1973)
15 L. Garcia-Iniguez, L. Powers, B. Chance, S. Sellin, B. Mannervilk, and A.S. Midvan: Biochemistry 23, 685 (1984)
16 V. Yachandra, L. Powers, and T.G. Spiro: J. Am. Chem. Soc. 105, 6596 (1983)
17 R. Jerome, G. Vlaic, and C.E. Williams: J. Physique-Lett. 44, L717 (1983)
18 R.A. Berner: Earth Planet. Lett. 18, 77 (1983)
19 M. Belli, A. Scafati, A. Bianconi, E. Buratini, S. Mobillo, C.R. Natoli, L. Palladino, A. Reale: Il Nuovo Cimento 2D, 1281 (1983)

EXAFS Studies on the c-Heme Environment in Native and Chemically Modified Cytochromes

A. Colosimo
Dept. of Exper. Medicine and Biochem. Sciences, Univ. of Rome
Tor Vergata, Italy
F. Andreasi Bassi
Inst. of Physics, Catholic University S.C., Rome, Italy
S. Mobilio
I.N.F.N. Laboratories, Frascati, Rome, Italy

The chemical environment of the c heme in low-molecular weight and water soluble cytochromes like cytochrome c is well known thanks to the X-ray diffraction studies carried out by Dickerson and coworkers (Ref.1) for both the oxidation states of the molecule and for a number of different species. Their conclusions, summarized in Table 1 as far as the interatomic distances of the first five shells surrounding the chromophores are concerned, have been confirmed and substantiated by various physical and biochemical methods, including XAS (see, for example, Ref.2).

Table 1. The chemical environment of heme-iron in heart cytochrome c (from Ref.1)

Shell	occupancy	Interatomic distance
1 N	(His 18)	1.97 (Å)
4 N	(Heme)	2.00
1 S	(Meth 80)	2.32
8 C_α	(Heme)	3.05
4 C_γ	(Heme)	3.43
8 C_β	(Heme)	4.25

The present paper refers to a series of EXAFS measurements carried out on the carboxymethylated form of cyt c (CM cyt c) as well as on the d-heme depleted form of cytochrome oxidasefrom Pseudomonas Aeruginosa taking advantage of: i) parallel investigations carried out on the native cytochrome c (NAT cyt c) and ii) extensive use of simulation techniques.

The carboxymethylated cytochrome c prepared according to Schejter and George (Ref.3) has been characterized spectroscopically and functionally by Brunori et al. (Ref.4). The latter authors suggested a reaction mechanism accounting for the kinetic and equilibrium behaviour of its reduced form in the reaction with CO and based on a proton induced shift between the alkaline and the acidic state of an amino acid residue in the heme pocket: at acidic pH (6) the prevailing conformation would have a pentacoordinated iron since Lys 79 (replacing Meth 80 as a distal ligand in CM cyt c) is only able to bind the iron in the dissociated state.

This is in good agreement with the results from our EXAFS spectra (Fig.1), in which the relevant differences between reduced NAT and CM cyt c are: 1) a narrower first peak in Fourier transform of the CM cyt c at pH 4.5 as compared to the NAT cyt c at pH 7.0, reflecting a higher level of chemical and structural homogeneity in the scatterers; 2) the presence, in the case of CM cyt c, of a new peak between those assigned to the heme nitrogens and Cα, which might be due to the S atom of Meth 80, here shifted at a longer distance from the iron than the 2.32 Å in the native form.

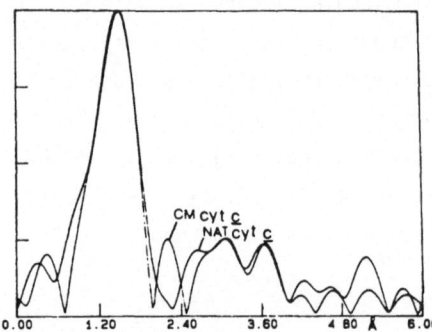

Fig.1. Fourier transforms of the EXAFS oscillations for native (NAT) and carboxymethylated (CM) cyt. c. The experimental data represented the average of at least three spectra run under identical conditions. The samples were prepared by addition to the lyophylized powder of NAT and CM cyt c of the appropiate buffer (K-Pho 0.1 M, pH = 7.0 and NaAcetate 0.1 M, pH = 4.5, respectively) up to a final concentration of 10 mM. The completely reduced state of the proteins was obtained by addition of minute amounts of dithionite. The EXAFS fluorescence spectra were recorded at T = 25°C and current intensity between 80 and 25 mA, using four photomultipliers in a semi-octahedral arrangement around a sample holder of three mm thickness. During the time required to run a single spectrum (between 60 and 80 min) no change occurred in the sample due to radiation damage, as checked by absorption spectroscopy in the VIS/ UV region. Data analysis was carried out with the help of a software package developed by one of the authors (S. Mobilio). The Fourier transforms have been performed in the range 2.5-9.5 Å^{-1} with a K^2 weight and a Hanning window function of 2 Å^{-1} width

Figure 2 shows the Fourier transforms obtained after simulating the EXAFS spectra of NAT cyt c by the method described in Refs. 5-6 and on the basis of the interatomic distances listed in Table 1. The satisfactory matching in the peak position and shape of the Fourier transforms between experimental and simulated spectra in the case of NAT cyt c enhances our confidence in the experimental and analytical procedures followed, while comparison between Figs. 1 and 2 provides, as a minimum estimate for the Fe-S distance in CM cyt c, the value of 2.7 Å.

Figure 3 shows the results of the Fourier analysis carried out over the sum of eight EXAFS spectra relative to the c heme environment in Pseudomonas cytochrome oxidase. This enzyme is characterized by the presence of two different hemes, of the c and of the d type, in two different (although unknown) environments, since the c heme only is covalently bound to the polypeptide chain (see Ref. 7 for a review). Taking advantage of this latter feature, selective and reversible splitting of the d heme has been achieved using a mild procedure which leaves unaltered the spectra properties of the c heme in the VIS/UV regions.

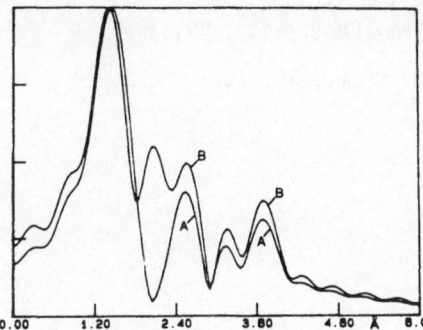 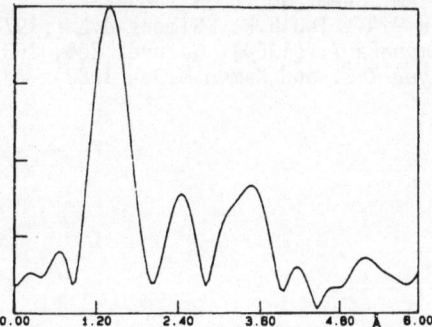

Fig.2. Fourier transforms of simulated EXAFS spectra for cytochrome c. EXAFS were simulated using the theoretical parameter of Refs. 5-6, on the basis of: the bond lengths listed in Tab. 1 (A); the same lengths above but a value of 2.7 Å for the Fe-S distance (B). The Fourier transforms have been calculated following the identical procedure described in Fig. 1.

Fig.3. Fourier transform of the EXAFS oscillations for the d Heme depleted form of Ps. cytochrome oxidase. Immediate after splitting of the d heme, the enzyme was centrifuged for 10 hs at $T = 0°C$ and 30×10^3 g, and the pellet used to fill the sample holder. All the other experimental conditions and technical precautions as those in Fig. 1.

The average distances of at least the first two shells can be defined, on the basis of our data, the relative figures being 2.05 ± 0.06 Å and 3.12 ± 0.06 Å. The significant difference in the latter one as compared to that assigned to the Cα in NAT cyt c (3.05 Å) indicates, in the case of Ps. oxidase, the possible contribution of some other scatterer situated at a noticeably longer distance. This is even more true for the 3rd peak, whose maximum is located around 4.05 Å and which is of clearly composite nature corresponding, as a first approximation, to the coalescence of the 4 Cγ and the 8 Cβ.

The chemical information avalaible up to now on Ps. cytochrome oxidase is limited to the amino acid composition and the sequence of the first 60 residues, the c heme-binding peptide. Assignment of the 6th coordination position of the iron to Meth 60, given by Meyer and Kamen (Ref. 8) on this basis alone, should be probably revised in the light of our data, although unequivocal definition of the chemical environment of the heme demands extensive use of suitable model compounds. This is a central topic in our present experimental plans.

References

1. Takano T. and Dickerson R.E. (1981), J. Mol. Biol., **153**, 95
2. Yuen C., Weissbluth M. and Labhardt A.M. (1977), SSRP Report n.77/06
3. Schejter A. and George P. (1965), Nature **206**, 1150
4. Brunori M., Wilson M.T. and Antonini E. (1972), J. Biol. Chem., **247**, 6076
5. Teo B.K., Lee P.A., Simons A.L., Eisenberger P. and Kincaid B.M. (1977), J. Am. Chem. Soc., **99**, 3854
6. Lee P.A., Teo B.K., Simons A.L. (1977), J. Am. Chem. Soc., **99**, 3856
7. Yamanaka T. (1964), Nature, **204**, 253
8. Meyer T.E. and Kamen M.D. (1982), Adv. Prot. Chem., **35**, 162

Structural Characterization of 3Fe Clusters in Fe-S Proteins by EXAFS

Robert A. Scott
School of Chemical Sciences, University of Illinois, Urbana, IL 61801, USA
James E. Penner-Hahn and Keith O. Hodgson
Department of Chemistry, Stanford University, Stanford, CA 94305, USA
H. Beinert
Institute for Enzyme Research, University of Wisconsin, Madison, WI 53706, USA
C. David Stout
Department of Crystallography, University of Pittsburgh,
Pittsburgh, PA 15260, USA

Fe EXAFS data have been collected on solution samples of unactivated beef
heart aconitase and a derivative of Azotobacter vinelandii ferredoxin I (Av
Fd I) in which the [4Fe-4S] cluster has been removed. Single crystal polarized EXAFS studies of native Av Fd I are also reported. The data have been
analyzed to determine the structural characteristics of the [3Fe-xS] cluster
present in each protein. Evidence is presented for the existence of a common
"compact" cluster structure in these two proteins (as well as Desulfovibrio
gigas Fd II). This is not in agreement with the "extended" [3Fe-3S] cluster
structure determined by x-ray crystallography on Av Fd I. Possible explanations for this discrepancy are discussed.

1. Introduction

Until about 1979, the field of Fe-S proteins seemed reasonably satisfied with
the assumption that all naturally-occurring Fe-S clusters would be found to
have either the [2Fe-2S] or [4Fe-4S] structures that had been modeled so well
by numerous synthetic analogs [1]. Thus, it came as something of a shock
when evidence was uncovered (primarily through Mössbauer spectroscopy) for
the existence of [3Fe-xS] clusters, now known to occur in several isolated
Fe-S proteins [2]. Since their initial discovery, it has been shown in a
number of cases that the [3Fe-xS] clusters can be transformed into [4Fe-4S]
clusters (and vice versa) in a relatively facile manner, suggesting the possibility that [3Fe-xS] clusters are artifactual remnants of protein isolation
techniques. We have taken the view that the fact that [3Fe-xS] clusters exist
in proteins makes them worthy of study, especially in light of the apparent
ability of some of these proteins to accomodate a structural rearrangement to
a [4Fe-4S] cluster.

Spectroscopic techniques such as Mössbauer can yield information about the
number of Fe atoms contained in a particular magnetically-coupled cluster,
but in order to ascertain metrical details (bond lengths, etc.), only two
techniques are applicable: x-ray crystallography and x-ray absorption
spectroscopy (XAS). Both of these techniques have been applied to [3Fe-xS]
cluster-containing proteins. Ferredoxin II of Desulfovibrio gigas (Dg Fd
II), a tetramer of identical subunits with one [3Fe-xS] cluster per subunit,
has been examined by Fe EXAFS in both oxidized and reduced states [3]. Fe
EXAFS has also been used to study beef heart aconitase, isolated (by traditional procedures) in an inactive [3Fe-xS] form [4]. Ferredoxin I from
Azotobacter vinelandii (Av Fd I) contains one [4Fe-4S] and one [3Fe-xS]
cluster as isolated and has been structurally characterized by both x-ray
crystallography (to 2.0 Å resolution) [5] and by Fe EXAFS [6]. Av Fd I
also appears to be one of the few examples for which results of x-ray
crystallography and EXAFS seem to be in direct and substantial conflict.

2. Experimental

Single crystal samples of Av̲ Fd I for the polarized EXAFS experiments were
grown and mounted as described in [5] and shipped from Pittsburgh to Stanford
for XAS data collection, after which they were confirmed to still diffract
showing that no radiation damage had occurred. The data were collected at
SSRL on (wiggler) beam line IV-2 under dedicated operation at 3.0 GeV and
~70 mA. Two crystals were examined, each at two different orientations for a
total of ~24 hours exposure time per crystal. The x-ray beam was collimated
to 2 x 2 mm and the crystals were positioned by use of a He/Ne laser directed
(in reverse) along the x-ray beam propagation path. The XAS data were collected by fluorescence excitation using a multi-element scintillation detector array [7] with Mn filters. The EXAFS data for each orientation of each
crystal consists of the average of ~20 scans lasting ~25 minutes each.

Curve-fitting analysis of the Fe EXAFS data was performed using a hybrid
empirical/theoretical technique described in detail elsewhere [8]. Briefly,
empirical amplitude and phase functions for each interaction are extracted by
complex backtransformation of the appropriate shell in the Fourier transform
(FT) of a structurally-characterized ("model") compound. For the Fe-S proteins discussed here, $(NEt_4)[Fe(SC_{10}H_{13})_4]$ was used as the model for Fe-S
interaction and $(NEt_4)_2[Fe_2S_2(S_2\text{-}o\text{-}xyl)_2]$ was used as the model for Fe-Fe
interaction. The appropriate E_0 for Fe-Fe was found to be 7130 eV and for
Fe-S, 7135 eV. The protein Fe EXAFS was then extracted assuming E_0 = 7130
eV, keeping ΔE_0 for Fe-Fe fixed at 0 eV, and varying ΔE_0 for Fe-S during the
optimizations.

3. Solution EXAFS

Figure 1a shows the FT of Fe EXAFS of unactivated aconitase [4]. The two main
peaks at ~1.7 and 2.3 Å correspond to Fe-S and Fe-Fe interactions, respectively. Fourier filtering these two peaks gives rise to the data shown in
Fig. 1b. Also shown is the best two-shell (Fe-S, Fe-Fe) fit of the filtered
data. The appearance of the aconitase FT is reminiscent of those observed for
[2Fe-2S] and [4Fe-4S] proteins and synthetic analogs [10] and is also very
similar to the FT observed for oxidized Dg̲ Fd II [3]. The Fe EXAFS of a derivative of Av̲ Fd I with the [4Fe-4S] cluster removed are reported elsewhere
[6]. The FT looks virtually identical to the aconitase data shown in Figure
1a. This type of FT (and filter) is a signature for a "compact" Fe-S cluster with Fe-S distances of 2.2 Å and Fe-Fe distances of ~2.7 Å.

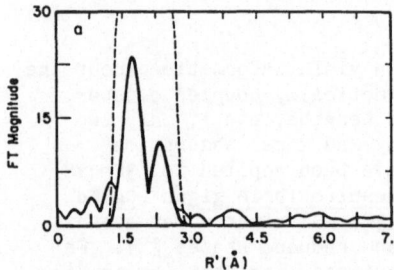 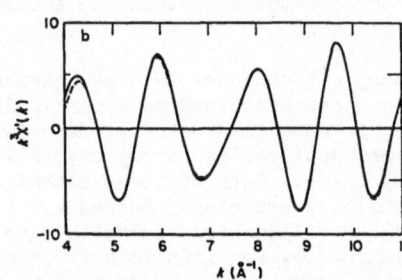

Figure 1. Fe EXAFS data on unactivated beef heart aconitase. (a) Fourier
transform (k^3-weighted k = 3.5-11.5 Å^{-1}) of Fe EXAFS. The dashed line
represents the filter window which gives rise to the filtered data in (b).
(b) the solid line is the result of Fourier-filtering the two main (Fe-S,
Fe-Fe) FT peaks in (a) and the dashed line is the best two-shell fit to the
filtered data (see Table 2).

4. Av Fd I Single Crystal Polarized EXAFS

The crystal structure results [5] on native (as isolated) Av Fd I suggest a [3Fe-3S] cluster structure with Fe-Fe distances of ~4.1 Å and with five cysteine sulfurs and one "oxo" (hydroxo or water) as terminal ligands (two per iron). The existence of a [4Fe-4S] cluster in addition to the 3Fe cluster in Av Fd I makes an EXAFS determination of the structure of the 3Fe cluster difficult in the native protein. However, the iron atoms of the [3Fe-3S] cluster define a plane and the Fe-Fe interactions of the [4Fe-4S] cluster should be nearly isotropic. Perhaps fortuitously, the [3Fe-3S] clusters of all four molecules in the unit cell have their Fe plane normals aligned within ± 15° of the crystallographic **c** axis. This prompted us to examine the Fe EXAFS of Av Fd I crystals in two orientations: one with the x-ray electric field polarized along the **c** axis (nearly **perpendicular** to the 3Fe plane) and the other with the electric field polarized 90° away from the **c** axis (nearly **parallel** to the 3Fe plane). The hope was that, if the 4.1 Å Fe-Fe interaction could be observed, it should have a very distinct polarization. The orientation-dependent FT's for one of two Av Fd I crystals examined are shown in Fig. 2 for one of the crystals. Again, the signature of a "compact" Fe-S cluster is observed, presumably attributable in large part to the [4Fe-4S] cluster. No peak is observed (above noise) in the region where a 4.1 Å Fe-Fe interaction would be expected. This is not surprising since the EXAFS amplitude falls off as R^{-2} and the room temperature Debye-Waller σ_{as} is expected to be quite large for the "extended" structure. What was not expected is the orientation-dependence of the amplitude of the Fe-Fe peak. Curve-fitting results were used to calculate the "polarization ratio" $N_\parallel/N_\perp$ for Fe-S and Fe-Fe peaks for both crystals (see Table 1). The polarization ratio for the Fe-S peak is expected and observed to be near unity assuming approximately tetrahedral coordination around each Fe. However, the Fe-Fe peak should only contain contributions from the [4Fe-4S] cluster and the polarization ratio is again expected to be near unity. The Fe-Fe polarization ratio is reproducibly much larger than expected. Two possible explanations present themselves: (1) The [4Fe-4S] cluster is significantly distorted giving rise to different resultant projections of Fe-Fe vectors in the two orientations; or (2) The [3Fe-xS] cluster contributes one or more 2.7 Å Fe-Fe vectors lying approximately in the plane defined by the Fe atoms of the [3Fe-3S] cluster.

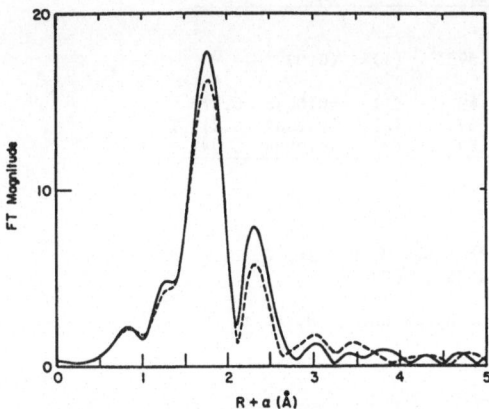

Figure 2. Orientation dependent Fourier transform (k^3-weighted, k = 2.0-12.0 Å^{-1}) of Fe EXAFS of a single crystal of Av Fd I. The solid line is the FT of data collected with the x-ray electric field (**e**) perpendicular to the crystal **c** axis (in the 3Fe plane). The dashed line is for **e** parallel to **c** (perpendicular to the 3Fe plane).

Table 1. Av Fd I polarized EXAFS curve-fitting results

	Fe-S R(Å)	N‖/N⊥ [a]	Fe-Fe R(Å)	N‖/N⊥ [a]
calculated[b]	2.31(3)	1.11	2.84(8)	0.90
observed (xtal 1)	2.26	1.07	2.73	1.60
(xtal 2)	2.26	1.14	2.74	1.75

[a]Ratio of the number of scatterers observed in the parallel orientation to the number observed in the perpendicular orientation.
[b]From D. Ghosh, S. O'Donnell, W. Furey, Jr., A. H. Robbins , C. D. Stout: J. Mol. Biol. **158**, 73 (1982).

5. EXAFS Curve-Fitting

Curve-fitting results for the aconitase Fe EXAFS data are shown in Table 2. In these fits, the number of scatterers (N_s) was fixed at some integer value and R_{as} (the Fe-S or Fe-Fe distance) and $\Delta\sigma_{as}^2$ (the relative mean square deviation in R_{as} compared to the model compound) for each shell were optimized. Three different fits were performed, varying the number of Fe neighbors from 1 to 3 (i.e., a 2Fe, 3Fe, or 4Fe cluster). Since a [2Fe-2S] cluster was used as a model for the Fe-Fe interaction, σ_{as}^2 for this model contains only the vibrational contribution (no static disorder) [10]. Thus, for Fe-S clusters of higher nuclearity, $\Delta\sigma_{as}^2$ would be expected to be zero or slightly positive. Using this criterion, the best fit is the one with N_s(Fe-Fe) = 2. This fit is also the one that gives the best goodness-of-fit value (f). Curve-fitting results on the [4Fe-4S]-depleted Av Fd I EXAFS are virtually identical [6]. The curve-fitting thus predicts a "compact" 3Fe cluster of very similar structure in both unactivated aconitase and [4Fe-4S]-depleted Av Fd I.

Table 2. Curve-fitting analysis of Fe EXAFS data for aconitase and model compounds

Compound/Protein	Fe-S[a] R_{as}	$\Delta\sigma_{as}^2$	ΔE_0	Fe-Fe R_{as}	N_s	$\Delta\sigma_{as}^2$	f[b]
	(Å)	(Å^2)	(eV)	(Å)		(Å^2)	
$[Fe(SR)_4]^{-}$[c]	(2.284)[d]	(0.0)	(5)	-	-	-	-
$[Fe_2S_2(S_2R)_2]^{2-}$[e]	-	-	-	(2.698)	(1)	(0.0)	-
Aconitase[f]	2.19	-0.0005	4.6	2.69	(1)	-0.0038	0.470
(unactivated)	2.20	-0.0007	3.7	2.69	(2)	+0.0004	0.217
	2.20	-0.0009	3.3	2.69	(3)	+0.0034	0.343

[a]N_s(Fe-S) was fixed at 4 for all fits.
[b]$f = \{ \sum_{i=1}^{N} [\underline{k}^3(\chi_{obs}^{(i)} - \chi_{calc}^{(i)})]^2/N \}^{1/2}$
[c]$(NEt_4)[Fe(SC_{10}H_{13})_4]$ was a gift from S. A. Koch - see M. Millar, J. F. Lee, S. A. Koch, R. Fikar: Inorg. Chem. **21**, 4106 (1982).
[d]Numbers in parentheses were not optimized.
[e]$(NEt_4)_2[Fe_2S_2(S_2\text{-}\underline{o}\text{-}xyl)_2]$ was a gift from K. S. Hagen and R. H. Holm. Structural data were taken from [1].
[f]EXAFS data taken from [5].

6. Discussion

The EXAFS evidence discussed above suggests that the 3Fe clusters in Dg Fd II, aconitase, and Av Fd I all have a very similar "compact" structure (with

~ 2.7 Å Fe-Fe distances). This result does not agree with the Av Fd I crystallographic determination of an "extended" [3Fe-3S] cluster (with ~4.1 Å Fe-Fe distances) [5]. There are at least four possible explanations for this discrepancy: (1) The 3Fe cluster changes from a "compact" structure in solutions of Av Fd I to an "extended" structure in Av Fd I crystals; (2) Removal of the [4Fe-4S] cluster results in rearrangement of the 3Fe cluster of Av Fd I from an "extended" to a "compact" structure; (3) The EXAFS data for Av Fd I is wrong or has been misinterpreted; (4) The x-ray crystallographic data for Av Fd I is wrong or has been misinterpreted.

Unfortunately, none of these explanations is very satisfying. Explanation (1) requires a substantial structural rearrangement of the Av Fd I 3Fe cluster upon crystallization. The single crystal polarized EXAFS experiment seems to observe a "compact" 3Fe cluster although the alternative of a significantly distorted [4Fe-4S] cluster cannot be ruled out. Resonance Raman studies of microcrystalline Av Fd I [11] seem to support a "compact" structure in the crystalline state as well. Explanation (2) is unsatisfactory since both magnetic circular dichroism and electron paramagnetic resonance spectroscopies indicate that the 3Fe cluster is unaffected by removal of the [4Fe-4S] cluster [6]. Explanation (3) would require either that our understanding of the EXAFS phenomenon in terms of single-scattering theory is completely wrong or that the FT peak at R' = 2.3 Å has been misassigned as the Fe-Fe peak. The argument must be that, for the "extended" [3Fe-3S] structure, another shell of (non-Fe) atoms gives rise to the 2.3 Å FT peak. In order to test this possibility, we have examined the Fe EXAFS of $(NEt_4)_3$-$[Fe_3(SPh)_3Cl_6]$, a trinuclear cluster with ~4.4 Å Fe-Fe distances [12]. Although this is not a precise model for the proposed [3Fe-3S] cluster, the FT showed only the expected Fe-(S,Cl) peak. Explanation (4) is also unsatisfactory. Re-refinement of the Av Fd I structure unfailingly results in an electron density map that cannot be made to fit a "compact" arrangement of Fe atoms. Thus, we do not currently have a satisfactory explanation for the difference in predicted [3Fe-xS] structures between x-ray crystallography and EXAFS.

One further point concerns the value of x in the "compact" [3Fe-xS] cluster. Careful iron and sulfide analyses on aconitase [4] have indicated that the probable value for x in this protein is 4 [4]. Two possible structures for [3Fe-4S] clusters have been suggested [2,4,11], one requiring three, the other four terminal cysteine ligands. As pointed out by JOHNSON, et al. [11], removal of a single bridging sulfide and concomitant binding of extra terminal ligands would convert a [3Fe-4S] cluster to a [3Fe-3S] cluster. It is possible that the conflicting results reported herein could have their basis in 3Fe-cluster lability, resulting in facile interconversion between "compact" and "extended" structures. The answer will have to await the future availability of structural data on other [3Fe-xS] proteins.

Acknowledgments
Jean-Luc Dreyer is thanked for assistance with preparation of the aconitase samples. Carl Cork is acknowledged for help in alignment of Av Fd I single crystal samples at SSRL. XAS work under RAS at Illinois is supported by an NIH Biomedical Research Support Grant (RR-07030). The XAS work was performed at SSRL, which is supported by the Department of Energy, Office of Basic Energy Sciences; and the NIH, Biotechnology Resource Program, Division of Research Resources.

References and Notes
1. J. M. Berg, R. H. Holm: in "Iron-Sulfur Proteins", T. G. Spiro, Ed. (Wiley, New York 1982); Chapt. 1

2.   For a review, see: H. Beinert, A. J. Thomson: Arch. Biochem. Biophys.
     222, 333 (1983)
3.   M. R. Antonio, B. A. Averill, I. Moura, J. J. G. Moura, W. H.
     Orme-Johnson, B.-K. Teo, A. V. Xavier: J. Biol. Chem. **257,** 6646 (1982)
4.   H. Beinert, M. H. Emptage, J.-L. Dreyer, R. A. Scott, J. E. Hahn, K. O.
     Hodgson, A. J. Thomson: Proc. Natl. Acad. Sci. USA **80,** 393 (1983)
5.   D. Ghosh, S. O'Donnell, W. Furey, Jr., A. H. Robbins, C. D. Stout: J.
     Mol. Biol. **158,** 73 (1982)
6.   P. J. Stephens, T.V. Morgan, F. Devlin, J. E. Penner-Hahn, K. O.
     Hodgson, R.A. Scott, C. D. Stout, B. K. Burgess: Proc. Natl. Acad. Sci.
     USA, submitted for publication.
7.   S. P. Cramer, R. A. Scott: Rev. Sci. Instrum. **52,** 395 (1981)
8.   R. A. Scott: Meth. Enzymol., submitted for publication.
9.   B.-K. Teo, P. A. Lee: J. Am. Chem. Soc. **101,** 2815 (1979)
10. B.-K. Teo, R. G. Shulman, G. S. Brown, A. E. Meixner: J. Am. Chem. Soc.
     101, 5624 (1979)
11. M. K. Johnson, R. S. Czernuszewicz, T. G. Spiro, J. A. Fee, W. V.
     Sweeney: J. Am. Chem. Soc. **105,** 6671 (1983)
12. This compound was a gift from K. S. Hagen and R. H. Holm. The struc
     tural parameters are from K. S. Hagen, R. H. Holm: J. Am. Chem. Soc.
     104, 5496 (1982)

Copper X-Ray Absorption Spectroscopy of Cytochrome c Oxidase

Robert A. Scott and James R. Schwartz

School of Chemical Sciences, University of Illinois, Urbana, IL 61801, USA

Stephen P. Cramer

Exxon Research and Engineering, Annandale, NJ 08801, USA

Beef heart mitochondrial cytochrome aa_3, purified by several different methods in several laboratories, has been examined by extended x-ray absorption fine structure (EXAFS) spectroscopy. Comparison of copper EXAFS and Fourier transforms (FT's) shows that, in spite of visual differences in the FT's among the different samples, the data of all the samples are the same, within experimental error. Curve-fitting results suggest the presence of 5-6 nitrogens (or oxygens or carbons) at ~1.96 Å and 2-3 sulfurs at ~2.32 Å per two coppers. In addition, it was found that x-ray induced reduction of the metal sites of cytochrome aa_3 is effectively nonexistent at or below ~200 K.

1. Introduction

Cytochrome c oxidase (ferrocytochrome $c:O_2$ oxidoreductase, EC 1.9.3.1) is a large (~150,000 daltons) membrane-bound enzyme of the mitochondrial respiratory chain [1]. The enzyme can be isolated in either dimeric or monomeric forms with two hemes and two coppers per monomer. The two hemes are both of the a type but are structurally and functionally distinct in the active enzyme, giving rise to the names a and a_3. For this reason, the mitochondrial cytochrome c oxidase is often referred to as cytochrome aa_3, a nomenclature that will be used herein. The two copper atoms are also distinct and will be referred to as Cu_A (the copper responsible for the g2 electron paramagnetic resonance (epr) signal) and Cu_B. In the respiratory chain, electrons are transferred from cytochrome c to cytochrome aa_3 which in turn reduces O_2 to H_2O in an energy-conserving step. The initial electron-accepting site is thought to be heme a (possibly in concert with Cu_A) and the O_2 reduction site is thought to consist of both heme a_3 and Cu_B as a magnetically-interacting pair. Until recent x-ray absorption spectroscopic (XAS) studies [2-8], very little was known about the detailed molecular structure of the metal active sites of cytochrome aa_3. Due to the size and membrane-bound nature of cytochrome aa_3, XAS would seem to provide the only means of obtaining such structural information, which presumably will improve our understanding of the function of the enzyme.

Extended x-ray absorption fine structure (EXAFS) studies on cytochrome aa_3 have been carried out by two groups with apparently disparate results. For example, POWERS, et al. [5,7] typically observe a single first-shell peak (at R' ≈ 1.5 Å) in the Fourier transform (FT) of Cu EXAFS of resting state cytochrome aa_3, whereas SCOTT, et al., have published data [6,8] indicating a split first-shell peak (with maxima at R' ≈ 1.4, 1.9 Å) in the Cu FT of resting state enzyme. Since these published data were collected on purified enzyme prepared by two different techniques, the question arises as to whether these differences in Cu FT's represent structural differences in cytochrome aa_3 samples prepared by different methods. To address this question, Cu EXAFS data have been collected on cytochrome aa_3 samples prepared by four different methods and in a number of laboratories. It is the major purpose of this report to show that any differences in Cu EXAFS among these preparations are within the experimental errors of the data.

2. Sample Preparation and Data Collection

Eight separate samples of purified resting state cytochrome aa_3 were examined. Table 1 summarizes the purification methods and origin of each sample. Published EXAFS data were collected on samples prepared by the Hartzell/ Beinert method [6,8] and the Yonetani method [5,7]. All XAS data were collected at the Stanford Synchrotron Radiation Laboratory on the wiggler beam line VII-3 under dedicated operation at 3.0 GeV and ~60 mA. Each data set consists of an average of ~15 20-minute scans of samples ~1.5-2.0 $m\underline{M}$ in Cu. Data were collected by fluorescence excitation techniques using a multi-element array of scintillation detectors and Ni filters [13].

3. Sample Integrity

The oxidation states of the metal sites in the cytochrome aa_3 samples examined in this study were characterized by epr spectroscopy before and after XAS data collection. The temperature history of each sample included initial epr at ~15 K, storage at 77 K, XAS data collection at either 190 K or 4 K, storage at 77 K, and final epr at 15 K. All samples were therefore never warmed above 190 K and four of the samples were never above 77 K. To detect any possible change in the oxidation state of the copper sites during XAS data collection, the Cu edge spectrum of each sample was monitored throughout data collection (~5-6 hours). As reported previously [8], no significant change in the copper edge occurred during data collection at or below ~200 K. In addition, we have never observed any significant change in the intensity of the g6 or g3 signals of samples with the temperature history described above.

X-irradiation does, however, generate radicals in such XAS samples, as noted by others [14,15]. We have observed two different types of epr signals in X-irradiated aqueous cytochrome aa_3 samples contained in lucite sample cells with Mylar windows. One type is the typical sharp "free radical" signal at g = 2.0. This signal can also be generated by X-irradiation of an identical cell containing only H_2O. We have not yet attempted to reproduce it using an empty cell. These "free radical" signals decay with time, disappearing after several days at 77 K without causing any change in the cytochrome aa_3 epr signals. Incubation of a sample at ~195 K for five hours also causes the disappearance of these signals without change in the

Table 1. Cytochrome aa_3 samples examined

Sample	Temperature (K)[a]	Purification Method [Reference]	Origin
Hb/U[b]	4,190	Hartzell/Beinert [9]	Urbana
HB/M	195	Hartzell/Beinert [9]	Madison[c]
HB/P	190	Hartzell/Beinert [9]	Pasadena[d]
vB/A	190	van Buuren [10]	Amsterdam[e]
K/B[b]	4	King [11]	Bern[f]
Y/E	4	Yonetani [12]	Essex[g]

[a]Temperature of XAS data collection; [b]Two samples were examined; [c]From the laboratory of H. Beinert; [d]From the laboratory of S. I. Chan; [e]This sample was a gift from B. F. van Gelder to H. Beinert; [f]From the laboratory of A. Azzi; [g]From the laboratory of M. T. Wilson

cytochrome aa₃ epr. However, warming the sample to room temperature for about one minute, and then refreezing causes loss of the "free radical" signal with a concomittant increase in the intensity of the g6 signal. We conclude that the species giving rise to the "free radical" epr signal can transfer reducing equivalents to the metal sites of cytochrome aa₃ but only at temperatures considerably above ~200 K (perhaps only in fluid solution). A second type of epr signal sometimes observed after X-irradiation is a very broad (full width of ~700 G at X-band) signal centered at g ≈ 3 (at X-band). We do not currently have an explanation for the origin of this signal.

4. Copper EXAFS Data Analysis

The Cu EXAFS was extracted from the averaged raw fluorescence data for each sample by standard techniques [16]. Fourier transforms of the Cu EXAFS of each sample are shown in Fig. 1. As can be seen in the data collected at ~190 K, some samples exhibit single peaks in the first shell (e.g., sample HB/Ua) and some show two resolved peaks (e.g., sample HB/M). In the data collected at 4 K, all four samples show two clearly resolved peaks with the higher-R' peak being significantly larger (relative to the same peak at 190 K). As has been shown previously [6,8], the first peak represents a shell of nitrogen (or carbon or oxygen) atoms at ~2.0 Å from the coppers, whereas the second peak represents a shell of sulfur atoms at ~2.3 Å from the coppers.

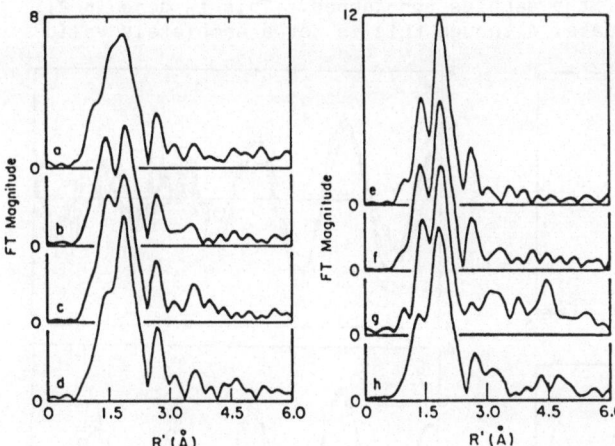

Figure 1. Fourier transforms (k_3-weighted, k = 3.0 – 13.0 Å⁻¹) of Cu EXAFS of eight resting state samples of cytochrome aa₃. The samples (see Table 1 for codes) are: (a) HB/Ua; (b) HB/M; (c) vB/A; (d) HB/P; (e) K/Bb; (f) K/Ba; (g) Y/E; (h) HB/Ub. (a)–(d) are at 190 K, (e)–(h) at 4 K.

In order to quantitate this further, the first two FT peaks were filtered out, giving rise to filtered EXAFS data as shown in Fig. 2. Two-shell (N,S) fits were then performed on filtered data for each sample. Empirical amplitude and phase functions representing Cu–N and Cu–S scattering were generated from EXAFS data on the structurally-characterized (model) compounds, [Cu(imid)₄]²⁺ (in frozen aqueous solution), and [Cu(14-ane-S₄)](ClO₄)₂ (as a solid), respectively, by complex Fourier backtransformation [17]. For each shell, the Cu–L (L = N,S) distance (R_{as}), the number of scatterers (N_S), the Debye–Waller relative mean square distance deviation (σ_{as}^2), and the shift in threshold energy (ΔE_0) need to be determined. For a particular shell, N_S and σ_{as}^2 are strongly positively correlated. To avoid problems, for each fit N_S was fixed for both shells and σ_{as}^2 was optimized. For a two-copper enzyme such as cytochrome aa₃, N_S may be half-integer (since the EXAFS is normalized on a "per copper" basis). For convenience, the total copper coordination number, N_S(N) + N_S(S), was also restricted to be 4.

An appropriate value for E_0 for each shell was estimated by extracting the empirical phase function (by complex backtransformation) from the model compound using various values for E_0 and then comparing with the theoretical phase function [18]. The E_0 giving the best match between empirical and theoretical phase functions was 9013 eV for Cu-N and 9007 eV for Cu-S. Thus, the Cu EXAFS of the cytochrome <u>aa</u>$_3$ samples was extracted assuming E_0 = 9013 eV and ΔE_0 for the Cu-N shell was fixed at 0.0 while ΔE_0 for the Cu-S shell was optimized.

The results of two-shell curve-fitting of the cytochrome <u>aa</u>$_3$ data reveal no significant differences in the R_{as} values for either shell nor in ΔE_0 for the Cu-S shell among the eight samples. The only differences among samples examined at a given temperature occur in $\Delta\sigma_{as}^2$ (measured relative to model compounds) for each shell. The question to be addressed involves whether the deviations in $\Delta\sigma_{as}^2$ values for a particular temperature are statistically significant. The major source of error in this study (as in most EXAFS work on spectroscopically dilute samples) is random noise in the original EXAFS data. This source of error is often disguised by the Fourier filtering of EXAFS data before curve-fitting. Fourier filtering is often required to filter out structural features that will not be included in the curve-fitting, but has the added effect of removing all high-frequency noise from the EXAFS. One way of assessing the effect of this random noise on the parameters calculated from curve-fitting is to compare the individual two-shell fits with the original EXAFS data for one of the samples considered. This is done in Fig. 3 for both the 190 K and 4 K data. Although this is not a completely valid

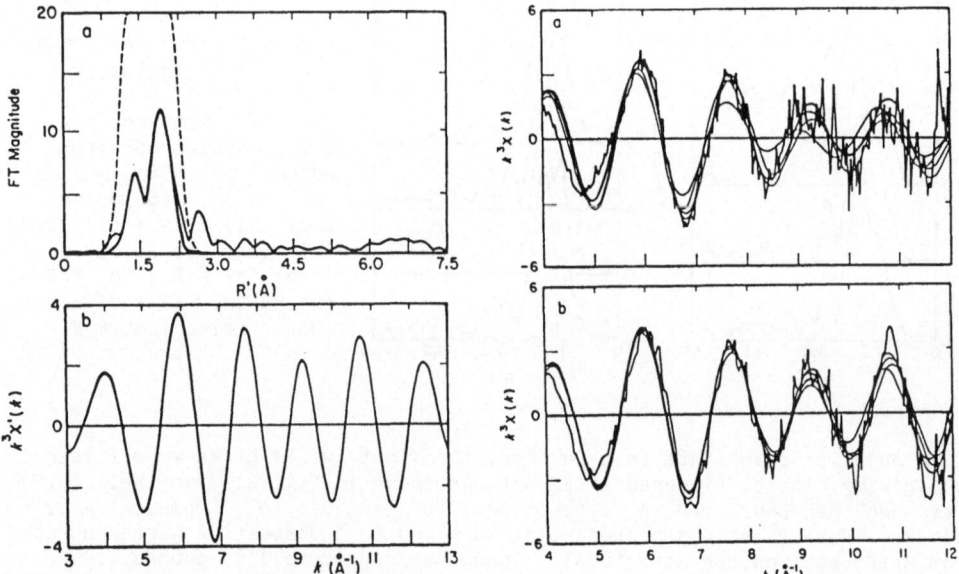

Figure 2. (left) Example of Fourier filter of first two shells of Cu EXAFS of sample K/Bb. The FT window (dashed line) of (a) covers the range from R' = 1.10-2.30 Å with half-Gaussian edges of width 0.20 Å and the result of filtering is shown in (b).
Figure 3. (right) Comparison of individual curve-fitting results (smooth curves) with the raw Cu EXAFS data for resting state cytochrome <u>aa</u>$_3$ sample. (a) Data collected at ~190 K: Fits are from samples HB/Ua, HB/M, HB/P, vB/A; raw data from sample HB/Ua. (b) Data collected at ~4 K: Fits are from sample K/Ba, K/Bb, Y/E, HB/Ub; raw data from sample K/Bb. In both the spread in the fits is very comparable to the noise in the data.

comparison (since only the two main FT peaks were used to construct the filtered data which gave rise to the fits), it can clearly be seen that the overall deviation among the fit spectra (at any particular k value) is comparable in magnitude to the noise in the original data. This is especially significant for the data collected at 190 K, since it shows that the fits (and, therefore, the EXAFS data) for the four samples examined are within experimental error the same, **even though the Fourier transforms are visually different.**

5. Discussion

Given the finding that the Cu EXAFS data of resting state cytochrome aa$_3$ collected at a given temperature are independent of the method of preparation (within experimental error), the curve-fitting results may be averaged together for each temperature and standard deviations calculated which yield information about the **precision** of the results. This is done in Table 2 for various values of N_s. The large negative values for $\Delta\sigma_{as}^2$ for the Cu-S shell are expected since the Cu EXAFS data of [Cu(14-ane-S$_4$)](ClO$_4$)$_2$ were collected at room temperature. The large positive values of $\Delta\sigma_{as}^2$ for the Cu-N shell most likely indicate a greater amount of static disorder of Cu-N(C,O) distances in cytochrome aa$_3$ as compared to [Cu(imid)$_4$]$^{2+}$ since a dynamic contribution should have shown up as a difference between 190 K and 4 K data. The best fits at 190 K occur for an average coordination (per two coppers) of six nitrogens and two sulfurs, whereas at 4 K, five nitrogens and three sulfurs fit best. The data does not support the existence of any short (~2.1 Å) Cu-S bonds, as required for the existence of a type 1 (blue) copper. In addition, as will be described in detail elsewhere [19], the peak observed at R' ≈ 2.7 Å in the Fourier transforms of Figure 1 can be conclusively assigned as an Fe atom, yielding a Cu$_B$-Fe$_{a3}$ distance of 3.0 Å, in conflict with the results obtained by POWERS, et al. [5], in which a distance of 3.8 Å was reported.

Table 2. Curve-fitting results for resting state cytochrome aa$_3$ Cu EXAFS

	Cu-N			Cu-S				
T	R_{as}	N_s	$\Delta\sigma_{as}^2$	R_{as}	N_s	$\Delta\sigma_{as}^2$	ΔE_0	f
(K)	(Å)		(Å^2)	(Å)		(Å^2)	(eV)	
191	1.96(2)	(2.0)	+0.0030(13)	2.30(1)	(2.0)	−0.0008(13)	−11(2)	0.360
	1.97(2)	(2.5)	+0.0042(10)	2.31(1)	(1.5)	−0.0026(12)	−14(2)	0.289
	1.98(1)	(3.0)	+0.0050(7)	2.33(1)	(1.0)	−0.0050(10)	−17(1)	0.245
4	1.94(1)	(2.0)	+0.0020(12)	2.29(1)	(2.0)	−0.0032(4)	−12(1)	0.394
	1.95(1)	(2.5)	+0.0035(11)	2.30(1)	(1.5)	−0.0048(4)	−14(1)	0.355
	1.97(1)	(3.0)	+0.0046(10)	2.31(1)	(1.0)	−0.0070(4)	−17(1)	0.398

The Cu EXAFS data presented here on resting state cytochrome aa$_3$ prepared by different procedures in different laboratories support the following conclusions: (a) Within experimental error, the first coordination spheres of the copper atoms are identical regardless of the method used to prepare the enzyme and in spite of visual differences in Fourier transform appearance; (b) XAS data collection below ~200 K avoids x-ray induced reduction of the metal sites of cytochrome aa$_3$, even though radical species are created; (c) A short (~2.1 Å) Cu-S interaction is not present, indicating that neither of the copper sites can be termed "blue" copper; and (d) The Cu$_B$-Fe$_{a3}$ distance is 3.0 Å.

Acknowledgments

The following are acknowledged for providing cytochrome $\underline{aa_3}$ samples for this study: H. Beinert, A. Azzi, S. I. Chan, and M. T. Wilson. D. B. Rorabacher is thanked for the sample of $[Cu(14\text{-}ane\text{-}S_4)](ClO_4)_2$ and T. Smith for the sample of $[Cu(imid)_4](NO_3)_2$. XAS work under RAS at Illinois was supported by NIH Biomedical Research Support Grant RR-07030. The work was done at SSRL which is supported by the Department of Energy, Office of Basic Energy Sciences; and the NIH, Biotechnology Resource Program, Division of Research Resources.

References

1. For a review, see B. G. Malmström: in "Metal Ion Activation of Dioxygen", T. G. Spiro, Ed. (Wiley, New York 1980); Chapt. 5
2. V. W. Hu, S. I. Chan, G. S. Brown: Proc. Natl. Acad. Sci. USA 74, 3821 (1977)
3. V. W. Hu, S. I. Chan, G. S. Brown: FEBS Lett. 84, 287 (1977)
4. L. Powers, W. E. Blumberg, B. Chance, C. H. Barlow, J. S. Leigh, Jr., J. Smith, T. Yonetani, S. Vik, J. Peisach: Biochim. Biophys. Acta 546, 520 (1979)
5. L. Powers, B. Chance, Y. Ching, P. Angiolillo: Biophys. J. 34, 465 (1981)
6. R. A. Scott, S. P. Cramer, H. B. Gray, R. W. Shaw, H. Beinert: Proc. Natl. Acad. Sci. USA 78, 664 (1981)
7. B. Chance, L. Powers, Y. Ching: in "Mitochondria and Microsomes", C. P. Lee, G. Schatz, G. Dallner, Eds. (Addison-Wesley, London 1981); p. 271
8. R. A. Scott: in "The Biological Chemistry of Iron", H. B. Dunford, D. H. Dolphin, K. N. Raymond, L. C. Sieker, Eds. (D. Reidel, Boston 1982); p. 475
9. Procedure I of C. R. Hartzell, H. Beinert, B. F. van Gelder, T. E. King: Meth. Enzymol. 53, 54 (1978)
10. Procedure III of C. R. Hartzell, H. Beinert, B. F. van Gelder, T. E. King: Meth. Enzymol. 53, 54 (1978)
11. Procedure II of C. R. Hartzell, H. Beinert, B. F. van Gelder, T. E. King: Meth. Enzymol. 53, 54 (1978)
12. T. Yonetani: J. Biol. Chem. 235, 845 (1960)
13. S. P. Cramer, R. A. Scott: Rev. Sci. Instrum. 52, 395 (1981)
14. B. Chance, P. Angiolillo, E. K. Yang, L. Powers: FEBS Lett. 112, 178 (1980)
15. G. W. Brudvig, D. F. Bocian, R. C. Gamble, S. I. Chan: Biochim. Biophys. Acta 624, 78 (1980)
16. R. A. Scott: in "Structural and Resonance Techniques in Biological Research", D. L. Rousseau, Ed. (Academic, New York 1984); vol. 2, Chapt. 4, in press.
17. P. A. Lee, P. H. Citrin, P. Eisenberger, B. M. Kincaid: Rev. Mod. Phys. 53, 769 (1981)
18. B.-K. Teo, P. A. Lee: J. Am. Chem. Soc. 101, 2815 (1979)
19. R. A. Scott, J. R. Schwartz, S. P. Cramer: in "Inorganic and Biochemical Perspectives on Copper Coordination Chemistry", Proceedings of the 2nd SUNY Albany Conference, in press.

Edge and EXAFS Studies of Cytochrome Oxidase

L. Powers

AT & T Bell Laboratories, Murray Hill, NJ 07974, USA

B. Chance

University of Pennsylvania, Philadelphia, PA 19104, USA

This terminal enzyme in the respiratory chain reduces O_2 to H_2O, a process involving $4H^+$, transfer of $4e^-$, charge separation across the membrane, and energy conservation. It contains four redox centers: the iron of cytochrome $\underline{a}$ (Fe_a) and its functionally associated copper (Cu_a) and on iron-copper bincular active site (Fe_{a_3}-Cu_{a_3}) which are magnetically coupled in the oxidized state. Fe_{a_3} is known to be the binding site for O_2 and CO in the reduced state. The structure-changes associated with its function, esp. the binuclear active site which cannot be observed by other spectroscopic techniques, is pivotal in our understanding of its mechanism.

Highly concentrated purified samples ($\sim 1.5mM$) were prepared from beef heart using the Yonetani method and from bacterial sources and characterized to assure maximum conversion to the desired derivatives [1,2]. Fluorescence data were collected with an array of plastic scintillation counters with appropriate filters and focusing lenses [2,3]. Since these samples are altered by the production of hydrated electrons and free radicals during X-ray exposure [4], the integrity of each sample was known and maintained during data collection. Hence, all data were collected at $-100^{\circ}C$ or lower and samples were simultaneously monitored by optical absorption spectroscopy and periodically by EPR spectroscopy [2,3].

The proximity of the metal atoms of the binuclear active site were uniquely identified in the first EXAFS data [5] using the following criterion[2]: each metal atom must be observed in the data of the other at the same distance having the same Debye-Waller contribution and the copper contribution must be absent on copper depletion of the enzyme [6]. This contribution was found to be $3.75 \pm .05$Å, a distance remarkably similar to the binuclear active site of hemocyanin [7]. Fe_{a_3} appears in the third shell of the copper EXAFS and Cu_{a_3} in the third shell of the iron EXAFS (Fig. 1) together with heme and proximal histidine contributions in both the oxidized and mixed valence formate derivatives where the binuclear site is magnetically coupled. Copper depletion leaves only the heme and proximal histidine contributions in the iron EXAFS [6] (Fig. 2).

Poised biochemical derivatives that selectively reduce Fe_a and Cu_a or the binuclear active site were first used in edge studies together with the fully oxidized and fully reduced derivatives where all four centers are oxidized or reduced to differentiate the contribution of the two copper atoms [8]. In addition, the 'site modeling' method was developed where the data of a model compound in the oxidized and reduced states were using to represent the contribution of a site [8,2]. The model for a site was chosen from consideration of all known biochemical and biophysical

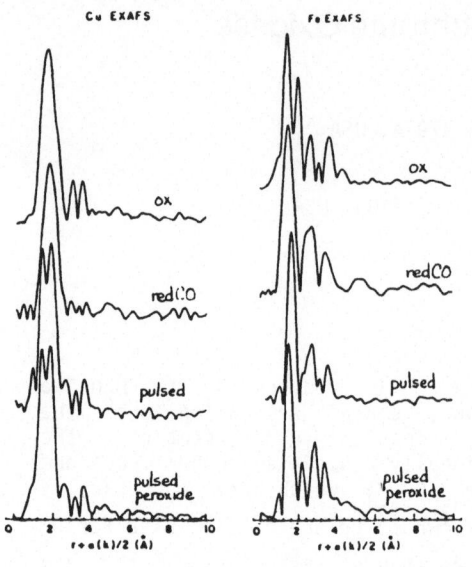

Cu EXAFS   Fe EXAFS

OX

redCO

pulsed

pulsed
peroxide

Fig.1. Fourier-transformed copper
and EXAFS data

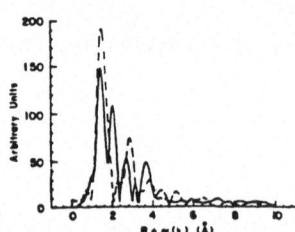

Fig.2. Fourier date for the oxidized
resting native oxidase (——) and the
oxidized copper depleted cytochrome
oxidase (---)

information and in some cases the model was used by other spectroscopic
techniques. Thus the data of the mixed valence formate derivative of
cytochrome oxidase would be compared to the <u>sum</u> of the model data for Cu_{a_3}
in the oxidized state and that for Cu_a in the reduce state. This method
involves <u>no fitting</u> , thus avoiding the pitfalls of those procedures
[9,10], esp. multiple solutions of Cu-N and Cu-S atom types, and requires
that the models be self-consistent. The two-atom type fitting procedure
averages all contributions into two average distances and types of ligands.
While neither method gives unique results, the fitting procedure does not
require self-consistency. In addition, all comparisons of the site
modeling method were required to be at least twice as good as the two atom
fitting procedures for each derivative to be acceptable.

From comparison of the copper edge spectra of the derivatives and with
that of model compounds using site modeling, the Cu_{a_3} contribution was
shown to be similar to that of stellacyanin, a Type I blue copper protein,
while that of Cu_a was more covalent and similar to models having N (or O)
and S as ligands [8]. The subsequent identification of a copper binding
site in cytochrome oxidase having an amino acid sequence nearly identical
to that of blue copper proteins [11] adds further support to these results.
The first shell filtered data of the copper EXAFS of the derivatives was
also analyzed using the site modeling method. The results are in agreement
with the edge studies with stellacyanin representing the Cu_{a_3} contribution
and Cu_a having three (or two) S at a normal Cu-S bond distance of
2.27 + .03Å and one (or two) N at 1.97 + .03Å (Fig. 3). Stellacyanin on
the other hand, has a short Cu-S (cysteine) distance of 2.19 + .03Å, two N
ligands, and a very long (2.84 ± .03Å) Cu - S(C, N, O) contribution [9].

Likewise, comparison of the iron EXAFS of the derivatives showed that
the structure of Fe_a did not change on reduction. This contribution was
represented by Im_2FeTPP in agreement with other spectroscopic techniques.
Fe_{a_3} in the reduced state is the binding site for O_2 and CO and this
contribution is similar to oxy and CO hemoglobin. In the oxidized state

Fe$_{a_3}$ is similar to met-hemoglobin but has a long (2.60 ± .03Å) Fe - S bond instead of O in the axial position that is clearly delineated in the second peak of the split first shell (Fig. 1). This long sulfur contribution is also present in the mixed valence formate derivative where Fe$_{a_3}$ is oxidized.

With Cu$_{a_3}$ - Fe$_{a_3}$ separated by 3.75 ± .05Å in the oxidized state, the sulfur is likely a bridging ligand and as such has an Fe$_{a_3}$-S-Cu$_{a_3}$ bond angle of ∿103°, a value typical of the sp^3 bonding required for a cysteine sulfur. Copper depletion leaves the Fe$_a$ site intact which is modeled by Im$_2$FeTPP but produces a species having a shorter Fe-S bond (2.28 ± .03Å) similar to that of cytochrome P-450 [6]. This sulfur has been identified as cysteine from EPR studies and lends support for a cystein sulfur bridging ligand. These results are summarized in Fig. 3 where error is ±.03Å unless specified and bars over distances indicate the average distance for the same kind of ligand in that center. It is worthwhile noting that recently a model compound for the Fe$_{a_3}$-S-Cu$_{a_3}$ active site has been synthesized and crystallography determination of the bond distances and angles have shown that they are very similar to those of Fig. 3 [12].

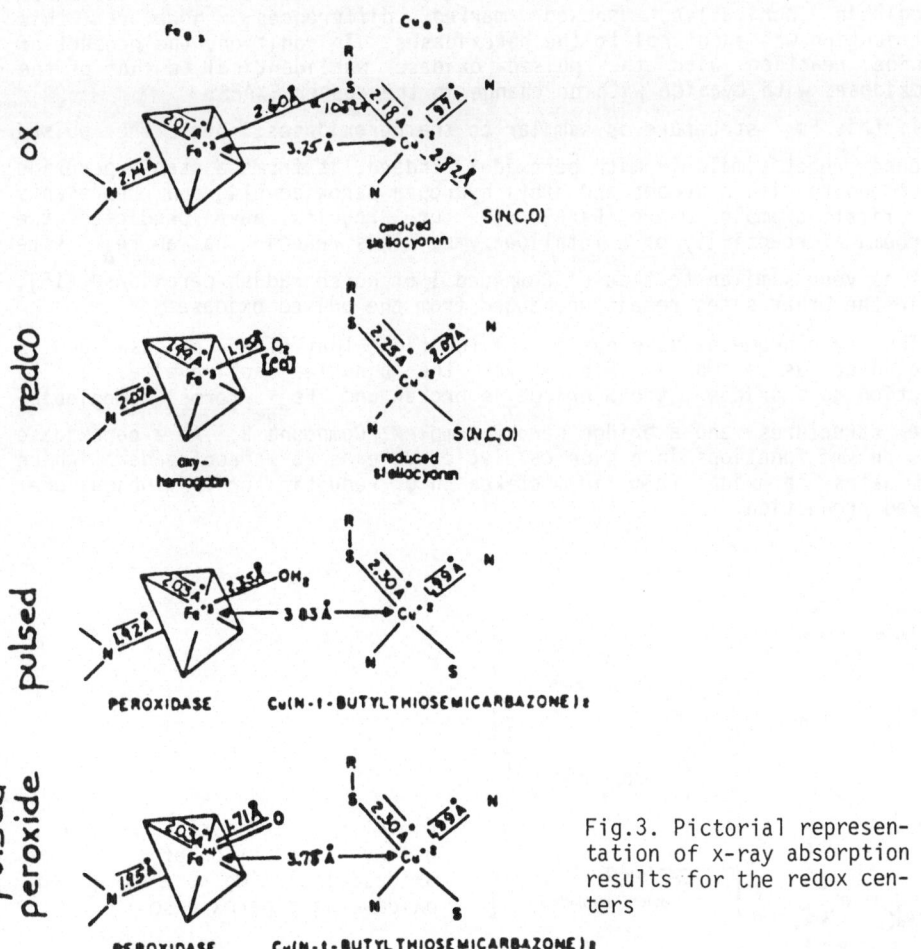

Fig.3. Pictorial representation of x-ray absorption results for the redox centers

Another oxidized derivative, 'pulsed' oxidase, has been observed during O_2 reduction which reacts more rapidly with cytochrome c and persists for sometime after substrate oxidation. EPR and optical data indicate that Fe_a and Cu_a are substantially unaltered from the resting oxidized derivative discussed above. Both the copper and iron EXAFS of the pulsed oxidase show the Cu_{a_3} - Fe_{a_3} binuclear active site distance [13] to be within the error of that of found for the resting oxidized derivative. Using site modeling, both Cu_a and Fe_a are likewise unchanged; however, significant changes are observed in the structure of the binuclear active site (Fig. 1). The short Cu_{a_3}-S (cysteine) distance has lengthened to a normal Cu-S distance and this site is modeled by Cu (N-t-butyl thiosemicarbazone)$_2$ (Fig. 3). The edge data indicate this model has a more planar geometry than the Cu_{a_3} model of the resting oxidized derivative, stellacyanin, in agreement with EPR results.

The bridging sulfur of the resting oxidized derivative that is found in the second peak of the split first shell of the iron EXAFS is missing in the pulsed oxidase (Fig. 1). Without this bridging ligand the site reacts more readily with cyanide and chemical intuition would suggest Fe_{a_3} might be a deoxy or met hemoglobin-like structure. Comparison with these hemoglobin derivatives showed marked differences; however, this contribution was identical to the peroxidases. In addition, the product of cyanide reaction with the pulsed oxidase was identical to that of the peroxidases with cyanide with no change in the copper EXAFS.

If this Fe_{a_3} structure is similar to the peroxidases, does the pulsed oxidase react similarly with peroxide? Indeed, it forms a stable peroxide intermediate with hydrogen and ethyl hydrogen peroxide [14] and represents the first example where EXAFS structure results have predicted the biochemical reactivity of a metalloenzyme. This reaction has an Fe_{a_3} site that is very similar to that of Compound I of horse radish peroxidase [15], while the other sites remain unchanged from the pulsed oxidase.

Thus cytochrome oxidase has a dual function; that of an oxidase and a peroxidase as shown in Fig 4 for the binuclear active site. In its function as a oxidase, the S bridge is broken and Fe_{a_3} forms hemoglobin-like structures and a bridge peroxy complex, Compound B. As a peroxidase this enzyme functions in a side catalytic cycle as a 'proofreader' which eliminates peroxide from the mistakes in O_2 reduction or from ubiquinone-linked production.

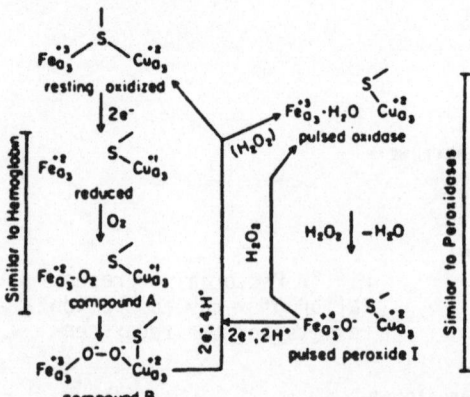

Fig.4. Proposed cyclic mechanism of oxygen reduction of cytochrome oxidase including role of pulsed oxidase as a peroxidase

The structures discussed above have been observed for cytochrome oxidase from bacterial sources in addition to beef heart. Although other sources such as yeast have identical optical and EPR spectra which suggest the Fe_a and Cu_a sites are unchanged, the EXAFS data show gross differences. This suggests that the structure of the binuclear active site of yeast cytochrome oxidase is different and is in agreement with the order of magnitude difference in the oxygen binding and reaction kinetics. In addition, different preparation methods of beef heart cytochrome oxidase also exhibit vastly different structures (Fig. 5) [16]. Variation is also observed from preparation to preparation with the same method as exhibited by the two different copper EXAFS data for preparations by the Hartzel-Beinert method. These different preparations and methods were investigated using cyanide and peroxide binding kinetics. Only the Yonetani method which was used for all the studies presented here gave preparations >85% homogeneous, with the other methods producing preparations having significant quantities of three distinctly different forms. Comparison of optical and EPR data for the methods as well as EXAFS data using the site modeling method indicate that the binuclear active sites are different in structure while Cu_a and Fe_a are not significantly altered.

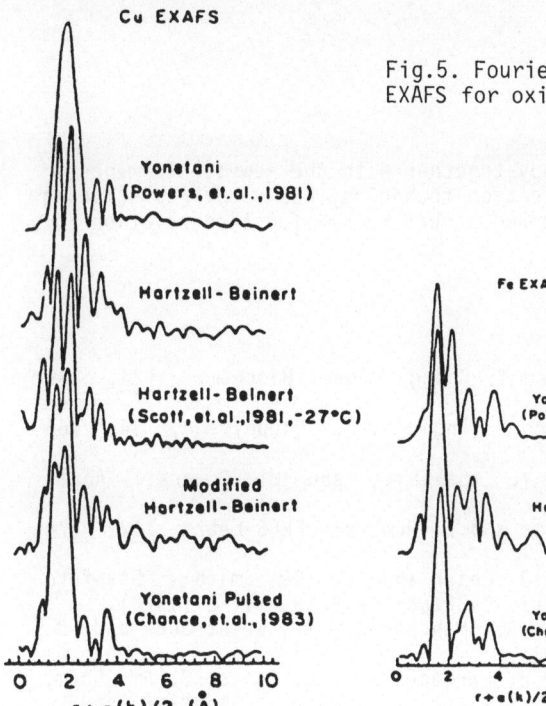

Fig.5. Fourier transforms of the Cu and Fe EXAFS for oxidized cytochrome oxidase

These results raise the question of how representative are the purified preparations of cytochrome oxidase in vivo. For this reason cytochrome oxidase was studied in submitochondrial particles which contain the respiratory chain intact within the mitochondrial membrane.

The results are shown in Fig. 6 and the resting oxidized derivative is identical to that of the Yonetani method having a sulfur-bridged active site. The copper EXAFS results also confirm this.

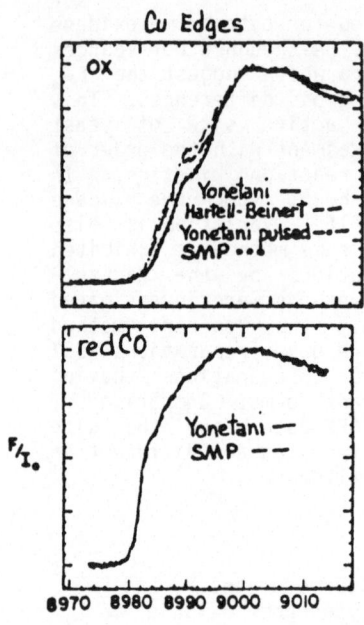

Cu Edges

Fig.6. Comparison of X-ray absorption edges in the oxidized and reduced states of cytochrome oxidase

Thus X-ray absorption spectroscopy together with the new developments in biochemical and fluorescence detection technology has uncovered striking new information about the mechanism and structure <=> function relation of this important metalloenzyme.

References

1. B. Chance, J. Moore, L. Powers and Y. Ching: Anal. Biochem. 124, 239 (1982).
2. L. Powers, B. Chance, Y. Ching and P. Angiolillo: Biophys. J. 34, 465 (1981).
3. B. Chance, W. Pennie, M. Carman, V. Legallais, and L. Powers: Anal. Biochem 124, 248 (1982).
4. B. Chance, P. Angiolillo, E. Yang, and L. Powers: FEBS Lett. 112, 178 (1980).
5. L. Powers, B. Chance, C. Barlow, J. Leigh and J. C. Smith: Stanford Synchrotron Radiation Laboratory Activity Report #78/10, VII-8 (1978).
6. L. Powers, B. Chance, Y. Ching, B. Muhoberac, S. Weintraub and D. Wharton: FEBS Lett. 138, 245 (1982).
7. J. Brown, L. Powers, B. Kincaid, J. Larrabee, and T. Spiro: J. Am. Chem. Soc. 102, 4210 (1980) G. Woolery, L. Powers, T. Spiro, M. Winkler, and E. Solomon: J. Am. Chem. Soc. (1982).
8. L. Powers, W. Blumberg, B. Chance, C. Barlow, J. Leigh, Jr., J. Smith, T. Yonetani, S. Vik, and J. Peisach: Biochem Biophys. Acta 546, 520 (1979).
9. J. Peisach, L. Powers, W. Blumberg, and B. Chance: Biophys. J. 38, 77 (1981).
10. P. Lee, P. Citrin, P. Eisenberger, and B. Kincaid: Rev. Mod. Phys. 53, 769 (1981).
11. G. Steffens and G. Buse: in Cytochrome Oxidase (ed. T. King, Y. Orii, B. Chance, K. Okunuki), Elsevier/North Holland, 79 (1979).

12. C. Schauer, K. Akabori, M. Elliott, and O. Anderson: J. Am. Chem. Soc. 106, 1127 (1984).
13. B. Chance, C. Kumar, L. Powers and Y. Ching: Biophys. J. 44, 353 (1983).
14. C. Kumar, A. Naqui, and B. Chance : J. Biol. Chem. 259, 2073 (1984).
15. B. Chance, L. Powers, Y. Ching, T. Poulos, I. Yamazaki, and K. Paul, Arch. Biochem and Biophys. in press.
16. A. Naqui, C. Kumar, Y. Ching, L. Powers, and B. Chance: Biochem. in press (1984).

EXAFS Studies of Non-Blue Copper Proteins: Superoxide Dimutase, Dopamine-Monooxygenase and Monoamine Oxidase

N.J. Blackburn
Department of Chemistry, University of Manchester, Institute of Science and Technology, P.O. Box 88, Manchester M60 10D, United Kingdom
S.S. Hasnain
Daresbury Laboratory, Warrington, Cheshire WA4 4AD, United Kingdom
P.F. Knowles
Department of Biophysics, University of Leeds, Leeds, LS2 9JT, United Kingdom

The copper containing enzymes superoxide dismutase, dopamine monooxygenase and plasma amine oxidase while functionally diverse are often classified together as 'non-blue' or 'type 2' on the basis of the similarity of their optical and EPR spectroscopic properties to those of simple mononuclear complexes[1]. X-ray absorption spectroscopy offers a further method of examining the structural relationships within this class of copper proteins. In this paper we summarise the results of EXAFS studies, which provide evidence of structural homologies and have identified imidazole coordination in all three enzyme systems. In the case of native superoxide dismutase where a crystal structure is available to 2Å resolution[2,3], comparison of EXAFS and crystallographic coordinates using molecular graphics has provided extra insights into the local structure of the active site.

1. Bovine erythrocyte superoxide dismutase. (SOD). This enzyme catalyses the dismutation of the superoxide anion according to the equation

$$2O_2^{-\bullet} + 2H^+ \longrightarrow O_2 + H_2O_2$$

Recent refinement of the crystal structure[2,3] has produced a clear view of the structure of the metal sites : the copper ligands, (N atoms of imidazole of His-44,-46,- 61 and - 118) show an uneven tetrahedral distortion from a square plane with the fifth axial co-ordination position occupied by solvent [4,5]. His -61 forms a bridge to the zinc, the other ligands of which are N atoms of His-69 and 78 and an oxygen of Asp-81. The geometry around zinc is tetrahedral with a strong distortion towards a trigonal pyramid having buried Asp-81 at the apex.

1.1 Copper site of the native enzyme. The interpretation of the EXAFS data for the copper site of (oxidised) bovine erythrocyte superoxide dismutase in aqueous phosphate buffer (0.010 M, pH 7.4) indicates an immediate coordination environment in close agreement with the crystallographic results, namely four imidazole groups with Cu-N = 1.99 Å and one solvent water molecule wth Cu-O = 2.24 Å as ligands to copper (Table 1, Fig.1). The Cu-N bond lengths are slightly shorter than the crystallographic value of 2.1 Å, determined as the average over the four copper sites of the asymmetric unit. However, the value determined by EXAFS agrees closely with Cu-N distances reported for similar Cu(II) imidazole complexes, especially [Cu(imid)$_4$] (NO$_3$)$_2$[6] the EXAFS of which has been successfully interpreted with 4 Cu-N distances of 1.98 Å as compared to the crystallographic value of 2.01 Å Furthermore, the Cu-N value obtained for the enzyme in aqueous solution is within 0.01 Å of that otained for the lyophillised derivative[7].
    The EXAFS interpretation of the aqueous enzyme also locates a single oxygen atom at 2.24 Å. Both crystallography[2] and water proton relaxation

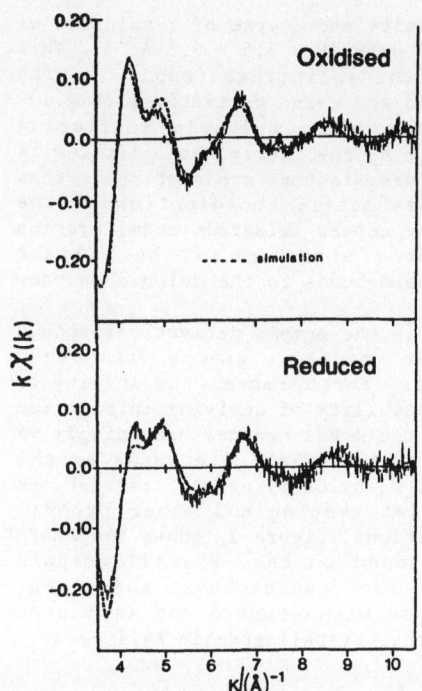

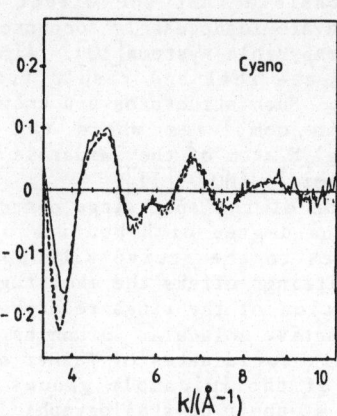

Fig.1. EXAFS associated with the copper K-edge of oxidised, reduced and cyano superoxide dismutase

Table. Parameters used to simulate the EXAFS associated with the copper K-edge of oxidised, reduced and cyano superoxide dismutase in aqueous phosphate buffer (0.01 M)

	OXIDISED			REDUCED			CYANO		
	Atom	$\sigma^2/(\text{Å})^2$	$R/(\text{Å})$	Atom	$\sigma^2/(\text{Å})^2$	$R/(\text{Å})$	Atom	$\sigma^2/(\text{Å})^2$	$R/(\text{Å})$
1st Shell	$4N_\alpha$	0.01	1.99	$3N_\alpha$	0.01	1.94	$3N_\alpha$	0.01	1.96
	1O		2.24				1C		2.00
							$1N_\alpha$		2.24
2nd Shell	$4C_\beta$	0.01	2.96	$3C_\beta$		2.98	$3C_\beta$	0.02	2.90
	$4C_\beta$		3.16	$3C_\beta$		3.08	$3C_\beta$		3.30
3rd Shell	$4N_\gamma$	0.01	3.90	$3N_\gamma$	0.01	3.90	$3N_\gamma$	0.025	3.80
	$4C_\gamma$		3.80	$3C_\gamma$		3.80	$3C_\gamma$		3.78

studies[4,5] provide strong evidence for the presence of solvent in an axial type position. The distance found for this oxygen as the fifth ligand is consistent with an axially coordinated water molecule in a five coordinate structure[8] but is too short for a CuN_4O_2 structure involving two axial water molecules, where $Cu-OH_2$ bond lengths are typically 2.6 Å[9].

An interesting feature of the analysis of the EXAFS of the Cu site of SOD is the need to split the $Cu - C_\beta$ distances by 0.2 Å (were C_β represents the C_2 and C_5 carbon atoms of the imidazole rings). The magnitude of the splitting is found to correlate with the intensity and degree of resolution of the low energy shoulder on the second EXAFS peak (k = 5.5 - 6.5 Å^{-1}). This effect can be seen by comparing the values of the splittings (table 1) with the EXAFS profiles of the oxidised, reduced and cyano derivatives (Fig.1). While it is possible that the effect results from multiple scattering processes which are inadequately accounted for by the theory, no splitting is apparent in tetrapyrrole systems[10]. A more satisfactory explanation is that the splittings are real and result from asymmetric coordination of the imidazole rings. Such structures are known for copper imidazole model systems particularly in complexes where the proton attached to the distant (noncoordinating) N atom of the imidazole group H-bonds to the anionic counter ion e.g. $\left[Cu\,(Himid)_6\right](NO_3)_2[11]$.

The magnitude of the splittings observed in the enzyme derivatives should then reflect the degree of H-bonding of the imidazole groups with other protein residues in the active site pocket. Furthermore, the ability to resolve the splittings offers the exciting possibility of deriving information on the orientation of the rings relative to the metal centre. Accordingly we have used interactive molecular graphics to fit the EXAFS distances to the crystallographic coordinates of Tainer et al[2] by carrying out translations and rotations of the imidazole groups whilst keeping all other protein residues fixed at their crystallographic positions. Figure 2 shows the EXAFS positions of the imidazole groups superimposed on the crystallographic positions for one particular orientation. The results have shown that asymmetrical coordination of imidazole is not unreasonable and is indeed observed for some rings at the resolution of the crystallographic refinement.

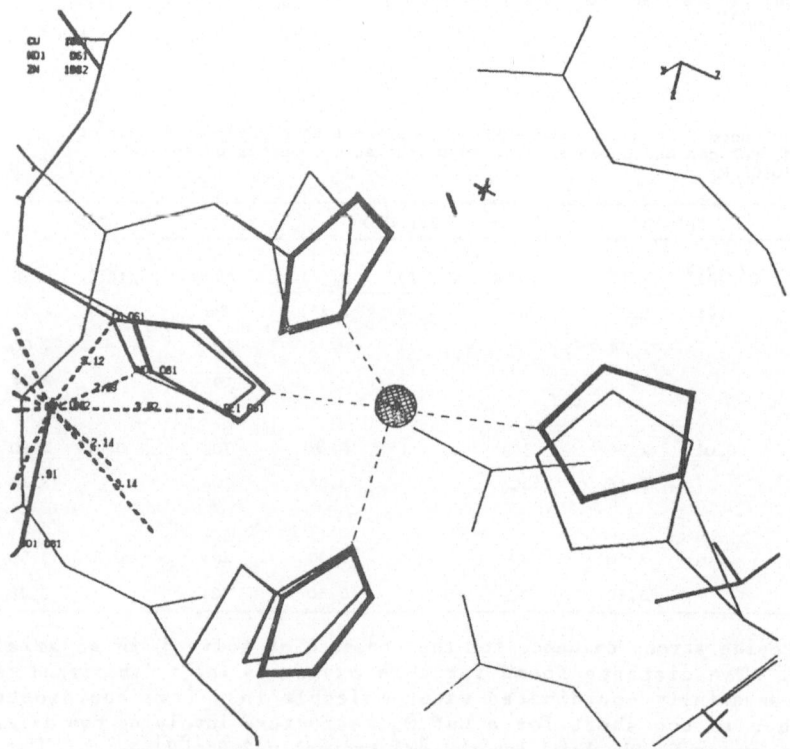

Fig. 2. EXAFS positions of imidazole rings (shown in heavy black) superimposed on the crystallographic coordinates of Tainer et.al.[2] for copper atom #1001. The data were generated using the graphics program FRODO.

1.2   Copper Site of the Reduced Enzyme. Features on the copper K-absorption edge and the X-ray Absorption Near Edge Structure (XANES) of oxidised and reduced bovine erythrocyte superoxide dismutase in aqueous phosphate buffer, show clear differences in relative intensity, suggesting a stereochemical change at the copper accompanying reduction.   The interpretation of the EXAFS for the reduced form of the enzyme has been most successful for three coordinate Cu(I) involving only imidazole coordination. However, and of more significance, the value of the Cu-N bond lengths of 1.94 Å in the reduced form of the enzyme is itself indicative of a coordination number lower than that in the oxidised form since reduction of the copper is expected to lengthen the Cu-N bonds, unless the coordination number is decreased.   Thus $[Cu^{II}(N\text{-methylimidazole})_4][BF_4]_2[12]$ and $Cu^I(N\text{-methylimidazole})_4[ClO_4][13]$ involve Cu-N distances of 2.006 and 2.054 Å, respectively. 3 coordinate Cu(I) complexes on the other hand exhibit Cu-N(pyridine) first shell distances in the range 1.89 - 1.93 Å (see refs 14 & 15 for a fuller discussion of this point).

The EXAFS results thus provide the first direct structural information on the copper site of reduced superoxide dismutase and are fully consistent with previous proposals based on redox titration data[16,17] and [113]Cd nmr spectra[18] that an imidazole group dissociates from the copper on reduction. EXAFS gives no indication as to which imidazole is lost although Fee[19] suggests that it is the bridging ligand to zinc.

1.3   Cyano - SOD.   The EXAFS of cyano SOD is not easy to interpret unambiguously but is best fit by three 'equatorial'type imidazole groups at 1.96 Å, CN^- at 2.00 Å and a more distant 'axial' imidazole at 2.24 Å (Fig.1, table 1).   These results support previous proposals based on other spectroscopic data that CN^- binding reorients the coordination sphere, such that one imidazole ligand is displaced from an essentially equatorial to an axial position with CN^- occupying the vacant equatorial site. The value of 2.00 Å is longer than the Cu(II)-CN distances documented by crystallography (although examples of Cu^{II}-cyano complexes are rare).   For example, distorted trigonal bipyramidal $[Cu(phen)_2CN](NO_3)H_2O[20]$ and $[Cu(bipy)_2CN](NO_3)2H_2O[21]$ have Cu-C = 1.935Å and 1.974Å respectively. However if coordinated CN in superoxide dismutase was also H-bonded to the protonated Arg-141 in the active site cavity, a lengthening of Cu-C bond would not be unexpected.   This proposal is in accord with the observation that chemical modification of arginine residues decreases the CN^- binding constant[22].   It is also consistent with a recent observation from this laboratory that evacuation or lyophilisation causes dissociation of the cyanide ligand from the enzyme as HCN, with a pH dependence which implicates a group with a pK_a >10.5 as the source of the proton.

2   Dopamine-ß- monooxygenase (DßM) and monoamine oxidase (MAO)
DßM catalyses the hydroxylation of dopamine to non adrenalin, an important step in the modification of transmitter molecules[25,26]

$+O_2 + 2H^+ + 2e^-$

$+ H_2O$

MAO catalyses the   oxidative conversion of biogenic amines to the corresponding aldehyde.

$$RCH_2NH_2 \; + \; O_2 \; + \; H_2O \; \longrightarrow \; RCHO \; + \; H_2O \; + \; NH_3$$

Due to the high molecular weight of these enzymes the copper is much more dilute (1 Cu per 77000 & 90,000 respectively for the preparations of DβM and MAO used in this study). The more limited quality of the data therefore renders our conclusions more tentative. However, the EXAFS study clearly shows that in each case Cu is coordinated to > 3 imidazoles and the structures of the catalytic sites in these enzymes are thus closely related. For dopamine monooxygenase best fits are obtained by 4 imidazoles at 2.01 Å and a solvent molecule at 2.3 Å. When combined with EPR data[25] these results are consistent with a square pyramidal site.

Acknowlegdements. We would like to thank the SERC for financial support and the Daresbury Laboratory for provision of facilities. We are especially grateful to Professor T. Blundell and Dr. Peter Linley of the Birkbeck College, University of London for use of the Molecular Graphics facility

References

1. R. Malkin and B.G. Malmstrom, Adv. Enz., 1970 33, 177.
2. J.A. Tainer, E.D. Getzoff, K.M. Beem, J.S. Richardson and D.C. Richardson, J. Mol. Biol., 1982, 181-217.
3. J.A. Tainer, E.D. Getzoff, J.S. Richardson and D.C. Richardson, Nature, 1983, 306, 284 - 286.
4. J.A. Fee and B.P. Gaber, J.Biol. Chem., 1972, 247, 60 - 65
5. N. Boden, M.C. Holmes and P.F. Knowles, Biochem J, 1979, 177, 303 - 309.
6. D.L.McFadden, A.T. McPhail, C.D. Garner and F.E. Mabbs, J. Chem. Soc. Dalton Trans. 1976, 47 - 52.
7. N.J. Blackburn, S.S. Hasnain, G.P.Diakun, P.F. Knowles, N. Binsted and C.D. Garner, Biochem. J. 1983, 213, 765 - 768.
8. K.D. Karlin, Y. Gultneh, J.G. Hayes and J. Zubieta, Inorg. Chem., 1984, 23, 521 - 523.
9. W. Vreudgdenhill, P.J.M.W.L. Birker, R.W.M. ten Hoedt, G.C. Verschoor and J. Reedijk, J. Chem. Soc. Dalton Trans., 1984, 429 - 432.
10. M.F. Perutz, S.S. Hasnain, P.J. Duke, J.L. Sessler and J.E. Hahn, Nature, 1982, 295, 535 - 538.
11. D.L. McFadden, A.T. McPhail, C.D. Garner and F.E. Mabbs, J. Chem. Soc., Dalton Trans. 1975, 263 - 268.
12. S.R..Acott, W. Clegg, D. Collison and C.D. Garner, personal communication.
13. W. Clegg, S.R. Acott and C.D. Garner, Acta Cryst. 1984, C.40, 768 - 769.
14. N.J. Blackburn, S.S. Hasnain, N. Binsted, G.P. Diakun, C.D. Garner, and P.F. Knowles, Biochem. J. 1984, 219, 985 - 990.
15. K.D. Karlin, J.C. Hayes, Y Gultneh, R.W. Cruse, J.W.McKown, J.P. Hutchinson and J. Zubieta, J.Am. Chem. Soc. 1984, 106, 2121 - 2128.
16. J.A. Fee and P.E. diCorletto, Biochemistry, 1973, 12, 4893 - 4899.
17. G.D. Lawrence and D.T. Sawyer, Biochemistry, 1979, 18 3045 - 3050.
18. D.B. Bailey, P.D. Ellis and J.A. Fee, Biochemistry, 1980, 19, 591 - 596.
19. J.A. Fee in Metal Ions in Biological Systems Vol.13, (Copper Proteins) (Sigel H.ed.) 1981, pp.259 - 298, Marcel Dekker, New York.
20. O.P. Anderson, Inorg. Chem., 1975, 14, 730 - 734.
21. S. Tyagi and B.J. Hathaway, J. Chem. Soc. Dalton Trans., 1983, 199 - 203.
22. O. Bermingham - McDonogh, D.M. De Freitas, A. Kumamoto, J.E. Saunders, D.M. Blech, C.L. Borders Jr., and J.S. Valentine, Biochem. Biophys.

Res. Comm. 1982, 108, 1376 – 82.

23. T. Skotland and T. Ljones, Inorg. Perspect. Biol. Med., 1979, 2, 152 – 180.

24. J.J. Villafranca, in Metal Ions in Biology 1981, Vol.2, (Copper Proteins) (ed. T.G. Spiro) pp. 264 – 281, John Wiley and Sons, New York.

25. N.J. Blackburn, D. Collison, J. Sutton and F.E. Mabbs, Biochem.J., 1984, 220. 447–454.

Light-Induced Changes in X-Ray Absorption (K-Edge) Energies of Manganese in Photosynthetic Membranes

David B. Goodin, Vittal K. Yachandra, Ron Guiles, R. David Britt, Ann McDermott, Kenneth Sauer, and Melvin P. Klein

Laboratory of Chemical Biodynamics, Lawrence Berkeley Laboratory
University of California, Berkeley, CA 94720, USA

1. Introduction

This report describes our observations of the participation of manganese in electron transport and oxidation equivalent storage by the photosynthetic oxygen evolution complex of higher plants. The oxidizing potential produced by photosynthetic charge separation is stored near the site of water oxidation in a membrane-bound complex that contains manganese. Broken, washed chloroplast preparations typically contain 4-6 Mn per PSII reaction center, and approximately 2/3 of this quantity is released by mild treatments that specifically inactivate O_2 evolution [1-3]. Recent studies with inside out thylakoids [4] and photosystem II (PSII) preparations [5] have given rise to a reevaluation of the Mn content per photosynthetic unit, but it remains clear that a direct relation exists between Mn and O_2 evolution.

Kinetic studies of the oxygen produced by pulsed light trains have given rise to a model for the accumulation of oxidizing equivalents [6] in which five intermediates, S_0-S_4, operate in a cyclic fashion. The first direct studies of the association of manganese with these intermediates has emerged from the observation of an EPR signal at low temperature in illuminated spinach chloroplasts [7]. The temperature dependence of the light-induced formation and subsequent decay of this signal indicates that it is correlated with the presence of the S_2 state of the oxygen evolving complex [8]. Treatments that affect oxygen evolution, including alkaline Tris washing, cause the disappearance of this signal [9]. It is significant that the signal is stable indefinitely at 77 K indicating that S_2 may be trapped for x-ray absorption studies.

The EPR signal was assigned to a manganese species on the basis of its hyperfine splittings, and simulations were consistent with either a weakly exchange coupled Mn(III,IV) binuclear or Mn(III,III,III,IV) tetra-nuclear manganese center [9,10]. These findings are in agreement with an earlier EXAFS study of whole chloroplasts which indicated a Mn-Mn interaction at 2.71 Å. This contribution is present in active chloroplasts and disappears upon Tris inactivation [11].

In this edge and EXAFS study, we use the newly characterized EPR signal as an S state indicator. The studies were performed using highly active detergent solubilized PSII sub-chloroplast preparations with a lower and presumably more homogeneous manganese content per reaction center than those available in earlier studies. The results demonstrate conclusively that manganese participates in the turnover of the oxygen evolution complex.

2. Materials and Methods

Preparation of oxygen-evolving PSII sub-chloroplast membranes from spinach was done as described by KUWABARA and MURATA [12]. Oxygen evolution was measured polarographically [13]. The samples had rates of oxygen evolution as high as 526 μmoles of O_2 (mg of Chl)$^{-1}$ h^{-1} and contained 3.6 Mn atoms per reaction center, assuming a photosynthetic unit of 250 Chl [14].

X-ray absorption edge spectra (XAES) were collected at the Stanford Synchrotron Radiation Laboratory, Stanford, CA., on wiggler beam line VII-3 (unfocussed, 1×10^{12} photons/s with 1 eV resolution) using a Si<111> double crystal monochromator during dedicated operation of the SPEAR storage ring, providing 40-80 mA electron beams at 3.0 GeV. EXAFS was measured in fluorescence mode using an NE104 plastic scintillation array similar to that described by POWERS et al. [15], equipped with Cr fluorescence filters and Soller slit assembly [16]. The filter consisted of 0.25 mm thick Be with Cr electro-deposited to a thickness of 0.013 mm. Energy calibration was maintained by simultaneous measurement of the strong pre-edge feature of $KMnO_4$ at 6543.3 eV [17].

X-ray absorption samples were prepared by layering pellets of spinach PSII membranes in 50% glycerol, 50 mM MES (pH 6.0), 15 mM NaCl and 5 mM $MgCl_2$ at approximately 8 mg Chl/ml (100-200 μM Mn) into narrow lucite sample cells with an open cavity of dimensions 5x30x2 mm. Dark adaptation, illumination and EPR measurements were carried out directly in these cells. EPR spectra were recorded on a Varian E-109 spectrometer. Samples were run at 10-12 K in an Air Products liquid helium cryostat at a microwave frequency of 9.12 GHz and a 100 KHz field modulation. During x-ray measurements, samples were suspended in a jacketted N_2 boil-off jet at 170 to 190 K in darkness.

3. Results and Discussion

We measured X-ray K-edge spectra of spinach PSII preparations in three different S states. Samples were prepared in the S_1 state by illumination for 60 s at 190 K followed by dark adaptation for 1 h at 273 K. This pre-illumination reduced the S_0 concentration, which would otherwise constitute about 25% of the oxygen evolving complex centers. Samples were prepared in state S_2 by first generating S_1 according to the above procedure followed by 60 s illumination at 190 K. This procedure results in the accumulation of state S_2 because reoxidation of the acceptor Q^- is blocked at this temperature [18] and thus only one oxidizing equivalent is transferred. An attempt was also made to generate the S_3 state by preparing S_2 as above, followed by warming the sample in the dark for 2 min to 260 K. This procedure was followed by illumination for 60 s at 220 K. Reoxidation of Q^- at 260 K is much faster than the decay of the S_2 state [8] so that brief warming to this temperature prepares the electron transport chain for another one-electron transfer.

These protocols gave samples containing the multi-line EPR signal for those prepared in S_2 but the signal was absent in S_1. The S_3 samples contained EPR signal intensities that were smaller by 30-60% compared to S_2.

The Mn K-edge spectra representing PSII membrane preparations in the states S_1, S_2 and S_3 are presented in Fig. 1. The S_2 edge inflection exhibits a shift to higher energy by 2.3 eV with respect to the S_1 edge

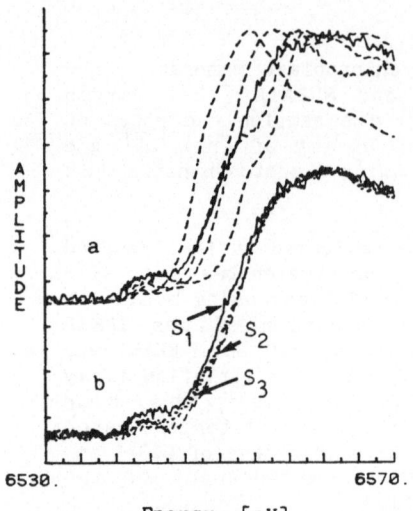

Figure 1. Mn K-edge fluorescence spectra of Model complexes and PSII preparations. In a, dashed spectra from left to right are: $Mn(II)(AcAc)_2$, $Mn(III)(AcAc)_3$, $Mn(III,IV)_2(Bipy)$, and $Mn(IV)O_2$. The solid line is spinach PSII in state S_1. In b are shown spinach PSII samples prepared in states S_1, S_2, or S_3. The S_1 spectrum is plotted with a solid line, the S_2 spectrum is plotted with a dashed line and the S_3 spectrum is plotted with a dotted line.

inflection. In addition, a pre-edge feature observed for the S_1 samples is not so intense for S_2. In spite of the significant reduction in EPR signal intensity, the S_3 sample edge appears to be identical to that observed for S_2.

The edge spectra of the PSII S_1 and S_2 samples showed a range of inflection point energies that were well outside the estimated 0.2 eV relative uncertainty. On each of three independent sets of measurements, the inflection point energies were well correlated with the relative magnitude of the multi-line EPR signal of the sample. One such correlation is presented in Fig. 2. These data indicate that the shifts in inflection position reflect the presence of a slightly different S state composition for each sample. In addition, each independent study gave differences in average edge positions as large as 1.9 to 2.5 eV between S_1 and S_2 samples.

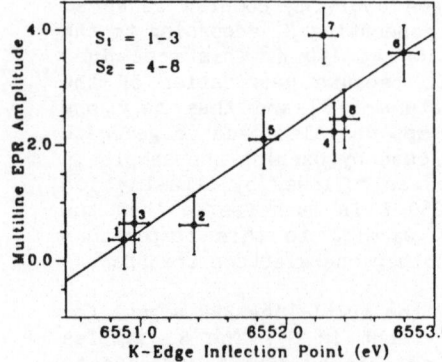

Figure 2. Correlation of Mn K-edge inflection with multi-line EPR signal intensity for various spinach PSII samples prepared in states S_1 and S_2.

From previous correlations of Mn K-edge inflection point energies [19], we expect an edge shift of approximately 2.3 eV per unit change in coordination charge. The simplest interpretation of the multi-line EPR signal associated with S_2 is that only one charge equivalent per P_{680} has been transferred. The observed S_1 to S_2 average K-edge shift of aproximately 2 to 2.5 eV is larger than one would expect if only one electron equivalent is removed from a total pool of 3.6 manganese atoms.

We offer two explanations for this observation. One pool of Mn might be involved in a structural role while another has a functional role. One-electron oxidation of the functional manganese center during the S_1 to S_2 transition may cause a conformational change altering the environment of the structural manganese and causing a change in the Mn K-edge energy. However, a large conformational change is not expected to occur at the illumination temperature. Therefore we propose that more than one charge equivalent is transferred from the Mn in the oxygen evolving complex under the illumination conditions used. This would occur if, for example, one oxidizing equivalent is stored on a PSII electron carrier other than manganese in S_1. Transition to state S_2 would then cause the removal of both reducing equivalents from the manganese center. This oxidation state change could also be accompanied by a conformational change. Based on the observation that samples prepared in S_3 exhibited a decreased EPR signal intensity, but had edge properties very similar to those in S_2, we speculate that the oxidation state of Mn in S_3 is similar to that of S_2, but is somehow chemically distinct.

A number of models for the oxygen evolving complex have included dimeric manganese species [20,11,21]. Based on the observation that state S_2 is paramagnetic, DISMUKES et al. [10] have proposed that it contains a Mn(III,IV) complex, and the edge properties observed in this work are consistent with that assignment. If it is true that two oxidizing equivalents are added to the manganese center accompanying the S_1 to S_2 transition, as discussed above, these observations would be consistent with the presence of a Mn(II,III) complex in S_1 and Mn(III,IV) in S_2. However, a Mn(III,III) complex which undergoes a large change in ligation and/or conformation in the S_1 to S_2 transition could also give rise to the observed properties for the S_1 state.

Our previous indications of a binuclear manganese complex in the oxygen evolution complex from difference EXAFS studies in whole chloroplasts has profound significance for the elucidation of the mechanism of oxygen evolution. The above demonstration of the generation and trapping of the states S_1, S_2 and S_3 under conditions of the EXAFS measurement will be a major advance in our studies of that mechanism. In addition, the development of photosystem II sub-chloroplast preparations that are more homogeneous in manganese content have simplified the data analysis and interpretation of our EXAFS results, without recourse to difference EXAFS (active minus Tris inactivated pool) used in earlier studies.

Figure 3 shows the fourier transform of the k-space data. The FT is dominated by a peak at $R+\alpha/2 \approx 1.5$ A with a small shoulder to higher R, and a second peak at $R+\alpha/2 \approx 2.1$ A. The fourier isolation of the two major peaks is shown in Fig. 4, along with the superimposed best fit to theoretical phases and amplitudes [22]. It should be noted that though the EXAFS data extend to higher photon energies, the presence of the iron K-edge at 7100 eV forces truncation of the data set. The results indicate a Mn dimer with a Mn-Mn distance of 2.75±0.03 A, with each Mn containing 5±1 N or O ligands at 1.99±0.03 A and 3±1 N or O ligands at distance of 1.66±0.03 A. This compares extremely well with our results from whole chloroplast difference EXAFS and closely resembles the published crystal structure parameters [23] for the di-μ-oxo tetrakis(2,2-bipyridine)dimanganese (III,IV) perchlorate.

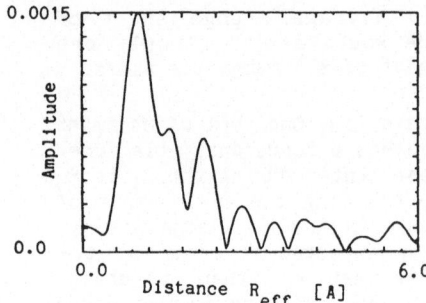

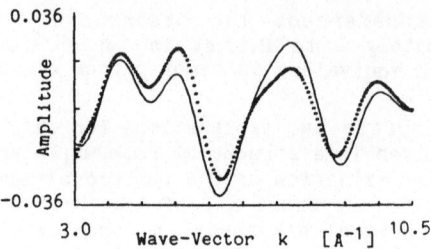

Figure 3. Fourier transform power spectrum of the k^6-weighted EXAFS data for spinach PSII prepared in state S_1.

Figure 4. Fit to the fourier filtered peaks between R_{eff} = 0.3 and 2.7 A for spinach PSII in state S_1. The fit utilized three scattering shells with parameter values of: 1.1 Mn atoms at 2.75 A, 5.0 N atoms at 1.99 A, and 3.4 O atoms at 1.66 A.

4. Conclusions

We have demonstrated that membrane associated manganese atoms participate in the light driven storage of oxidizing equivalents of the photosystem II oxygen evolution complex. It was observed that manganese is oxidized, possibly by more than one equivalent in the transition from S_1 to S_2. The state S_3 appears to be remarkably similar in its edge properties to that of S_2 thus implying that the oxidation state of Mn in state S_3 is similar to that in state S_2. EXAFS data for PSII particles prepared in state S_1 provide a direct measurement of the center which agrees well with our earlier proposal of a bridged binuclear cluster.

5. Acknowledgements

Sun Un and Sue Dexheimer are acknowledged for help in data collection. Dr. John Dini of the Lawrence Livermore National Laboratory prepared the Be fluorescence filters. This work was supported by the Office of Basic Energy Sciences, Biological Energy Research Division of the U.S. Department of Energy under Contract No. DE-AC03-76SF00098 and by NSF Grant PCM82-16127. Synchrotron radiation facilities were provided by the Stanford Synchrotron Radiation Laboratory which is supported by the U.S. Department of Energy, Office of Basic Energy Sciences, and by the National Institutes of Health, Biotechnology Resource Program, Division of Research Resources.

6. References

1. G.M. Cheniae and I.F. Martin (1970) Biochim. Biophys. Acta 197, 219-239.
2. R.E. Blankenship and K. Sauer (1974) Biochim. Biophys. Acta 357, 252-266.
3. R.E. Blankenship, G.T. Babcock, and K. Sauer (1975) Biochim. Biophys. Acta 387, 165-175.
4. R. Mansfield and J. Barber (1982) FEBS Lett. 140, 165-168.
5. L.E.A. Henry, B. Lindberg-Møller, B. Andersson, and H.-E. Akerlund (1982) Carlsberg Res. Commun. 47, 187-198.

6. B. Forbush, B. Kok, and B. McGloin (1971) Photochem. Photobiol. 14, 307-321.
7. G.C. Dismukes and Y. Siderer (1981) Proc. Natl. Acad. Sci. 78, 274-278.
8. G.W. Brudvig, J.L. Casey and K. Sauer (1983) Biochim. Biophys. Acta 723, 366-371.
9. O. Hansson and L.-E. Andreasson (1982) Biochim. Biophys. Acta 679, 261-268.
10. G.C. Dismukes, K. Ferris, and P. Watnick (1982) Photobiochem. Photobiophys. 3, 243-256.
11. J.A. Kirby, A.S. Robertson, J.P. Smith, A.C. Thompson, S.R. Cooper, and M.P. Klein (1981) J. Amer. Chem. Soc. 103, 5529-5537.
12. T. Kuwabara and N. Murata (1982) Plant Cell Physiol. 23, 533-539.
13. D.B. Goodin (1983) Ph.D. Thesis, Univ. of California.
14. G.T. Babcock, D.F. Ghanotakis, B. Ke, and B.A. Diner (1983) Biochim. Biophys. Acta 723, 276-286.
15. L. Powers, B. Chance, Y. Ching, and P. Angiolillo (1981) Biophys. J. 34, 465-498.
16. E.A. Stern and S.M. Heald (1979) Rev. Sci. Instrum. 50, 1579-1582.
17. D.B. Goodin, K.-E. Falk, T. Wydrzynski, and M.P. Klein (1979) 6th Annual Stanford Synchrotron Radiation Laboratory Users Group Meeting, SSRL Report No. 79/05, 10-11.
18. A. Joliot (1974) Biochim. Biophys. Acta 357, 439-448.
19. J.A. Kirby, D.B. Goodin, T. Wydrzynski, A.S. Robertson, and M.P. Klein (1981) J. Amer. Chem. Soc. 103, 5537-5542.
20. K. Sauer (1980) Accts. Chem. Res. 13, 249-256.
21. G. Renger (1978) Photosynthetic Water Oxidation (H. Metzner, ed.) Academic Press, London, 229-248.
22. B.-K. Teo and Lee, P.A. (1979) J. Amer. Chem. Soc. 101, 2815-2832.
23. P.M. Plaksin, R.C. Stoufer, M. Mathew, and G.J. Palenik (1972) J. Amer. Chem. Soc. 94, 2121-2124.

EXAFS Investigations of Zinc in 5-Aminolevulinic Acid Dehydratase

Eileen M. Wardell and C. David Garner
Department of Chemsitry, University of Manchester
Manchester M13 9PL, United Kingdom
Manfred Schlösser and Detmar Beyersmann
Department of Biology and Chemistry, University of Bremen
D-2000 Bremen 33, Fed. Rep. of Germany
S.S. Hasnain
Science and Engineering Research Council, Daresbury Laboratory
Daresbury, Warrington WA4 4AD, United Kingdom

5-Aminolevulinic acid dehydratase (ALAD, 5-aminolevulinate hydrolyase, EC 4.2.1.24) catalyzes the synthesis of the pyrrole porphobilinogen from 2 molecules of 5- aminolevulinic acid. The enzyme from bovine liver has a molecular weight of 280.000 daltons and consists of eight subunits of 35.000 daltons each [1]. Lysine, hystidine and cysteine have been identified as amino acid constituents of the enzyme protein involved in the active site [2]. Two cysteine residues essential for the enzyme activity are rapidly oxidized by air, and the maintenance of the active state requires the presence of a thiol [3]. The isolated enzyme protein contains 2-6 g.atoms of zinc per octamer and it binds a maximum number of eight zinc ions per octamer with high affinity [4]. The function of zinc in ALAD is not clear. Several findings imply a direct catalytic function for zinc whilst other results support the assumption that zinc stabilizes the active structure of the enzyme protein, without directly interacting with the substrate.

The ligands of zinc in ALAD have not been identified in any previous investigation. However, coordination by cysteine has been implied. Thus, Bevan et al [5] showed that Zn^{2+} did not bind to the enzyme, unless the essential sulphydryl groups of the enzyme were first reduced, and that alyklation of sulphydryl groups resulted in a reduction in the binding of Zn^{2+} to the enzyme.

Hence, it was considered to be highly desirable to investigate the nature of the zinc site(s) in various forms of ALAD by EXAFS. Four forms of the enzyme were investigated and details of the zinc environment elucidated in each case. The data was obtained using the fluorescence EXAFS facility at the Daresbury's synchrotron radiation source. SRS was operating at 1.8 GeV with an average current of 90 mA. A channel-cut monochromator Si111 was used.

Figure 1 shows the near edge and EXAFS spectra of the native (reduced), air-oxidised, air-oxidised plus substrate, and native plus inhibitor (2-bromo-3(5-imidazolyl) proprionic acid) forms of ALAD. The corresponding spectra of the first three of these forms are very similar in their overall appearance suggesting that each involves a similar environment for the zinc. However, the near edge and EXAFS profiles observed for the inhibitor-bound form are markedly different from corresponding profiles for the other forms, clearly indicating that the binding of the inhibitor produces major changes in the zinc environment. Detailed analyses of the EXAFS have confirmed these preliminary conclusions.

The best agreement between the experimental and calculated EXAFS for the native form of ALAD (Fig.2) was obtained with zinc coordinated by four ligands. Three of these are sulphur atoms at 2.28(2) Å and one is a lower Z atom at the substantially shorter distance of 1.92(5) Å (if it is nitrogen) or 1.89(5) Å (if it is oxygen). On the basis of the EXAFS simulation, we favour nitrogen as the

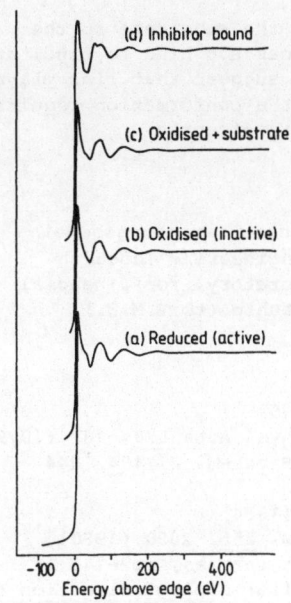

(d) Inhibitor bound

(c) Oxidised + substrate

(b) Oxidised (inactive)

(a) Reduced (active)

-100 0   200   400
    Energy above edge (eV)

Fig.1. Comparison of the XANES and EXAFS associated with the zinc K-edge of various forms of ALAD: (a) native (reduced); (b) air-oxidised; (c) air oxidised plus substrate; (d) native plus inhibitor

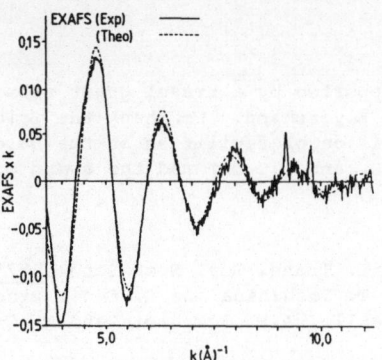

Fig.2. EXAFS and the interpretation, with 3 Zn-S of 2.28 Å and 1 Zn-N of 1.92 Å, of native (reduced) ALAD

fourth ligand but cannot be definitive. Reaction of the enzyme with the inhibitor produced significant changes in the EXAFS, the nature of which are consistent with coordination by ca. three sulphur atoms at 2.22(2) Å, a nitrogen at 1.93(5) Å, and an imidazole group, presumably from the inhibitor, with Zn-N of 2.14(5) Å. Inactivation of the enzyme by air-oxidation of essential sulphydryl groups affects the zinc environment and the corresponding EXAFS has been successfully interpreted with three sulphur atoms at 2.25(2) Å, a low Z atom (as found in the native form but slightly closer) at 1.85(5) Å, plus a fifth ligand at 2.08(5) Å (if nitrogen) or 2.02(5) Å (if oxygen). Binding of the substrate to the air- oxidised form produces very slight changes in the zinc environment and the EXAFS is consistent with the presence of a nitrogen (or oxygen) atom at ~ 2.36 Å, beyond an inner coordination sphere virtually identical to that in the air-oxidised form, and interpreted by three sulphur atoms at 2.27(2) Å and two nitrogens, one at 1.85(5) Å and one at 2.08(5) Å.

Therefore, zinc in native ALAD appears to be four-coordinate and ligated by the sulphur atoms and one nitrogen (or oxygen) atom. Tetracoordination of zinc in ALAD by four amino acid residues would be expected to be more compatible with a structural rather than a catalytic role for this metal. The identification of an imidazole group bound to zinc, following the addition of the inhibitor, fits well with the evidence from kinetic studies that the binding of the inhibitor is a function of the zinc content of the enzyme [7].

The third important result of the EXAFS investigation is that the oxidation of ALAD by air does not involve one of the three cysteine ligands of zinc and, therefore, affects other sulphydryl groups. Nevertheless, oxidation changes the co-ordination at the zinc and this has been interpreted as involving the ligation of an additional oxygen or nitrogen atom. This change is attributed to a conforma-tional shift of the protein, induced by oxidation of the sulphydryl groups essen-tial for the activity of the enzyme which are probably located at or near the active centre. This result demonstrates that zinc protects its sulphydryl ligands, but not the other essential sulphydryl groups, from oxidation by air. The slight

137

change in the zinc environment induced by the binding of the substrate to the oxidized (inactive) enzyme suggests that the substrate does not bind to zinc, at least in the oxidised form of the enzyme. Therefore, we suggest that zinc plays a structural role in ALAD, enabling the enzyme to adopt the conformation required for activity, without binding and activating the substrate directly

Acknowledgements

This work was supported by a travel grant from the Deutsches Forschungsgemein-schaft to Detmar Beyersmann. We thank the Science and Engineering Research Council for provision of facilities at the Daresbury Laboratory, for financial support (to S.S.H. and C.D.G.) and the award of a studentship (to E.M.W.)

References

1 D. Shemin: Phil. Trans. Roy. Soc. Lond. B273, 109 (1975)
2 I. Tsukamoto, T. Yoshinaga and S. Sano: Biochem. Biophys. Acta 570, 167 (1979)
3 A.M. del C. Batlle, A.M. Ferramola and M. Grinstein: Biochem. J. 104, 244 (1967)
4 R. Sommer and D.Beyersmann: J. Inorg. Bochem. 2, 131 (1984)
5 D.R. Bevan, P. Bodlaender and D. Shemin: J. Biol. Chem. 255, 2030 (1980)
6 J.F. Chlebowski and J.E. Coleman: Zinc and its role in enzymes, pp2-140 in Sigel, H. (ed.), Metal ions in biological systems, Vol.6: Biological action of metal ions, (New York, Marcel Dekker, Inc., 1976)
7 D. Beyersmann and M. Cox: Biochim. Biophys. Acta, (1984) in press

Influence of Phosphoproteins on Calcification

J.C. Irlam and D.W.L. Hukins
Department of Medical Biophysics, University of Manchester
Manchester M13 9PT, United Kingdom
C. Holt
Hannah Research Institute, Ayr, KA6 5HL, Scotland
S.S. Hasnain
SERC Daresbury Laboratory, Warrington WA4 4AD, United Kingdom

This paper presents a brief report of our progress in investigating the phases of calcium phosphates precipitated in vitro in the presence of the phosphoproteins casein and phosvitin with emphasis on the application of EXAFS to characterising the precipitates. Phosphoproteins have been found in all the sites of normal physiological and pathological calcification in the body where they have been sought; as a result it is widely believed that they act as sites for the heterogeneous nucleation of calcium phosphates from solution [1]. We have chosen to use casein (from milk) and phosvitin (from eggs) because, unlike the other phosphoproteins, they are available in adequate quantities for our experiments. A wide range of calcium phosphates occurs in normal and pathological deposits - their structures may resemble those of highly crystalline, well characterised minerals such as hydroxyapatite, $Ca_5(PO_4)_3OH$, and brushite, $CaHPO_4.2H_2O$, or they may be less well defined with little evidence for long-range order.

In our experiments calcium chloride and sodium phosphate were slowly mixed at carefully controlled pH either with or without a protein present; X-ray diffraction patterns and EXAFS spectra were recorded from the resulting precipitates. Mixing was carried out over a 12 hour period to prevent aggregation of proteins by calcium ions and the pH was adjusted to either 5.4 or 7.1 by addition of sodium hydroxide solution - all under computer control. The final concentration of calcium and phosphate was the same as in cow's milk since casein was to be used as one of the phosphoproteins. Unfractionated casein was prepared from cow's milk, phosvitin was purchased from Sigma Chemical Co. and β-lactoglobin, a protein which contains no phosphate groups, was used as a control. X-Ray diffraction patterns were recorded using a Debye-Scherrer camera and nickel-filtered copper K_α radiation; where appropriate the patterns were compared with the ASMT powder file and patterns from well characterised standards. EXAFS spectra were recorded above the Ca K edge in the transmission mode at the Synchrotron Radiation Source of the SERC Daresbury Laboratory under similar conditions to those described previously [2].

X-Ray diffraction patterns from samples precipitated in the presence of phosphoproteins consisted only of a few diffuse rings; other precipitates, where either β-lactoglobulin or no protein at all were present, had X-ray diffraction patterns characteristic of crystalline phases. Fig. 1 compares some typical results. At pH 7.1 (no protein or β-lactoglobulin present) poorly crystalline hydroxyapatite was precipitated: at pH 5.4 the precipitate was brushite - as expected [3]. Patterns from samples precipitated in the presence of phosphoproteins were inadequate for identifying the phases present.

EXAFS spectra showed that the local environment of calcium was the same in the presence or absence of phosphoproteins; thus it appears that

Fig. 1   X-Ray diffraction patterns of precipitates formed at pH 5.4 with no
protein present (top) and in the presence of casein (bottom)

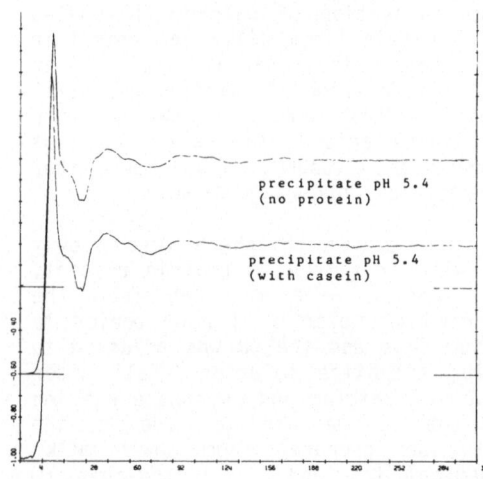

precipitate pH 5.4
(no protein)

precipitate pH 5.4
(with casein)

Fig. 2 EXAFS spectra of precipitates formed at pH 5.4 with no protein
present (top) and in the presence of
casein (bottom)

phosphoproteins influence the extent or degree of crystallisation without
changing the chemical type of calcium phosphate precipitated. For example
at pH 5.4 the EXAFS spectrum from the precipitate with casein is virtually
identical to that from the precipitate with no protein, identified by X-ray
diffraction as brushite - see Fig. 2.   Further experiments need to be
performed, especially those involving chemical analysis.   But two important
points can be made.   One: EXAFS, which provides information on the local
environment of calcium, complements X-ray diffraction, which is more
sensitive to long-range order, for characterising calcium phosphates. Two:
our results are consistent with phosphoproteins acting as sites for the
nucleation of calcium phosphate microcrystals so that, at least until all
these sites are occupied, larger crystals with long-range order are not
formed.

Acknowledgements

We thank the staff of Daresbury Laboratory for help and SERC for support,
including a studentship for J.C.I.

140

References

1. M.J. Glimcher: J.Dent.Res. B58, 791 (1979)
2. C. Holt, S.S. Hasnain & D.W.L. Hukins: Biochim. Biophys. Acta
   719, 299 (1982)
3. P-T. Cheng & K.P.H. Pritzker: Calcif. Tissue Int. 35, 596 (1983)

Structural Study of Arthropathic Deposits

J. Harris and J.S. Shah

Department of Physics, University of Bristol, Bristol BS8 7TL, United Kingdom

S.S. Hasnain

Science and Engineering Research Council, Daresbury Laboratory
Daresbury, Warrington WA4 4AD, United Kingdom

1 Introduction

Destructive joint arthropathics and connective tissue diseases are often associated with the presence of crystalline deposits. One such deposit is hydroxyapatite; a mineral frequently observed in the articular and intra-articular cartil-
age of patients suffering from acute synovitis or more progressive arthropath
ics, for example osteoarthritis. Other, more extensive deposits are common in
connective tissue diseases, such as systemic sclerosis.

Although these deposits are believed to be impure apatitic phases the principle impurity being carbonate, CO_3^{2-}, little has been done to comprehensively
characterise them in terms of their degree of crystallinity and impurity
content. Consequently, the atomic scale mechanisms involved in apatite
deposition and maturation are poorly understood.

A preliminary characterisation of pathological and synthetic apatite depos
its, of varying crystallinity and CO_3^{2-} content, using x-ray powder diffraction,
infra-red spectroscopy and EXAFS techniques has therefore been made. In this
way, it is hoped that the structural changes associated with crystallinity and
CO_3^{2-} content may be understood.

2 Method

X-ray diffraction was conducted using a standard Debye-Scherrer powder camera;
each sample being placed in a 0.5 mm diameter lindemann tube and exposed to
nickel filtered Cu $K\alpha$ radiation at 35 keV, 40 mA for four hours.

Infra-red spectra were obtained using a Perkin Elmer 577 spectrophotometer
over the wavelength range 3600 - 200 cm^{-1}. Samples were prepared by mixing ca
2 mg with 20 mg dry spectroscopically pure KBr, grinding this mixture to a fine
powder and compressing it into a thin disc under a pressure of 10 MN/m^2.

Specimens were prepared for EXAFS by grinding to a uniform fine powder and
mounting between strips of sellotape. Spectra above the Ca K-edge were recorded
in transmission mode at the Daresbury Synchrotron Radiation Source, operating at
2 GeV with an average current of 90 mA. A silicon 220 monochromator was employed and the ion chamber for detection of the incident and transmitted flux optimised to give attenuation of 20% and 60% respectively.

3 Results

Infra-red spectroscopy revealed additional absorption peaks at 1460, 1420 and
870 cm^{-1} in both the pathological deposits and the __carbonated__ synthetic samples.

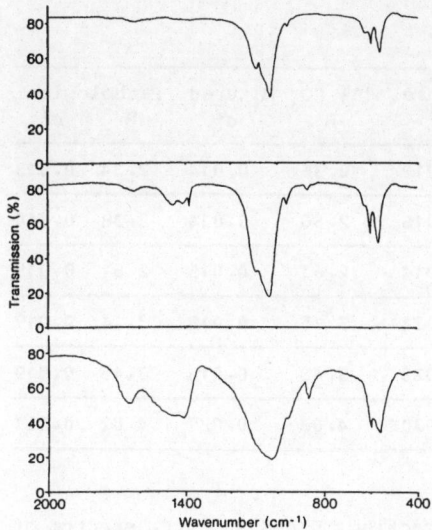

Fig.1.   IR spectra of hydroxyapatite
(top) carbonated apatite (middle) and
pathological deposit (bottom)

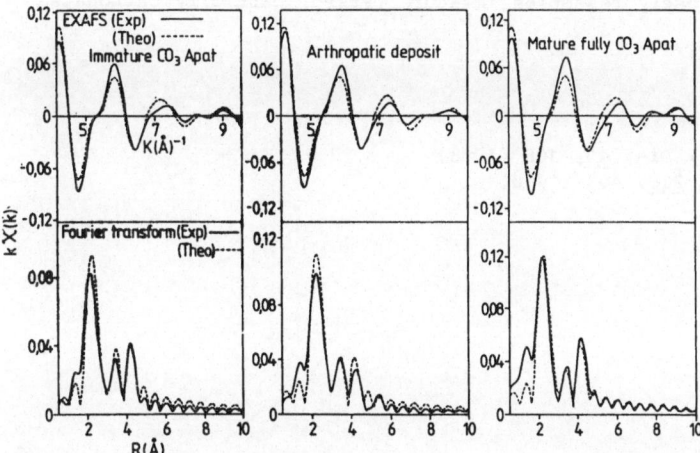

Fig.2.   Fourier filtered and simulated EXAFS spectra of (a) immature carbonated
apatite (b) arthropathic deposit and (c) mature 5-6% carbonated apatite

These peaks are characteristic of carbonate substituted into the phosphate site
of the apatite lattice [1].

X-ray diffraction of the samples showed the 'a' and 'c' lattice parameters
vary with carbonate content, in agreement with the results of LEGROS [2].   It
was also observed that the degree of crystallinity degrades with increasing
carbonate content.

K weighted EXAFS spectra also displayed changes with varying CO_3^{2-} content
and structural disorder within the unit cell.   Figure 2 presents the theoretical and experimental EXAFS data for the arthropathic deposit, and partially
(Fresh) and fully carbonated apatite.   Table 1 summarises the results of a detailed analysis of EXAFS data.

Table 1

Shell No.	Atoms	1.5% CO₃ Matured R σ²		2% CO₃ Immature R σ²		10% CO₃ Matured R σ²		Pathological R σ²	
		R	σ^2	R	σ^2	R	σ^2	R	σ^2
1	3 O	2.36	0.014	2.34	0.012	2.36	0.014	2.34	0.013
2	2 O	2.37	0.014	2.44	0.016	2.36	0.014	2.38	0.014
3	2 O	2.60	0.018	2.64	0.014	2.61	0.015	2.61	0.015
4	2.5P	3.15	0.020	3.17	0.021	3.15	0.018	3.14	0.019
5	2.5P	3.51	0.019	3.48	0.023	3.43	0.018	3.45	0.019
6	3 Ca	4.03	0.020	4.04	0.030	4.03	0.027	4.02	0.031

In conclusion, a comparison of x-ray diffraction, EXAFS and I.R. spectra of both pathological deposits and synthetic carbonated apatites shows that pathological apatite most closely resembles immature (Fresh) partially carbonated apatite.

References

1. J.S. Shah: Ann. Rhem. Dis, 42, 168 (1983)
2. R.S. Legros: Nature, 206, 403 (1965)

Environment of Calcium in Biological Calcium Phosphates

S.S. Hasnain

Science and Engineering Research Council, Daresbury Laboratory
Daresbury, Warrington WA4 4AD, United Kingdom

1 Introduction

An understanding of calcification process is a fundamental problem in biology
 and therefore a knowledge of the structure of biological calcium phosphates
is of crucial importance. Due to the 'amorphous' nature of the biological
calcium phosphates, x-ray diffraction techniques have been of limited use.
Over the last six years EXAFS and XANES at the Ca K-edge ($\sim$ 3.1 Å) have been
used to probe the structure and chemical nature of calcium in a number of
biological systems. Until recently, these studies were primarily based on
qualitative comparison of spectra and their Fourier transforms. Recently, a
self-consistent method has been developed to obtain accurate quantitative
structural information using the EXCURVE program based on sigle scattering
theory of LEE and PENDRY [1]. Phase shifts were calculated using the MUFPOT
program. The examples chosen are for the following biological systems: (a)
bone, tooth enamel and arthropathic deposits, (b) milk and (c) binding of Ca
to DNA and polydA.polydT.

A) Bone, ACP and HAP

Initial EXAFS studies [2] showed that the environment of Ca in matured bone
is not the same as that in hydroxyapatite but is like the Ca environment in
partially matured amorphous calcium phosphate. It was also [3] shown that
the environment of Ca changes with increasing age (maturation) in the same
way as the amorphous calcium phosphate matures in the presence of water.
EXAFS analysis of hydroxyapatite was started with the crystal structure data
as the initial parameter. From Fourier transform and back-transform it was
established that contributions from atoms (around the Ca atom) within $\sim$ 5 Å
needed to be considered. Figure 1(a) shows the experimental and simulated
spectra for hydroxyapatite. This simulation was transferred to amorphous
calcium phosphate (Fourier-filtered 1.0 - 4.5 Å) and matured bone. The final
simulation for ACP and bone are shown in Figs.1(b) and 1(c) with their
Fourier transforms. Parameters used for simulation are given in Table 1. We
note that the first two shells (coordinated oxygen) increase slightly in
going from HAP to ACP while shell 3 (2 O) increase by $\sim$ 0.09 Å and shell 5
(2.5 P) decrease by $\sim$ 0.09 Å also σ^2 for the 6th shell (3(a)) increases nearly three-fold.

Tooth Enamel and HAP

An insight in the mineralization process of the calcified tooth tissue requires an exact knowledge of the nature of the mineral. A number of models
have been proposed. Recently DRIESSENS and VERBEECK [4] have proposed a
three-phase model for the mineral in tooth enamel and dentine. They proposed
that both enamel mineral consists of a magnesium Whitelockite, a sodium and
carbonate containing apatite and a slightly carbonated hydroxyapatite approxi

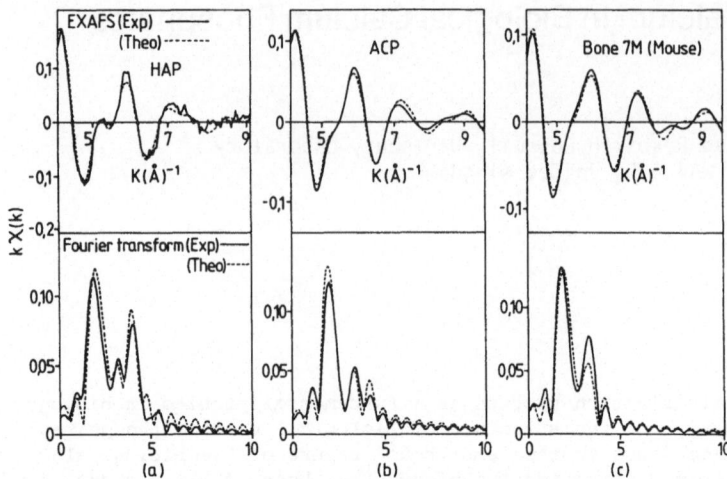

Fig.1(a) Experimental and simulated spectra for hydroxyapatite Fourier filtered and simulated spectra for ACP (1(b)) and bone (1c).   Parameters are
given in Table 1

Table 1

Shell No.	Atoms	HAP		ACP		BONE	
		R	σ^2	R	σ^2	R	σ^2
1	3 O	2.34	0.014	2.35	0.014	2.33	0.012
2	2 O	2.35	0.014	2.35	0.014	2.36	0.013
3	2 O	2.54	0.018	2.63	0.018	2.63	0.017
4	2.5P	3.13	0.025	3.16	0.017	3.13	0.014
5	2.5P	3.52	0.017	3.47	0.017	3.43	0.014
6	3 Ca	4.01	0.011	3.99	0.027	3.96	0.037

mately in the ratio 1:1:3. A comparison of the EXAFS spectra for enamel,
$Ca_{8.5}$ $Na_{1.5}$ $(PO_4)_{4.5}(CO_3)_{2.5}$, magnesium Whitelockite and a slightly carbonated
hydroxyapatite are shown in Fig.2; also included in the figure is the spec-
trum of OCP. It is clear that the calcium environment in enamel is very similar to that in a slightly carbonated hydroxyapatite and pure hydroxyapatite.
A detailed analysis of the EXAFS spectra of HAP and enamel gave almost identical structure differing little from the slightly carbonated apatite.
Figure 3 shows the spectra with theoretical simulation for enamel and slightly carbonated apatite. Table 2 compares the structural parameters used for
these simulations. We note again that inner oxygen shells change only slightly. Thus, EXAFS results do not support the three-phase model but favour a
hydroxyapatite or slightly carbonated hydroxyapatite structure of the tooth
enamel. This is consistent with the x-ray diffraction [5] study of tooth
enamel. X-ray diffraction pattern of tooth enamel is similar to that of

146

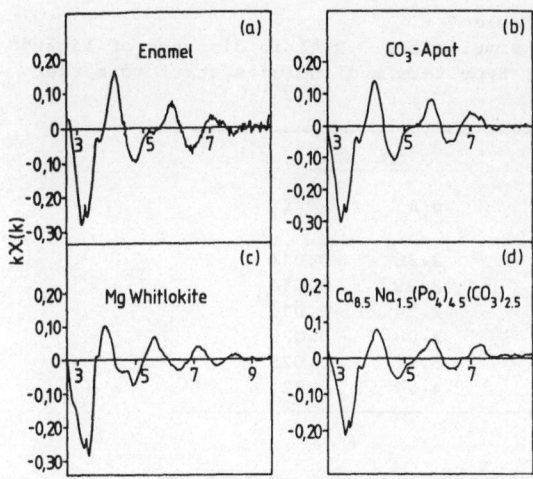

Fig.2. EXAFS spectra of (a) tooth enamel, (b) a slightly carbonated apatite, (c) magnesium whitlockite and (d) sodium and carbonate containing apatite

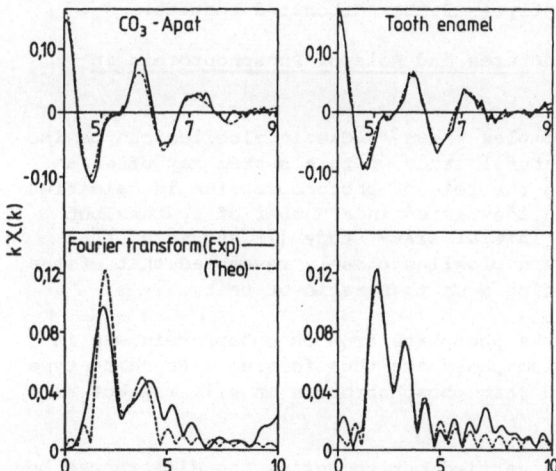

Fig.3. Experimental and simulated spectra for a slightly carbonated apatite and tooth enamel. Parameters used for simulation are given in Table 2

hydroxyapatite with the crystallographic a-axis varying between 9.44 - 9.45 Å compared to pure single phase HAP (9.44 Å).

Arthropathic Deposits

There are many diseases of the joints and periarticular connective tissues where chronic destructive changes and/or inflammation is associated with growth of microcrystals. Some of these crystals are believed to be hydroxy-apatite. Apatite deposition is frequently associated with the condition known as periarthritis. Recently, SHAH et al [6] have shown, using the pow-der diffraction technique, that the lattice parameters for periarthritic deposits are different to that of HAP. Infra-red spectra of these deposits show additional peaks at 870 cm^{-1}, 1450 cm^{-1} and 1510 cm^{-1} which are known to be associated with the presence of CO_3^{2-} group.

Table 2

Structural parameters used for EXAFS simulations. R(Å) is distance of ligands from Ca and $\sigma^2(Å)^2$ - the Debye Waller type term and includes static disorder (ΔR^2)

Shell/Atom	Enamel		CO_3-APat	
	R(Å)	$\sigma^2(Å)^2$	R(Å)	$\sigma^2(Å)$
3 O	2.34	0.014	2.36	0.014
2 O	2.35	0.014	2.35	0.014
2 O	2.56	0.016	2.54	0.016
2.5 P	3.15	0.018	3.08	0.025
2.5 P	3.55	0.016	3.57	0.025
3 Ca	4.05	0.035	4.04	0.024

EXAFS measurements of these pathological samples have been made along with HAP, mature, immature, slight and heavy substituted CO_3 apatite. Detailed analysis of these results discussed elsewhere in the proceeding (Harris et al) show that the Ca environment in arthropathic deposit to be most like a partially matured slightly (~ 2-3%) carbonated apatite.

B) Ca in Milk: Brushite-type Structures and Role of Phosphoprotein in Calcification

Calcium phosphate-phosphoprotein complex known as Casein micelles can be isolated as such and therefore a structural study of this system may offer a unique possibility in understanding the role of phosphoproteins in calcified tissues. Phosphoproteins have been identified in a number of systems but their specific role is not known. Initial EXAFS study [7] suggested that environment of calcium atom in casein micelles closely resembled that of the Ca atom in Brushite mineral suggesting a Ca to P ratio of unity.

This can be reconciled only if the phosphate from phosphoprotein was incorporated into the inorganic calcium phosphate thus forming a Brushite-type lattice. Thus Holt et al suggested that phosphoprotein in milk may act as the nucleation site.

Detailed EXAFS analysis has been carried out to define the differences between the calcium environment in casein milk and brushite mineral. Crystal structure of brushite was used as the starting parameters. For brushite nearly all atoms up to 6 Å needed to be incorporated while for the Casein micelles a simulation without the ~ 6 Å shell (of Ca atoms) was obtained. This difference could also be established from the fourier filtering of Brushite-data with or without the 6 Å and a comparison of the experimental EXAFS data for the two systems which are identical except for the high k shoulder in the three beat regions. This difference may arise from the lack of a long range order (crystallinity) in the case of Casein micelles.

In a recent experiment Calcium phosphates have been precipitated in vitro with and without proteins. β-lactoglubin, a non-phosphoprotein had no effect on the precipitate formed (its x-ray diffraction pattern and EXAFS were identical to that of precipitate formed in the absence of protein). Precipitates formed in the presence of phosphoproteins (Casein and phosvitin) had no extensive crystal lattice as shown by powder diffraction but the local order, verified by EXAFS remained the same. This observation further supports sug-

gestions that the phosphoproteins act as nucleation sites for the calcium
phosphate.

Calcium salts of natural DNA and polydA.polydT

The DNA molecule is a polyanion, and to retain a stable double-helical structure it needs (in the absence of interaction with cellular proteins) the
proximity of water molecules and countercations (eg metal ions). If it is
effectively dried or there is a marked counterion deficit, DNA denatures.
If, however, the water or counterion content is changed within certin limits,
one can observe in the general case, multiple conformational transitions
between different double-helical structures. Besides, solid-state DNA reacts
to changes in relative humidity (r.h), ie in the water content of the specimens, by different patterns of conformational behaviour depending on the kind
of counterion bound to it. In order to understand the mechanisms underlying
the conformational transitions in double-helical DNA, it is important to
understand how the cations and at least the nearest water molecules are
disposed around it.

The preliminary EXAFS experiments were carried out on film of the Ca salts
of the synthetic polynucleotide polydA.polydT at 95, 81 and 76% r.h. and for
the Ca salt of chicken erythrocyte DNA at 81% r.h. (approximately 43% GC
pairs). These EXAFS spectra show beat patterns typical of calcium phosphate,
particularly to those of monetite and brushite, with whom the spectrum of
Ca-polydA.polydT is compared in Fig.4. From the detailed analysis of EXAFS
data, it is clear that the Ca^{2+} ion is in fairly close proximity (within 4 Å)
to a number of phosphorous atoms. This is in contradiction with the recently
proposed model which assumes a close coordination between the cations and the
nitrous base atoms deep inside the polynucleontide molecule, so that the distance to the nearest phosphorous atoms must not be less than 5 Å. Instead,
the EXAFS results suggest that the Ca^{2+} ions are, for the most part, located
at the periphery of individual polydA.polydT (or DNA) molecules, possibly
serving as intermolecular links.

Acknowledgements

The above work was done in collaboration with a number of colleagues particularly with Drs. D.W.L. Hukins, Carl Holt and Mr. J. Irlam, Dr. J. Shah and

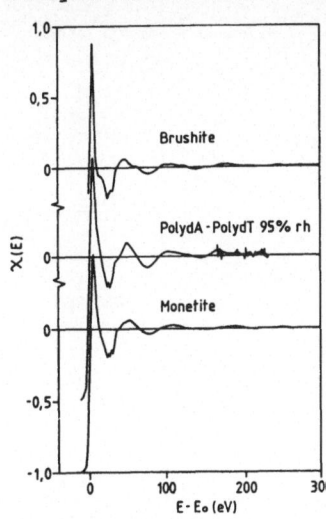

Fig.4. XANES and EXAFS spectra
of Ca-polydA.polydT and minerals
of Brushite and monetite.

Mr. J. Harris, Dr. Skuratovskii, Professor Driessens and Dr. Verbeeck. The author is thankful to Dr. David Hukins for many stimulating discussions on the subject. Science and Engineering Research Council is acknowledged for its help. The experimental results reported here were obtained at the EXAFS station 7.1 on SRS except data reported in [2] was obtained at L.U.R.E.

References

1 P.A. Lea and J.B. Pendry: Phys. Rev. $\underline{B11}$, 2795 (1975)
2 R.M. Miller, D.W.L. Hukins, S.S. Hasnain and P. Lagarde: Biochem. Biophys. Res. Comm. $\underline{99}$, 102 (1981)
3 N. Binsted, S.S. Hasnain and D.W.L. Hukins: Biochem. Biophys. Res. Com. $\underline{107}$, 89 (1982)
4 F.C.M. Driessens and R.M.H. Verbeeck: Bull. Soc. Chim. Belg. $\underline{91}$, 573 (1982)
5 F.C.M. Driessens: Bull. Soc. Chim. Belg. $\underline{89}$, 663 (1980)
6 J.S. Shah: Ann. Rheum. Dis. $\underline{42}$, 68 (1983)
7 C. Holt, S.S. Hasnain and D.W.L. Hukins: Biochemica Biophysica Ata $\underline{719}$, (1982)

Examination of Rhodium Carboxylate Anti-Tumor Agents Complexed with Nucleosides by EXAFS Spectroscopy

N. Alberding, N. Farrell, and E.D. Crozier

Department of Physics, Simon Fraser University, Burnaby, B.C. Canada V5A 1S6
and
Universide Federal de Minas Gerais, Instituto de Ciencias Exatas
30 000-Belo Horizonte, MG, Brasil

1. Introduction

Rhodium carboxylates form anti-tumor agents [1] which bind specifically to adenine and its nucleotides and nucleosides [2]. A clear understanding of the exact mechanism of action of the rhodium carboxylates has been hampered because they are difficult to crystallize and it has not been possible to elucidate their structure-function relationship with x-ray crystallography. Indirect spectroscopic evidence has suggested that the rhodium binds at the purine N_7 site in adenine nucleoside systems [3,4]. The present EXAFS study was undertaken to determine the rhodium binding site in $[Rh_2Ac_4(adenosine)]$. H_2O (I), $[Rh_2Ac_4(corycepin)]$.H_2O (II), and $[Rh_2Ac_4(ara-A)]$.H_2O (III). An unambiguous determination of the N_7 site was not possible using only the Rh K-edge EXAFS. Subsequently additional structural evidence was sought from the Rh and Br EXAFS of $[Rh_2Ac_4(8-Br-adenosine)]$.$H_2O$ (IV).

2. Experimental and Results

X-ray absorption spectra were measured in the transmission mode on Beamline IV-1 at the Stanford Synchrotron Radiation Laboratory. Spectra were recorded at 22°C and -185°C on complexes I-IV, Rh_2butyrate.$2CH_3OH$ (V), and the reference compounds $[Rh_2Ac_4(caffeine)_2]$ (VI) and 8-Br-adenosine (VII).

The preparation of the complexes and specific details of the EXAFS analysis will be presented elsewhere [5]. Briefly the EXAFS interference function $\chi(k)$ was approximated by the theoretical expression for the case of single scattering, valid in the limit of low disorder. Effects of shake-up, shake-off [6], non-Gaussian pair distribution functions [7], and multiple scattering [8] were not pertinent in the present study. The similar chemical structure (oxidation state, ligation) at the Rh site of the complexes is evident in the $k\chi(k)$ shown in Fig. 1. The high k region of $k\chi(k)$ is dominated by Rh whose backscattering amplitude $f(k,\pi)$ is large, peaking at $10A^{-1}$, whereas the $f(k,\pi)$ of the lighter elements coordinated to Rh are small at high k. This fact facilitated an accurate determination, ±0.005A, of the Rh-Rh bond lengths from the phases; the phases were obtained by suitable Fourier filtering and the difference between the unknown and the reference compound VI, of known crystal structure [4], was fit to a linear function over the range $9-17A^{-1}$. The Rh-Rh bond lengths of I, II, III and IV are the same as VI, 2.395A, within experimental error. The r(Rh-Rh) of $[Rh_2Bu_42CH_3OH]$, however, is 0.026A shorter. Curve-fitting to a two-shell model of one Rh and four O surrounding the central Rh atom indicates that changes in the disorder parameter σ^2(Rh-O) and minor changes in r(Rh-O) produce the change in the beat structure at $9A^{-1}$ observed between VI and I.

A search was made for Rh-purine ligation. In the reference complex VI, Rh binds to the N_9 site of the purine with a bond length 0.08A shorter than

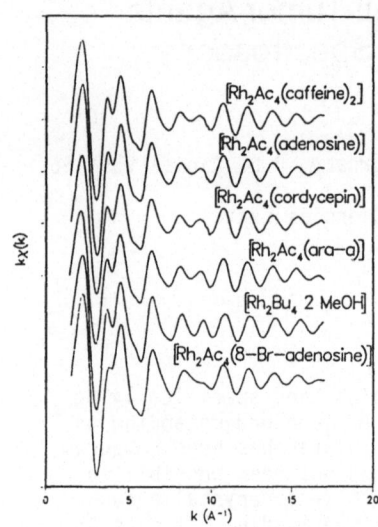

Fig. 1. EXAFS spectra of the Rh K-edge of six Rh-carboxylate compounds. The tick marks on each spectrum are at the same k value and reveal the shorter bond length of $[Rh_2Bu_4 \cdot 2CH_3OH]$ relative to the others.

[Rh$_2$Ac$_4$(caffeine)$_2$]

[Rh$_2$Ac$_4$(adenosine)]

[Rh$_2$Ac$_4$(cordycepin)]

[Rh$_2$Ac$_4$(ara-a)]

[Rh$_2$Bu$_4$ 2 MeOH]

[Rh$_2$Ac$_4$(8-Br-adenosine)]

$k\chi(k)$

k (A^{-1})

r(Rh-Rh) [4]. Comparable distances are expected for Rh binding to the purines of the other complexes. However, a three-shell fit with O, N and Rh did not improve the goodness of fit obtained with the two-shell model. Evidently the stronger signal from the four O atoms and the Rh atom masks the contribution of the purine ligation.

The Fourier transforms of the Br edge spectra are shown in Fig. 2. The known crystal structure of (VII) [10] enabled the following peak identifications: peak 1 with C_8, peak 2 with N_7 and N_9, peak 3 with C_4 and C_5, peak 5 with C_6 and N_3 and possibly some contributions from atoms on the ribose moiety, peak 4 with $O_{2'}$, $C_{2'}$, and the heterocyclic oxygen. Peak 4 is more sensitive to temperature as expected for backscattering from atoms on the sugar in comparison with those on the purine. The syn-conformation of the sugar with respect to the purine ring can be recognized by the existence of this peak. Upon complexation with rhodium acetate, peaks 1, 2, 3 and 5 retain the same identification. Peak 4, however, is considerably reduced and a shoulder appears on the lower side of peak 3 at 3A. The shoulder is not due to Rh backscattering as confirmed by the absence of any identifiable Br signal in the Rh spectrum. The absence of peak 4 implies a movement of the sugar to the anti-conformation. The movement can be calculated to bring several atoms, e.g., heterocyclic oxygen, into conjunction around 3-3.7Å which could produce this shoulder. This strongly implies binding at N_1. The alternative N_7 binding should give an observable Rh-Br distance of approximately 3.6Å.

3. Conclusion

The Rh-edge EXAFS spectra did not permit assignment of the Rh binding site in the non-brominated adenosine species I-III. Indeed the spectra showed no features which could obviously be assigned to structure outside the Rh-carboxylate complex. The spectra did, however, enable determination of Rh-Rh and Rh-O bondlengths and Debye-Waller factors for complexes I-IV, quantities which have not been put into evidence by x-ray crystallography. In the brominated complex IV it is suggested that Rh binds to the N_1 site of the purine and that complexation of Rh induces a syn- to anti-conformation change at the nucleoside.

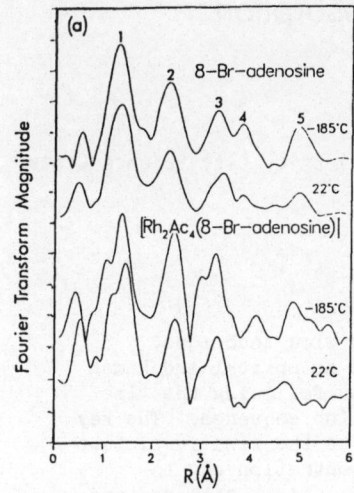

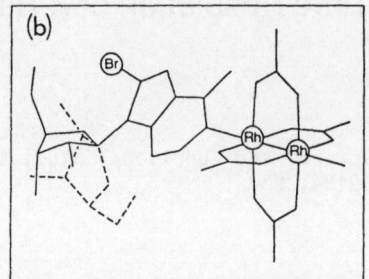

Fig. 2a. The Fourier transforms of the Br K-edge spectra.
Fig. 2b. A possible structure of Rh$_2$-acetate bound at N$_1$ to 8-Br-adenosine in the anti-conformation. The dashed lines show the syn-conformation.

4. Acknowledgements

The authors wish to thank A.J. Seary for technical assistance. The research was partially funded by a grant from N.S.E.R.C., Canada. The SSRL is supported by the D.O.E. through the Office of Basic Energy Sciences and the N.I.H. through the Biotechnology Resource Program in the Division of Research Resources.

5. References

1. J.L. Bear, H.B. Gray, L. Rainen, I.M. Chang, R. Howard, G. Serio and A.P. Kimball: Cancer Chemother. Rep. 59, 611 (1975).
2. L. Rainen, R.A. Howard, A.P. Kimball and J.L. Bear: Inorg. Chem. 14, 2752 (1975).
3. N. Farrell: J. Inorg. Biochem. 14, 261 (1981).
4. K. Aoki and H. Yamazaki: J.C.S. Chem. Comm. 186 (1980).
5. N. Alberding, N. Farrell and E.D. Crozier: J. Am. Chem. Soc. (1984).
6. E.A. Stern, B.A. Bunker and S.M. Heald: Phys. Rev. B 21, 5521 (1980).
7. E.D. Crozier and A.J. Seary: Can. J. Phys. 59, 876 (1981).
8. N. Alberding and E.D. Crozier: Phys. Rev. B 27, 3374 (1983); B.K. Teo: J. Am. Chem. Soc. 103, 3990 (1981); J.J. Boland, S.E. Crane and J.D. Baldeschweider: J. Chem. Phys. 77, 142 (1982).
9. S.S. Tavale and H.M. Sobell: J. Mol. Biol. 48, 109 (1978).

Limitations to Time Resolution of X-Ray Absorption Spectroscopy

B. Chance

Institute for Structural and Functional Studies, University City Science Center
Philadelphia, PA 19104, USA

X-ray absorption spectroscopy is essentially a slow technique, particularly when compared with "rapid" methods such as optical and Raman spectroscopy. Thus, it is not a preferred method for following the time course of intermediate compounds in a chemical reaction sequence. The key information that X-ray absorption spectroscopy adds to the kinetic picture is the structure of intermediate compounds. The concentration can be appropriately maximized by studies usually employing optical methods, and in fact, monitoring of the maximization of concentrations of intermediate compounds during the X-ray absorption spectroscopic study may be essential. The EXAFS methods described below are not appropriate to follow the time course of a chemical change (for which they are extremely inefficient) but to determine the structure of the labile intermediates of which the kinetic studies have afforded a maximum of concentration of a single species. Higher accuracies than those required here are needed if multiple species are simultaneously present in the reaction system.

TABLE I	Temperature Range (°K)	Signal Count rate(sec^{-1})	Time Resolution (sec)	Observation Time, ΔT (sec)	Repetition Frequency (f) (Hz)	Counts/sec = sig.ΔT.f	Time for 400 pts at 10^6 counts	efficiency of beamline usage (%)
1) Relaxation method Laser Photolysis liquid/solid	4-300	10^6	10^{-6}	10^{-6}	10	10	4×10^7	0.001
2) Optical Pump Xenon Flash Photolysis liquid/solid	4-300	10^6	10^{-6}	5×10^{-3}	2	10^4	4×10^3	1.0
3) Flow Method Regenerative Flow liquid	240-300	10^6	10^{-3}	0.5	1	5×10^5	8×10^2	50
4) Cryokinetics Trapped states solid	4-200	10^6	1	1	1	10^6	400	100

Relaxation methods, flow methods and cryokinetics (see Table) (1) afford time resolution of chemical reactions. Relaxation methods have been used by a variety of investigators [Sandstrom (2), Mills (3), Huang (4)].

1. Relaxation methods. Laser photolysis (5) activates some biochemical reactions. At a time resolution of a usec and a laser repetition frequency of 10 Hz, 4×10^7 sec is required for an EXAFS spectrum of 400 data points at 10^6 counts per point. The efficiency of beam time utilization is 0.001% relative to a static scan (3.4).

2) Long flash optical pumping. Under the same conditions, a time resolution of 1 usec over the interval of the pump flash (5 msec) and a repetition frequency of 2 Hz requires 4×10^3 sec and the efficiency is 1% (6.7). A current design is illustrated in Fig. 1.

3) Regenerative flow method. This method is applicable to most chemical reactions (8). With a time resolution of a msec, an observation time of 0.5 sec, a repetition frequency of 1 Hz, the EXAFS scan requires 2×10^3 sec and the efficiency is 50%. A current design is illustrated in Fig. 2. (The efficiency is independent of the time measured.) Thus,

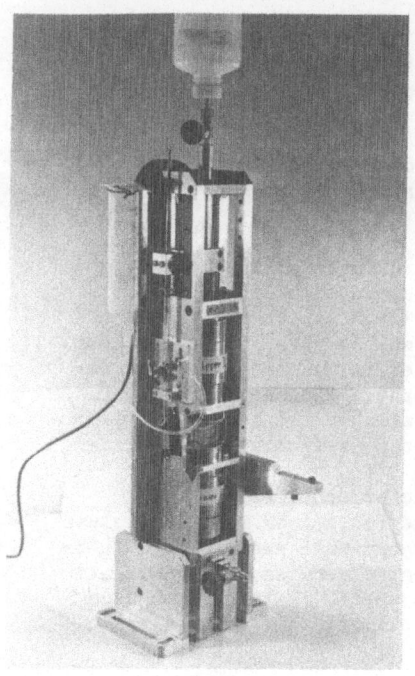

Fig. 1. Long flash optical pumping system providing for "illumination of optically thin samples with X-ray absorption pathway through four (or more) for achievement of adequate X-ray absorption thickness. Each flow-through cell is provided with dual Xenon flash illumination on both sides, and X-ray absorption spectroscopic monitoring occurs through appropriate filters. The flow rate is appropriate to maintain a temperature rise of less than 4° during the flash interval

Fig. 2. Modification of observation chamber of regenerative flow apparatus for X-ray absorption spectroscopy. The mylar walled chamber is moved forward to obtain an adequate solid angle for sensitive recording of X-ray fluorescence in the K region

observation of the structural state of liquid samples during light activation in a pumped system or after mixing in a flow apparatus offers worthwhile advantages in minimizing onerous requirements of beam time required by laser flash photolysis or indeed any impulse perturbation (T-jump, E-jump, P-jump, etc.) (9).

4. Cryokinetics in the liquid phase require cryo-protectants (alcohols, glycols, glycerols) that allow chemical mixing at temperatures from -30 to -90°. Usually the stability of intermediates formed in cryokinetic conditions is sufficient to permit their further trapping at low temperatures at which they are indefinitely stable and the efficiency rises to 100% (10). A diagram of an apparatus for cryokinetic studies that we

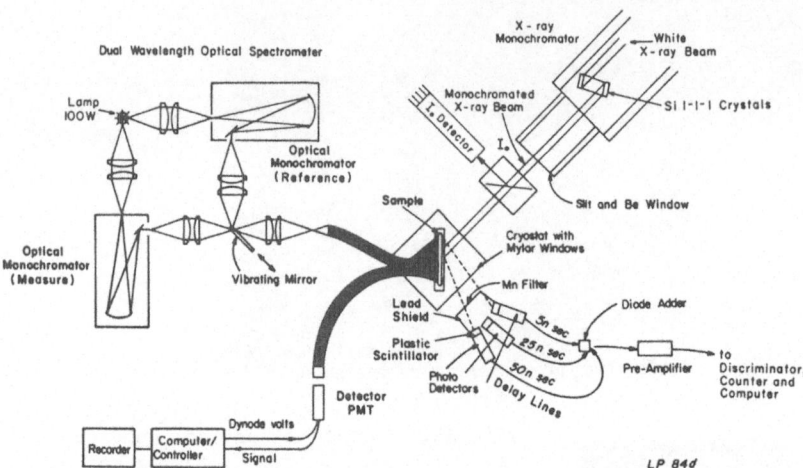

Fig. 3. Schematic diagram of the coupling of X-ray and optical
illumination of sample in cryostat for low temperature kinetic studies.

have used is illustrated in Fig. 3. If kinetics are desired the
temperature is adjusted so that the kinetics are slow compared to the scan
duration (400 sec).

Cryokinetics in the solid phase (11,12) also affords efficient data
collection and both direct photolysis and flash pumped photolysis
techniques are usable with frozen solutions (13) or with crystals (14,15)
with the caveat that the repetition frequency may be governed by the time
for recombination or reprocessing of the samples. Thus reaction kinetics
becomes an important parameter in determining the efficiency of beam time
utilization. If the structures are unstable states at the lowest
temperature possible, the optically pumped system seems most appropriate.

Time-resolved diffraction study presents special problems of
perturbation. Electrical stimulation of muscle (16,17) or laser photolysis
of MbCO (13) are employed. In the latter case, crystals small enough to be
completely converted to Mb* in a single flash give very small signals and
the efficiency as calculated here will be very small. Lattice disorders
caused by low temperatures make the use of cryokinetic approaches less
effective in this case.

SUMMARY

For high efficiency of beam time utilization (100%), which often means
a feasible experiment, prolongation of the lifetime of labile intermediates
is achieved by cryokinetic approaches, many of which bring the time profile
for biochemical reactions within the time range of EXAFS data accumulation
(for concentrated samples, 1 sec/point for 400 points) and for more rapid
reactions, the trapped state method affords even longer data collection
intervals. In the liquid phase the regenerative flow method has an ultimate
time resolution of about 50usec (9) particularly when sharply focused beams
(1mm diameter) can be employed. The fluid volume requirements for the
stopped flow method are much less, but the method (1) does not prolong the

156

lifetime of intermediates, as in all relaxation methods, which may well require flow systems because higher repetition rates and higher beam intensities can cause heating and X-ray damage of the sample. Flash pumped systems using high power Xenon lamps and a slowly flowing liquid at diminished temperatures seems to be a method which has efficiencies of 1% currently that can readily approach 20%. It is predicted that the methods described here and others that will complement them may be a principal initiative of X-ray absorption spectroscopy in the next decade. Current progress is indicated by the description of the design, construction and tests of typical apparatuses appropriate for X-ray absorption spectroscopy of intermediates in chemical reactions.

REFERENCES 1. Rapid Mixing and Sampling Techniques in Biochemistry (Chance. B., Gibson, Q.H., Eisenhardt, R., Lonberg-Holm, K.K., eds.) Academic Press, NY, 1964. 2. Sandstrom, D.R. & Lytle (1980) SSRL Rpt. 80/01, pp. VII-45. 3. Huang, H.W., Teng, T.Y. & Wang, X.F. (1984) SSRL Report 84/01, pp. IX-39-IX-40 4. Mills, D.M., Lewis, A., Harootunian, A., Juang, J. & Smith, B. (1984) Science 223:811-813. 5. DeVault, D. & Chance. B., Biophys J. 6:825-847 (1964). 6. Chance, B., Kumar, C., Korszun, R., Khalid, S., Legallais,, V. Penney, W. & Sorge, J. (1983) 10th Ann. Users Group Mtg. Rpt 83/03, 18-32. 7. Chance. B., Korszun, Z.R., Kumar, C., Zhou, Y., Powers, L. & Winick, H. (1984) SSRL Report 84/01, pp. IX-3. 8. Chance. B., (1955) Faraday Soc.Disc., 20:205-216. 9. Chance, B., DeVault, D., Legallais, V., Mela. L. & Yonetani, T.(1967) in Fast Reactions and Primary Processes in Chemical Kinetics (Vth Nobel Symposium, S. Claesson, ed)., Almquist & Wiksell, Stockholm, 437-468. 10. Chance. B., Graham, N. & Legallais (1975) Anal. Biochem. 67:552-579. 11. Chance, B., DeVault, D., Tasaka, A. & Thornber, J.P. (1979) in Tunneling in Biological Systems (Chance, B., DeVault, D., Frauenfelder, H., Marcus, S.eds) Academic Press, NY, pp. 387-403. 12. Douzou, P., Siriex, R. & Travers, F. (1970) Proc. Natl. Acad. Sci., USA 66:787-792. 13. Chance, B., Powers, L. & Fischetti, R. (1983) Biochem. 22:3820-3829. 14. Anderson,, N.M., Reed, T.A. & Chance, B., (1970) Biophys. Soc. Abstr. 320a & Ann. N.Y. Acad. Sci. 174, 189-182. 15. Bartunik, H., et al (1982) EMBL Research Rpt. 16. Rosenbaum, G. (1979) "The use of synchrotron radiation for x-ray structure analysis in biology," Ph.D. Thesis Rupprecht-Karl-Univ.of Heidelberg. 17. Huxley, H., Faruqi, A.R. Kress, M., Simmons, R.M., Maeda, Y. & Bordas, J. (1982) EMBL Res.Rpt.p.232.

SUPPORT - Partial research support by NIH Grants GM31992, GM33165, RR01633, HL18708, HL31909. Work partially done at SSRL (Projects 660B, 659B, 632B) supported by NSF through Div.Matl.Res. & NIH through BRRP,DRR,DOE. & CHESS (Project 54).

An Integrating Method of Time-Resolved EXAFS Measurement *

Huey W. Huang

Department of Physics, Rice University, Houston, TX 77251, USA

One of the advantages of using EXAFS as a structure probe is its high event rate. With a beam of 10^{12} photons-s^{-1}-ev^{-1}, it is feasible to measure microsecond- or submicrosecond-resolved transmission EXAFS spectra. However, one would need integrating detectors for such measurements. Even in the fluorescence mode, dilute biological samples often yield strong enough fluorescence that an integrating detector is needed in order to take the full advantage of intense synchrotron radiation (SR). For example a SR pulse of 10^6 photons (of 7.2 Kev) incident on a haemoglobin (Hb) solution of 0.5 mM concentration in Fe yields $(1/3) \times 10^3$ Fe K-shell fluorescence photons. A detector subtending 5% of 4π at the sample (which is easily achievable with a 1" detector) would intercept about 20 photons. Since these photons are practically inseparable in time, the conventional counting method would have to sacrifice a factor of 20 in the detecting solid angle in order to measure the variations in absorption. As the SR laboratories continue to develop more powerful sources, the need for integrating detectors will undoubtedly intensify. In this paper I will describe an integrating method for measuring 100 μs-resolved EXAFS of photolyzed haemoproteins [1].

The basic idea is to measure, at a given X-ray energy, a time sequence of EXAFS points following sample excitation, and repeat the process at as many different energies as required. The necessity of measuring at one energy a time is a drawback compared to the energy dispersive method. However the latter is not applicable to fluorescence EXAFS. The innovation we made is in the method of measuring an EXAFS point over a short interval of time, henceforth denoted as τ or the time-resolution. For the type of applications under consideration and given the intensities of available SR beams ($\sim 10^{12}$

---

*This work was supported in part by the Office of Naval Research Contract No. N00014-76-C-0273 and the Robert A. Welch Foundation. The work performed at SSRL was also supported by the Department of Energy, Office of Basic Energy Sciences; the National Science Foundation, Division of Materials Research; and the National Institutes of Health, Biotechnology Resource Program, Division of Research Resources.

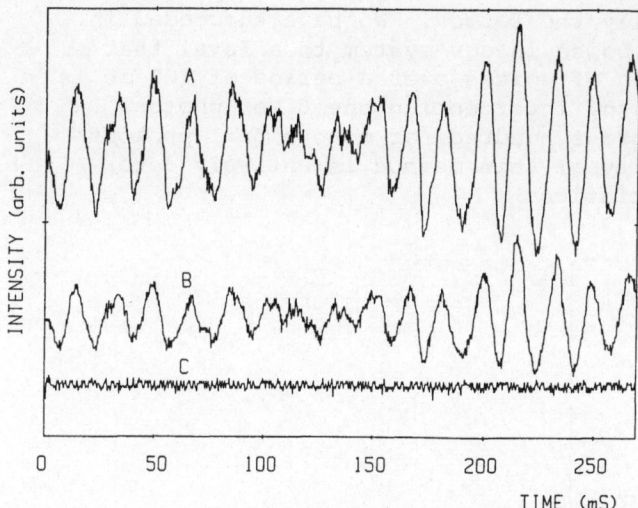

Fig. 1. 100 μs-resolved SR intensity of the SSRL beamline VII-3 at 7.0 Kev. I_o(B) and I(A) were measured in two consecutive ion chambers. The 60 HZ sinusoidal modulation is about 4 %. C is the background. The Fourier transform is shown in Fig. 4.

photons-s^{-1}-ev^{-1}), 100 μs is a practical (lower) limit for τ if a stationary (versus e.g. rapid flow) sample is used.

In doing time-resolved experiments, one should be aware that the intensity of a SR beam is, in general, not steadfast even over a short period of time (Fig. 1). If one uses the counting method, the measurement of incident intensity, I_o, presents as much a problem as the measurement of fluorescence intensity, I. Ignoring variations (Fig. 4 at end) in I_o can easily lead to 1% errors in absorption spectra. We measured both I_o and I with integrating method. We used an ion chamber to monitor I_o and a NaI detector to monitor I. The output of each detector is accumulated in an integrator described in [2,3] during the integration periods. The timing of integrations is controlled in such a way that the same length of interval τ always covers the same number of SR pulses. This is achieved by using the SR pulse clock as a timer. A counter is used to count the pulses arriving from the SR clock. At a chosen time t after a photolysis flash is triggered, the counter is reset to zero. Thereafter when the counter counts one, the integrators are enabled. When the counter counts a preset number N (N-1 = τf, where f is the frequency of the SR pulse clock), the integrators are turned off. The integrated signals are read off via a 14-bit A/D converter.

A potential problem for an integration method is its accumulative effect on noise, since the latter is integrated along with signal. Therefore special care needs to be taken to screen or suppress the noises due to electronics and undesir-

able radiation, especially the former. We have succeeded in
limiting the electronic noise in our system to a level that
the effect of integration of noises over a period of 100 μs is
equivalent to the effect of intercepting one 6 Kev photon.
Since we typically integrate hundreds or more signal photons
in 100 μs, the sensitivity of this method is entirely determined by the photon statistics.

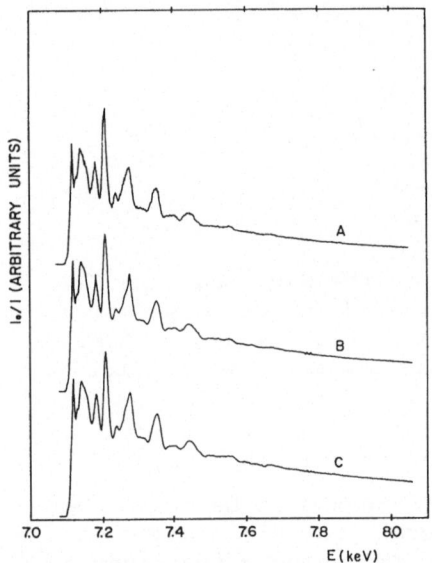

Fig. 2 Transmission EXAFS on
iron foil with $\tau = 100$ μs (A),
$\tau = 50$ ms (B) and $\tau = 1$ s (C). I_o
$I_o \sim 10^{11}$ photons-s^{-1}, $I \sim 10^9$
photons-s^{-1}

As an example we show in Fig. 2 transmission EXAFS of an
iron foil with $\tau = 100$ μs, 50 ms and 1 s. In this case ion
chambers were used for both I_o and I. The monochromatized incident beam I_o was about 10^{11} photons s^{-1} and $I \sim 10^9$ photons
s^{-1}. The signal to noise ratios (S/N) of the spectra are consistent with the numbers of photons per data point. The $\tau =$
100 μs spectrum is quite adequate for determining the interatomic distances of the first three shells.

To determine the minimum number of signal photons per data
point (N/pt) necessary for studying small structural changes in
hemoproteins, we show in Fig. 3 the EXAFS spectra of carboxy-
myoglobin (MbCO) at various levels of statistical accuracy.
The study of the 4°K metastable state Mb*CO by CHANCE et al.
[4] indicated that N/pt $\gtrsim 10^6$ is necessary for EXAFS analysis.
For studying the near edge spectral changes N/pt $\sim 10^5$ seems to
be sufficient. On the other hand, one expects to integrate
about 10^3 signal photons in 100 μs for hemoprotein samples,
(see below). Therefore, until the intensity of SR increases by
two orders of magnitude above the present level, iterated measurements at the same point are required.

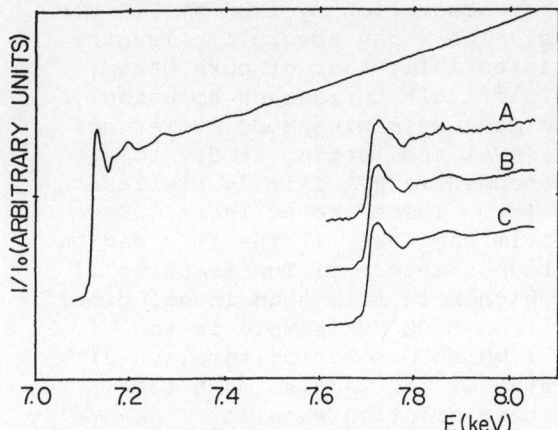

Fig. 3 The EXAFS of MbCO measured in the fluorescence mode
with τ = 100 μs integration. The long spectrum is the result
of 10^4 iterations (7.5×10^5 photons/pt). The short spectra are:
A: 1.9×10^4 photons/pt, B: 7.5×10^4 photons/pt and C: a portion of the long spectrum.

MbCO and carboxyhaemoglobin (HbCO) are perhaps the most often
used samples for studying haemoprotein dynamics because of their
high apparent quantum yields of photolysis [5]. However the
competing experimental requirements for X-ray absorption and
photolysis pose a serious problem for sample preparation--the
former requires the sample to have a high iron concentration
and the latter requires the opposite. We tried to overcome
this difficulty by sample segmentation, which is accomplished
by spacing out a stack of thin samples in the direction of
incident X-ray; each of the samples is thin enough for photolysis, and together they provide sufficient X-ray absorption.
Table I shows the X-ray and optical absorption lengths of MbCO

Table I

	C (mM)	Fe abs. (%)	X-ray abs. length (mm) (7.2 Kev)	Optical abs. length (mm) 423 nm	542 nm	Minimum thickness (mm)
MbCO	5	0.70	0.72	0.011	0.14	
in	1	0.15	0.71	0.054	0.71	$\simeq 0.5$ mm
buffer	0.5	0.076	0.71	0.11	1.4	
	0.1	0.015	0.71	0.54	7.1	
MbCO-	(max.					
PVA	1.5)					
dry	1	1.4	0.70	about 50% of		$\simeq 0.02$ mm
film	0.5	1.1	0.57	above		
	0.1	0.32	0.35			

samples and the percent of X-ray absorption by iron at 7.2 Kev
at various MbCO concentrations. The X-ray absorption lengths
of MbCO buffer solutions are essentially that of pure water,
i.e. 0.7 mm. Therefore it is difficult to segment solution
samples. A thin sample can be made by mixing MbCO buffer solution with polyvinyl alcohol (PVA) and letting it dry to a
film. The property of MbCO embedded in PVA film is similar to
that of MbCO in frozen buffer [6]. There are at least three
advantages in using MbCO-PVA film samples. 1) The film can be
made much thinner than a solution sample. 2) The fraction of
X-ray absorption due to Fe is higher in film than in solution.
For example the signal from a 1 mM MbCO-PVA sample is ten
times as large as that from a 1 mM MbCO solution sample. 3)
MbCO embedded in PVA is very stable. It is also much more
resistant to radiation damage than solution samples.

The kinetics of rebinding of CO to Mb after photodissocia-
tion has been described in detail in [6]. For example it is
believed that below 150°K the photodissociated CO does not
leave the heme pocket. Its recombination time, say at 85°K,
is about 1 s. Therefore 100 µs-resolved EXAFS is ideal for
studying the structural relaxation of the metastable state
Mb*CO. The structural changes in photolyzed Hb are more
complicated and interesting. For example 100 µs-resolved EXAFS
of photolyzed HbCO at 85°K may shed light on the important
question of whether a tilting occurs in the proximal histidine
during the R-T transition [5]. So far our spectrometer has not
been used extensively due to limited SR beam time allocations.
Including instrumental testings, we have tried the experiment
of photolyzed MbCO three times, once at CHESS in 1982 and twice
at SSRL in 1983. 85°K 100 µs-resolved measurements reproduced
the spectrum shown in Fig. 3 due to failure in sample photolysis.

References
    1. H.W. Huang, W.H. Liu, T.Y. Teng and X.F. Wang: Rev.
          Sci. Instrum. 54, 1488 (1983)
    2. H.W. Huang, W.H. Liu and J.A. Buchanan: Nucl. Instrum.
          Methods 205, 375 (1983)
    3. W.H. Liu, X.F. Wang, T.Y. Teng and H.W. Huang: Rev. Sci.
          Instrum. 54, 1653 (1983)
    4. B. Chance, R. Fischetti and L. Powers: Biochemistry 22,
          3820 (1983)
    5. J.M. Friedman, T.M. Scott, R.A. Stepnoski, M. Ikeda-
          Saito and T. Yonetani: J. Biol. Chem. 258, 10564
          (1983)
    6. R.H. Austin, K.W. Beeson, L. Eisenstein, H. Frauenfelder
          and I.C. Gunsalus: Biochemistry 14, 5355 (1975)

Note added:

Fig. 4 shows the frequency analysis of the noise spectrum A in Fig. 1. Note that the modulations at 54 Hz and 60 Hz give rise to 6 Hz beats in Fig. 1.

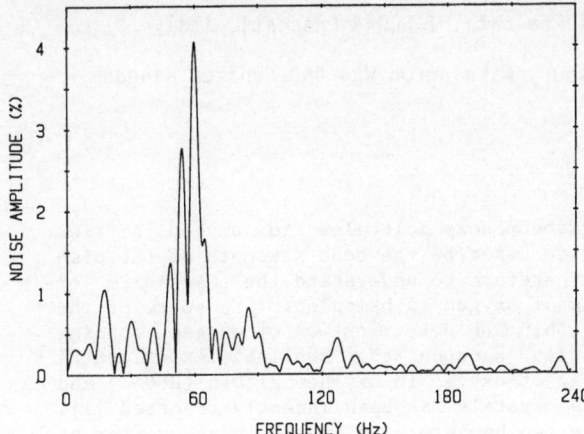

Fig. 4

O$_2$ and CO Bonding Geometry in Heme-Proteins in Solution Investigated by XANES

A. Congiu-Castellano, A. Bianconi, M. Dell'Ariccia, A. Giovannelli
Dipartimento di Fisica, Università "La Sapienza", I-00185 Roma, Italy
E. Burattini
CNR-INFN, Laboratori Nazionali di Frascati, I-00044 Frascati, Italy
P.J. Durham
SERC, Daresbury Laboratory, Daresbury Warrington WA4 4AD, United Kingdom

1 Introduction

The bonding angle of oxygen and carbonmonoxy molecules in oxygen carrier hemoproteins is a key parameter to describe the bond strength of the dia-tomic molecules to iron atom and therefore to understand the mechanism of reversible bonding and releasing of oxygen in hemoglobin. In spite of the large number of studies of haemoglobin the determination of oxygen bonding angle in the proteins in solution escapes the available experimental methods. A different bonding angle of oxygen in oxy-hemoglcbin (HbO$_2$) and in oxy-myoglobin (MbO$_2$) single crystals has been recently reported [1]. The different bonding angle in the two hemoproteins which have different biological roles, transport and storage of oxygen molecules respectively, but similar local structure at the iron site has renewed the interest in this problem. We report the first application of the XANES (X-ray Absorp-tion Near Edge Structure) spectroscopy to obtain information on the oxygen bonding geometry in the protein in solution (close to the "in vivo" situa-tion) which cannot be studied by diffraction methods. The appealing aspect of XANES to study this problem is that the multiple scattering resonances of the photoelectron emitted at the iron site in the 10-50 eV energy range depend on the relative atomic positions of neighbour atoms and not only on the first order radial distribution function like EXAFS (Extended X-ray Ab-sorption Fine Structure), therefore it is a direct structural probe of bonding angles [2-4].

2 Experimental

The XANES measurements were performed at the Frascati "wiggler" beam line using synchrotron radiation monochromatized by a Si(111) channel-cut single crystal; the protein concentration ranged from 4 to 10 mM in heme. Optical spectra of unirradiated and irradiated samples were used to control the possible radiation effects. The sample prepared at pH=7.2 was kept at constant temperature +10 C during the experiment. Hemoglobin sample was obtained from human adult blood and myoglobin was obtained from Sigma sperm whale myoglobin.

3 Results and discussion

Figure 1 shows the XANES of oxy-myoglobin and oxy-hemoglobins of adult human and carp fish. The zero of the energy scale is fixed at the Fe metal K-edge. There are no differences between the XANES spectra over a range of 50 eV within the experimental noise. The comparison between the XANES spectra of HbCO and MbCO, shown in Fig. 2 shows clearly a difference in the intensity of peak C between the two spectra.

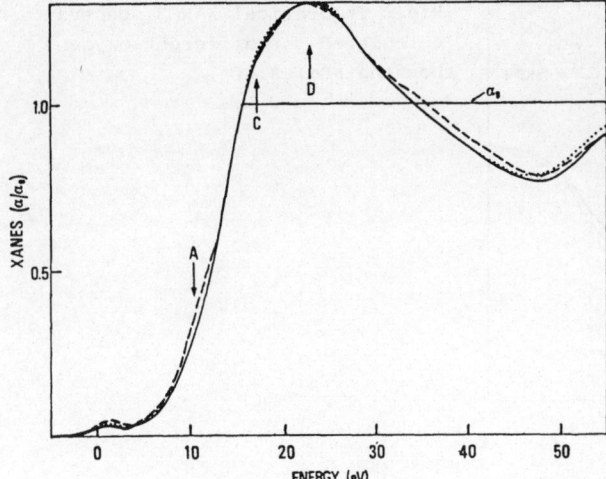

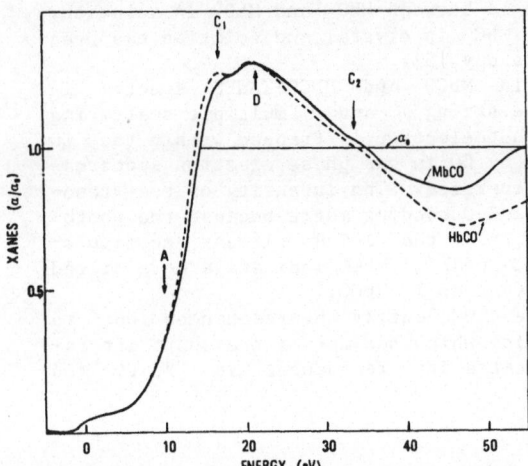

Fig.1 Experimental XANES spectra of oxy-myoglobin (solid line), human oxy-hemoglobin (dashed line) and carp fish oxy-hemoglobin (dotted line).

Fig.2 Experimental XANES spectra of HbCO (dashed line) and MbCO (solid line).

The similarity between the XANES of oxy-hemoproteins indicates similar local atomic arrangement near the Fe site in these proteins and therefore similar oxygen orientation in the protein in solution. This is in opposition with the situation in crystals where the oxygen bonding angle is different [1].

The XANES multiple scattering calculations have been performed starting from the coordinates of the atoms of MbO_2 and HbO_2 crystals. The multiple scattering resonances have been calculated for a photoelectron emitted by the Fe atom and reflected and transmitted by three shells of neighbour atoms forming a cluster of 29 atoms including the 4 nitrogens and 20 carbon atoms of the heme 3 atoms of the proxymal hystidine and the two oxygen atoms. The calculations are in good agreement with the energy position of the experimental features and only in qualitative agreement concerning the lineshape of the spectra. A large change of the XANES spectra for the variation of the oxygen bonding angle Fe-O-O between $115°$ and $156°$, as it has been found in MbO_2 and HbO_2 crystals, is predicted by the calculations, mainly at the peak C in Fig. 3. Because we do not observe in the XANES spec-

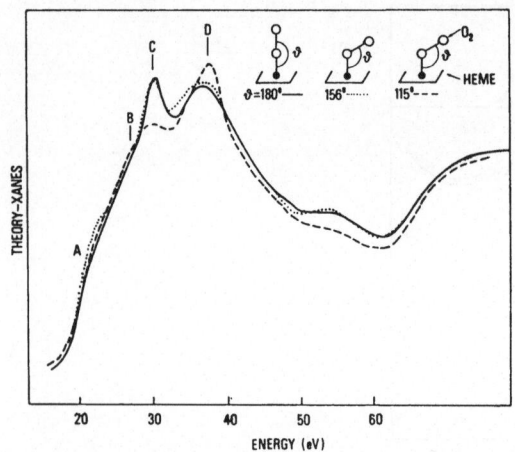

Fig.3 Theoretical XANES spectra calculated for different oxygen bonding angles.

tra a variation in this energy range we find no evidence of such a large variation of the oxygen bonding angle between HbO_2 and MbO_2 in solution. Evidence for different C-O bonding in MbCO in crystal and solution has been found by infrared vibrational spectroscopy [5].

Finally the structures C_1 and C_2 in MbCO and HbCO-XANES spectra in Fig. 2 have been assigned [3,4] to the strong π and σ multiple scattering resonances of the CO molecule. The photoelectron is trapped within the CO molecule like in electron scattering by CO in gas phase electron spectroscopy experiments at positive kinetic energies. The intensity of the resonances is strongly dependent on the Fe-C-O bonding angle because the photoelectron is strongly reflected toward Fe in the Fe-C-O colinear configuration. The weaker intensity of the C_1 peak in MbCO indicate a more tilted CO orientation (a smaller Fe-C-O angle) than in HbCO.

In conclusion we have shown that we can identify the resonances due to CO and O_2 in the XANES of heme-proteins which because of the short interatomic distance give strong multiple scattering resonances at ~20 eV and ~35 eV, above Fe-metal K-edge.

References

1 B.Shaanan: Nature 296, 683 (1982) and J. Mol. Biol. 171, 31 (1983).
2 A.Bianconi, M.Dell'Ariccia, P.J.Durham and J.B.Pendry: Phys. Rev. B26, 6502 (1982).
3 P.J.Durham, A.Bianconi, A.Congiu-Castellano, S.S.Hasnain, L.Incoccia, S.Morante and J.B.Pendry: EMBO Journal 2, 1441 (1983).
4 A.Bianconi: in "EXAFS and Near Edge Structure" Springer Series in Chemical Physics, Vol.27 (Berlin, 1983) pag.118.
5 M.W.Makinen, R.A.Houtchens, W.S.Coughey: Proc. Nat. Acad. Sci. 776, 6042 (1979).

XANES: One Electron Multiple Scattering Resonances and Splitting of Final State Configurations at Threshold

A. Bianconi

Department of Physics, University of Rome "La Sapienza"
I-00185 Roma, Italy

The energy range of the X-ray absorption near edge structure (XANES) and the definition of the X-ray absorption threshold have been recently object of discussion. Here a definition of the absorption threshold is proposed and the energy range of XANES is discussed.The different physical processes present in the XANES range will be resumed and the role of multiple scattering resonances in the angular resolved Fe XANES of myoglobin single crystal will be shown.
The breakdown of one electron approximation is observed only at threshold of correlated electronic systems where a mixing of localized orbitals with delocalized orbitals of the ligands occurs in the valence band.

INTRODUCTION

XANES (X-ray absorption near edge structure) is actually a new tool for local structure studies. XANES probes the unoccupied states at a selected site and of selected angular momentum.Different physical excited states can be distinguished according to the description of the final state wavefunction of the excited photoelectron. Here, first, a definition of the energy range of XANES , second,a short classification of one-electron excitations is given.
 Finally the effects beyond the one electron approximation will be described i.e. the splitting of final state configurations of passive electrons.The suppression of the strong "shake up" satellites, observed in core level X-ray photoelectron spectra (XPS), has been observed in XANES of rare-earth insulating compounds.

1.THE ABSORPTION THRESHOLD

It is possible to define three different absorption thresholds in XANES.
a) The absorption threshold $E(1)$:at the energy of the lowest energy core transition.
b) The step absorption edge $E(2)$:at the energy of half maximum of the absorption jump.
c) The continuum threshold $E(0)$:at the energy threshold of excitations in the continuum. The $E(0)$ value in non metals cannot be obtained directly from XANES but is determined from XPS and optical data.
In metals the continuum threshold $E(0)$ and the absorption threshold $E(1)$ are at the same energy i.e. at the excitation energy to the Fermi level E_F .Fig.1 shows the L_3 and L_1-XANES of palladium/1/ and fig.2 shows the $M_{4,5}$ absorption threshold measured by total yield on the Frascati grasshopper beam-line/2/, where the contiuum threshold $E(0)$, obtained by XPS data, is indicated by arrows. Because of selection rules the absorption cross section can be enhanced or suppressed for excitations close to the Fermi

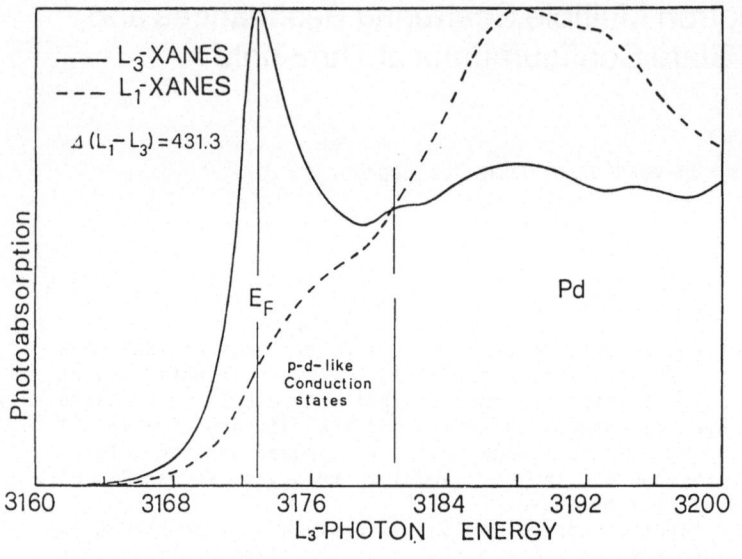

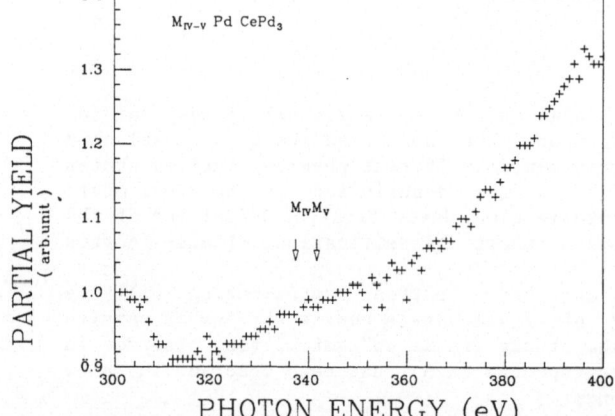

Fig.1
L_3 and L_1 XANES of Pd metal.
E indicates the position
of the Fermi level i.e. of
E(0)

Fig.2
$M_{4,5}$ absorption edge of Pd
in $CePd_3$ measured by total
yield at the Frascati grasshopper beam line. The arrows
indicate the position of E(0)
determined by photoemission
data during the same experimental run

level.In Pd at the Fermi level the l=1 (p components) density of states is
a step function and the E(0) threshold in the L_1XANES is at the first
inflection point;the l=2 (d components) density of states has a spike at
the Fermi level and the E(0) threshold in the L_3XANES is at the maximum of
the white line;the l=3 (f components)local density of states is so weak
near the Fermi level that there is a complete suppression of the
absorption cross section jump at the E(0) threshold.In this case the step
absorption edge E(2) is at about 40 eV above E(0) as shown in fig.2.In
conclusion E(2) can be defined in the spectra, but it has not a general
physical meaning.

In non metals there are bound states below the continuum threshold,
threfore E(0) and E(1) are not at the same energy.
In atoms and molecules E(0) is defined as the threshold for excitations in
the vacuum and is well determined from XPS data. E(1) is the energy of the
first bound state which in the atomic spectra ,reported in fig.3,are at
few eV below E(0).

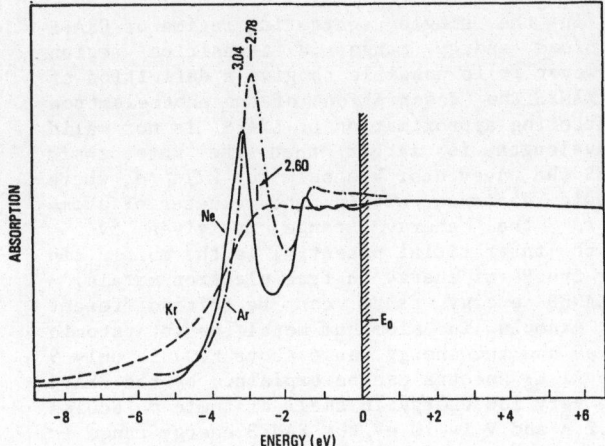

Fig.3
K-XANES of atoms .The
spectra are aligned at the
same continuum threshold E(0).
The E(1) threshold is at the
energy of the first member of
the Rydberg series.

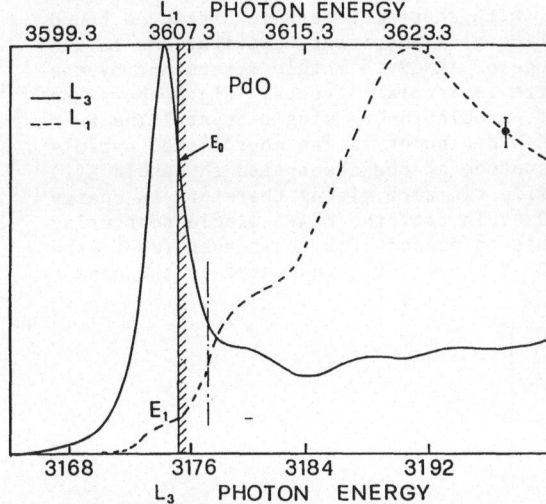

Fig.4
L_3 and L_1 -XANES of PdO. E(0)
has been determined from XPS
and optical data. /3/

In insulating crystals E(0) is given by the energy of excitations at the
bottom of the conduction band in a one-electron scheme. Fig.4 shows the L_1
and the L_3 XANES of PdO /3/. The continuuum threshold E(0) has been
determined by joint XPS and optical data and the two spectra have been
aligned at the same E(0). The first bound state at the absorption
threshold E(1) is at about 1 eV below E(0). Again because of dipole
selection rules the absorption cross section is enhanced at E(1), giving a
white line in the L_3 XANES and a very weak peak in the L_1 XANES which
are due to the same final state. This is a common spectral feature of K and
L_1 edges of transition metal compounds ,where, because of selection rules
the step absorption edge E(2) is some eV above the first excited state
and the first weak excited state , to mostly d-like states , has the
spectroscopical name of pre-peak.

2.THE HIGH ENERGY LIMIT OF XANES

The high energy part of XANES is determined by multiple scattering
resonances of the photoelectron in the continuum. Increasing the energy of

the photoelectron we enter in the single scattering regime of EXAFS oscillations. There is a broad energy range ,a transition region /4/,between the two regimes.However it is possible to give a definition of a minimum energy range of XANES.The description of the photoelectron wavefunction by the single scattering approximation of EXAFS is not valid where the photoelectron wavelength is larger than the interatomic distances.Threfore in the XANES the wavevector k should be k $\lesssim \pi$/d, where d is the shortest interatomic distance within the cluster of atoms determing the XANES.Therefore the energy range is given by E(eV)=151/$d(Å)^2$-V ,where V is the interstitial potential in the muffin tin description of the potential or the Fermi energy in free electron metals. Using this definition the XANES energy range can be quite different depending on the system.For example in aluminum metal the interatomic distance is 2.8 Å, V is 11.7 eV and the energy range above E(0) is only 9 eV.It is well known that the Al L_3 spectra can be explained by the EXAFS single scattering theory up to very low energy.In small diatomic molecules like hydrocarbons, where d=1.2 Å and V is 10 eV, the XANES energy range is 90 eV and EXAFS oscillations are not observed in many cases .In the case of biological molecules,and in many inorganic compounds, there are molecular groups of low Z elements, with short interatomic distances bound to the central absorbing atom, which give important contributions to the XANES.In this case the energy range of XANES is mainly determined by the short interatomic distances of the molecular groups. Fig.5 shows the angular resolved Fe K- XANES of myoglobin-CO single crystal.The Fe-N distances are 1.98 Å and the C-N distances in the coordinated pyrrols are 1.35 Å,the 3d atomic-like resonance at the absorption threshold E(1) is at about 13 eV above the muffin tin zero,giving therefore an energy range for myoglobin XANES of 70 eV.In fact the EXAFS single scattering theory using spherical waves is able to account for the structures in the experimental spectrum only from 60 eV above E(1) indicated as the peak P_1 in fig.5./5/

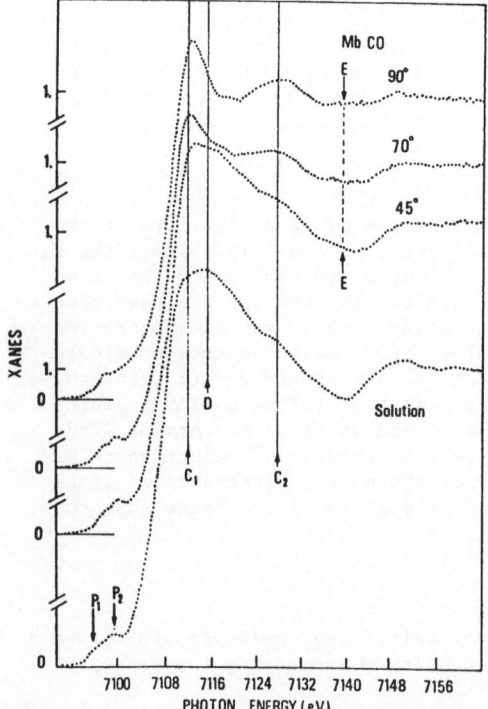

Fig.5
Angular resolved XANES of myoglobin-CO single crystal at different angles of the electric field E with the $\hat{c}$-axis of the crystal.At 90° $\hat{E}$ is at 28° off the heme normal.Decreasing the angle, the $\hat{E}$ direction moves toward the heme plane.The lowest cu- rve is the solution spectrum.

170

Bunker and Stern /6/ have recently discussed the contribution of the multiple scattering in the XANES energy range but they do not take account of the important second shell contributions which determine a larger range in many inorganic and biological compounds. We estimate in $KMnO_4$ a XANES energy range of about 45 eV above the absorption threshold E(1), at the pre-peak energy, where the strong absorption maxima are, in fact, before the onset of EXAFS oscillations.

3.ONE-ELECTRON EXCITATIONS IN THE XANES RANGE

We can distinguish several type of one-electron excitations in different systems.
     a)Rydberg states: In atomic spectra,as it is shown at the K-edges of rare gases in fig.3 a series of bound Rydberg states appear below the continuum threshold E(0) over a range of 2 - 3 eV. In molecules the Rydberg states are very weak and difficult to be detected.
     b)Bound valence states. In molecular K and L_1 spectra,bound states appear in the energy range of 10 eV below E(0) which are due to molecular orbitals unoccupied in the ground state which are deepen by the core hole Coulomb attraction.In insulators where the core hole is not completly screened by valence electrons , and where there are localized states at the site of the absorbing atom , like in transition metal compounds,bound molecular-like valence states appear below E(0).In fig.4 the bound states at both L_3 and L_1 edges of PdO can be described as the first unoccupied orbitals of the cluster formed by Pd and its first O coordination shell.In this case, because of the large contribution of Pd d-orbitals, the white line at the L edge has a large atomic character and it can be described as an atomic-like bound resonance.
     c)Local density of states near the Fermi level. In metals the core hole is well screened by the conduction electrons in the valence band and no bound states are present.The XANES in the first 10 eV above the Fermi level probes the local and partial density of electronic states of the system.
     d)Multiple scattering resonances. At kinetic energies of the photoelectron higher than 10 eV , the inelastic scattering with valence electrons at binding energies lower than 10 eV becomes very important and the mean free path of the photoelectron decreases to few Å.The elastic scattering is mainly due to bound electrons at higher binding energy,which generally do not participate to the chemical bond and are in an atomic-like potential.Therefore the XANES structures above 10 eV due to inteference effects on the photoelectron wavefunction determined by elastic scattering with neighbour atoms can be described by multiple scattering in a muffin tin potential within a small atomic cluster.There is experimental evidence that the multiple scattering resonances depend mainly on atomic positions/7/. XANES calculations using different approaches e.g.: multiple scattering formalism in the real space /8,9/,X_α multiple scattering /10,11/ ,and cross section calculations,starting from band structures of metals, in the k space /12/, have reached the same conclusion that the structures above about 10 eV are mainly determined by the geometry of the arrangement of neighbour atoms.

The multiple scattering formalism is in fact able to calculate the XANES of very complex systems without long range order.We have recently shown that also the XANES of a protein,hemoglobin,can be predicted by XANES calculation based on crystallographic data of atomic coordinates./13/ Recently also the angular resolved XANES of myoglobin-CO single crystal have been calculated for the photon polarization in the plane of the heme

171

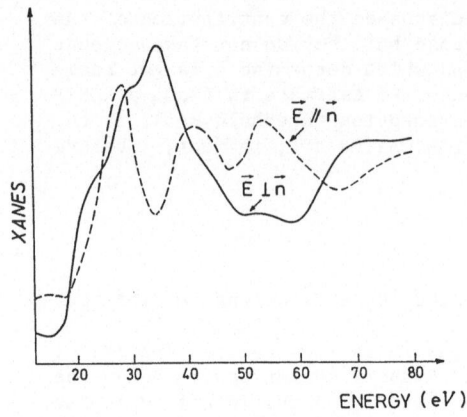

Fig.6 Fe K–XANES calculations of the myoglobin–CO polarized spectra. The parameters of theory were fixed from calculations of the $Fe(CN)_6$ spectra and the atomic positions from diffraction data. The energy scale is at the muffin tin zero. The Fe-3d resonance (peak P_1 in experiment) is at about 13 eV.

$(\hat{E} \perp \hat{n})$ and for the polarization normal to heme plane $(\hat{E} /\!/ \hat{n})$ toward the CO molecule bound to Fe, which has been assumed to be along the normal. The results of the calculations, shown in fig.6 show a large difference for the two polarizations. The experimental angular-resolved XANES of myoglobin–CO single crystal shown in fig.5 /5/ shows in fact a good agreement with the theory. The peaks C_1 and C_2 are determined by multiple scattering resonances due to the CO molecule which are enhanced when the electric field of the X ray beam is close to the normal. The features due to CO in the XANES spectrum of the solution can be assigned by comparison with the angular resolved spectra of the crystal and the variaton of the bonding angle of CO on the heme can be determined in different heme-proteins. /14/

4. SUPPRESSION OF SHAKE UP SATELLITES IN XANES AND SPLITTING OF FINAL STATE CONFIGURATIONS

In the one-electron approximation for core excitations the valence electrons are considered in be fixed in a **single** configuration. For transitions of the photoelectron to a completly unoccupied orbital the passive elecrons are assumed in be fully relaxed in the final state potential with the core hole in the absorbing atom. This description is in good agreement with a large amount of experimental data. However it is well known that in the photoionization of a core level in XPS spectra the sudden creation of a core hole induces excitations of the passive valence electrons giving shake up or shake down satellites. /15/ A peak on the high energy side of the multiple scattering resonance in $MnCl_2$ /16/ has been assigned to a shake up peak in analogy with XPS data.
We have investigated a series of insulating compounds of lanthanum which exhibit very strong shake up or shake down satellites/15/in the 3d XPS core spectra which are as strong as the main line. The shake down peaks are due to excitations of the valence electrons to the 4f orbitals which are completly empty in the ground state but because of the Coulomb interaction with the core hole are deeper in the final state. Fig.7 shows the L_3 XANES of La compounds which in the XPS spectra obtained from the same samples present shake down peaks at about 4 eV from the main line and of similar intensity. /17/The spectra are dominated by a bound state due to a 2p $\rightarrow$ 5d transition. No evidence of a satellite is found. /18/

This result is in agreement with a recent experiment on the XANES of metallic rare earth palladium alloys where a low energy satellite

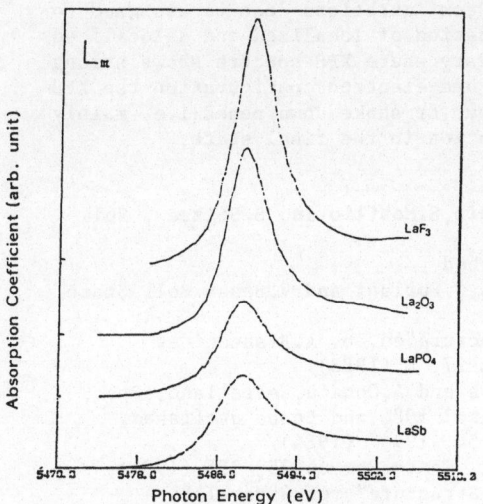

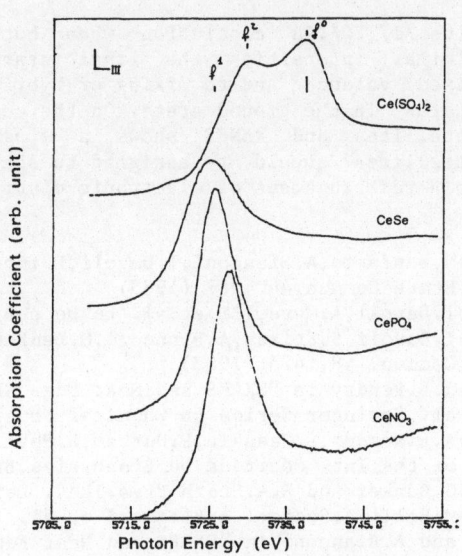

Fig.7 L XANES of lanthanum com-
pounds.No shake down or
shake up satellites are
observed.

Fig.8 L XANES of CeSO$_4$ upper curve,
and of other Ce compounds.
The splitting of the CeSO$_4$
white line in the three com-
components 4f^o,4f^2 and 4f^1
is observed.The other compounds show the single
4f^1 configuration.

,corresponding to the shake down peak in XPS core level spectra,has been
found to be present in LaPd$_3$ but its intensity is 12 % of the main
one-electron line while in XPS is 70% of the main line./19/Further studies
of the Pd L$_3$ XANES /17/allow us to assign the presence of the low energy
satellite in the XANES to hybridization of the Pd 5d orbitals with La 4f
orbitals in the occupied valence band.The strong satellites in XPS spectra
are due to rearrangement of valence electron to screen the core hole on
the absorbing atom.In XANES the core is screened by the excited
photoelectron in a bound state (in insulators)or in a well localized 5d
orbitals,threfore the Coulomb repulsion between the photoelectron and the
valence screening electron suppress the shake up or shake down satellite
due to transfer of charge on the absorbing atom.
Although shake up peaks are suppressed the splitting of the final state
configurations,not predicted by the one-electron approximation can be
observed in XANES.This occurs in correlated electronic systems where a
localized orbital is degenerate or hybridized with a delocalized orbital
derived by the same atom, like is the case for valence fluctuating systems
/20/ or there is mixing of localized (4f or 3d) orbitals of an atom with
delocalized orbitals derived by its ligands /21,17,18/. In these cases the
splitting of final state configurations in the XANES is due to the
different Coulomb interaction with the core hole for the valence
configurations with different occupation of the localized orbitals.In fig.8
the L$_3$ XANES of Ce insulating compounds are shown.Only in the case of CeSO$_4$
a 4f^o compound where a large mixing of Ce(4f) and O(2p) orbitals occurs in
the valence band the one electron white line is splitted in three
components corresponding to occupations of the 4f state in the final state
4f^1 ,4f^2 ,4f^o. In this case a similar splitting is found in the XPS core

173

line./17,20/.In conclusion when both XPS and XANES show satellites with similar intensities the final state configurations can be assigned to mixed valence and to mixing or hybridization of localized and delocalized states in the ground state. On the contrary where XPS spectra shows strong satellites and XANES shows a single one-electron configuration the XPS satellites should be assigned to shake up or shake down peaks i.e. mainly to a rearrangement of electronic distribution in the final state.

1) M.Benfatto,A.Bianconi,I.Davoli,L.Incoccia,S.Mobilio and S.Stizza , Sol. State Commun.46, 367 (1983)
2) I.Davoli,A.Marcelli et al. to be published
3) I.Davoli,S.Stizza,A.Bianconi,M.Benfatto,C.Furlani and V.Sessa Sol. State Commun. 48, 475 (1983)
4) J.B.Pendry in "EXAFS and Near Edge Structure"ed. by A.Bianconi et al.,Springer Series in Chemical Physics 27, 4 (1983)
5) A.Bianconi,S.Hasnain,P.Durham,S.Phyllips and A.Congiu Castellano, Proc. of the Int. Congress on Biophysics,Bristol 1984 and to be published.
6) G.Bunker and E.A.Stern Phys. Rev. Lett.52, 1990 (1984)
7) M.Belli,A.Scafati,A.Bianconi et al. Sol. State Commun.35, 355 (1980);
   and A.Bianconi in "EXAFS and Near Edge Structure" ref.4 pag.118;
   and Appl.of Surface Science 6,392 (1980) ;
   and in "EXAFS,XANES,SEXAFS and their applications"ed. R.Prins
   and D.Koningsberger,J.Wiley New York 1984
8) G.N.Greaves,P.J.Durham,G.Diakun,and P.Quinn Nature 294,139 (1981)
9) A.Bianconi,M.Dell'Ariccia,P.J.Durham and J.B.Pendry Phys. Rev. B26,6502 (1982)
10) C.R.Natoli,D.K.Misemer,S.Doniach and F.W.Kutzler Phys.ReV. A22, 1104 (1980)
11) F.W.Kutzler,D.F.Ellis,T.I.Morrison,G.K.Shenoy et al.Sol. State Commun. 46,803 (1983)
12) J.E.Muller,O.Jepsen and J.W.Wilkins Sol. State Commun. 42, 365 (1982)
13) P.J.Durham,A.Bianconi,A.Congiu-Castellano,A.Giovannelli,S.Hasnain, L.Incoccia, S.Morante and J.B. Pendry EMBO Journal 2, 1441 (1983)
14) A.Congiu Castellano,A.Bianconi, M.Dell'Ariccia, A.Giovannelli, E.Burattini and P.J.Durham on the proc. of the 3rd Int. Conf.on EXAFS Stanford 1984, Springer Verlag,Berlin 1984
15) G.Wendin "Breakdown of one)electron pictures in photoelectron spectra" in Structure and Bonding ,Springer Verlag ,,Berlin ,1981
16) E.A.Stern Phys. Rev. Lett. 49 ,1353 (1982)
17) A.Marcelli Thesis Univ. of Rome "La Sapienza" 1984,and A.Marcelli et al. Int. EXAFS Conf.,Stanford ,1984 ,Springer Verlag,to be published
18) A.Marcelli, A.Bianconi et al. J.of Magnetism and Magnetic Materials to be published
19) A.Bianconi,A.Marcelli, I.Davoli, S.Stizza and M.Campagna Sol. State Commun. 49, 409, (1984)
20) A.Bianconi,S.Modesti,M.Campagna,K.Fisher and S.Stizza J. of Physics C 14, 4737 (1981)
21) A.Bianconi Phys. Rev. B26, 2741 (1982)

Part IV Catalytic Systems and Small Metal Clusters

EXAFS Studies of Supported Bimetallic Cluster Catalysts

G.H. Via, G. Meitzner, and J.H. Sinfelt

Exxon Research and Engineering Company, Annandale, NJ 08801, USA

R.B. Greegor and F.W. Lytle

The Boeing Company, Seattle, WA 98124, USA

Bimetallic cluster catalysts represent an important class of materials in the field of catalysis from both a fundamental and technological point of view. Significant interest in these systems resulted from the demonstration that various combinations of metals, even pairs which were not miscible in the bulk, produced substantial alterations in the selectivity of competing chemical reactions. Thus, one of us (JHS) [1] showed that the addition of copper to ruthenium and osmium resulted in an increase in the cyclohexane dehydrogenation to benzene relative to hydrogenolysis to methane. This observation stimulated considerable activity in the field since it offered the hope that catalytic structures could be fabricated that produced selective reaction path-ways for complex systems. Technological implementations of bimetallic catalysts have occurred in the petroleum industry in at least two instances, both related to the reforming of hydrocarbons for motor fuel. Thus, supported catalysts based on Re-Pt [2] and Ir-Pt [3,4] are in commercial use in petroleum refineries in place of supported Pt catalysts because of improvements in activity, yield, product distribution, and catalyst activity maintenance.

Very recently, Klabunde and Imizie [5] published a study which may also have a similar stimulating effect on the field. They showed that manganese-cobalt particles prepared by a special solvent dispersion technique had a much higher activity for the hydrogenation of alkenes than similarly prepared cobalt particles, although manganese has little activity for this reaction. Thus, bimetallic systems offer significant opportunities for designing catalysts for specific applications, and understanding the structure of these catalysts and how the structure changes with environment is an important aspect of developing this field. We have been interested for some time in the structure of supported bimetallic clusters, and we have used EXAFS extensively to study these materials [6,7].

In our studies of bimetallic cluster catalysts, we have emphasized two types of systems. One consists of elements from Group VIII and Group IB of the periodic table, while the other consists of elements from within Group VIII alone. Ruthenium-copper, osmium-copper, and rhodium-copper clusters are examples of the first type, while platinum-iridium and rhodium-iridium are examples of the second. The clusters are commonly dispersed on a refractory oxide carrier such as silica or alumina at a concentration of 1-2 wt.% and have the feature that the surface atoms constitute a significant fraction of the total atoms present in the cluster. Because of the low concentration and high state of metal dispersion in these materials, conventional structural tools such as x-ray diffraction provide limited information on the nature of the metal-metal interaction. EXAFS, on the other hand, has been particularly useful in dealing with this question.

The analysis of EXAFS data for structural information ((N, σ, R)) generally proceeds by use of Fourier transform and non-linear least squares techniques [8]. In the case of highly dispersed metal catalysts, we have been most interested in information on metal atoms in the first coordination shell, and for bimetallic clusters this shell will contain, in general, two types of atoms. Thus, the experimental first-shell EXAFS function for a bimetallic cluster catalyst will contain contributions of the type A-A and A-B where A and B are the two different species and A is the absorbing atom. Since EXAFS is element specific, it is possible to independently obtain information on the pairs B-B and B-A by examining an absorption edge for the B species. By combining the information derived from these two independent experiments, one can devclop a complete picture of the average structure of the bimetallic system. It is possible to use K edge spectra for both experiments, L edge spectra for both experiments, or a combination of both K and L edge spectra.

The osmium-copper system is an example of a system in which both K (copper) and L (osmium) edge spectra are used to study the structure of the supported clusters [9]. These two metals show very limited bulk miscibility, but the catalysis data and hydrogen chemisorption in conjunction with electron microscopy data [10] indicate that they interact strongly. In the EXAFS investigation silica supported clusters of osmium and copper in atomic ratio 1:1 were examined at 100 K following in situ reduction in flowing hydrogen at 700 K. Reference data obtained on pure copper and osmium at 100 K were used in carrying out the analysis.

Table 1 - EXAFS results on Os-Cu/silica system

Bond Type	Bond Distance, Å	Percentage In 1st Shell
Cu-Cu	2.55 (2.556)*	49%
Cu-Os	2.68	51%
Os-Cu	2.68	17%
Os-Os	2.68 (2.705)*	83%

* Values for pure metals

The results from this analysis, shown in Table 1, indicate that the osmium atoms in the copper-osmium clusters are coordinated predominantly to other osmium atoms, while the copper atoms are coordinated about equally to both copper and osmium atoms. This result is very similar to our findings in the silica supported copper-ruthenium system [11,12]. Furthermore, we find that the total coordination number about osmium (osmium + copper) is approximately 12 while the total coordination number about copper (copper + osmium) is approximately 9. The values obtained for the various interatomic bond distances are also given in Table 2 and compared with the corresponding values for the pure metals. It should be noted that osmium metal has the hexagonal close packed structure, and the metal distance given in the table is the average of the interatomic distance (2.735Å) in a hexagonal layer and the distance of closest approach (2.675Å) between two atoms in adjacent hexagonal layers [13]. The value for the Os-Os pair in the catalyst is very close to this smaller value for pure osmium metal, and the value for the Cu-Cu pair in the catalyst is also very close to the corresponding value for pure copper metal. The value for the Os-Cu pair in the catalyst is very similar to the value for the Os-Os pair. These results, in conjunction with the results on total coordination number and average atomic composition about copper and osmium atoms in the catalyst, indicate that the supported osmium-copper clusters are composed of osmium rich cores with copper atoms segregated in the surface layer. This picture is very similar to the model proposed for supported ruthenium-copper clusters [12]. Additional evidence in support of this model comes from experiments involving EXAFS measurements of the catalysts after exposure to oxygen in helium [9]. These results show that the metal-metal bonding associated with copper is affected by the oxygen to a much greater extent than that associated with osmium, which is readily understood if the copper is concentrated at the surface of the osmium-copper clusters.

Table 2 - EXAFS results of Rh-Cu/silica system

Bond Type	Bond Distance, Å	1:1 (Rh:Cu) Percentage In 1st Shell	2:1 (Rh:Cu) Percentage In 1st Shell
Cu-Cu	2.62 (2.556)*	50%	44%
Cu-Rh	2.64	50%	56%
Rh-Cu	2.64	21%	8%
Rh-Rh	2.68 (2.690)*	79%	92%

* Values for pure metals

The osmium-copper and ruthenium-copper systems represent extreme cases with regard to the immiscibility of copper with the other two metals. In addition, both osmium and ruthenium have hcp structures while copper is fcc. In contrast to this, in the rhodium-copper system both metals are fcc, and copper shows partial miscibility in rhodium at catalytically interesting conditions [14]. We have investigated this system at two different atomic ratios on silica [15], and the results are summarized in Table 2.

The results for the rhodium-copper system are quite similar to those for the ruthenium-copper and osmium-copper systems indicating that copper concentrates in the surface of the rhodium-copper cluster. The data for the 1:1 atomic ratio rhodium-copper cluster, when compared with the data for ruthenium-copper or osmium-copper cluster, do, however, show a higher 1st shell copper concentration about the non-copper component. This presumably reflects the increased tendency for copper to dissolve in rhodium relative to ruthenium or osmium. In addition, we note that the Cu-Cu bond distance in the rhodium-copper cluster is considerably longer than in metallic copper (2.62 vs. 2.556Å) and close to the value of 2.64Å for the Rh-Cu distance. It is also much longer than the Cu-Cu bond in Ru-Cu clusters (2.58Å) or Os-Cu clusters (2.55Å), again indicating that copper is more compatible with rhodium than with ruthenium or osmium. Finally, when the atomic ratio of Rh:Cu is changed from 1:1 to 2:1, we observe changes in 1st shell composition which are at least directionally what one would expect. Thus, the concentration of copper about rhodium drops from 21% to 8% and the concentration of rhodium about copper increases from 50% to 56%. These small changes appear reasonable when one considers that these are highly surface segregated structures.

We have also investigated the structures of bimetallic clusters formed from iridium and rhodium on both silica and alumina [16]. This is an example of a Group VIII – Group VIII pair, as well as an example of a pair which shows complete bulk miscibility. Thus, iridium and rhodium are both fcc, they form a complete series of bulk solid solutions, and their nearest neighbor interatomic distances are very close (2.714 vs. 2.690Å) [13]. Our results for 1:1 atomic ratio clusters supported on both silica and alumina are given in Tables 3A and 3B.

Table 3-A - EXAFS results on Ir-Rh/silica

Bond Type	Bond Distance	Percentage In 1st Shell	Total N_1
Rh–Rh	2.72 (2.690)*	40%	
			10
Rh–Ir	2.72	60%	
Ir–Rh	2.72	27%	
			11
Ir–Ir	2.72 (2.714)*	73%	

Table 3-B - EXAFS results on Ir-Rh/alumina

Bond Type	Bond Distance	Percentage In 1st Shell	Total N_1
Rh–Rh	2.71 (2.690)*	60%	
			5
Rh–Ir	2.71	40%	
Ir–Rh	2.71	43%	
			7
Ir–Ir	2.75 (2.714)*	57%	

* Values for pure metals

In the case of silica, we see a 1st shell composition pattern which is rather similar to ruthenium-copper, osmium-copper, and rhodium-copper. In this case, however, we see that rhodium is segregated in the surface of the iridium-rhodium cluster, since iridium shows a high fraction of iridium near neighbors whereas rhodium shows roughly equal numbers of rhodium and iridium near neighbors. That is, rhodium is similar to copper in the Group VIII – Group IB examples. Wong, et. al. [17], also report rhodium surface enrichment in iridium-rhodium/silica clusters based on CO and H_2 chemisorption measurements. We also note that all of the interatomic bond distances are equal within the experimental error.

In the case of alumina-supported clusters, the interatomic bond distances are all equal except for the Ir-Ir distance which is some 0.04Å larger. This difference is probably within the experimental error of the measurement due to the low signal/noise for the iridium data in this set. The 1st shell composition pattern also suggests surface segregation of rhodium, but the tendency seems much less in this case. That is, iridium-rhodium/alumina clusters appear to be much more homogeneous compositionally than iridium-rhodium/silica clusters. The other major difference between the silica-supported clusters and the alumina-supported clusters is cluster size, as reflected in the 1st shell coordination number (N_1) given in the table. Thus, the silica-supported clusters have an average coordination number on the order of 10.5 whereas alumina-supported clusters have an average coordination number of around 6. This also suggests that differences in cluster topology may exist between the two systems, with the silica clusters being more "3-D" in character while the alumina clusters are more "raft-like" in character. Whether the change in compositional homogeneity is due to cluster size, cluster shape, or cluster-support interaction is not clear at this point.

In conclusion, we have investigated the structure of a number of bimetallic cluster catalysts with significant variation in the bulk miscibility characteristics of the bimetallic pair. The systems include osmium-copper, rhodium-copper, and iridium-rhodium which were discussed above, and ruthenium-copper [11,12] and iridium-platinum [18] which were not discussed here. In all cases, we have found a strong tendency for surface segregation of one of the components in the system. In general, the surface-segregated component is the component with the lowest surface energy. In the case of iridium-rhodium, the degree of surface segregation depends on either the cluster size, cluster shape, or cluster-support interaction, and this suggests that the detailed compositional distribution within a cluster may vary with cluster environment. This observation may have important catalytic consequences.

1. J.H. Sinfelt: J. Catal., 29, 308 (1973).
2. R.L. Jacobson, et al: Proc. Amer. Petrol. Inst. Div. Refining, 49, 504 (1969).
3. J.H. Sinfelt: U.S. Pat. 3,953,368 (1976).
4. J.H. Sinfelt and G.H. Via: J. Catal., 56, 1 (1979).
5. K.J. Kalbunde and Y. Imizu: J. Am. Chem. Soc., 106, 2721 (1984).
6. J.H. Sinfelt, G.H. Via and F.W. Lytle: Catal. Rev. - Sci. Eng., 26 (1), 81 (1984).
7. F.W. Lytle, G.H. Via and J.H. Sinfelt: "X-Ray Absorption Spectroscopy: Catalyst Applications," in Synchrotron Radiation Research (H. Winick and S. Doniach, eds.), Plenum, New York, 1980, pp. 401-424.
8. G.H. Via, J.H. Sinfelt and F.W. Lytle: J. Chem. Phys., 71, 690 (1979).
9. J.H. Sinfelt, G.H. Via, F.W. Lytle and R.B. Greegor: Ibid., 75, 5527 (1981).
10. E.B. Prestridge, G.H. Via and J.H. Sinfelt: J. Catal., 50, 115 (1977).
11. F.W. Lytle, G.H. Via and J.H. Sinfelt: Am. Chem. Soc., Div. Pet. Chem. Prepr., 21 (2), 366 (1976).
12. J.H. Sinfelt, G.H. Via and F.W. Lytle: J. Chem. Phys., 72, 4832 (1980).
13. International Tables for X-Ray Crystallography, Vol. III (T.H. MacGillavry, G.D. Riech, and K. Lonsdale, eds.), Kynoch Press, Birmingham, England, 1962, pp. 278, 282.
14. M. Hansen: Constitution of Binary Alloys, 2nd ed. (McGraw-Hill, New York, 1958).
15. G. Meitzner, G.H. Via, F.W. Lytle and J.H. Sinfelt: J. Chem. Phys., 78, 882 (1983).
16. G. Meitzner, G.H. Via, F.W. Lytle and J.H. Sinfelt: Ibid., 78, 2533 (1983).
17. T.C. Wong, L.F. Brown, G.L. Haller and C. Kemball: J. Chem. Soc., Faraday Trans., 1 71, 519 (1981).
18. J.H. Sinfelt, G.H. Via and F.W. Lytle: J. Chem. Phys., 76, 2779 (1982).

Co-Mo Hydrodesulfurization Catalysts Studied by EXAFS

B.S. Clausen, H. Topsøe, and R. Candia

Haldor Topsøe Research Laboratories, DK-2800 Lyngby, Denmark

B. Lengeler

Institut für Festkörperforschung, Kernforschungsanlage Jülich
D-5170 Jülich, Fed. Rep. of Germany

1 Introduction

Hydrotreating reactions, and in particular the hydrodesulfurization (HDS) reaction, are among the most important catalytic processes. This has, of course,
inspired massive research efforts in order to understand the structural and
chemical form of the elements in the catalysts used in such reactions (i.e.
sulfided Co-Mo/Al_2O_3 and Ni-Mo/Al_2O_3 catalysts). Great efforts have been
made to establish possible connections between such structural properties and
the various catalytic functions. Despite the numerous studies, greatly di-
verging views on these properties exist (see e.g. [1, 2]). It has recently
been shown that the promotion of the HDS activity of both supported and un-
supported Co-Mo catalysts can be related to the presence of the Co promoter
atoms in a so-called Co-Mo-S phase [3, 4]. The characterization of the Co-
Mo-S phase was carried out by use of Mössbauer emission spectroscopy [5]
and EXAFS at both the Co and Mo K edges [6-8]. These studies indicated that
Co-Mo-S can be considered as Co atoms located at the edge positions of an
MoS_2-like structure. Supporting evidence of this picture has recently been
given by infrared and electron microscopy studies [9].

The first EXAFS studies of Co-Mo/Al_2O_3 catalysts after typical sulfiding
[6] showed that the Mo atoms have six nearest neighbour sulfur atoms at a
bond distance equal to that of well-crystallized bulk MoS_2. Furthermore, it
was found that the nearest Mo-Mo interatomic distance is the same for the
catalysts and the model MoS_2 sample, although this second shell gives a
smaller contribution to the EXAFS compared to the MoS_2 sample (in MoS_2
molybdenum has 6 sulfur atoms in the first coordination shell and 6 molybdenum atoms in the second shell). Since the local surroundings of the Mo
atoms in the catalysts are quite similar to MoS_2, it was concluded that the
catalysts studied have the Mo atoms present in MoS_2-like structures [6].
EXAFS studies of other sulfided Co-Mo or Ni-Mo catalysts seem to be in
qualitative agreement with this conclusion [10-12].

The reduced second shell contribution to the EXAFS of the catalysts compared to MoS_2 was interpreted [6] as being due to the presence of very small
(~10 Å) domains or crystallites of MoS_2. A similar explanation has also
been advocated by other investigators [11, 12]. In the present paper we will
discuss in more detail why this interpretation is likely. In addition, we
will also present results which show that the different preparation parameters
may have a profound influence on the structure of the catalysts.

2 Experimental

The Mo/Al_2O_3 and Co-Mo/Al_2O_3 catalysts were prepared as described in Ref. [3]
using the incipient wetness impregnation method. After calcination in air at
500°C for 2 h, the oxidized catalysts were activated by sulfiding in a 2% H_2S

in H_2 gas mixture at $400^\circ C$ for 4 h. A reference material of well-crystallized MoS_2 was obtained from Riedel-de Haën AG.

The EXAFS experiments were performed at DESY in Hamburg using the synchrotron radiation from the DORIS storage ring and the EXAFS set-up at RÖMO at HASYLAB. The X-rays, which were emitted by electrons in the storage ring, were monochromatized by two Si(220) single crystals. The beam intensity was measured before and after passing through the sample by use of two ionization chambers containing one atmosphere of Ar. The sulfided catalysts were studied *in situ* by placing self-supporting wafers of pressed catalyst powder in specially designed cells equipped with X-ray transparent windows [7]. The EXAFS was extracted from the experimentally recorded X-ray absorption spectra following the method described in detail in Ref. [13].

3 Results

Figure 1 shows the Fourier transforms of the EXAFS$\cdot k^2$ for well-crystallized MoS_2, a sulfided Mo/Al_2O_3 catalyst containing 4% Mo by weight, and a sulfided Mo/Al_2O_3 catalyst containing 8.6 wt% Mo. The transforms for the two catalysts show essentially only the presence of two distinct peaks, one located at about 1.95 Å, and the other at about 2.85 Å. The locations of these peaks are very close to those of MoS_2 and also the heights of the first shell peaks are quite similar. The second shell peak is, however, reduced in height (most strongly for the low loading catalyst) compared to MoS_2. In order to obtain the interatomic distances and the coordination numbers of the first and second neighbour shells in the two catalysts, the phase and amplitude functions of the absorber-scatterer pair Mo-S (first shell) and Mo-Mo (second shell) of MoS_2 have been used to fit the Fourier filtered EXAFS. The structural parameters obtained from these fits are given in Table 1.

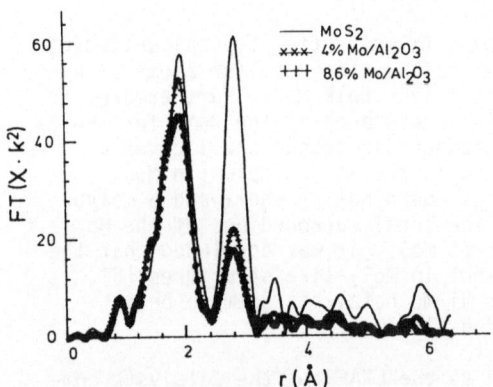

Fig. 1. Absolute magnitude of the Fourier transforms of EXAFS$\cdot k^2$ above the Mo K edge for well-crystallized MoS_2, sulfided Mo/Al_2O_3 (4 wt%) catalyst, and sulfided Mo/Al_2O_3 (8.6 wt%) catalyst.

The effect of changing the sulfiding temperature from $400^\circ C$ to $700^\circ C$ for a $Co-Mo/Al_2O_3$ (Co/Mo = 0.3) catalyst is shown in Fig. 2. Again, it is observed that the first shell of the high temperature sulfided catalyst is essentially identical to the MoS_2 sample, whereas the second shell peak is reduced somewhat in height. The results of the least squares fits of the two shells are given in Table 1.

Table 1. Bond lengths R, coordination numbers N, and Debye-Waller corrections $\delta\sigma^2$ obtained by fitting the Fourier filtered Mo EXAFS of
sulfided catalysts

	1st Shell			2nd Shell		
	R(Å)	N[1]	$\delta\sigma^2$	R(Å)	N[2]	$\delta\sigma^2$
Mo/Al$_2$O$_3$(4% Mo)	2.41	6.9	0.0034	3.13	2.2	0.0011
Mo/Al$_2$O$_3$(8.6% Mo)	2.41	7.1	0.0023	3.15	3.2	0.0017
Co-Mo/Al$_2$O$_3$(T_{sulf}=400°C)[3]	2.41	7.1	0.0023	3.15	3.0	0.0024
Co-Mo/Al$_2$O$_3$(T_{sulf}=700°C)[3]	2.41	7.8	0.0016	3.15	4.5	0.0015

1) Good quality fit is also obtained by constraining N to being equal to 6.
2) Fit is very poor when N is constrained to being equal to 6.
3) 8.6 wt% Mo and Co/Mo = 0.30 (atomic ratio).

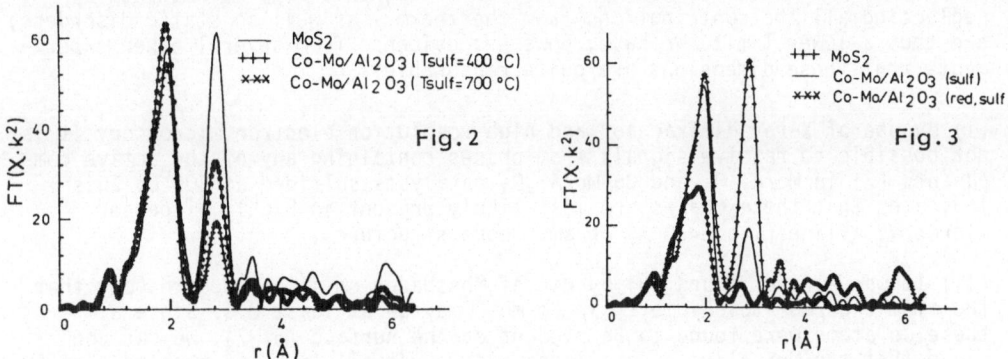

Fig. 2. Fourier transforms for well-crystallized MoS$_2$, Co-Mo/Al$_2$O$_3$ (Co/Mo =
0.3) catalyst sulfided at 400°C, and the same catalyst sulfided at
700°C.

Fig. 3. Fourier transforms for well-crystallized MoS$_2$, sulfided Co-Mo/Al$_2$O$_3$
(Co/Mo = 0.3) catalyst, and the Co-Mo/Al$_2$O$_3$ (Co/Mo = 0.3) catalyst
reduced at 400°C before being sulfided.

From Fig. 3. it is seen that a reduction treatment (at 400°C in H$_2$) prior
to the sulfiding has a strong influence on the catalyst structure. Not only
the second shell, but also the first shell of the prereduced catalyst are
significantly different from both MoS$_2$ and the non-reduced catalyst. Attempts
to fit the nearest neighbour shell by use of the phase and amplitude functions
of MoS$_2$ were not successful.

4 Discussion

The heights of the peaks in the Fourier transforms are related to the coordination number but may be influenced by thermal (Debye-Waller smearing),
as well as static disorder. Thus, in order to obtain accurate coordination
numbers, it is important to be able to extract these different contribu-

tions. Thermal and static disorders with a Gaussian distribution can be accounted for through the Debye-Waller term, $\exp(-2\sigma^2 k^2)$. However, the presence of non-Gaussian disorder and the occurrence of several different interatomic distances in a coordination shell being too similar to give resolved peaks can make the determination of reliable coordination numbers a most difficult task.

The fact that in all the catalysts studied we observe a reduction in the peak height of the second shell compared to well-crystalline MoS_2 may be a result of some of the above effects. However, an estimated lower coordination number in the second shell may also arise if the MoS_2-like structure present in the catalysts consists of very small crystallites with a large fraction of atoms located at the surface. The latter situation is interesting since in this case EXAFS may provide information on the dimension of the layered MoS_2 crystallites parallel to the basal planes (the Mo atoms in the second shell are located in the same basal plane as the absorbing Mo atoms). In other words, the 'edge dispersion' can be estimated. This parameter is of course very important in view of the structure of Co-Mo-S which is supposed to consist of small MoS_2 crystals with Co at the edges.

The dimensions of the MoS_2-like crystallites, which may be obtained by neglecting all the contributions from the thermal as well as static disorders, are thus a lower limit. We have, however, evidence from several other experiments that these dimensions are quite reasonable:

(i) By use of X-ray diffraction and high resolution electron microscopy it is not possible to resolve signals from phases containing any of the active components [2] in Mo/Al_2O_3 and $Co-Mo/Al_2O_3$ catalysts sulfided at $400^{\circ}C$. This indicates that these phases are most likely present in highly dispersed microcrystalline (i.e. <30 Å) or amorphous structures.

(ii) It was earlier found [3] by use of Mössbauer emission spectroscopy that the atomic Co/Mo stoichiometry in Co-Mo-S may be at least 0.5. Since all these Co atoms were found to be present on the surface [5, 7], we can conclude that the MoS_2-like crystallites are very small (i.e. of the order of 10 to 20 Å). Evidence of an edge stoichiometry of Co/Mo $\sim$0.8 has recently been given by use of analytical electron microscopy [9].

(iii) Finally, oxygen chemisorption measurements [14, 15] on similar sulfided Mo/Al_2O_3 and $Co-Mo/Al_2O_3$ catalysts suggest that the Mo phase is present in a high degree of dispersion.

Although it may not be possible from EXAFS alone to determine whether disorder or small crystallites are responsible for the reduced height of the second shell peak, these data, when combined with results obtained by a number of other methods, give strong indications of the presence of very small crystallites of MoS_2. From the data in Table 1, it can be estimated that an Mo/Al_2O_3 or $Co-Mo/Al_2O_3$ catalyst sulfided at $400^{\circ}C$ has about 90% of the Mo atoms located at the edges. This corresponds to a crystallite (or domain) size of $\sim$10 Å. If the Mo loading is decreased to about half, an apparent decrease in the crystallite size is observed (Fig. 1 and Table 1), giving a higher edge dispersion. Upon increasing the sulfiding temperature to $700^{\circ}C$, the second shell coordination number seems to increase (Fig. 2 and Table 1), which suggests larger crystallites and, therefore, a lower edge dispersion. The effect on the crystallite size of increasing the sulfiding temperature is in agreement with electron microscopy results which clearly show layered structures due to MoS_2 for the catalysts sulfided above $700^{\circ}C$ [16].

The fact that a very strong dependence of preparation parameters on the structure has been observed for Co-Mo HDS catalysts has made direct comparison difficult between the different studies reported in the literature. The influence reported above of the metal loading and the sulfiding temperature on the crystallite size are just two examples. Also the specific activation procedure is important for the resulting structures in the sulfided catalysts. For example, by reducing the Co-Mo/Al_2O_3(Co/Mo = 0.3) catalyst in hydrogen before sulfiding, the structure of the catalyst is strongly modified as evidenced from the EXAFS results reported in Fig. 3. The quite low amplitude of the first shell peak indicates that the Mo atoms are coordinated to less than six sulfur atoms. Furthermore, the second shell peak is found to be quite different from that of both MoS_2 and the catalyst sulfided in the usual manner. Not only the height but also the position of the peak is changed. Thus, it is clear that by reducing the catalyst before sulfiding the local surroundings around the Mo atoms are different from those in MoS_2.

5 Conclusion

In the present paper it has been discussed how a comparison of EXAFS data with results obtained by other methods can be used to estimate the crystallite dimension and thus the edge dispersion of the MoS_2-like phase in various Co-Mo HDS catalysts. A lower limit of the crystallite dimensions is estimated from the average coordination number of the second shell around the absorbing Mo atoms. Specifically, it is found that an increase in the Mo loading and the sulfiding temperature both lead to an increase in the crystallite size and thus a decrease in the edge dispersion.

Acknowledgments

We are grateful to HASYLAB for offering beam time on the synchrotron radiation facility of DESY and for providing access to the EXAFS spectrometer at ROMO.

References

1 F.E. Massoth: in 'Advances in Catalysis and Related Subjects', Vol. 27, p. 265, Academic Press, New York/London, 1978
2 H. Topsøe: in 'Surface Properties and Catalysis by Non-metals: Oxides, Sulfides and Other Transition Metal Compounds', p. 326, Reidel, Dordrecht (1983)
3 C. Wivel, R. Candia, B.S. Clausen, S. Mørup, and H. Topsøe: J. Catal. 68, 453 (1981)
4 R. Candia, B.S. Clausen, and H. Topsøe: J. Catal. 77, 564 (1982)
5 H. Topsøe, B.S. Clausen, R. Candia, C. Wivel, and S. Mørup: J. Catal. 68, 433 (1981)
6 B.S. Clausen, H. Topsøe, R. Candia, J. Villadsen, B. Lengeler, J. Als-Nielsen, and F. Christensen: J. Phys. Chem. 85, 3868 (1981)
7 B.S. Clausen, B. Lengeler, R. Candia, J. Als-Nielsen, and H. Topsøe: Bull. Soc. Chim. Belg. 90, 1249 (1981)
8 B.S. Clausen, H. Topsøe, R. Candia, and B. Lengeler: in 'Catalytic Materials: Relationship between Structure and Reactivity', ACS Symposium Series, San Francisco, California, p. 71, 1983
9 H. Topsøe, N.-Y. Topsøe, O. Sørensen, R. Candia, B.S. Clausen, S. Kallesøe, and E. Pedersen: in 'Symposium on Role of Solid State Chemistry in Catalysis', ACS Meeting, Washington, D.C., p. 1252, 1983

10 H. Harnsberger: Priv. Communications
11 T.G. Parham and R.P. Merrill: J. Catal. 85, 295 (1984)
12 M. Boudart, J.S. Arrieta, and R. Dalla Betta: J. Am. Chem. Soc. 105,
        6501 (1983)
13 B. Lengeler and P. Eisenberger: Phys. Rev. B21, 4507 (1980)
14 J. Bachelier, J.C. Duchet, and D. Cornet: Bull. Soc. Chim. Belg. 90, 1301
        (1981)
15 W. Zmierzcak, G. MuraliDhar, and F.E. Massoth: J. Catal. 77, 432 (1982)
16 R. Candia, O. Sørensen, J. Villadsen, N.-Y. Topsøe, B.S. Clausen, and
        H. Topsøe: To be published

In Situ EXAFS of CoMo/γ-Al$_2$O$_3$ Catalysts During Hydrodesulfurization of Benzothiophene

M. Boudart, R. Dalla Betta, K. Foger, and D.G. Löffler

Department of Chemical Engineering, Stanford University,
Stanford, CA 94306, USA

1 Introduction

The treatment of oil and coal liquefaction products with hydrogen at high pressures and temperatures over CoMo/Y-Al$_2$O$_3$ catalysts to remove organic sulfur in the form of gaseous H$_2$S is a well-established technology. The recent application of EXAFS to the study of the catalyst has had considerable impact in the understanding of the structure of the active phase. Most authors agree today that the catalyst exhibits a MoS$_2$-like structure with Co atoms probably located on the edges [1,2,3]. However, all published studies have been carried out with the catalyst sample in a non-reacting atmosphere under ambient conditions of pressure and temperature. The assumption that the structure of the active phase does not change during cooling down and depressurization after reaction is implicit in the analysis of the results.

The in situ study of a catalyst while a chemical reaction is occurring is a long-sought goal in heterogeneous catalysis. In the present contribution we study the structure of a CoMo/Y-Al$_2$O$_3$ catalyst during the hydrodesulfurization of benzothiophene under pressures and temperatures similar to those found in industrial practice.

2 Experimental

The catalyst used had been prepared by Boudart et al. [2] and contained 2.3 weight % Co and 7.55% Mo supported on Y-Al$_2$O$_3$. The powder was compacted to form a 2.5-cm-diameter wafer. The reaction was carried out in a high pressure cell shown in Fig.1. Key features are the convection baffles and cooling jackets which allow operation of the catalyst at 523 K and 7.3 MPa while keeping the 508-µm-thick beryllium windows at room temperature. The liquid reactant, a 7.7 mole % solution of benzothiophene in decalin flowed downwards over the catalyst wafer. The gas phase was a mixture of 2% H$_2$S in H$_2$ flowing at a rate of 5 ℓ/h STP. After an initial sulfidation of the catalyst the liquid flow was started and kept at a rate of 6 mℓ/h for approximately 4 h while EXAFS spectra were taken continuously. The flow of liquid rectant was then stopped and the cell was depressurized and cooled down to room temperature. A wafer made of a mixture of MoS$_2$ and Y-Al$_2$O$_3$ was used as model compound.

3 Results and Discussion

All EXAFS spectra were taken at SSRL in transmission mode around the Mo K-edge (20 kev). The data analysis procedure used has been described in

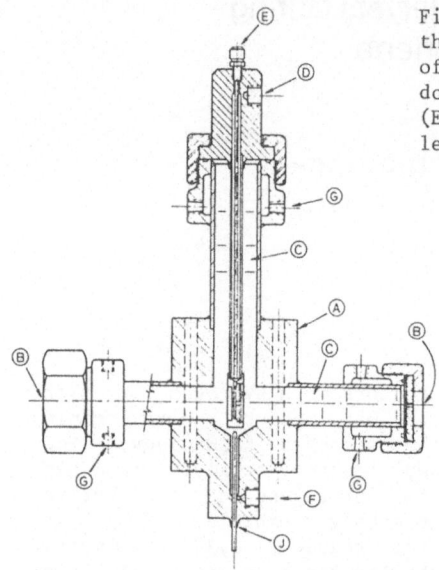

Figure 1. Cross section of the side view of the high pressure cell. (A) main body made of 321 stainless steel; (B) beryllium windows; (C) convection baffles; (D) gas inlet; (E) liquid inlet; (F) gas and liquid outlet; (G) cooling jackets; (J) thermocouple

detail in the literature [4]. Scattering contributions to the absorption coefficient from atoms other than Mo were substracted by extrapolating the pre-edge region into the post-edge region. Then the threshold energy E_0 was specified as the inflection point of the absorption edge and the data were transformed into electron momentum (k-space). To obtain the pure EXAFS oscillations, the smooth background was removed by fitting a polynomial spline in six regions. The resulting interference functions were then multiplied by a k^3 weighting factor and Fourier transformed in the range 40-130 nm^{-1}, to provide a qualitative picture of the structural data in the form of the radial distribution functions (RDFs).

Amplitude and phase-shift functions for the catalysts and model compound were calculated. A non-linear least squares fitting routine was used to adjust the functions for the catalysts to those of the model compound by varying the interatomic distance R, the coordination number N and the Debye-Waller coefficient σ. The results obtained are shown in Table 1. No changes in the structure parameters during and after reaction are evident.

Table 1. Structure Parameters for CoMo/γ-Al$_2$O$_3$ catalysts.

P	T	N_{Mo-S}	N_{Mo-Mo}	R_{Mo-S}	R_{Mo-Mo}	σ_{Mo-S}	σ_{Mo-Mo}
7.3 MPa	523 K	6.3	4.4	240 pm	314 pm	19 pm	6.8 pm
0.1 MPa	RT	6.0	3.8	240 pm	314 pm	11 pm	3.5 pm

The RDFs for the catalyst and model compounds at room and reaction temperatures shown in Fig. 2 present essentially the same features, suggesting that the Mo atoms in the catalyst are in a MoS$_2$-like phase. The peaks in the model compound have been identified following Partham and Merrill [3] as caused by first and second Mo-S and Mo-Mo shells, and third Mo-Mo shell. All peaks decrease with temperature due to Debye-Waller smearing.

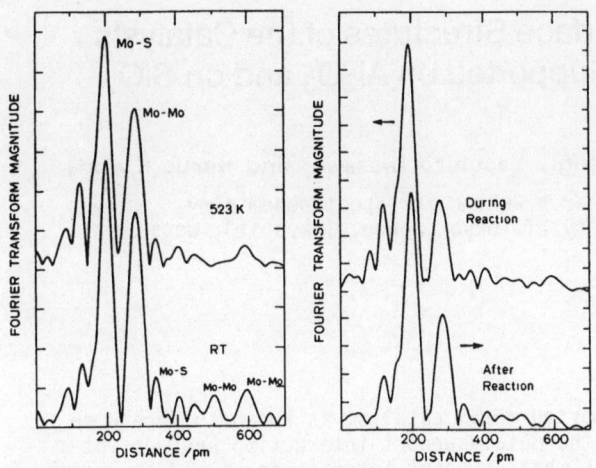

Figure 2. Fourier transforms of the Mo K-edge of a) MoS$_2$ at room temperature (RT) and at reaction temperature (523 K); b) CoMo/Y-Al$_2$O$_3$ catalyst during reaction at 523 K and 7.3 MPa under ambient conditions of pressure and temperature after reaction.

The RDFs for the catalyst during and after reaction present the same features, indicating that the structure of the Mo environment seems not to be affected by depressurization and cooling. The first Mo-Mo peak on the catalyst is much smaller relative to the Mo-S peak than the model compound, as consequence of the lower coordination number in the second shell.

We may conclude that the CoMo/Y-Al$_2$O$_3$ catalyst, when examined during or immediately after reaction, presents a MoS$_2$-like active phase with a degree of crystallinity enough to yield third Mo-Mo shell scattering. No changes in the Mo environment are observed as as consequence of cooling down and depressurization.

4 References.

1. B. S. Clausen, H. Topsoe, R. Candia, J. Villadsen, B. Lengeler, J. Als-Nielsen and F. Cristenesen: J. Phys. Chem. 85, 3868 (1981).
2. M. Boudart, J. Sanchez-Arrieta, and R. Dalla Betta: J. Am. Chem. Soc. 105, 6501 (1983).
3. T.G. Partham, and R.P. Merrill: J. Catal. 85, 295 (1984).
4. P. A. Lee, P.H. Citrin, P. Eisenberger, and B.M. Kincaid: Rev. Mod. Phys. Part 1 53, 769 (1981).

EXAFS Studies on the Surface Structures of the Catalysts Derived from $Ru_3(CO)_{12}$ Supported on Al_2O_3 and on SiO_2

Kiyotaka Asakura, Nobuhiro Kosugi, Yasuhiro Iwasawa, and Haruo Kuroda

Department of Chemistry and Research Center for Spectrochemistry
Faculty of Science, The University of Tokyo, Hongo, Tokyo 113, Japan

1. Introduction

The catalytic activity of a supported metal catalyst is known to depend on the nature of the support. But the metal-support interaction has not yet been well understood, because the metal-support interaction is hidden under much pronounced metal-metal interaction. In the cases of ordinary supported metal catalysts, it is hard to derive information on the metal-support interaction from the analysis of EXAFS data. In a carbonyl-derived catalyst, metal clusters are highly and uniformly dispersed over the support (less than 10 Å in diameter); therefore, this kind of catalyst may be suitable for the study of the metal-support interaction [1].

Ru catalysts which are prepared by use of $Ru_3(CO)_{12}$ as the precursor show different catalytic behaviors depending on the supports. The $Ru/\gamma-Al_2O_3$ prepared from $Ru_3(CO)_{12}$ produces C_2-C_4 hydrocarbon from $CO + H_2$, whereas Ru/SiO_2 prepared from the same precursor is active for the formation of methane [2]. We investigated the catalysts $Ru/\gamma-Al_2O_3$ and Ru/SiO_2 prepared from $Ru_3(CO)_{12}$ by means of EXAFS, and found the difference of the metal-support interaction between the two catalysts.

2. Experimental

The Ru catalysts were prepared by immersing $\gamma-Al_2O_3$ and SiO_2 in the pentane solution of $Ru_3(CO)_{12}$ under high-purity Ar flow. The Ru content was 2 wt.% for $\gamma-Al_2O_3$ and 1.75 wt.% for SiO_2. The carbonyls supported on Al_2O_3 and on SiO_2 were thermally decomposed by heating at 473 K and 400 K, respectively, and then the reduction with H_2 was performed at 673 K and 200 Torr for an hour. The samples were transferred into EXAFS cells without exposing to air. EXAFS measurements were carried out on the EXAFS Beam Line 10B at Photon Factory in National Laboratory for High Energy Physics (KEK-PF).

3. Results and Discussion

The Fourier transforms of $k^3 \cdot \chi(k)$ of Ru-K EXAFS data are given in Fig. 1. We carried out curve-fitting analysis by using phase shifts and amplitude functions derived empirically from model compounds [1]. Summarizing all the results from the analysis of EXAFS data, we propose structural models for the catalysts in Fig. 2. After the thermal decomposition of $Ru/\gamma-Al_2O_3$, the Ru-Ru bond is cleaved and the monomeric Ru carbonyl species is anchored by the surface oxygen of Al_2O_3 as shown in Scheme 1. Two peaks observed in Fig. 1b correspond to Ru-O (surface oxygen) and Ru-C bonds and Ru-O (carbonyl oxygen) bonds, respectively. After the H_2 reduction, the Ru-Ru peak is found with the distance of about 2.3 Å together with the Ru-O (surface oxygen) peak

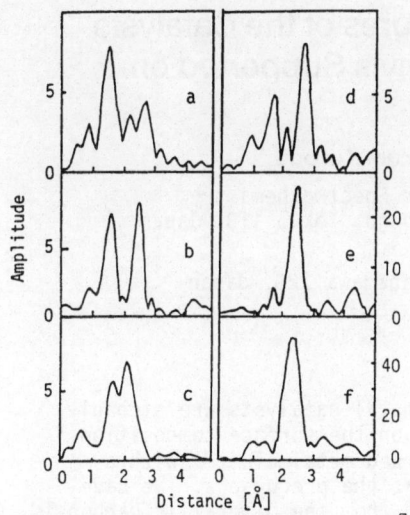

Fig.1. Fourier transforms of $k^3 \cdot \chi(k)$ of Ru K-edge EXAFS data
(a) $Ru_3(CO)_{12}$ crystalline powder
(b) thermal decomposition on γ-$A\ell_2O_3$
(c) H_2 reduction on γ-$A\ell_2O_3$
(d) thermal decomposition on SiO_2
(e) H_2 reduction on SiO_2
(f) Ru metal

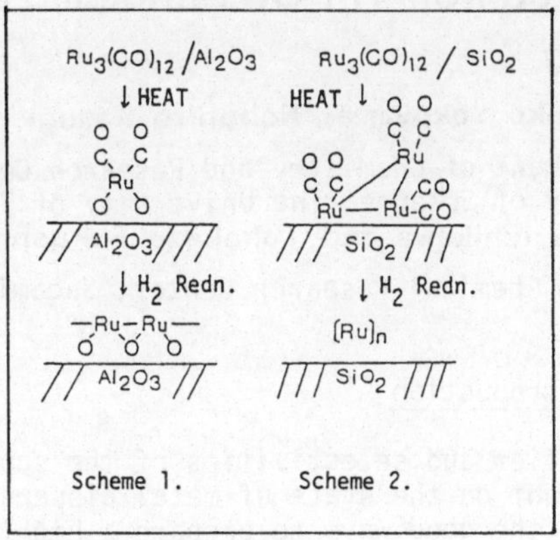

Fig.2. A structural model for the catalysts derived from $Ru_3(CO)_{12}$

in the Fourier transform as shown in Fig. 1c, but a Ru-O (carbonyl) peak is not found. This is consistent with the result of the volumetric observation that the total amount of CO evolved during the two treatments was 12 per Ru_3. The bond lengths of Ru-Ru and Ru-O (surface) are determined to be 2.62 ± 0.03 Å and 2.17 ± 0.05 Å, respectively. This strong metal-support interaction through the surface oxygen keeps the Ru cluster well dispersed and cationic as shown in Scheme 1.

On the other hand, the interaction between Ru and SiO_2 is weak. After the thermal decomposition, partially decarbonylated species are formed as shown in Scheme 2, where the Ru-Ru bond length is 2.76 ± 0.03 Å and no contribution from the surface oxygen is found. After the H_2 reduction, rather large Ru clusters with the coordination number 5.0 ± 0.5 are formed.

In conclusion, the metal-support interaction is much stronger in Ru/γ-$A\ell_2O_3$ catalyst than in Ru/SiO_2 catalyst. Even in the reducing atmosphere, Ru particles are highly dispersed on γ-$A\ell_2O_3$. This causes the multiple adsorption of CO and reduces the hydrogenation activity, and increases the selectivity for the formation of C_2-C_4 hydrocarbon [2].

References

1. K. Asakura, N. Kosugi, Y. Iwasawa, and H. Kuroda, submitted to J. Chem. Soc. Chem. Commun.
2. K. Asakura et al., to be published

EXAFS Studies on the Surface Structures of the Catalysts Derived from Rh-Co Bimetallic Carbonyls Supported on Al$_2$O$_3$

Toshihiko Yokoyama, Nobuhiro Kosugi, and Haruo Kuroda

Department of Chemistry and Research Center for Spectrochemistry
Faculty of Science, The University of Tokyo, Hongo, Tokyo 113, Japan
Masaru Ichikawa and Takakazu Fukushima

Sagami Chemical Research Center, Sagamihara, Kanagawa 229, Japan

1. Introduction

Activities and selectivities of the supported metal catalysts are strongly dependent on the state of metal dispersion and on the surface composition. One of the good ways to prepare a highly dispersed metal cluster with a uniform composition is to use metal carbonyls as the precursors. We have studied the structures of the catalysts derived from the bimetallic carbonyls, Rh$_x$Co$_{4-x}$(CO)$_{12}$ (x = 0,1,2,4), by the analysis of the Rh-K and Co-K EXAFS spectra.

2. Experimental

The catalysts were prepared by the method described elsewhere [1]; the start-ing carbonyls were synthesized after MARTINENGO et al. [2], and identified by IR spectroscopy. The total content of the supported metals was about 4 wt.% in all cases. The x-ray absorption measurements were carried out at room temperature on samples sealed with pure He and/or N$_2$ in the specially designed sample cells, by use of an EXAFS spectrometer of Beam Line 10B of Photon Factory in National Laboratory for High Energy Physics (KEK-PF). EXAFS spectra were also measured on the catalysts which were prepared by impregna-tion of the support with an aqueous solution containing RhCℓ_3 and CoCℓ_2 (1:1 or 1:3 mole ratio).

3. Results and Discussion

Examples of the Fourier transforms of $k^3 \cdot \chi(k)$, which were calculated by use of the Rh-K EXAFS data over the region of k 4-14 Å^{-1}, are shown in Figs. 1 and 2. After the reverse-Fourier transformation for the region of the peaks corresponding to Rh-Rh and Rh-Co bonds, we carried out the curve-fitting analysis by using the empirical parameters derived from the analysis of the EXAFS data of Rh$_4$(CO)$_{12}$ and RhCo$_3$(CO)$_{12}$ [3]. The results of the curve-fit-ting analysis are given in Table 1. When the dehydrated γ-Al$_2$O$_3$ was just im-pregnated with Rh$_2$Co$_2$(CO)$_{12}$ the coordination numbers (N) and bond lengths (R) are almost the same as those obtained for the powder of the original carbonyl. This seems to indicate that the Rh-Co cluster in the original carbonyl molecule is retained in the state supported on γ-Al$_2$O$_3$. When the carbonyl-impregnated specimen is first exposed to air and subsequently reduced with hydrogen at 400°C, the coordination numbers are nearly the same as those for the original carbonyl, indicating that Rh-Co bimetallic cluster skeleton of the carbonyl is retained even in the treated catalyst. Also in the case where RhCo$_3$(CO)$_{12}$ was used as a precursor, the resultant catalyst retains RhCo$_3$ cluster skeleton. On the other hand, when the catalysts are prepared by impregnating Al$_2$O$_3$ with a solution of Rh and Co salts, relatively large metal particles with a Rh-rich bulk composition are formed after reduction with H$_2$.

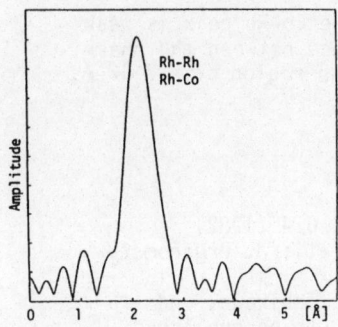

Fig.1. Fourier transform of k^3.
$\overline{X(k)}$ (Rh-K) for $Rh_2Co_2(CO)_{12}$/
$A\ell_2O_3$ reduced with H_2 at 400°C

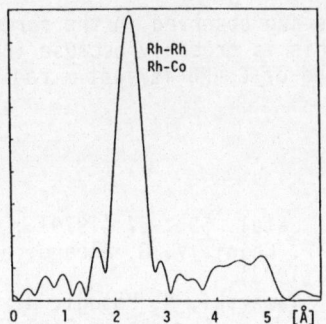

Fig.2. Fourier transform of k^3.
$\overline{X(k)}$ (Rh-K) for Rh-Co (1:1)/
$A\ell_2O_3$ reduced with H_2 at 400°C

Table 1. Results of the curve-fitting analysis for the Rh-K EXAFS data

Sample	Condition	Rh-Rh		Rh-Co	
		N	R [Å]	N	R [Å]
$Rh_4(CO)_{12}$	crystalline powder	(3)	2.73	—	—
$RhCo_3(CO)_{12}$	crystalline powder	—	—	(3)	2.62
$Rh_2Co_2(CO)_{12}$	crystalline powder	1.3	2.77	1.9	2.61
$Rh_2Co_2(CO)_{12}/A\ell_2O_3$	untreated	0.8	2.81	2.0	2.60
$Rh_2Co_2(CO)_{12}/A\ell_2O_3$	heated at 200°C	—	—	2.2	2.64
$Rh_2Co_2(CO)_{12}/A\ell_2O_3$	redn. with H_2	0.7	2.69	2.0	2.55
$RhCo_3(CO)_{12}/A\ell_2O_3$	untreated	—	—	1.0	2.60
$RhCo_3(CO)_{12}/A\ell_2O_3$	heated at 200°C	—	—	1.5	2.56
$RhCo_3(CO)_{12}/A\ell_2O_3$	redn. with H_2	—	—	2.7	2.60
Rh-Co (1:1)/$A\ell_2O_3$	redn. with H_2	8.7	2.70	3.3	2.57
Rh-Co (1:3)/$A\ell_2O_3$	redn. with H_2	7.6	2.70	3.5	2.58

In the Fourier transform for Co-K EXAFS data, peaks corresponding to Co-O
bonds are observed both for the cluster-derived catalysts and for the salt-
derived ones, as demonstrated in Figs. 3 and 4. This indicates, in agreement
with the results from the Rh-K EXAFS data, that in the former Co atoms are
attached to the surface oxygen of the support and in the latter are concentrated at the surface of the metal particles. However, no peaks correspond-

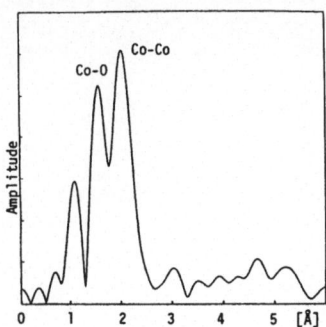

Fig.3. Fourier transform of k^3.
$\overline{X(k)}$ (Co-K) for $RhCo_3(CO)_{12}$/
$A\ell_2O_3$ reduced with H_2 at 400°C

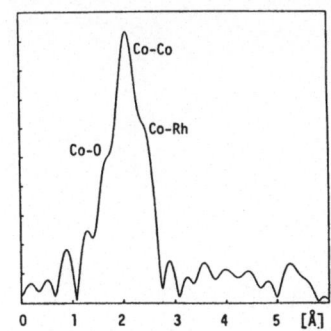

Fig.4. Fourier transform of k^3.
$\overline{X(k)}$ (Co-K) for Rh-Co (1:1)/
$A\ell_2O_3$ reduced with H_2 at 400°C

193

ing to Co-Rh bonds are observed in the former, and the Co-Rh peak is weak in the latter. This is probably because the difference between the phase shifts of Co-Co and of Co-Rh is just π radian over the region of k 7-12 Å.

References

1. M. Ichikawa: J. Catal. 59 , 67 (1979) ; Chemtech , 674 (1982)
2. S. Martinengo, P. Chini, V. G. Albano, and F. Cariati: J. Organomet. Chem. 59 , 379 (1973)
3. T. Yokoyama, K. Yamazaki, N. Kosugi, H. Kuroda, M. Ichikawa, and T. Fukushima: J. Chem. Soc. Chem. Commun. (1984) , to be published

EXAFS Investigations of Industrial Reforming Catalysts: Pt-Re or Rh on Gamma Alumina at the Platinum L_3 Edge

D. Bazin, H. Dexpert, and P. Lagarde

LURE, Bat. 209C, UPS, F-91405 Orsay, France

J.P. Bournonville

IFP, BP 311, F-92406 Rueil-Malmaison, France

Reforming is a large area of interest in heterogeneous catalysis. The basic material is made of small platinum particles (one nm or less in diameter) supported onto a light oxide (typically alumina of high specific area: 200 m^2g^{-1}). All the different chemical routes taken during the preparation procedures go through three steps from the precursor impregnation to the active metal final stage: drying, calcination and reduction. Using a furnace similar to that related by LYTLE (1) we have already described these stages in an in situ study of a pure platinum on gamma alumina catalyst (2). Bimetallic cases, which are close to industrial materials, have been considered in the same way and this work summarises the quantitative results we got at the platinum L_3 edge for the dried and calcined steps. Two systems, where the added metal is either rhenium or rhodium, are investigated.

1 EXPERIMENTAL
These systems have been studied for different metal ratios, the total weight remaining lower than 2%. The particles size is checked by electron microscopy and chemisorption measurements. The impregnation salt is first the hexachloroplatinic acid to which either ammonium perrhenate or rhodium trichloride is added. The EXAFS data are taken through a Si(220) or (400) channel cut or Si(311) double crystal monochromator using the DCI ring operating at 1.72 or 1.85 Gev. Data coming from low metal concentrations have been collected and Fig. 1 is an example of the signal to noise ratio

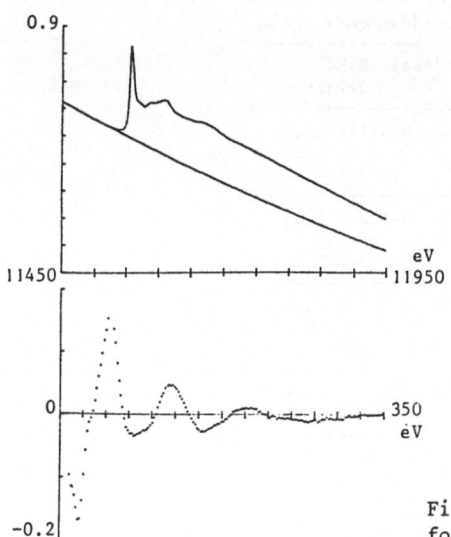

Figure 1 : Platinum L_3 edge data example for a Pt-Rh/Al$_2$O$_3$ catalyst (0.47% wt Pt)

obtained for a 0.5% weight of platinum. Experimental phases and amplitudes
values are extracted from standards (H_2PtCl_6, PtO_2) and transferred to the
samples which are considered within a 40-330 eV bandwith in each case (the
Re L_2 edge rises only 400 eV beyond the Pt L_3 absorption). The calculated
and experimental inverse Fourier are k^3 weighted and fitted using a two
shells least square procedure. They are taken valuable when the difference
is within 10^{-3} and 10^{-4}, all parameters (R,N,σ,E) being simultaneously
calculated except the photoelectron mean free path which is fixed to the
value taken for the standard. Figure 2 shows the agreement we are looking
for in any case.

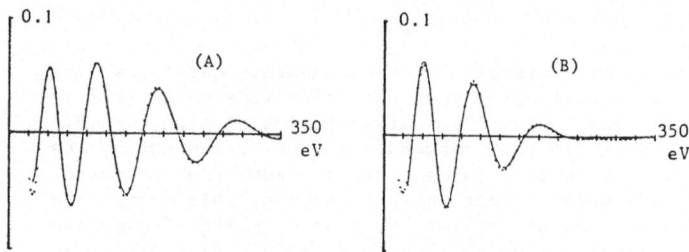

Figure 2 : Fit examples of two bimetallic catalysts at the dried stage :
(A) : Pt-Rh/Al$_2$O$_3$ (Rh/Pt=0.6;0.47%Pt) (B) : Pt-Re/Al$_2$O$_3$ (Re/Pt=0.5;1.35%Pt)
Dotted line=experiment,full line=model

2 RESULTS and DISCUSSION
The results are listed in Table 1. The comments are as follow :
  1- Only one shell is extracted : no metal-metal bonds are detected on
the radial distribution fonction (Fig.3);
  2- Pt-Re : as we reported qualitatively in a recent work (3), the
platinum environment of the bimetallic Pt-Re at the drying step (383 K) is
already that taken by the monometallic catalyst at the calcined stage (803
K). Oxygen atoms surround the heavy atoms very soon, at low temperature,
and therefore the calcination does not bring significant changes, the O/Cl

Table I: Fit results
N=Atoms number, R=Distance (Å), σ =Debye-Waller variation (au), E=Energy
variation (eV), r=Experimental-calculated difference (x10^{-5})

| | Dried Stage (383K) | | Calcined Stage (803K) | |
	Chlorine	Oxygen	Chlorine	Oxygen
2.0%Pt	N=6.1 R=2.31 σ =0.08 E=1.5	N=0.7 R=2.02 σ =0.05 r=4	N=2.1 R=2.31 σ =0.10 E=1.6	N=5.1 R=2.02 σ =- 0.06 r=14
Re/Pt=2.0 0.65%Pt	N=2.1 R=2.35 σ =0.00 E=5	N=6.4 R=2.03 σ =0.00 r=19	N=2.1 R=2.35 σ =0.11 E=6.5	N=6.2 R=2.04 σ =0.00 r=70
Re/Pt=0.5 1.35%Pt	N=1.6 R=2.36 σ =0.02 E=5.7	N=6.1 R=2.04 σ =0.00 r=50	N=1.2 R=2.36 σ =-0.06 E=7.4	N=6.3 R=2.03 σ =0.09 r=100
Rh/Pt=1.8 0.47%Pt	N=4.3 R=2.35 σ =0.09 E=4.7	N=4.3 R=2.04 σ =0.01 r=14	N=2.7 R=2.27 σ =0.00 E=5	N=4.3 R=2.10 σ =0.00 r=89
Rh/Pt=0.6 0.47%Pt	N=6.4 R=2.29 σ =0.00 E=3.2	N=2.7 R=2.22 σ =0.10 r=40	N=2.1 R=2.31 σ =0.10 E=0.8	N=5.3 R=2.02 σ =0.09 r=97

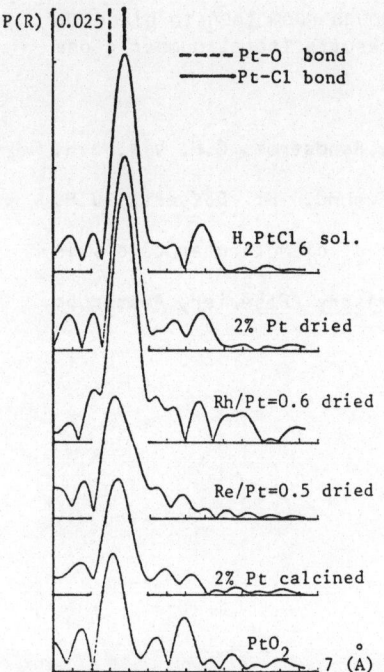

P(R) |0.025|

- - - - Pt-O bond
———Pt-Cl bond

H_2PtCl_6 sol.

2% Pt dried

Rh/Pt=0.6 dried

Re/Pt=0.5 dried

2% Pt calcined

PtO_2
.7 (A)

Figure 3 : Fourier transforms of
some mono and bimetallic catalysts

ratios remaining close to 3. A small tendency to decrease the chlorine number (2 to 1.5) seems related with an increase of the platinum content. This could be specific to the rhenium precursor which enriches in oxygen the platinum vicinity. Another hypothesis is a link with the oxidised species of the carrier surface : these bimetallic catalysts are known to delay efficiently the sintering, and therefore to interact more with the alumina.

3- Pt-Rh : this ambiguity is lifted off with the results got in the Pt-Rh phases. As the added constituant is now totally chlorinated ($RhCl_3$), the oxygen atoms still detected at low temperature are assigned to be from the support. The O/Cl ratio depends there strongly on the Rh/Pt one, so that when there is less rhodium (Rh/Pt = 0.6) the platinum neighbourhood is very close to the monometallic case. Increasing the rhodium percentage and the temperature brings these bimetallic catalysts nearer to the situation obtained in the pure Pt case, the final oxygen number remaining smaller (one must remember that we have there highly chlorinated phases).

Some odd Pt-Cl or Pt-O distances are found and one explanation could be that our fit procedure involves only a two shells interaction applied on a too narrow Hanning window.

4- The total number of atoms is always greater than six. As the Pt-O and Pt-Cl distances (around 2.02 and 2.31 Å) correspond exactly to the ionic radii sum, it seems that there is not enough space to place these supplementary atoms in the direction of the triangular faces of the Pt(Cl,O) octahedron. But as nuclear quadripolar resonance measurements show(4), the bond is partially ionic in these complexes where Pt is therefore farless from having a 4+ charge.

Within the Pt-O and Pt-Cl fixed distances, the result is a different equilibrium between the respective sizes of platinum and its ligands. A value of 1.05 Å for the platinum radius allows enough room then to place up to 14 atoms in its vicinity. In this view, our results (total number close to 8) are quite reasonable.

1 F.W. Lytle, R.B. Gregor, E.C. Marques, D.R. Sandstrom, G.H. Via, J.H. Sinfelt : J. Chem. Phys. 70, 4849(1979)
2 P. Lagarde, T. Murata, G. Vlaic, E. Freund, H. Dexpert, J.P. Bournonville : J. of Catalysis 84, 333(1983)
3 H. Dexpert, P. Lagarde, J.P. Bournonville : J. of Molecular Catalysis 25, 347(1984)
4 R.B. Heslop, P.L. Robinson : Inorganic Chemistry (Elsevier, Amsterdam 1969)

EXAFS Studies of Intermetallic Compounds Used as Heterogeneous Catalysts

V. Paul-Boncour, A. Percheron-Guegan, and J.C. Achard
Chimie Métallurgique des Terres Rares, C.N.R.S., F-92190 Meudon, France
J. Barrault and A. Guilleminot
Laboratoire de Catalyse Organique, Faculté des Sciences
F-86022 Poitiers, France
H. Dexpert and P. Lagarde
LURE, Bat. 209C, UPS. F-91405 Orsay, France

1 INTRODUCTION

Recently a number of studies have demonstrated that intermetallic compounds may be used as the precursors of new metallic catalysts (1,2). Intermetallics of the $LaNi_5$ series, extensively studied for hydrogen storage, exhibit an appreciable activity in catalytic reactions, particularly in carbon monoxyde or hydrocarbon hydrogenation. The nickel substitution by another transition element implies a significant variation of catalytic properties (3,4).

Characterization of the solids by CO chemisorption, X-ray powder diffraction and microprobe analysis has evidenced a progressive transformation during the catalytic reaction, since part of the intermetallic disappeared and was replaced by nickel metal and rare earth oxide or hydride depending on the type of reaction.

In this work we observe the extraction of nickel particles from $LaNi_5$, $LaNi_4Fe$ and $LaNi_3Mn_2$ using X-ray absorption analysis in the following reactions :

1)–Carbon oxyde hydrogenation $CO + H_2 \longrightarrow CH_4$, CO_2, $C_nH_{2n,2n+2}$
2)–Ethane hydrogenolysis $C_2H_6 + H_2 \longrightarrow 2\ CH_4$
3)–Propene hydrogenation $C_3H_6 + H_2 \longrightarrow C_3H_8$

2 EXPERIMENTAL

The powdered samples were mixed with boron nitride under purified argon atmosphere to avoid oxydation.

The data were collected at room temperature using the synchroton radiation provided by DCI at LURE. Ring energy was 1,72 Gev, beam current typically started at 250mA. Si (220) or (311) monochromators have been used. EXAFS spectra were collected at Ni-K edge. The other elements data (transition metals and rare earth) have been registered only within the edge region.

3 RESULTS and DISCUSSION

A 12.5 μm nickel foil has been used as model compound in order to determine experimental phases and amplitudes for Ni-Ni pairs and eventually for Ni-Mn and Ni-Fe pairs in case of substituted compounds. For all these analyses the mean free path was kept constant and equal to the nickel one.

Before catalytic reaction, the intermetallics radial function P(R), Fourier transform of the $k^3\chi(k)$ data, shows only one single peak (Fig. 1). EXAFS analysis of this main peak leads for $LaNi_5$ to a mean value of 6.3 Ni at 2.485 Å which is very close to the cristallographic value of 7.2 Ni at 2.48 Å. After catalytic reactions, the $LaNi_5$ (Fig.2), $LaNi_4Fe$ (Fig.3) and $LaNi_3Mn_2$ (Fig.4) Fourier transforms are compared to the starting intermetallic (a) and metallic nickel (b). Exept $LaNi_5$ after (CO,H_2) reaction (Fig.2b), a main peak increase is clearly seen. This increase is

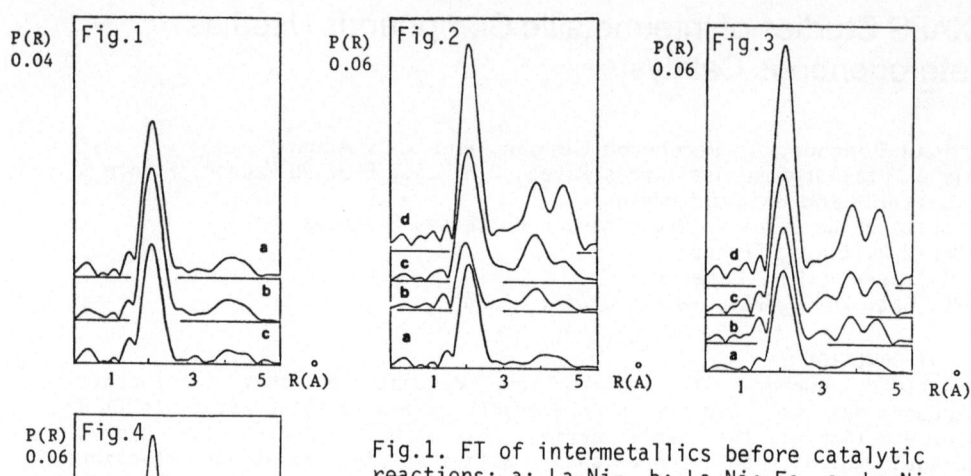

Fig.1. FT of intermetallics before catalytic reactions: a: La Ni₅, b: La Ni₄ Fe, c: La Ni₃ Mn₂

Fig.1. FT of intermetallics before catalytic reactions: a: La Ni_5, b: La Ni_4 Fe, c: La Ni_3 Mn_2

Fig.2. FT of La Ni_5: a: before reaction, b: after CO,H_2, c: after C_2H_6, d: Ni reference

Fig.3. FT of La Ni_4 Fe: a: before reaction, b: after C_2H_6, c: after CO,H_2, d: Ni reference

Fig.4. FT of La Ni_3 Mn_2: a: before reaction, b: after C_3H_6, c: after CO,H_2, d: Ni reference

Table I

EXAFS results for the intermetallic compounds before and after catalytic reactions

Compound	Reaction	1st Shell			3rd Shell			4th Shell		
		N	$R[\overset{\circ}{A}]$	$\sigma[\overset{\circ}{A}^{-2}]$	N	$R[\overset{\circ}{A}]$	$\sigma[\overset{\circ}{A}^{-2}]$	N	$R[\overset{\circ}{A}]$	$\sigma[\overset{\circ}{A}^{-2}]$
$LaNi_5$	before	6.3	2.48	0.04						
$LaNi_5$	CO, H_2	5.9	2.49	0.05	6.	4.33	0.025	1.5	5.	0.015
$LaNi_5$	C_2H_6	9.	2.49	0.04	13.	4.31	0.	3.7	4.99	0.015
$LaNi_4Fe$	before	6.4	2.48	0.04						
$LaNi_4Fe$	C_2H_6	7.1	2.49	0.04	7.	4.33	0.	4.	4.95	0.015
$LaNi_4Fe$	CO, H_2	10.7	2.50	0.05	16.	4.34	0.03	5.2	5.	0.015
$LaNi_3Mn_2$	before	6.8	2.51	0.06						
$LaNi_3Mn_2$	C_3H_6	9.	2.54	0.06	8.5	4.44	0.03	3.2	5.09	0.03
$LaNi_3Mn_2$	CO, H_2	10.3	2.48	0.04	14.7	4.32	0.03	6.5	4.95	0.015
Ni metal		12	2.49	0.	24.	4.31	0.	12.	4.97	0.

accompanied by the growth of two new peaks corresponding to the third (4.31 Å) and fourth (4.97 Å) Ni metal distances.

First shell analysis is not enough to separate metallic nickel from intermetallic contribution in the transformed compounds. Therefore we have also analysed the third and fourth Ni shells. Each peak was filtered in the $k^3X(k)$ Fourier transform, backtransformed in k space and analysed with a fitting procedure. All the EXAFS results are listed on table 1.

LaNi$_5$ appears to have a different behaviour from substituted compounds. After CO,H_2 conversion we observe for LaNi$_5$ a decrease of the first shell coordination number compared to the starting metallic, instead of an increase as for the other compounds. This result, confirmed by the small coordination numbers of the third and fourth shell, may be due to small Ni particles contribution.

Another surprising result is the greater distance values for LaNi$_3$Mn$_2$ after C_3H_6 reaction. The Ni-Mn distance being close to 2.59 Å instead of 2.49 Å for Ni pairs, this means that an important percentage of Ni-Mn bonds exists in the compound.

From all these analyses the nickel extraction rate is on the way to be more clearly seen. Accurate fits involving the third and fourth shells are undertaken from data collected at liquid helium temperature. Therefore the chemical behaviour of decomposing intermetallics are hoped to be more understood.

1 H. Imamura, W.E. Wallace : J. Phys. Chem. 84, 3145 (1980)
2 T.E. Fisher, S.R. Kelemen, R.S. Polizotti : J. Catal. 69-2, 345 (1981)
3 J. Barrault, D. Duprez, A. Percheron-Guegan, J.C. Achard : J. Less Common Metals, 89, 237 (1983)
4 J. Barrault, D. Duprez, A. Guilleminot, A. Percheron-Guegan, J.C. Achard : Applied Catalysis, 5, 99 (1983)

Mn Valence and Mn-O Bond Length in La$_{0.7}$A$_{0.3}$ MnO$_3$ Perovskite Catalysts (A = Ca, Sr, Ba, and Pb)

Wang Qi-wu

Chemistry Department, State University of New York, Albany NY, USA

Joe Wong

General Electric Corporate Research and Development, P.O. Box 8
Schenectady, NY 12301, USA

1. Introduction

Catalysts commonly used for oxidation of CO and hydrocarbons are those containing Pt, Pd and other noble metals. They are used for purification of
exhaust gases from automobiles and chimneys of electric power plants, for
oxidation of hydrocarbons as well as combustion promoters in the regenera-
tion of cracking catalysts. The noble metals are very expensive, and rare-
earth (RE) elements are very abundant in China. Therefore, the investigation of RE catalysts as substitutes for the noble metals is obviously of
technological and scientific significance.

It was found that the catalytic activity of RE perovskites of the type
La$_x$A$_{1-x}$MnO$_3$, where A is a divalent cation such as Ca, Sr, Ba or Pb, can be
related to crystal parameters [1], electrical conductivity and magnetic
properties [2]. In this study, the bonding and local atomic structure of
the catalytic active Mn sites in a series of La$_{.7}$A$_{.3}$MnO$_3$ perovskite materials are investigated using a combined XANES and EXAFS spectroscopy. MnO,
Mn$_2$O$_3$, Mn$_3$O$_4$ and MnO$_2$ of known chemical structures are used to model the Mn
sites in these perovskite catalysts.

2. Experimental

The perovskite catalysts were prepared by first precipitating the mixed
carbonates with a NH$_4$OH/NH$_4$HCO$_3$ solution from the corresponding mixed metal
nitrate solutions of predetermined stoichiometry. The carbonate precipitates formed in this was were filtered, washed, dried and decomposed at
350°C. The products obtained were ground and mixed with a little graphite,
pelletized and calcinated at 900 – 1100°C. The calcined catalysts were
divided into appropriate portions for catalytic activity and x-ray absorption measurements.

Mn K-edge absorption spectra were recorded with the C2 spectrometer at
CHESS (Cornell High Energy Synchrotron Source) with CESR (Cornell Electron
Storage Ring) operating at an electron energy of ~ 5.2 GeV and injection
current of ~ 15 mA. The x-ray beams from CESR was monochromatized with a
channel-cut Si(220) crystal and detuned to 50% to minimized harmonic contents at the Mn K-edge. All spectra were recorded at room temperature in
the transmission mode. Spectral specimens were prepared by mixing fine
powders (-400 mesh) of the materials with Duco^R cement and casting the mull
into films between two microscope slides.

3. Results and Discussion

In Figs. 1 and 2 the Mn K-edge spectra in a series of manganese oxides and
the RE perovskites are shown. The results for the oxides are in good

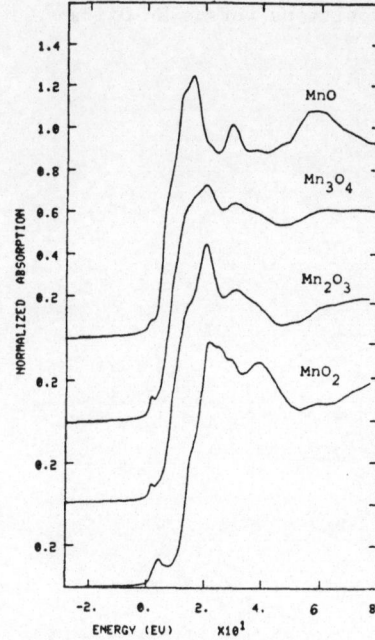

Fig. 1   Mn K-edge XANES
Spectra of Manganese Oxides

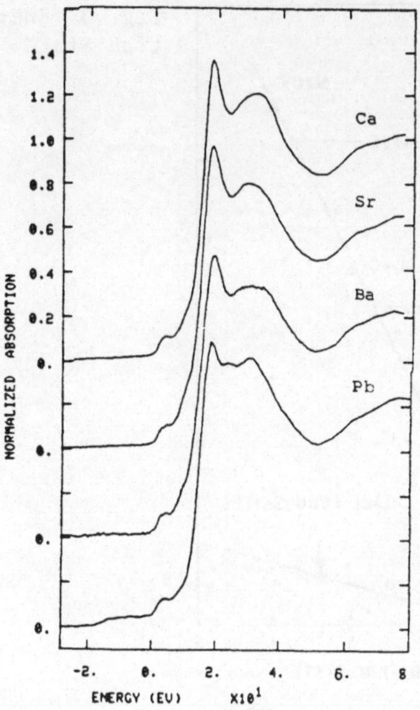

Fig. 2   Mn K-edge XANES
Spectra of RE Perovskites

agreement with those of Belli et al [3]. The zero of energy in these spectra is taken at the first inflection point in the derivative spectrum of
manganese metal at 6539.0 eV. The threshold of K-photoionization as given
by the first inflection point in the derivative spectra and position of the
K-edge are plotted versus formal oxidation state of Mn in the oxides. The
corresponding threshold and K-edge energy positions of Mn in the perovskite
catalysts, when plotted in Fig. 3, are found to be located between those of
Mn_2O_3 and MnO_2, indicating a mixed Mn^{3+}/Mn^{4+} valence state in the
perovskites. Compared with $LaMnO_3$ in which Mn is trivalent, it is clear
that the mixed Mn^{3+}/Mn^{4+} valence in the $La_xA_{1-x}MnO_3$ catalysts is induced by
a partial substitution of divalent cations for the trivalent La host ions
in order to maintain an overall charge neutrality. The oxidation of CO and
hydrocarbons on mixed oxide catalysts is strongly related to the labile
state of the valence of manganese in the catalytic reaction [1].

Table 1   Mn-O Bond Distance and Comparison with d-Spacing of the Strongest
Diffraction Line

Catalyst	Mn-O Bond Distance (Å) (±0.02Å)	d-Spacing (Å)
$La_{.7}Ca_{.3}MnO_3$	1.92	2.76
$La_{.7}Sr_{.3}MnO_3$	1.94	2.77
$La_{.7}Ba_{.3}MnO_3$	1.96	2.77
$La_{.7}Pb_{.3}MnO_3$	1.97	2.78

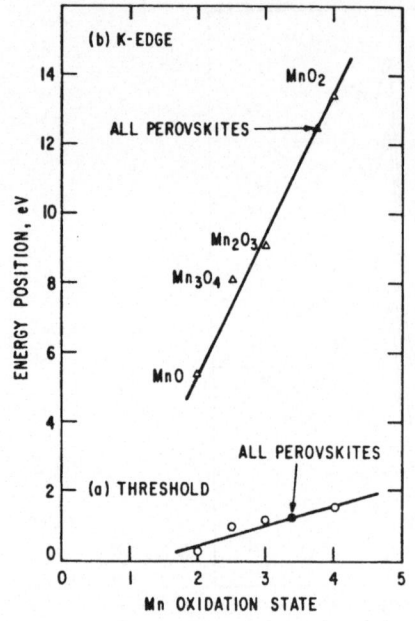

Fig. 3 Energy Positions versus Mn Oxidation State

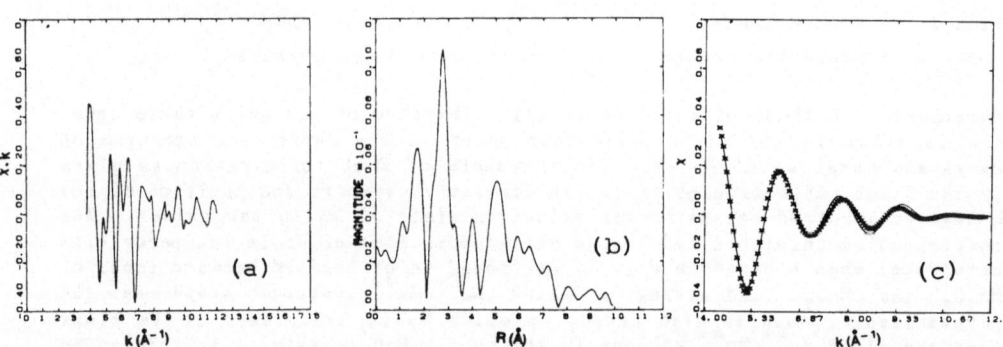

Fig. 4 (a) Normalized EXAFS, (b) Fourier Transform of (a), and (c) Experimental(line) and simulated(points) first-shell inverse for MnO

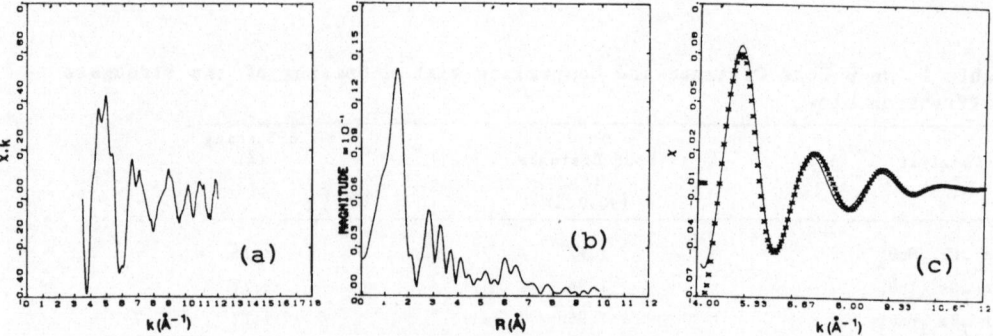

Fig. 5 (a) Normalized EXAFS, (b) Fourier Transform of (a), and (c) Experimental(line) and simulated(points) first-shell inverse for La$_{.7}$Sr$_{.3}$MnO$_3$

The Mn-O bond distance in these perovskites was determined from the inverse EXAFS signal from the first shell in radial structure function in the region 0-2.3Å. MnO having a known NaCl structure [4] was used as model to determine a self-consistent empirical Mn-O phase shifts by fitting. This Mn-O phase shift from MnO was then transferred to the perovskites to determine the Mn-O distances. Figure 4 and 5 show EXAFS data typified by MnO and La$_{.7}$Sr$_{.3}$MnO$_3$. In Table 1 the Mn-O bond distance in these RE perovskites are given. The increase in the MnO separation with divalent cation substitution correlates with an increase of the d-spacing of the strongest line in the diffraction pattern [1].

Catalytically, the Sr-containing perovskite is the most active in this series [1]. Indeed, investigation of Sr-bearing the Pb-bearing perovskites catalysts have been most widely reported in the catalyst literature.

4. Acknowledgment

We are grateful for experimental opportunities at CHESS which is supported by NSF.

5. References

1. Wang Oi-wu, Rong Jing-fang, Lin Pei-yan and Shan Shao-chun, Kexue Tong-
        bao, 25, 907 (1980).
2. E. G. Vieland, J. Cat. 32, 415 (1974).
3. M. Belli, A. Scafati, A. Bianconi, S. Mobilio, L. Palladino and A.
        Reale and E. Burattini, Solid State Comm. 5, 355 (1980).
4. G. Wyckoff, "Crystal Structures" Vol. 1, 2nd Edition. John Wiley (1963)
        p. 88.

Catalyst Preparation Procedure Probed by EXAFS Spectroscopy: Co/TiO$_2$

Yasuo Udagawa and Kazuyuki Tohji

Institute for Molecular Science, Okazaki, Aich 444, Japan

Akifumi Ueno, Takashi Ida, and Shuji Tanabe

Toyohashi University of Technology, Toyohashi, Aichi 440, Japan

1. Introduction

It is well known that activity and selectivity of supported
metal catalyst strongly depends on the method of the catalyst
preparation procedure employed. It has been believed that the
size of metal particles on a support makes a difference in
selectivity ; hence the development of techniques to control
the metal particle size has been needed. One of the present
authors found that the hydrolysis of a solution of metal
alkoxide, the alkoxide method, gives catalysts with small
metal particles [1].
   The purpose of the present study is to find out why the
alkoxide method is superior to the more conventional
impregnation method in preparing small metal particles. EXAFS
is best suited for such a study because the local structure
change around the metal atom can be studied at every step of
the catalyst preparation procedure. A result on Ni/SiO$_2$ system
has already been reported [2], and that on Co/TiO$_2$ system is
described here.

2. Experimental

The detail of the catalyst preparation procedure by the
alkoxide method has been described elsewhere [1]. In short, it
consists of the following five steps : 1. reaction of Co(NO$_3$)$_2$
with Ti(i-C$_3$H$_7$O)$_4$ in ethylene glycol, 2. formation of a gel by
hydrolysis, 3. drying at 110 [C], 4. calcination in air at 450
or 700 [C], 5. reduction. The impregnation method consists of
impregnation, drying, calcination,and reduction.
   The in-house laboratory EXAFS spectrometer employed here
has been reported [3], and further development will be
discussed in this conference. With the receiving slit width
of 0.2 [mm], the resolution of about 5 eV HWHM was obtained.
The regular data aquisition time was about 2 hours, collecting
over one million counts at each data point. Sometimes over ten
million counts were accumulated by an overnight measurement.

3. Result

A major difference in the local structure of the two
preparation methods appears in EXAFS at the calcination stage.
Figure 1 shows the EXAFS spectra of the samples prepared by
the alkoxide method and by the impregnation method, both
calcined at 450 [C], as well as that of the reference compound
Co$_3$O$_4$. It is apparent from Fig.1 that the spectrum of the
sample by the impregnation method is the same as that of Co$_3$O$_4$

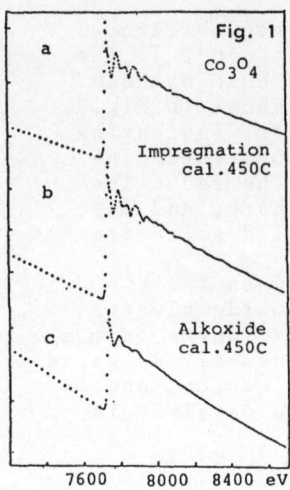

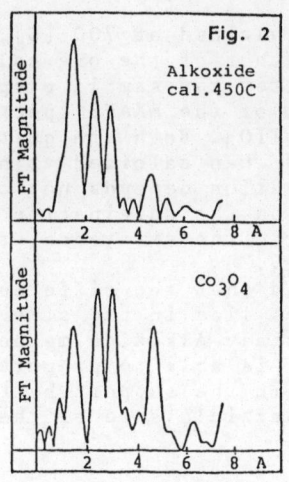

Figure 1. EXAFS spectra of Co_3O_4 and of the calcined samples.

Figure 2. Fourier transforms of a:Fig.1c and b:Fig.1a.

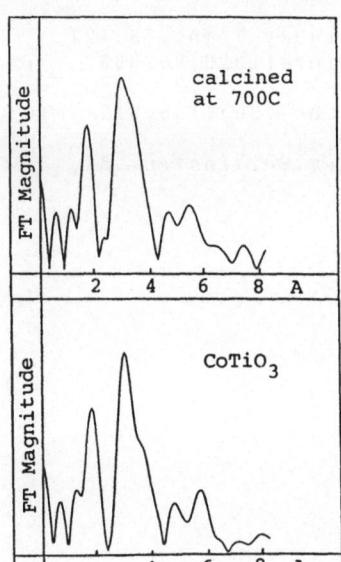

Figure 3. Fourier transforms of the sample calcined at 700 C and of $CoTiO_3$.

powder, indicating that Co_3O_4 crystallites are formed. On the other hand, the spectrum of the sample by the alkoxide method is not the same as those of reference compounds CoO, Co_3O_4, and $CoTiO_3$. A radial structure function of this material is shown in Fig.2, together with that of Co_3O_4. In order to make the comparison meaningful, the same k range (3.6-12.5 [1/A]) was used for the calculation. Although the overall structure looks different, the position of every peak coincides exactly. The difference between these two can be rationalized if one takes the surface effect into account ; Co_3O_4 clusters are formed also in the calcined material by the alkoxide method, but they are so small that the number of atoms at distant shells decreases rapidly with distance. A model calculation shows that the cluster size is as small as about 10 A.

If the sample is calcined at 700 [C], the spectrum becomes quite different from that of the one calcined at 450 [C]. The spectra are the same for the sample prepared by both methods. The Fourier transform of the EXAFS spectrum is shown in Fig.3, as well as that of $CoTiO_3$. Both are quite similar, indicating that $CoTiO_3$ is formed when calcined at high temperature. The particle size distribution depends not only on the reduction temperature but also on the calcination temperature, and the reason is attributed to the existence of the solid state reaction that makes $CoTiO_3$.

It can be concluded that the differences between the two preparation procedures lies in the size of the oxide clusters in the calcined material. Alkoxide method, which starts from a reaction in solution, is able to prepare homogeneously dispersed small oxide clusters in the support by the calcination, and as a result, the metal particle size of the reduced catalyst is small.

References
1. A.Ueno,K.Suzuki,Y.Kotera;J.Chem. Soc.Faraday Trans.79,127 (1983), S.Takasaki,S.Tanabe,A.Ueno,Y.Kotera;ibid,80,803 (1984).
2. K.Tohji,Y.Udagawa,S.Tanabe,A.Ueno;J.Am.Chem.Soc.106,612 (1984).
3. K.Tohji,Y.Udagawa,T.Kawasaki,K.Masuda;Rev.Sci.Instrum.54, 1482(1983).

Dynamical Studies of Catalytic Systems Using Dispersive X-Ray Absorption Spectroscopy

D.E. Sayers
Physics Department North Carolina State University,
Raleigh, NC 27695-8202, USA
D. Bazin, H. Dexpert, and A. Jucha
LURE, Bat. 209 C, Université Paris Sud, F-91405 Orsay, France
E. Dartyge, A. Fontaine, and P. Lagarde
Lab. Physique des Solides Bat. 510, F-91405 Orsay, France

I. Introduction

During the last several years the feasability of performing x-ray absorption measurements in the dispersive mode using synchrotron radiation sources has been demonstrated [1]. Recently, JUCHA, et al. at LURE have reported [2] the development of an improved photodiode detector which has the capability of acquiring data in times as fast as 4 msec. Using this system a wide variety of experiments have been performed to test the feasability of doing dispersive x-ray absorption experiments including dynamical studies on catalysts, conducting polymers and solutions, high pressure experiments and dispersive reflected EXAFS (REFLEXAFS) measurements to study interfaces. In this paper, the first time dependent studies of the reduction under H_2 of catalytic systems are reported. Pt/Al_2O_3 and $Pt-Rh/Al_2O_3$ catalysts were used for these first studies since it has been shown that the Pt L_3 edge changes systematically with the reduction of the Pt from Pt^{4+} to Pt°. [3]

II. Experimental

A) Samples.

Two catalyst samples were studied. The first is a monometallic $Pt/\gamma-Al_2O_3$ containing ~ 1 wt% of Pt. The catalyst was prepared by the impregnation technique which involved contact of the high surface area support ($\gamma-Al_2O_3$ with a surface area of approximately 200 m^2/g) with a solution of the metal salt (in this case $H_2Pt\,Cl_6$). The mixture is then dried in air at 110°C for about 50 hrs, leaving agglomerates of Pt bonded with oxygen and chlorine in contact with the support. When reduced, the particle size in this system is <15Å as measured by chemisorption.

The second system is a bimetallic $Pt-Rh/\gamma-Al_2O_3$ catalyst containing ~ 2 wt% Pt and 1 wt% Rh with a Pt/Rh ratio of 1.8. The sample was prepared in the same manner as described above for the first system except that an appropriate amount of $RhCl_3$ was combined with H_2PtCl_6 to give the desired ratio of Pt/Rh. The average particle size of this system is approximately the same as for sample 1.

B) Method.

Data was collected on LURE station D16 using a 7 cm long triangular Si_{111} monochromator. The curvature of the crystal was adjusted and calibrated using a Pt foil to cover about a 300 eV interval about the Pt L_3 edge at 11560 eV. The sample to detector distance was ~ 1.5 m and was chosen so that the desired energy band pass illuminated most of the 1024 pixels of the Recticon array. The calibration of the pixels versus energy is done by comparing the Pt foil spectrum obtained in the dispersive mode with a Pt foil spectrum taken with a standard EXAFS apparatus. The resulting energy resolution of this configuration is about twice as large as the resolution obtained with a conventional apparatus using a flat Si_{111} monochromator. The reason for this additional resolution is thought to be due to the horizontal width (~ 7 cm) of the beam in DCI but further studies of the factors effecting resolution are being made. The effect of the increased resolution is to broaden and reduce the amplitude of the white line but still results in spectra of sufficient quality to follow changes in the white line as a function of reduction of the catalyst. Data were taken with DCI operating in the parasitic mode at 1.56 GeV and a typical current of 100 mA.

Samples were mounted in a controlled environment furnace which has been described previously[4] and encapsulated in boron nitride sample cells through which flowing H_2 could be passed. Temperature was controlled to about ± 1°C by an automated temperature controller using a thermocouple mounted in the boron nitride cell as the temperature indicator.

The experiment consisted of first optimizing the sample position and then heating a sample under flowing H_2 at uniform rate (10°C/min) taking spectra every 5 or 10°C until the temperature region over which the reduction occurs is identified. Then a fresh sample was mounted and the sample was quickly heated (~ 10°C/min) in air to the desired temperature. The hydrogen flow was begun and a series of spectra were recorded at this temperature as a function of time until the reaction has gone to completion as indicated by a reduction of the white line at the Pt L_3 edge. For these experiments this time was typically 20 minutes.

III. Results.

The results of one such series of measurements for Pt/Al_2O_3 at 210 C are shown in fig. 1a. Curve A was taken soon after the hydrogen flow was begun and is essentially identical to the room temperature spectrum, indicating no changes had occurred in the system during the initial heating. Curves B-E are samples of the subsequent spectra taken over a period of 16 minutes. The acquisition time for each spectrum was 26.4 sec. The reduction of the system can clearly be seen by the decrease of the white line at the Pt L_3 threshold and by the changes of the EXAFS oscillations in the vicinity of 50 eV above the edge.

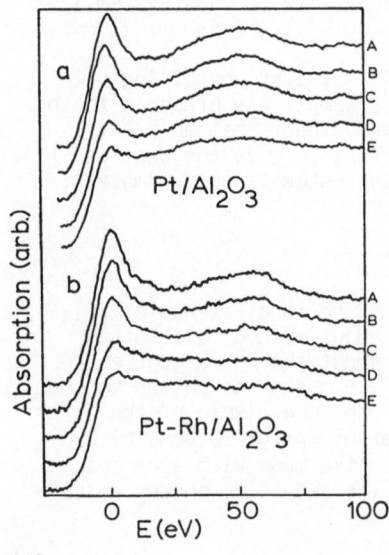

Fig. 1a) Changes of the Pt L_3 edge of a Pt/Al_2O_3 catalyst at 210°C under H_2 as a function of time. Curve A is taken as t = 0 as discussed in the text. Curve B – E are taken at times of 5.4 min, 9.8 min, 12.6 min and 15.7 min respectively relative to curve A.

Fig. 1b) Similar curves as in Fig. 1a for a $Pt-Rh/Al_2O_3$ catalyst at 85°C. The acquisition times are at t = 0, 5.0 min, 6.5 min, 11.0 min and 15.9 min for curves A – E.

The corresponding experiment was performed for the Pt-Rh/$A\ell_2O_3$ catalyst; however, the temperature in this case was $85^{\circ}C$. This lower reduction temperature relative to the pure Pt catalyst is expected because of the higher affinity of Rh for oxygen, thus making it easier to reduce the Pt. The results are shown in fig. 1b. with a similar range of times to those of fig. 1a. The acquisition time for these spectra was 16 s. Again the reduction of the system is evident. These results represent only a part of the studies done to date. Other measurements have been done at other temperatures on the direct thermal reduction of these systems when heated in air, and on subsequent oxidation.

The times scales shown in Fig. 1 are only relative, since the onset of reduction may be delayed from the actual start of the H_2 flow by an induction time, which may depend on a variety of experimental factors. For these curves t = 0 was typically taken to be after the induction time but before any noticable change in the spectrum has occurred.

Conclusions

These measurements clearly show that dynamical changes of catalytic systems undergoing reduction or oxidation reactions can be followed on the time scale of minutes. A more detailed analysis of these data is currently underway to see to what extent models of reduction or kinematical parameters can be extracted from this type of experiment. It is also expected that improvements in the optics and detector systems and the utilization of the next generation of intense synchrotron sources will greatly increase the amount of information which can be obtained from studies of this type and reduce the time scales of change which can be followed down to the vicinity of 10 msec.

References

1. A. M. Flanck, A. Fontaine, A. Jucha., M. Lemonnier, C. Williams: J. de Physique Lettres 43, L315 (1982)

2. A. Jucha, D. Bonin, E. Dartyge, A. M. Flanck, A. Fontaine, D. Raoux, NIM, to be published.

3. F. W. Lytle, J. Catal. 43, 376 (1976).

4. P. Lagarde: EXAFS and Near Edge Structure, ed: A. Bianconi, L. Ineoccia, S. Stipich (Springer Verlag, Berlin, 1983) p. 294.

An EXAFS Study on the Influence of CO Absorption on the Structure of Small Rhodium Clusters Supported on γ-Al$_2$O$_3$ or TiO$_2$

D.C. Koningsberger

Laboratory for Inorganic Chemistry and Catalysis, Eindhoven University of Technology, P.O. Box 513, NL-5600 MB Eindhoven, The Netherlands

INTRODUCTION

Controversies exist in literature about the structure and oxidation state of highly dispersed Rh/Al$_2$O$_3$ catalysts. Some authors conclude on the basis of CO infrared data that rhodium monatomatically dispersed as RH^{1+} ions (1-3). However, electron microscopy studies shows the presence of small metallic rhodium particles (4-5).

The carbon-oxygen stretching frequence of CO chemisorbed on Rh/Al$_2$O$_3$ (1-3) and Rh/TiO$_2$ (6) have been extensively studied by infrared spectroscopy. One surface species consisting of two CO molecules bound to one surface rhodium (bands at 2095 and 2027 cm^{-1}, representing the symmetrical and anti-symmetrical modes, respectively) is solely present on catalysts with a low rhodium loading. This surface species is assigned to CO adsorbed on isolated Rh^{1+} cations (3), since the wavenumbers closely correspond to those observed for the bridged [Rh^{1+}(CO)$_2$Cl]$_2$ dimer and do not shift in frequency with increasing coverage. At high rhodium loadings two CO bands are observed, one around 2060 cm^{-1} and another broad band between 1800 and 1900 cm^{-1}. The former band has been interpreted as arising from a CO molecule on top of a rhodium surface atom of a metal crystallite with the latter broad band due to a CO molecule bridged between neighbouring Rh surface atoms (3).

Rh/Al$_2$O$_3$ catalysts with low rhodium loading showing solely the rhodium dicarbonyl species have been studied with electron microscopy also. However, these studies showed the presence of small metallic rhodium particles. To clarify these contradicting results our group performed EXAFS studies on the influence of CO chemisorption on the topology of small rhodium clusters supported on γ-Al$_2$O$_3$ (7,8,9) and on TiO$_2$ (10). In this paper we will summarize the results and discuss the influence of the support and the cluster-size on the type of species which are formed during CO chemisorption.

EXPERIMENTAL

The preparation of a 0.6 wt% Rh/γ-Al$_2$O$_3$, a 1 wt% Rh/γ-Al$_2$O$_3$ and a 1 wt% Rh/TiO$_2$ has been extensively described in (8), (9) and (10) respectively. All catalysts were prepared by incipient wetting of high surface area supports with an aqueous solution of RhCl$_3$, dried and subsequently reduced under flowing H$_2$ at high temperature.

Hydrogen chemisorption measurements resulted in H/M values of 1.7, 1.65 and 1.5 for the Rh(0.6)/Al$_2$O$_3$ (8), Rh(1)/Al$_2$O$_3$ (9) and Rh(1)/TiO$_2$ (10), respectively.

EXAFS spectra of the rhodium K-edge were recorded in-situ at liquid nitrogen temperature on X-ray beamline I-5 at S.S.R.L., Stanford University with ring energies of 3 GeV and ring currents between 40-80 mA.

RESULTS AND DISCUSSION

The EXAFS spectrum of the reduced Rh(0.6)/Al$_2$O$_3$ catalyst shows oscillations
due to rhodium nearest neighbours (7), which proves the existence of rhodium
metal crystallites. The Rh-Rh coordination parameters as given in table I
for the reduced catalyst were obtained by making use of rhodium-rhodium
phase- and amplitude corrected Fourier transforms (11).

Table I Coordination parameters (N numbers, R distance, $\Delta\sigma^2$ Debeye Waller
Factor, difference with reference compound) for the 0.6 wt%
Rh/γ-Al$_2$O$_3$ catalyst. (Accuracies N$\pm$10 - 20%, R$\pm$0.5 - 1%, $\Delta\sigma^2\pm$10-30%)

TREAT MENT	COORDI- NATION	RH(0.6)/γ-AL$_2$O$_3$		
		N	R	$\Delta\sigma^2$ x10^{-3}
REDUCTION	RH-RH	3.7	2.68	5
ADMISSION OF CO AT 296 K	RH-RH	-	-	-
	RH-O	3.1	2.12	3
	RH-(C≡O)	1.8	1.85	7
	RH-(C≡O)	1.8	3.00	7

A 3μ-thick rhodium metal foil was used as reference for the phase and backscattering amplitude of a rhodium-rhodium absorber-scatterer pair. The Rh-
Rh coordination number (N=3.7) points to very small metal crystallites.

Adsorption of CO at room temperature leads to a complete disappearance of
the Rh-Rh EXAFS oscillations (7). This means that the metal-metal bonds in
very small supported rhodium metal clusters are disrupted by CO adsorption
at room temperature. The EXAFS results explain the seeming contradiction
between the results obtained from CO infrared studies and from high-resolut-
ion Electron Microscopy. The EXAFS technique shows that this contradiction
is only apparent, since CO adsorption changes the structure of the small
metal clusters completely.

Analysis of the EXAFS spectrum obtained for the Rh(0.6)/Al$_2$O$_3$ after CO
chemisorption has been carried out using phase corrected Fourier transforms
with Rh$_2$O$_3$ as reference for the Rh^{n+}-O^{2-} and [Rh^{1+}(CO)$_2$Cl]$_2$ for the Rh-
(C≡O) absorber-scatterer pairs respectively (12). The results of the EXAFS
data analysis, which is described in detail in (12), are summarized in table
1. It has been found that after CO chemisorption rhodium is coordinated by
two CO molecules. The coordination distances between rhodium and the carbon-
oxygen ligands of the carbonyl group are equal to the distances as found in
the Rh(CO)$_2$Cl monomer. A Rh-O coordination could also be analysed with three
oxygen neighbours at a distance of 2.12Å. These oxygen neighbours arise most
probably from the support, implying that the Rh(CO)$_2$ species is absorbed to
the support with three Rh-O bonds. A coordination distance of 2.12 Å points
to a valence state of rhodium higher than 0 and lower than 3+ (R=2.05 Å for

Rh$_2$O$_3$). The infrared results strongly suggest a valence state 1+ for rhodium in this species, which is further supported by the results of XPS measurements (12).

EXAFS oscillations due to rhodium-rhodium bonds can still be detected after CO chemisorption on rhodium particles with a notable size. This is illustrated in Fig. 1 for a Rh(1)/Al$_2$O$_3$ and a Rh(1)/TiO$_2$ catalysts.

The results of the data analysis carried out for the CO chemisorbed Rh(1)/Al$_2$O$_3$ (9) and Rh(1)/TiO$_2$ (10) catalysts are given in table II.

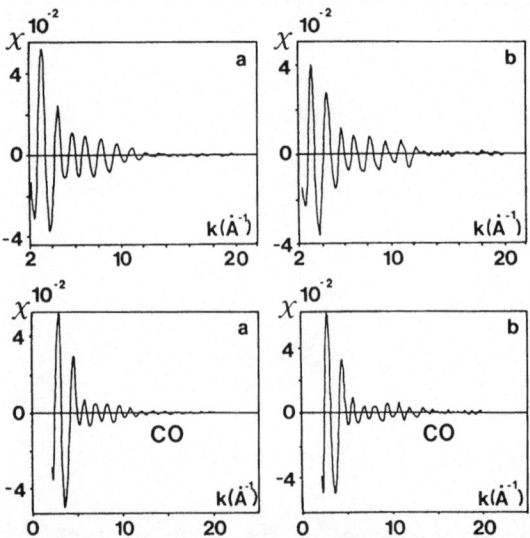

Figure 1 EXAFS spectra of Rh(1)/Al$_2$O$_3$ (a) and Rh(1)/TiO$_2$ (b) catalysts after reduction and after CO chemisorption at RT

Table II Coordination parameters (N number, R distance, $\Delta\sigma^2$ Debeye Waller factor, difference with reference coumpound) for the 1 wt% Rh on γ-Al$_2$O$_3$ or TiO$_2$ catalysts. (Accuracies N±10 - 20%, R±0.5 -1%, $\Delta\sigma^2$ +10 ± 30%)

TREATMENT	COORDINATION	RH(1)/γ-AL$_2$O$_3$			RH(1)/TIO$_2$		
		N	R	$\Delta\sigma^2$ x10^{-3}	N	R	$\Delta\sigma^2$ x10^{-3}
REDUCTION	RH-RH	5.9	2.68	5	5.9	2.66	7.4
ADMISSION OF CO AT 296 K	RH-RH	1.6	2.68	4	1.6	2.68	4
	RH-O	2.2	2.12	5	2.1	2.12	3
	RH-(C≡O)	1.5	1.85	5	1.5	1.83	4
	RH-(C≡O)	1.5	3.00	5	1.5	3.00	5

An EXAFS function for the CO chemisorbed Rh(1)/TiO$_2$ sample has been calculated (10) with the coordination parameters shown in table II using the phases and backscattering amplitudes obtained from the reference compounds. This function and its Rh-O phase corrected Fourier transform are given in Fig.2 with dotted lines. A good agreement exists as well as in k-space as in r-space between the experimental data and the results obtained from the data analysis.

The influence of the clustersize upon the formation of a specific surface species during CO chemisorption at RT is clearly demonstrated by our EXAFS results. The Rh^{1+}(CO)$_2$ is solely created on very small clusters. Rhodium catalysts with a somewhat lower dispersion still showed the presence of some metallic rhodium crystallites after CO chemisorption. Apparently, the larger particles present in these catalysts cannot be broken up by CO adsorption. These particles give rise to the linear and multicentre Rh-CO species as mentioned in the introduction.

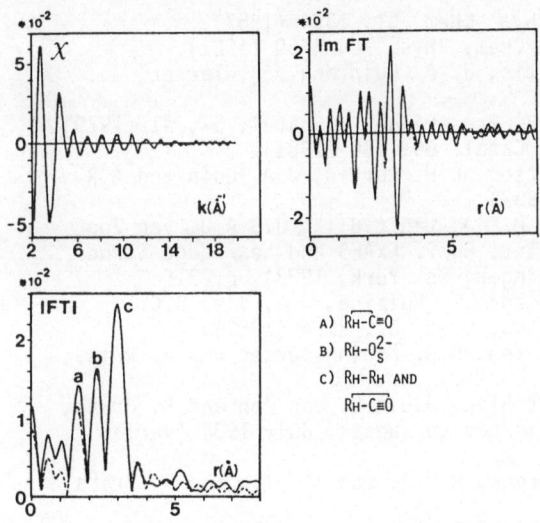

A) $\overline{Rh-C\equiv O}$
B) $Rh-O_s^{2-}$
C) $Rh-Rh$ AND
   $\overline{Rh-C\equiv O}$

Figure 2 Experimental EXAFS data (solid line) and calculated EXAFS$_2$(Rh-C≡O+ Rh-O+Rh-Rh) (dotted line) and corresponding transforms (k^3, Rh-O phase corr, k$_{min}$=3.4 Å^{-1}, k$_{max}$=9.2 Å^{-1}) for Rh(1)/TiO$_2$catalyst

The structure of the surface Rh^{1+}(CO)$_2$ can be directly derived from the coordination parameters as given in Table I. However, the coordination numbers for the Rh^{1+}(CO)$_2$ species found in Table II are averaged over all rhodium atoms and should be corrected for the amount of rhodium that remains metallic. Reversely, the fraction metallic rhodium can be estimated by using the fact that the real Rh-(C≡O) coordination number in the geminal dicarbonyl species is 2 and the expected coordination with support oxygen ion is 3. The fraction metallic rhodium is related to the real (N_R) and measured (N_m) coordination number by f=1-N_m/N_R. This fraction is found to be 25% for both the Rh(1)/Al$_2$O$_3$ and the Rh(1)/TiO$_2$ catalysts using the values given in Table II.

The coordination distance for the Rh-O$_s$ (O$_s$ support oxygen) bond is the same for γ-Al$_2$O$_3$ and TiO$_2$ supported catalysts. Also the Rh-(C≡O) coordination is not influenced by the support, indicating that also on TiO$_2$ the adsorption of CO has to be oxidative leading to Rh^{1+} valence state. In summary , the support determines the particle size distribution, which in turn influences the amounts of geminal dicarbonyl and linear/multicentre CO species formed during CO adsorption.

ACKNOWLEDGEMENT

This work was done at S.S.R.L. (Stanford University), which is supported by the Department of Energy,the National Science Foundation and the National Institute of Health. The author gratefully acknowledges the assistance of the SSRL staff and thanks all members of our EXAFS group for their indispensable support. He would like to thank (ZWO) also for supplying a travel grant R (71-24).

REFERENCES

1. A.C. Yang and C.W. Garland, J. Phys. Chem. 61, 1504 (1957).
2. R.R. Cavanagh and J.T. Yates, J. Chem. Phys. 74, 4150 (1981).
3. C.A. Rice, S.D. Worley, C.W. Curtis, J. A. Guin and A.R. Tarrer, J. Chem. Phys. 74, 6487 (1981).
4. D.J.C. Yates, L.L. Murell and E.B. Prestridge, J. Catal. 57, 41 (1979).
5. F.W. Graydon and M.D. Langan, J. Catal. 69, 180 (1981).
6. S.D. Worley, C.A. Rice, G.A. Mattson, C.W. Curtis, J.A. Guin and A.R. Tarrer, J. Chem. Phys. 76, 20 (1982).
7. D.C. Koningsberger, T. Huizinga, H.F.J. van't Blik, J.B.A.D. van Zon, R. Prins and D.E. Sayers, Proc. Int. Conf. EXAFS and Near Edge Structures, Frascati, Italy 1982 (Springer, New York, 1983) p. 310.
8. H.F.J. van 't Blik, J.B.A.D. van Zon, T. Huizing, J.C. Vis, D.C. Koningsberger and R. Prins, J. Phys. Chem. 87, 2264 (1983).
9. H.F.J. van 't Blik, J.B.A.D. van Zon, D.C. Koningsberger and R. Prins, J. Mol. Catal. 25, 379 (1984).
10. D.C. Koningsberger, H.F.J. van 't Blik, J.B.A.D. van Zon and R. Prins, Proc. 8th Int. Congr. on Catalysis Berlin (West), July 1984 (Verlag Chemie, Basel, 1984) p.V-123.
11. J.B.A.D. van Zon, D.C. Koningsberger, H.F.J. van 't Blik and R. Prins, J. Chem. Phys. 80, 3914 (1984).
12. H.F.J. van 't Blik, J.B.A.D. van Zon, T. Huizinga, J.C. Vis, D.C. Koningsberger and R. Prins, submitted to J. Amer. Chem. Soc.

An X-Ray Absorption Study of Gold/Y-Zeolite

M. Boudart[1] and G. Meitzner[2]

Department of Chemical Engineering, Stanford University
Stanford, CA 94305, USA

1. INTRODUCTION

There have been few investigations of catalysis by highly dispersed supported gold. In a previous study of gold supported on SiO_2, gold particles with mean diameter near 2.5 nm were shown to catalyze the reaction between H_2 and O_2 at an appreciable rate per exposed gold atom at 284 K [1]. As far as we know, the supported gold crystallites in those samples were the smallest to have been reported. The successful preparation of platinum clusters of 1 nm diameter in the supercages of a Y-zeolite [2] prompted us to try to prepare clusters of gold in the same support, to determine their structure by X-ray absorption spectroscopy (XAS) and to study the rate of reaction between H_2 and O_2 on these clusters.

This paper reports the preparation and study of gold clusters in the supercages of a Y-zeolite.

2. EXPERIMENTAL

The precursor to the samples of gold in Y-zeolite, abbreviated Au/Y, was prepared by ion-exchange [2] between a zeolite and the complex $[Au(en)_2]Cl_3$ (en=ethylene diamine) made from $HAuCl_4 \cdot 3H_2O$ (Baker Analyzed Reagent, Table 1). The ion-exchange gave a product that had about 13 wt % gold, corresponding to the exchange of about 30 cationic sites per unit cell. The ion-exchanged precursor was evacuated during 2 h at room temperature and maintained in flowing O_2 (volumetric flow rate 1200 h^{-1}) at 390 K for 4 h. It was then cooled to room temperature and evacuated for 0.5 h. Finally the samples were exposed to flowing H_2 for 0.5 h at room temperature before heating to the final temperature of reduction for 4 h.

Table 1. Elemental analysis of Baker Analyzed
Reagent grade $HAuCl_4 \cdot 3H_2O$.

Element:	Fe	Co	Ni	Ru	Rh	Pd	Ag	Os	Ir	Pt
Conc./ppm:	2200	<13	18	<13	<16	<13	26	<26	663	607

X-ray absorption spectra (XAS) were collected at SSRL using a controlled atmosphere cell that was described previously [3]. The structure of the gold L_{III} absorption edge has been discussed [4,5,6]. Gold(III) is distinguished from other oxidation states of gold by a conspicuous absorp-

---

1   To whom queries concerning this paper should be addressed.
2   Present address: Exxon Research and Engineering Co.,
      Annandale, New Jersey 08801

tion edge resonance. Gold(I) exhibits a small edge resonance in some
environments possibly because of a p→s transition or the s-d hybridization
of bonding orbitals, and gold(0) has no apparent edge resonance. Thus the
absorption edge gave information on the state of reduction of the gold.
The theory and methods of analysis of extended X-ray absorption fine structure (EXAFS) have been reviewed [7,8]. The technique was used previously
with gold supported on SiO_2, Al_2O_3, and MgO [9,10]. Interatomic distances,
R, were determined for the samples of unknown structure from comparison of
the sample and reference radial structure functions (RSF). Average
coordination numbers, N, and Debye-Waller factors, $\Delta\sigma^2$, were obtained from
analysis of inverse Fourier transforms by the ratio method described previously [11].

The average diameter of gold crystallites in some samples was estimated
with the Scherrer equation from the integral broadening of the Au(111) and
(200) powder diffraction lines.

3. Results and Discussion

Gold crystallites larger than 10 nm resulted when $[Au(en)_2]^{3+}/Y$ was heated
above 410 K in H_2, O_2, or vacuum, as previously reported for other samples
of supported gold [12]. Treatment of the precursor in O_2 at 390 K enabled
the intermediate to be reduced to gold metal in H_2 at room temperature.

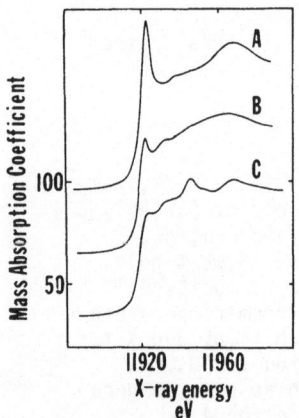

Figure 1. X-ray absorption
spectra in the vicinity of the
Au L_{III} absorption edge

A) $[Au(en)_2]Cl_3$

B) $[Au(en)_2]^{3+}/Y$, treated in
O_2 at 393 K;

C) Large gold crystallites on
Y-zeolite

X-ray absorption edge spectra of the $[Au(en)_2]^{3+}/Y$ precursor, of the
intermediate, and of gold metal, are shown in Fig. 1. The treatment in O_2
decreased the area of the sharp threshold resonance characteristic of
Au(III). The absorption edge of the intermediate treated in O_2 resembled
that of gold(I) in $KAu(CN)_2$ [4] and in complexes with 2 phosphorous or
sulfur ligands [13]. The radial structure functions of a series of samples
are shown in Fig. 2. Treatment of the precursor in O_2 did not change the
nearest neighbor (nitrogen ligand) distance, and no peak attributable to Au
neighbors appeared in the RSF. But the O_2 treatment diminished by one half
the magnitude of the nearest neighbor peak. Evidently each gold atom was
stripped of two nitrogen ligands, without changing the Au-N distance.
This is consistent with the tendency of Au(III) to form square planar complexes and of Au(I) to form linear complexes, in either case with a Au-N
distance between 200 pm and 205 pm [14]. Table 2 is a compilation of
structural parameters from EXAFS describing the reference materials and a
series of intermediate and reduced samples.

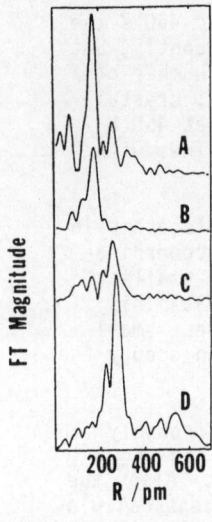

Fig. 2 Radial structure functions from EXAFS. All are plotted against the same scale of magnitude.
(A) [Au(en)$_2$]Cl$_3$;

(B) [Au(en)$_2$]$^{3+}$/Y, treated in O$_2$, 393 K;

(C) Sample B, reduced in H$_2$, room temperature, 0.5 h;

(D) Large gold crystallites on Y-zeolite.

FT Magnitude

200 400 600
R /pm

Table 2. Structural parameters from EXAFS. Spectra were collected at room temperature in O$_2$ (first 5 rows) or H$_2$.

Sample (Treatment)	Coordination Number (±20%)	Bond length /pm(±3)	$\Delta\sigma^2$ /pm^2
[Au(en)$_2$]Cl$_3$	4.0 (Au–N)	201.	0.
[Au(en)$_2$]$^{3+}$/Y	4.6 (Au–N)	201.	9.
(O$_2$,383K)	2.6 (Au–N)	201.	4.
(O$_2$,383 K)	2.0 (Au–N)	201.	0.
(O$_2$,413 K)	11.0 (Au–Au)	281.	10.
Bulk Au^o	12.0 (Au–Au)	288.	0. (σ^2=60.)

[Au(en)$_2$]$^{3+}$/Y, O$_2$, 393 K, followed by:

(H$_2$, RT)	6.5 (Au–Au)	273.	27.
(H$_2$,332 K)	7.7 (Au–Au)	273.	22.
(H$_2$,373 K)	9.3 (Au–Au)	277.	26.
(H$_2$,373 K)	10.4 (Au–Au)	287.	7.

The absorption edge spectra from gold clusters produced by reduction in H$_2$ at room temperature, and from large Au crystallites, are similar. In the RSF of the sample reduced at room temperature, little or nothing remains of the peak due to the Au–N distance, and the principal peak resembles that of the bulk metal but has lower magnitude. Hence, the gold appears to have been reduced to the zero-valent state. The XRD of NaY-zeolite and of samples in which the gold was not reduced were indistinguishable. The sample of Au/Y reduced under the mildest conditions exhibited a diffuse background but no diffraction lines characteristic of

219

gold metal. In the range of reduction temperatures from 350 K to 450 K the apparent crystallite size was 3-4 nm and did not change significantly, but the integral area of the diffraction line, proportional to the number of diffracting crystallites, increased sharply. Starting at 450 K, crystallite size increased with temperature. After reduction for 4 h at 450 K there was no further increase in the areas of the diffraction lines with more severe reduction.

From the formulae given by Van Hardeveld and Hartog [15], gold atoms in a cubo-octahedron with diameter of only 1.2 nm have an average cooordination number of about 8. The XAS indicates that the gold in the smallest clusters had an average coordination number near 6 (see Table 2). This argument, together with the diffuse diffraction of X-rays by these small clusters, suggest that these clusters were sufficiently small to occupy the supercages of the Y-zeolite.

The small clusters grew very easily to 3-4 nm crystallites, probably on the surface of the zeolite. This is not surprising since gold has shown little resistance to crystallite growth on other supports [12]. Also, the melting temperature of gold is only 1336 K and according to recent calculations [16] is further depressed to as low as 600 K in clusters with size between about 1 nm and 2.5 nm. By gravimetry, Yates [17,18] has demonstrated similar behavior of Zn, Cd, and Hg after deposition in Y-zeolite by ion-exchange. These metals were removed rapidly from the interior of the zeolite crystals at low temperature upon exposure to H_2. The gold particle size calculated from EXAFS and from XRD results agrees only for samples reduced near 300 K and above 450 K. The EXAFS indicates the more likely average coordination number, while the XRD includes only particles large enough to give reasonably sharp diffraction lines. Hence the discrepancy arises from the existence of a bimodal distribution of gold particle sizes following reduction at intermediate temperatures.

4. CONCLUSIONS
By the use of sufficiently mild conditions of reduction it has been possible to prepare gold clusters inside the supercages of a Y-zeolite. The reduction proceeded in two steps, through a gold(I) intermediate. The growth of the gold clusters into larger crystallites on the external surface of the zeolite crystals indicates a high mobility of the clusters that may be related to their possible low melting point, as well as the lack of a chemical interaction between gold and the zeolite framework. Perhaps more stable gold clusters can be made in acidic Y-zeolites as a result of metal-support interactions already demonstrated in the case of Pt clusters in an acidic Y-zeolite [2].

1 Y.L. Lam, J. Criado, and M. Boudart, Nouv. J. Chim. 1, 461 (1977).
2 R.A. Dalla Betta and M. Boudart, Proc. 5th Int. Cong. Catal., vol. 2, J. Hightower, ed., (Amsterdam, North Holland 1973), p. 1329.
3 R. Weber, Ph.D. Thesis, Stanford University, 1983.
4 F.W. Lytle, P.S.P. Wei, R.B. Greegor, G.H. Via, and J.H. Sinfelt, J. Chem. Phys. 70, 4849 (1979).
5 P. Gallezot, R. Weber, R.A. Dalla Betta, and M. Boudart, Z. Naturforsch. 34a, 40 (1979).
6 J.A. Horsley, J. Chem. Phys. 76, 1451 (1982).
7 P.A. Lee, P.H. Citrin, P. Eisenberger, and B.M. Kincaid, Rev. Mod. Phys. 53, 769 (1981).
8 F.W. Lytle, G.H. Via, and J.H. Sinfelt, "X-ray Absorption Spectroscopy: Catalyst Applications," in Synchrotron Radiation Research, H. Winick and S. Doniach, eds. (Plenum, New York 1980), pp. 401-424.

9   I.W. Bassi, F.W. Lytle, and G. Parravano, J. Catal. 42, 139 (1976).
10 G. Cocco, S. Enzo, G. Fagherazzi, L. Schiffini, I.W. Bassi, G. Vlaic,
    S. Galvagno, and G. Parravano, J. Phys. Chem. 83, 2527 (1979).
11 D.E. Sayers, E.A. Stern, and F.W. Lytle, Phys. Rev. Lett. 35, 584
    (1975).
12 W.N. Delgass, M. Boudart, and G. Parravano, J. Phys. Chem. 72, 3563
    (1968).
13 R.C. Elder, K.G. Tepperman, M.K. Eidsness, M.J. Heeg, C.F. Shaw, and
    N. Schaeffer, ACS Symp. Ser. 209, 385 (1983).
14 P.G. Jones, Gold Bull. 14, 102 (1981); 14, 159 (1981).
15 R. Van Hardeveld and F. Hartog, Surf. Sci. 15, 189 (1969).
16 J. Ross and R.P. Andres, Surf. Sci. 106, 11 (1981).
17 D.J.C. Yates, J. Phys. Chem. 69, 1676 (1965).
18 D.J.C. Yates, Chem. Eng. Prog. 63, 56 (1967).

Noble Metal Small Cluster EXAFS

S. Mobilio and E. Burattini

INFN-Laboratori Nazionali di Frascati, I-00044 Frascati, Italy

A. Balerna, E. Bernieri, P. Picozzi, A. Reale, and S. Santucci

Istituto di Fisica dell'Università dell'Aquila, I-67100 L'Aquila, Italy

1. Introduction

Many experimental and theoretical works (1) have been and are being performed on the structural properties of small metal clusters in order to investigate possible differences from the corresponding bulk metal structure. Attention is generally devoted to the following points: 1) if there is or not a contraction in the lattice parameter with decreasing cluster size; 2) if there is or not a structural transition from the bulk crystallographic structure to icosahedron which should be for thermodynamical and dense packing considerations the most stable structure for a few atoms system.

From an experimental point of view the answer to the first question seems to be dependent on the clusters preparation method and on the interactions with the substrates. Metal clusters supported on weakly interacting substrates and prepared using physical methods, like thermal evaporation, generally show a contracted lattice parameter though there are large discrepancies on the numerical values found for the contractions. On the other hand, chemically prepared clusters, like reduced supported catalysts, never show appreciable contractions. Therefore great attention must be given to fully understand the physical origin of such different behaviours.

With regard to the structural changes, no definite evidence has been found to ascertain or exclude experimentally the occurrence of the icosahedral structure for small metal clusters.

This work reports an EXAFS structural characterization of evaporated Au metal clusters supported on a weakly interacting substrate, like mylar, in order to study the cluster intrinsic structural properties. The presence of lattice parameter contractions, the non existence of icosahedron-like structure and the decrease of the cluster Debye temperature due to the softening of the phonon spectrum are the main results here reported.

2. Experimental

Samples were prepared in vacuum, by evaporating on a 6 μm mylar film multiple layers of gold and mylar until the optimum metal thickness for X-ray measurements was achieved.

Gold and mylar depositions were both controlled with a quartz-crystal detector. Discontinuity of the multilayer samples has been checked with optical transmission measurements (2). Cluster size distributions were obtained using electron microscopy analysis. Their mean diameter ranges from 11 Å to 50 Å.

The X-ray absorption spectra have been recorded on the Au L_3 edge at both liquid nitrogen temperature (LNT) and room temperature (RT). Measurements have been done at the Frascati National Synchrotron Radiation Laboratory using the wiggler beam line, whose monochromator was equipped with a Si(111) channel-cut crystal.

3. Results and discussion

The X-ray absorption spectra have been analyzed using a standard procedure (3) in order to obtain the values of the interatomic distances (R_1), the mean coordination

Table I. EXAFS results obtained using k-space analysis for the first and second coordination shells. Gold bulk was used as reference compound. The Debye-Waller factors are the total values. Values for gold bulk were obtained from Ref.(4).

D(Å)	R_1(Å)	R_2(Å)	R_2/R_1	N	σ^2_{RT}(Å^2)	σ^2_{LNT}(Å^2)
BULK	2.88	4.07	1.41	12.0	4.9×10^{-3}	1.7×10^{-3}
42.5	2.86	4.03	1.41	11.0	6.7×10^{-3}	2.4×10^{-3}
30.0	2.85	4.03	1.41	10.6	7.0×10^{-3}	2.6×10^{-3}
24.0	2.85	4.02	1.41	10.0	8.6×10^{-3}	4.2×10^{-3}
20.0	2.83	4.01	1.42	9.6	10.7×10^{-3}	5.5×10^{-3}
15.0	2.82	3.99	1.42	9.4	10.0×10^{-3}	---------
11.0	2.81	3.95	1.41	---	---------	---------

numbers (N) and the Debye-Waller factors (σ^2) for the first shells and the interatomic distances (R_2) for the second coordination shells (Table I).

It can be seen in Fig. 1 that as the cluster size decreases there is a decrease of the interatomic distances of both the first and second coordination shells. These contractions are due to the comprehensive force exerted by the surface stress on the small good clusters. Using a macroscopic liquid drop model, the contractions can be described by the relation:

$$\Delta R = -\frac{4}{3} K R_b f \frac{1}{D} \tag{1}$$

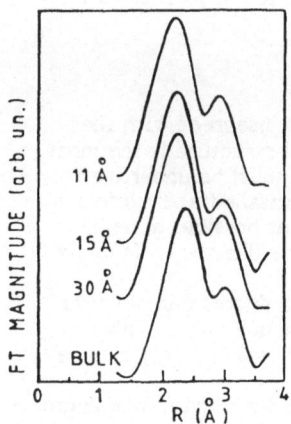

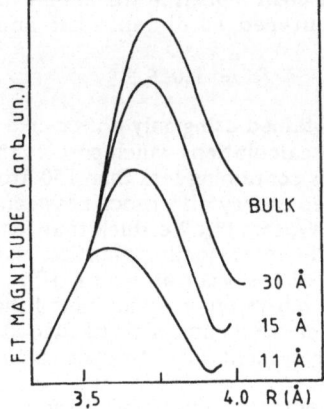

Fig. 1a. Fourier transform of some films and bulk EXAFS spectra relative to the first coordination shell. The transformations have been performed in the k range 3–11 Å^{-1} using a gaussian window and a k weight.

Fig. 1b. Fourier transform of some films and bulk EXAFS spectra relative to the second coordination shell using the same conditions of Fig. 1a.

223

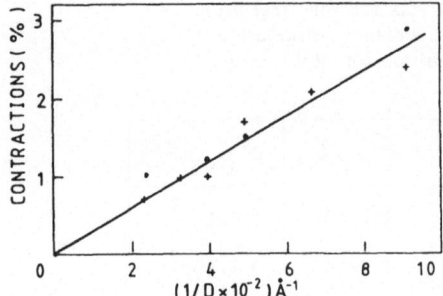

Fig. 2. Plot of the first (+) and second (.) neighbour distances contraction percentages vs. inverse cluster mean diameter. Using eq.(1) the f value obtained is 3464 ± -415 dyn/cm in good agreement with Ref.(8).

where f is the surface stress, K is the bulk comprenssibility, R_b is the metal bulk interatomic distance and D is the cluster mean diameter. A good agreement between this model and the experimental data is shown by the linear behaviour reported in Fig. 2.

Great attention was given to the possible presence of asymmetry effects in the radial distribution function, which can result in apparent nearest-neighbour contractions (5,6). The origin of such effects can be dynamical or static. Dynamical asymmetry effects can be excluded because the cluster total phase, obtained by EXAFS data analysis was found to be independent from temperature. On the other hand the presence of asymmetry static effects can be excluded "a posteriori" since the analysis with a gaussian function gives results for R, N and σ^2 which are completely explained by the increase surface to volume ratio. Moreover the good agreement between our data and the ones obtained using electron diffraction (7,8) excludes further the presence of asymmetry effects.

From Table I it can be seen that the ratio R_2/R_1 is for all samples equal to $\sqrt{2}$, as expected for an fcc structure: it is thus possible to exclude changes from the fcc structure to the icosahedron one also for the 11 Å clusters. In order to exclude this further on, the first shell EXAFS data were fitted using a single shell model (fcc-like structure) and a double shell model. In the icosahedron, in fact, the first fcc shell is split in two shells centered at distances R and R' related by:

$$R' = 1.056 \ R .$$

Good fits were obtained using only an fcc-like model. This result disagrees with the thermodynamical calculations which predict that the icosahedral structure is the most stable for clusters containing less than 150 atoms (D ≤ 15 Å). It must be underlined that the calculated energy difference between the fcc and the icosahedral structure is only about 0.02 eV/atom (9). We think that for this reason it is not possible to neglect contributions to the interatomic potential, given by larger effects like the shift of the clusters 5-d band of the order of ~1 eV (10).

Coordination numbers reported in Table I show a decrease with decreasing clusters size due to the increased number of surface atoms respect to the bulk ones. Values obtained are in agreement with the ones calculated by Mason (10) using highly symmetrical geometries.

Debye-Waller factors evaluated at RT and at LNT increase as the cluster size decreases. This increase in cluster disorder is mainly dynamical in origin due to the increased surface to volume ratio and to the higher mobility of the surface atoms. As reported in Ref. (2), using the Debye approximation it is possible to calculate the clusters Debye temperatures, Θ_D (Table II).

The decrease in clusters Debye temperatures is due to the higher mobility of the surface atoms with respect to the bulk. LEED measurements determined gold surface Debye tem-

Table II. Debye temperatures
for some of the samples studied.

D(Å)	Θ_D(K)
42.5	150
30.0	148
24.0	148
20.0	140

peratures for the different Miller planes: such temperatures are lower than the bulk
one (11). Assuming that the clusters Debye temperatures approach the surface one a-
veraged on all the possible surface planes, we obtained for gold:

$$\Theta_{D \text{ cluster}} = 135 \text{ K} \simeq 0.8 \ \Theta_{D \text{ bulk}}$$

in good agreement with data reported in Table II.

This work clearly shows that for gold clusters non-interacting with the substrate,the-
re is a real presence of contractions in the lattice parameters and there are no structural transitions from the bulk fcc structure to icosahedron. Attention will be given in
the future to the structural behaviour of gold clusters using different substrates and
preparation methods.

References

1. I.D.Morokhov, V.I.Petinov, L.I.Trusov and V.F.Petrunin, Sov. Phys. Usp. 24, 295
   (1981).
2. A.Balerna, E.Bernieri, P.Picozzi, A.Reale, S.Santucci, E.Burattini and S.Mobilio,
   to appear.
3. S.Mobilio, F.Comin and L.Incoccia, Frascati report LNF-82/19 (1982).
4. R.B.Gregor and F.W.Lytle, Phys. Rev. B20, 4902 (1979); E.Sevillano, H.Mevth and
   J.J.Rehr, Phys. Rev. B20, 4908 (1979).
5. P.Eisenberger and G.S.Brown, Solid State Commun. 29, 481 (1979).
6. S.Mobilio, L.Incoccia, in Proceedings of Workshop on EXAFS data analysis in
   disordered systems, Parma 1981, to be published on Nuovo Cimento.
7. F.W.C.Boswell, Proc. Phys. Soc. (London) 64A, 465 (1951).
8. C.Solliard and M.Flüeli, Proceedings of 3rd Intern. Symp. on Small particles and
   inorganic clusters, Berlin 1984, pag. 160.
9. M.Gordon and F.Cyrot-Lackmann, Surf. Sci. 80, 159 (1979).
10. M.G.Mason, Phys. Rev. B27 748 (1983).
11. D.P.Jackson, Surf. Sci. 43, 431 (1974).

EXAFS Studies of Transition Metal Carbonyl Clusters

N. Binsted, S.L. Cook, and J. Evans

Department of Chemistry, The University, Southamton SO9 5NH, United Kingdom

G.N. Greaves

SERC Daresbury Laboratory, Daresbury, Warrington, WA4 4AD, United Kingdom

1. Introduction

Transition metal carbonyl clusters are a growing area of chemistry, with
relevance to dispersed metal catalysts [1]. This study is concerned with
assessing the application of EXAFS to structural characterisation of these
complexes. Complications arise from the occurrence of co-existing
chemically different sites and the necessity to identify back-scattering
from remote metal sites in order to characterise a cluster skeleton. We will
consider three aspects:
  (a) the characterisation of osmium skeletons,
  (b) studies on supported osmium cluster catalysts, and
  (c) extension to lighter metal systems.
Brief communications of some of this work have been published [2,3].

2. Experimental

The X-ray absorption spectra were recorded on beam line 7 of the SRS at the
Daresbury Laboratory. Silicon (220) (order sorting and channel cut) and
(111) monochromators were used for Os L(III) and Co K edge spectra
respectively, all of which were   recorded in transmission. Calibration
and background subtraction routines from the SRS Program Library [4] were
employed. Analysis involved curve and Fourier transform simulations using
a modified version of the program EXAFS [5], and latterly using the least
squares spherical wave procedures previously described [6,7] and enacted in
the program EXCURVE [8].

3. Results

(a) Characterisation of osmium skeletons

The EXAFS associated with the Os L(III) edge spectra of four cluster
complexes with different skeletons were analysed. The structures of these
complexes, $[Os_3(CO)_{12}]$ (1), $[N(PPh_3)_2]_2[Os_6(CO)_{18}]$ (2), $[Os_6(CO)_{18}]$ (3), and
$[N(PPh_3)_2]_2[Os_{10}C(CO)_{24}]$ (4), are presented in Fig. 1. Initially, curve
fitting was carried out using plane-wave routines and ab initio phase shifts
from the Daresbury Laboratory Data Base [9]. These had been calculated for
elemental structures with the exception of oxygen, which was based on
silica. As illustrated for complex (2) in Fig. 2, close fits of the
experimental EXAFS were obtained from 200-750 eV above the edge. The Os-C
and Os-Os shell radii were estimated to within 0.02 Å of the mean distances
obtained by X-ray diffraction [10]. The precision of the Os-C determination
for complex (2) by diffraction was low and probably accounts for the larger
discrepancy in that case. However, the Os...O non-bonded distance was
consistently underestimated by ca. 0.15 Å. Spherical wave analyses using
the same phase shifts improved the fit of the EXAFS curve at low energies

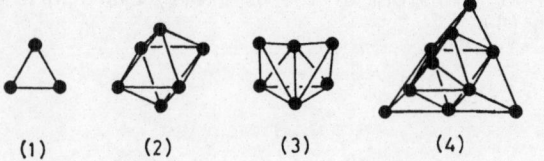

(1)   (2)   (3)   (4)

Fig. 1   Metal skeletons in complexes (1) - (4)

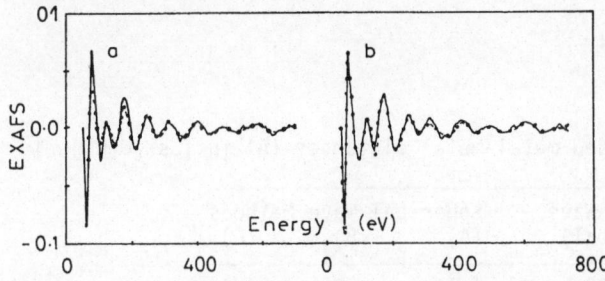

Fig. 2 Fits of the Os L(III) edge EXAFS of the $[Os_6(CO)_{18}]^{2-}$ anion in (2), (a) plane wave and (b) spherical wave

(>35 eV), as illustrated in Fig. 2. The two procedures yielded very similar shell radii (Table 1), but the spherical wave method afforded higher and more realistic Debye Waller α factors ($\alpha = 2\sigma^2$, σ^2= mean square variation in bond length). Accordingly, spherical wave routines were adopted throughout the remainder of this work.

Table 1 Comparison of plane and spherical wave results for $[Os_6(CO)_{18}]^{2-}$

| Shell | Distances ($\overset{\circ}{A}$) | | | α values ($\overset{\circ}{A}{}^2$) | |
	Diffraction	Plane	Spherical	Plane	Spherical
Os-C	1.83	1.90	1.90	0.0003	0.004
Os-Os	2.86	2.85	2.82	0.003	0.009
Os--O	(2.93)	2.87	2.87	0.0003	0.020

Differentiation between cluster skeletons in higher nuclearity clusters (6 or more metal atoms), however, can only be achieved using remote metal shells. The Fourier transforms of the EXAFS of each of the larger three clusters (2), (3) and (4), showed peaks attributable to Os...Os separations (Fig. 3). Three fitting procedures were tested on the decanuclear cluster (4) to estimate these distances, with the results shown in Table 2. Matching the peaks in the Fourier transforms provided a better estimate of these remote shells than a manual minimisation of the residual on the EXAFS curve (Method A). Slightly closer agreements were obtained by matching the Fourier transforms corrected using the first shell phase shifts (Method C) than without any correction (Method B).

227

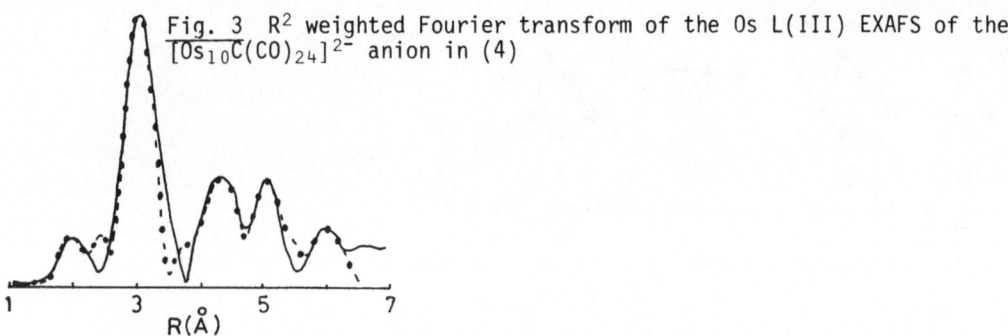

Fig. 3 R^2 weighted Fourier transform of the Os L(III) EXAFS of the $[Os_{10}C(CO)_{24}]^{2-}$ anion in (4)

Table 2 Estimation of non-bonded metal-metal distances (Å) in $[Os_{10}C(CO)_{24}]^{2-}$

Type	Diffraction (idealised)	Plane Wave (global)	Spherical Wave Methods (A)	(B)	(C)
(1)	4.01	4.08	4.11	4.08	4.05
(2)	4.91	4.82	5.19	4.87	4.86
(3)	5.67	5.77	6.09	5.80	5.74

(b) Characterisation of supported osmium cluster catalysts

Two types of supported trinuclear osmium clusters have been investigated. Firstly, direct interaction of $[Os_3(CO)_{12}]$ with high surface area alumina has been shown to form a trinuclear species (5) initially which has been formulated as $[Os_3(H)(CO)_{10}(O-)]$ [11]. The second type is prepared by interacting $[Os_3(CO)_{12}]$ with a ligand functionalised oxide e.g. $HS(CH_2)_3SIL$, and has been identified as $[Os_3(H)(CO)_{10}\{S(CH_2)_3SIL\}]$ (6) on infrared spectroscopic evidence [12]. In neither case has the geometry of the supported cluster been established.

A series of model compounds of the general form $[Os_3(CO)_{10}(X)(Y)]$ (X = Y = H (7); X = H, Y = OMe (8); X = Y = OEt (9); X = H, Y = SPr-n (10)) were studied to assess the parameters to be used for the supported species. Analysis was carried out using the least squares procedures in the program EXCURVE. The results are presented in Table 3. Several non-crystallographic models were used to identify the sensitivity of EXAFS towards various structural aspects. For example, Fig. 4 demonstrates the optimised fits for a 3 shell model {Os-C, Os-Os and Os...O(carbonyl)} for complex (9) and the improvements made using the 5 shell form expected from the structure in the crystal [13]. The two additional features viz. the bridging oxygen atoms and an isosceles Os_3 triangle with one long Os...Os distance give a significant improvement in the fit.

Analysis of the splined EXAFS curve of the alumina supported species clearly showed the presence of one or more bridging oxygen atoms. Two models of the types $[Os_3(H)(CO)_{10}(O-)]$ and $[Os_3(CO)_{10}(O-)_2]$ give very similar fits. However the latter model required Debye Waller factors which were consistent with those in the model compounds for the bridging oxygen atoms and is therefore favoured. The similarity of the EXAFS of the model and supported thiolate bridged cluster indicated a very similar local geometry and this was confirmed in the curve fitting results.

228

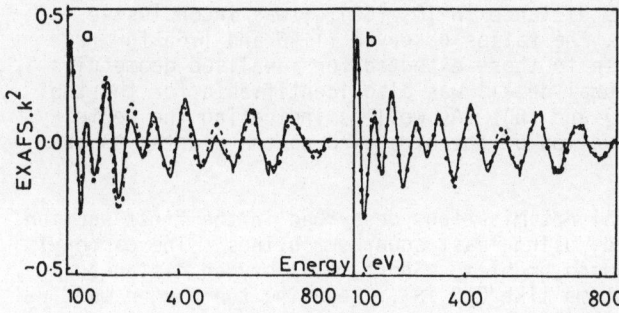

Fig. 4 Fits of the k^2 weighted EXAFS of $[Os_3(CO)_{10}(\mu\text{-}OEt)_2]$, (a) 3 shell and (b) 5 shell models

Table 3 EXAFS derived distances and Debye Waller factors (α) for models for the oxide supported species (X-ray diffraction distances in parentheses)

Compound	Distances ($\overset{o}{A}$) and α ($\overset{o}{A}{}^2$)			
	Os-O	α	Os-Os	α
(7)	-	-	2.83(2.82)	0.010
			2.64(2.68)	0.013
(8)	2.24(2.10)	0.016	2.81(2.82)	0.012
(9)	2.24(2.09)	0.019	2.82(2.82)	0.009
			3.08(3.08)	0.011
(5), 1 bridge	2.25	0.006	2.84	0.009
(5), 2 bridges	2.24	0.016	2.84	0.006
			3.13	0.014

(c) Extension to light metal systems

It is important to establish whether remote metal shells can be detected for elements lighter than osmium. The complex $[Co_3(CO)_9(CH)]$ (11) has been used as a model complex [14]. Again a close fit of all but the Co...O shell was obtained using ab initio phase shifts. A more complex example is provided by $[Co_4(CO)_{10}(PPh)_2]$ (12), which contains a Co...Co separation of 3.69 Å across a Co_4 rectangle [15]. In addition to the shells due to the bonded atoms and the carbonyl oxygen, evidence was obtained for the remote metal shell. Distances of 3.63(2) and 3.736(2) Å were derived by fitting the k^3 weighted experimental EXAFS and Fourier filtered data (3.3-3.8 Å) respectively.

4. Discussion

The spherical wave analysis of the higher nuclearity clusters indicated that detail of cluster skeletal geometry was obtainable. In the anion $[Os_{10}C(CO)_{24}]^{2-}$, there are three Os...Os separations with $r(Os...Os)/r(Os-Os)$ ratios of 1.414, 1.732 and 2.0 for an idealised geometry. The observed ratios were 1.44, 1.73 and 2.05. Radius ratios can also be used to distinguish between the skeletons of the two hexanuclear clusters since the evidence for the longer Os...Os distance in $[Os_6(CO)_{18}]$ was inconclusive. For the first non-bonded shell, the ratios observed (1.46 and 1.65 for (2) and (3) respectively) were close to those expected for idealised geometries (1.414 and 1.63). Some structural detail was also identifiable for the two supported triosmium species (5) and (6). As well as indicating the geometry of the Os_3 triangle, the observation of the atom binding the cluster to the surface is encouraging.

The objectivity of the manual optimisations described in the first section of the work have been confirmed, using least squares routines. The carbonyl oxygen atoms remain an outstanding problem. Short metal-oxygen distances have also been observed in systems like CaO [8]. Reducing the oxygen muffin-tin radius in ab initio phase shift calculations afforded an improvement in the metal-oxygen distance, but a match of crystallographic separations was not achieved for physically realistic radii. Empirical phase shift adjustment, which was successful in CaO and related systems [16], has given a close match of the crystallographic distances in $[Co_3(CO)_9(CH)]$ and $[Os_3(CO)_{12}]$. A diminution in the Debye Waller factors, due to the simple refinement procedure, can be offset by refining Z+1 or Z+2 element phase shifts. Improvements in this procedure are underway. The omission of multiple scattering contributions may contribute to the short metal-oxygen distances, since these are expected to be strong for the linear M-C-O units. Calculations using a least squares multiple scattering routine are also underway.

Acknowledgements

We are grateful to our colleagues, Professor P B Wells, Mr P Worthington and Dr P R Raithby for their assistance, and Johnson Matthey for the loan of OsO_4.

1.   B F G Johnson: Transition Metal Clusters (Wiley, New York, 1980).
2.   S L Cook, J Evans, G N Greaves, B F G Johnson, J Lewis, P R Raithby, P B Wells and P Worthington: J Chem Soc, Chem Commun, 777 (1983).
3.   S L Cook, J Evans and G N Greaves: J Chem Soc, Chem Commun, 1287 (1983).
4.   E Pantos: Daresbury Laboratory Preprint, DL/SCI/P346E (1982).
5.   S J Gurman: Daresbury Laboratory Technical Memorandum, DL/SCI/TM21T (1980); C D Garner and I Ross: unpublished results.
6.   P A Lee and J B Pendry: Phys Rev B, 11, 2795 (1975).
7.   S J Gurman, N Binsted and I Ross: J Phys C, Solid State Phys, 17, 143 (1984).
8.   N Binsted: unpublished results.
9.   E Pantos and G D Firth: Daresbury Laboratory Preprint: DL/SCI/P346E (1982).
10.   M R Churchill and B G DeBoer: Inorg Chem, 16, 878 (1977); R Mason, K M Thomas and D M P Mingos: J Am Chem Soc, 95, 3802 (1973); M McPartlin, C R Eady, B F G Johnson, J Lewis: J Chem Soc, Chem Commun, 883 (1976); P F Jackson, B F G Johnson, J Lewis, M McPartlin, and W J H Nelson: J Chem Soc, Chem Commun, 224 (1980).
11.   R Psaro, R Ugo, G M Zanderighi, B Besson, A K Smith and J M Basset: J Organomet Chem, 213, 215 (1981); B Besson, B Moraweck, A K Smith, J M Basset, R Psaro, A Fusi and R Ugo: J Chem Soc, Chem Commun, 569 (1980).
12.   J Evans and B P Gracey: J Chem Soc, Dalton Trans, 1123 (1982).
13.   V F Allen, R Mason and P B Hitchcock: J Organomet Chem, 140, 297 (1977).
14.   P Leung, P Coppens, K R McMullan and T F Koetzle: Acta Cryst, B37, 1347 (1981).
15.   R C Ryan and L F Dahl: J Am Chem Soc, 97, 6904 (1975).
16.   N Binsted, C M B Henderson and G N Greaves: Prog Experimental Petrology, in press (1984).

EXAFS and Mössbauer Study of Small Metal Clusters Isolated in Rare-Gas Solids

P.A. Montano
Department of Physics, West Virginia University, Morgantown, WV 26506, USA
G.K. Shenoy and T.I. Morrison
MST Division, Argonne National Laboratory, Argonne, IL 60439, USA
W. Schulze
Fritz-Haber Institut der Max-Planck-Gesellschaft, D-1000 Berlin

1. Introduction

The determination of the interatomic separation in small metal clusters is
of great importance. For instance, the knowledge of this parameter permits
one to carry out theoretical calculations on small clusters to predict their
experimentally measured properties. There have been several electron diffraction studies to determine the lattice parameter for various cluster
sizes [1]. However, a simple interpretation of the electron diffraction
data for small clusters is often unreliable and this obscures the determination of the atomic separation [1]. On the other hand, EXAFS provides a
unique approach to such measurements on small clusters.

We have undertaken a general program of isolating small metal clusters in
rare-gas solids and measuring the atomic separation in them using EXAFS.
It should be emphasized that the small cluster will not interact with the
rare-gas supports and hence will not disturb the electronic and geometric
properties of atoms in such isolated clusters. In this summary report, such
data is presented for clusters of Cr and Fe isolated in neon and those for
Ag clusters in argon. In the case of Fe clusters isolated in argon and
xenon, we have also carried out the measurements of the isomer shift and the
magnetic hyperfine field at Fe-57 nucleus using the Mössbauer effect.

2. Experimental

The metal clusters were prepared starting with an atomic beam of Cr or Fe,
codeposited with Ne onto an ultrapure Al foil kept at 4.2 K. The desired
dilution of 3d atoms in Ne was achieved by carefully monitoring the deposition rates using 6.3 keV X-ray and 14.4 keV gamma-ray absorption in the
sample. Samples containing a higher concentration of 3d atoms would contain
a broader range of cluster sizes compared to lower concentrations [2]. At
the lowest concentration (0.1 at.%) one forms majority of dimers of metal
atoms. In the case of Ag, the clusters of well-defined size between 25 and
140 Å were isolated in Ar matrix by the gas aggregation technique [3].

All X-ray absorption and fluorescence spectra were measured at the
Stanford Synchrotron Radiation Laboratory. Analysis of the EXAFS data was
carried out using standard procedure [2]. Mössbauer spectra of Fe-57 were
measured using conventional electromechanical drives on samples produced
under conditions identical to those used for X-ray studies.

3. Results and Discussion

In Fig. 1, the separation between metal atoms in Cr and Fe clusters is
plotted as a function of atomic concentration. In general there is a strong
reduction in the near-neighbor (nn) separation on decreasing the metal con-

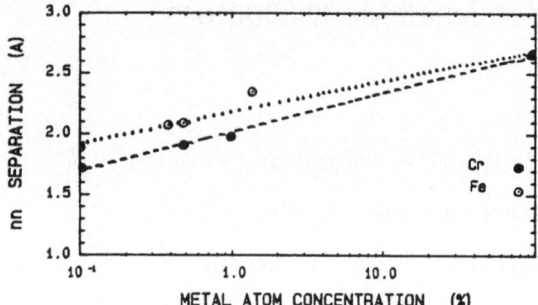

Fig. 1. nn separation vs. metal atom concentration for Cr and Fe in Ne.

centration or the cluster size. At the lowest concentration of 0.1 at.% of
Fe, the sample primarily contained dimers (Fe_2) which was unambiguously
identified from Mössbauer spectroscopy [4]. The contraction in the case of
Cr is far more dramatic and the dimer separation of 1.70 ± 0.02 Å agrees well
with that deduced from optical measurements.

The changes in the cluster size influence their physical properties as
well. To illustrate this, we have measured the Mössbauer isomer shift and
the magnetic hyperfine field at Fe-57 nucleus in Fe clusters isolated in Ar
and Xe solids. The results are shown in Fig. 2. Concurrent with the decrease in the nn separation with the cluster size, both the isomer shift and
the hyperfine field show major changes under about 10 at.%. The isomer
shift is a measure of the total electron density and the hyperfine field of
the spin density at the nucleus. These changes arise not only from the
changes in the nn separation with cluster size, but also owing to a transition from a bond description of the electronic structure in a dimer to an
electronic band picture in larger clusters. The present measurements of the
nn separation could be the starting point for detailed calculation of the
electronic properties of these small clusters.

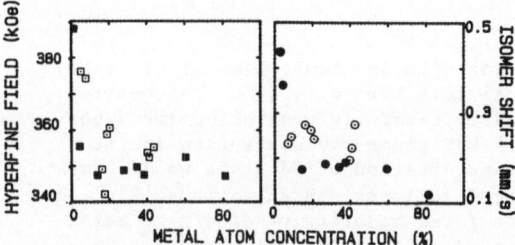

Fig. 2. Isomer shift and magnetic field for Fe-57 in Ar (open) and Xe
        (full).

The gas aggregation technique [3] used for the case of separating Ag
clusters in Ar matrix defines the size of the clusters more uniquely. The
cluster sizes have been measured from electron micrographs obtained during
the cluster preparation. In Fig. 3 the variation in the nn separation with
the cluster size is shown. The contraction, R, was analyzed using a simple
stress model [1] described by $R = -2fRk/3r$ where k is the compressibility, R
is the nn separation in Ag metal, f is the surface stress, and r is the
cluster radius. The dashed line in Fig. 3 is the fit using experimental
values for k, R, and f (=2296 dyn/cm). The agreement between the model
calculations and the experiment is very good.

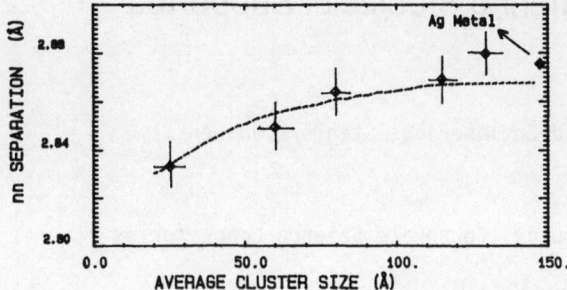

Fig. 3. nn separation vs. cluster size for Ag in Ar.

This work was supported by the U.S. Department of Energy and Max Planck Gesellschaft. We also would like to thank the staff at SSRL.

References

1. H.J. Wasserman and J.S. Vermaak: Surf. Sci. 22, 164 (1970); 32, 165 (1972).

2. H. Purdum, P.A. Montano, G.K. Shenoy, and T.I. Morrison: Phys. Rev. B 25, 4412 (1983).

3. P.A. Montano, G.K. Shenoy, T.I. Morrison, and W. Schulze: Phys. Rev. B (in press).

4. P.A. Montano, P.H. Barrett, and H. Micklitz: Mössbauer Effect Methodology, Plenum Press, NY) 10, 245 (1976).

Differential Anomalous Scattering Studies of Supported Pt Clusters

S.S. Laderman

Department of Materials Science and Engineering, Stanford University
Stanford, CA 94305, USA

K.S. Liang and J.H. Sinfelt

Exxon Research and Engineering Company, Corporate Science Laboratories
Annandale, NJ 08801, USA

Using the differential anomalous scattering technique, the diffuse diffraction pattern associated with small Pt clusters has been successfully separated from the strong and broad Bragg scattering of an η-Al_2O_3 substrate. At 5 weight %, highly dispersed Pt clusters, reduced in H_2, appear to be face centered cubic microcrystals. Examination of the data from an air exposed sample shows that changes in structural correlations beyond first neighbors brought about by changes in chemical environment appear clearly with this method.

The differential anomalous scattering (DAS) technique is a means of obtaining x-ray diffraction patterns with EXAFS-like atomic species selectivity [1]. We are applying DAS to better understand the structure of Pt clusters supported on alumina. To our knowledge, this is the first time the diffraction pattern of highly dispersed Pt clusters has been separated from the signal of a strongly scattering polycrystalline substrate and normalized. The experimental technique, analysis method and results are presented below.

A sample of 5.0 weight % Pt on η-Al_2O_3 was prepared as previously described [2]. Hydrogen adsorption measurements showed the Pt was about 70% dispersed. X-ray data were collected at room temperature with the sample in air and also after reduction in dry H_2 at 220 degrees C for 60 minutes. The reduced sample was maintained in a dry H_2 atmosphere throughout data collection. The data were collected with a two-circle diffractometer at wiggler station IV-3 at the Stanford Synchrotron Radiation Laboratory during dedicated running. The two crystal Si(220) monochromator was detuned to limit harmonic contamination of the incident beam. Patterns were collected using x-ray energies of 9 eV, 14 eV and 99 eV below the absorption maximum (white line) at the L_{III} edge of Pt. The incident intensity, I_0, was monitored with a scintillation detector and the scattered intensity, I, was measured with an intrinsic Ge detector. Total count rates in each detector were kept below 40,000 cps to insure a linear response. Pulse discrimination removed harmonics and noise in I_0 and I and the L_α fluorescence in I. After averaging individual scans, 1200 points between $k = (4\pi/\lambda)\sin\theta = 1$ and 11 Å^{-1} with total counting times of up to one minute per point were obtained at each energy.

Figure 1a displays the total diffraction pattern, I/I_0, for the reduced sample at 99 eV below the white line. At 9 eV below the white line, the sample absorption, the instrumental efficiency and the Pt scattering factor are all different. The changes in intensity of the strongest Al_2O_3 reflections are primarily due to the first two effects. Patterns from these two energies were put on the same relative scale by best matching the two strongest Al_2O_3 peaks. The low energy scan minus the scaled high energy scan appears in Fig. 1b. A similar difference of data from the air exposed sample appears in Fig. 1c. For that, data from 99 and 14 eV below the white line were used.

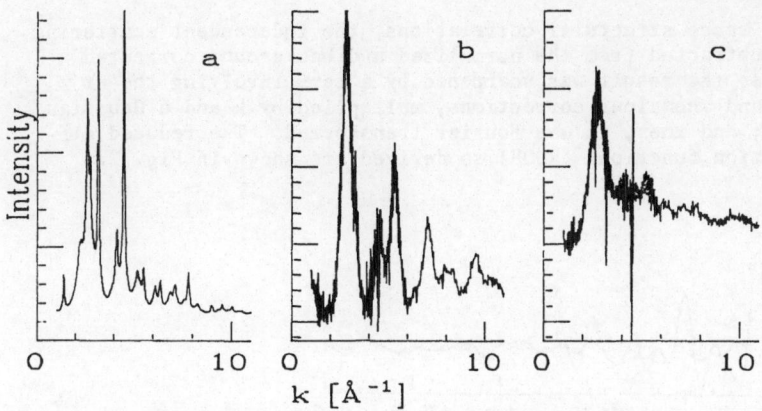

Fig. 1. (a) Sample scattering, (b) difference in H_2, (c) difference in air. The ordinate limits are 0.0 to 1.75 in (a), -0.025 to 0.060 in (b) and (c).

The broad peaks from the reduced sample may be indexed as face centered cubic (FCC) Pt peaks while the air exposed sample appears amorphous. The difference in background level is not related to the difference in structure. The L_β fluorescence present in scans at 9 eV below the white line appears clearly as a negative background, as seen from the reduced sample. This is very small in the case of the air exposed results because at 14 eV below the white line, the fluorescence is very small.

To normalize the data, the level of the flat fluorescence background and the instrumental constant are needed. Methods developed for amorphous materials [1] did not work well. The crystallinity of the substrate makes absolute normalization of the original patterns difficult. We derived integral relations and high k matching criteria for the difference pattern, but the normalization constants so derived were very sensitive to the upper limit of integration because of the strong oscillations present at the highest angles. A more definite result was obtained by using a structure-dependent but self-consistent observation. The widths of the peaks from the reduced sample suggest an average cluster size near 15 Å diameter. For a range of models of such clusters, calculations we performed show that between 1.5 and 2.0 $Å^{-1}$, a minimum with value zero appears in the scattering. Such minima appear in Figure 1b and 1c. We assume that these minima are at the fluorescence level. That the minimum appears near zero height in the air exposed pattern, where the background due to fluorescence is known to be very small, adds assurance that the assumption is valid. Forcing the high k scattering to oscillate about the independent scattering is the second relation needed to find the two unknowns. This procedure may underestimate the magnitude of the fluorescence background, resulting in an overestimate of coordination numbers. There is, also, uncertainty in scaling the high k scattering. We currently estimate the uncertainty in normalization to be about 20%.

To scale the data, the anomalous scattering corrections to the Pt scattering factor are needed. They were obtained from inversion of transmission EXAFS data [3]. At 99, 14 and 9 eV below the white line, the real corrections were -12.4, -17.7 and -19.3 electrons while the imaginary corrections were 3.92, 4.42 and 5.17 electrons, all of them independent of k. At the energies studied, chemical effects are expected to be very small [4].

235

To see the real space structural correlations, the independent scattering contribution was subtracted from the normalized and background corrected difference patterns; the result was sharpened by a term involving the Pt scattering factor and anomalous corrections, multiplied by k and a Gaussian convergence factor; and then, it was Fourier transformed. The reduced differential distribution functions (RDDF) so derived are shown in Fig. 2.

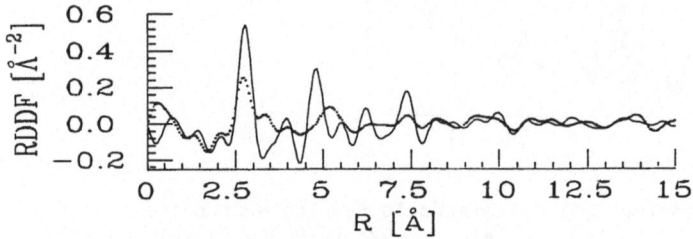

Fig. 2. Distribution function for sample in H_2 (solid) and in air (dots).

The reduced sample is well modelled as an FCC microcrystal. Each of the first 7 FCC shells appears clearly while each experimentally derived local maximum between 2.70 and 8.00 Å corresponds to such a shell. Assuming only Pt-Pt correlations, the area of the first peak corresponds to 11 atoms and that of the second to 6 atoms. Taking the 20% uncertainty into account, these agree well with FCC cluster values. The first peak in the air exposed case is much lower, consistent with the oxidized nature of the cluster. The peak at 5.17 Å is at a minimum in the distribution from the H_2 reduced sample, showing a clear difference in the higher shell structures.

We are pursuing a detailed structural interpretation of the data. We expect, also, to extend the measurements to additional catalyst preparations and chemical environments.

We thank M. Boudart for the sample cell and A. Bienenstock, K. F. Ludwig and F. W. Lytle for helpful discussions. This work was partially done at SSRL, which is supported by the DOE, Office of Basic Energy Sciences, and the NIH, Biotechnology Resource Program, Division of Research Resources.

References

1. P. H. Fuoss, P. Eisenberger, W. K. Warburton and A. Bienenstock: Phys. Rev. Lett., 46, 1537 (1981)
2. J. H. Sinfelt: Annu. Rev. Mater. Sci., 2, 641 (1972).
3. L. Wilson, W. K. Warburton and K. F. Ludwig, to be published
4. P. H. Fuoss and A. Bienenstock: "X-ray Anomalous Scattering Factors - Measurements and Applications", in: Fabian, D. J., Kleinpoppen, A. and Watson, L. M., eds., Inner-Shell and X-ray Physics of Atoms and Solids (Plenum, New York 1981)

Part V **Surface Structure**

New Perspectives in SEXAFS

F. Comin

Istituto di Struttura della Materia del CNR, I-00044 Frascati, Italy

1. *Introduction*

In the past years SEXAFS has made significant contributions to the study of ordered overlayers on substrates. This has been possible by exploiting two powerful characteristics of the SEXAFS technique: atom specificity and bond orientation sensitivity. The intrinsic anisotropy of an ordered interface makes full use of the polarization dependence of the photoabsorption cross section, whereas the (ad)atom specificity inherently provides surface sensitivity. Nevertheless, such SEXAFS studies of ordered overlayers do not utilize another important feature of EXAFS spectroscopy, i.e., the short-range character which makes it possible to study disordered systems. Recently, results[1] from reactive interfaces have shown that SEXAFS can determine site geometries in disordered systems *without* the assistance of polarization dependence. This is a marked departure from previous studies on ordered overlayers where the polarization dependence has played an important role for structure determination.

Another point is the inherent surface sensitivity by atom-specific discrimination in SEXAFS experiments. Dropping this convenience directly addresses the problem of solving structural problems from *clean* surfaces. It will be shown here that a proper choice of detection schemes can make SEXAFS an invaluable complementary technique in the structural study of such surfaces. As one example, evidence of surface reordering from clean amorphized silicon will be given. In this experiment low-k SEXAFS data are used to provide a direct structural probe on a system in which the lack of long-range order makes other surface structural experiments infeasible.

2. *A Solid-State Reaction: Ni on Si(111).*

The Ni-Si system has particular importance not only for its wide use in silicon technology, but also as a prototype for studying solid-state reactions and Schottky barrier formation. Despite intense efforts using

238

a broad range of experimental techniques, no satisfactory model for the room-temperature (RT) nucleation of nickel silicide had been proposed. The lack of long-range order, even in the first stages of deposition, precluded most structural probes from following the formation process. The short-range nature of SEXAFS, however, is ideally matched for this study.[1]

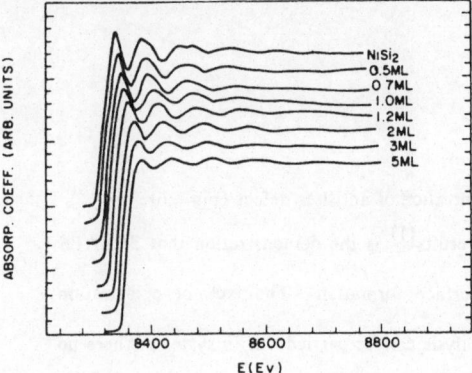

FIG. 1. Ni K-edge absorption data from bulk NiSi$_2$ and from Ni deposited on Si(111) at RT for different coverages. The K-edge jumps of the latter are normalized to that of NiSi$_2$. Energy scales are shifted relative to 5-ML data for clarity (ML, monolayer).

In Fig. 1 Ni KLL Auger yield data from the model compound NiSi$_2$ are compared with those obtained from increasing RT coverages of Ni on Si(111). Standard analysis procedures confirm the similarity between NiSi$_2$ and the 0.5 ML Ni/Si(111) system, with the first coordination shell around Ni being composed of 6-7 Si atoms at a distance of 2.37 ± .03 Å, in close agreement with the 2.34 Å Ni-Si bond in NiSi$_2$. With these results the chemisorption site can be unambiguously determined to be the sixfold hollow *between* the first and second Si layers. To accommodate the observed Ni-Si bond length, the three surface silicon atoms must expand outward, breaking the Si-Si interlayer bond. The resulting configuration corresponds to one (111) monolayer of cubic NiSi$_2$ with the bonds along the [111] direction partially unsaturated. Any other proposed nucleation site has been ruled out by extensive computer simulations and by comparisons with other experimental results. The determined chemisorption site naturally accounts for a variety of other measurements[2] and forms a basis for understanding the observed 180° rotation of the epitaxial layer which results from light annealing of low RT coverages of Ni.[3] At higher coverages, SEXAFS data exhibit a monotonic decrease of intensity and a modification of the envelope function, while the first shell bond length remains essentially unchanged. This behavior is compatible with progressive Ni substitution of the Si atoms within an NiSi$_2$ overlayer structure (antisite defects). The Ni-Ni and Ni-Si phase shifts differ by π, so that each substituted atom contributes antiphase, leading to an apparent decrease of the coordination number.[4]

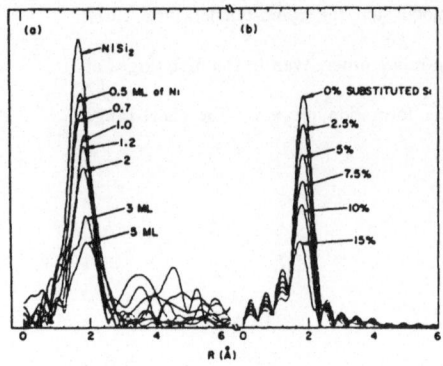

FIG. 2. (a) Fourier transforms of data shown in Fig. 1. Higher-frequency peaks at $R > 3$ Å are due primarily to noise. (b) Fourier transforms of simulated data assuming the $NiSi_2$ structure with varying amounts of Ni substituting Si sites. The 0% data have been normalized to the 0.5-ML data in (a). Side peaks are due to truncation errors.

Computer simulations have been used to estimate the concentration of antisites defect (Fig. 2b).

In addition to the other conclusions drawn from these results[1] is the demonstration that SEXAFS can be effectively applied to the study of reactive interface formation. The lack of polarization dependence is not a limiting factor, and the structural analysis can be carried out in systems where no other structural probe is applicable.

3. SEXAFS From Clean Surfaces: Amorphized Silicon.

Dealing with clean surfaces removes the surface sensitivity offered by the atom selectivity of SEXAFS from an adsorbate. Total and partial yield detection modes cannot be used because they probe far inside the surface (even secondaries with low escape depth originate well within the bulk). However, elastically scattered Auger electrons with low escape depth and detected at shallow takeoff angles can be used to probe only the surface contribution. Almost every element has low escape depth Auger transitions: for Si, the LVV Auger electrons at ~90 eV can be used. Their escape depth of ~5 Å, combined with shallow takeoff angles, can in principle provide probing depths as low as <2 Å. Clean single crystal Si samples with different crystal orientations were prepared following usual UHV procedures to obtain ordered crystalline surfaces.[6] Amorphous silicon was deposited on top of these surfaces by electron gun evaporation. "Amorphized" samples were obtained by heavily sputtering with 2 kV rare gas ions. Si KLL and Si LVV Auger yield absorption data were collected from the crystalline, amorphous, and amorphized samples with a geometry that provided probing depths of ~4 Å for the LVV electrons and ~30 Å for the KLL. Analysis was carried out on the Si K-edge absorption data in the range of 10-90 eV following standard EXAFS procedures. This range is not normally applicable to single scattering analysis procedures, even though theoretical work (presented elsewhere in this volume) suggests the possibility of extending the usable EXAFS range to energies very close to threshold. It is not the aim of the present work to demonstrate the validity of such an approach. Rather, since we always deal with the same atomic species, Si, questions of phase and amplitude transferability are unimportant and so we can use this low-k EXAFS regime to test the degree of order of the sample. In fact, a noticeable advantage of the low-k region is the reduced Debye-Waller damping, allowing departure from medium range order (MRO) (as tested from the ratio between the 1st and the 2nd and 3rd Fourier peak magnitudes[5]) to be easily detected. Compared with data obtained using Si KLL electrons, that obtained with the more surface sensitive LVV electrons from the same (ion-damaged) amorphized Si were found to exhibit a marked enhancement of MRO. The reference sample, (evaporated) amorphous Si, did not show any enhancement. This is a clear signature of *reordering* of the topmost layer of amorphized Si. In Fig. 3 the raw absorption data for crystalline and amorphized Si are shown. In Fig. 4 the Fourier transforms of analogous data clearly show the additional crystalline content in the LVV surface sensitive yield.

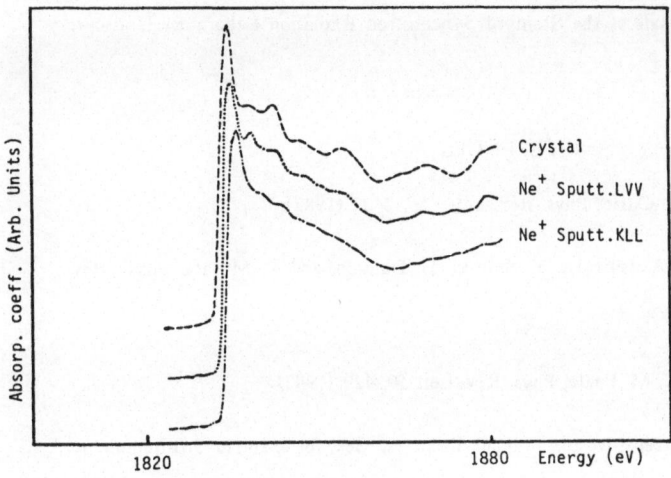

Fig. 3. K edge absorption data of Silicon

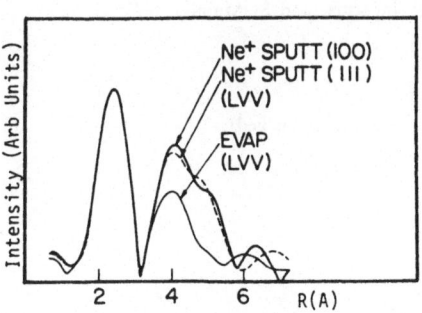

Fig. 4. Fourier transform ofSi K edge data in the range 10-90 eV. The curve for evaporated Si is indistinguishable from the KLL of Ne^+ sputtered sample

Besides the direct implications of these results on the characterization of amorphous systems is the demonstration that a proper choice of detection scheme and geometry allows SEXAFS to be obtained from *substrate surfaces either in the presence or absence of chemisorbed species.* The unimportance of long-range order in such SEXAFS experiments allows almost any system to be investigated with these techniques. Future experiments will include the complete EXAFS range, extending beyond that portion of the data where photoelectrons enter the low-energy Auger window.

4. *Conclusion*

It has been shown that new applications are open to the SEXAFS technique. Clean and disordered surfaces can be studied , with the possibility of addressing such long-standing problems as surface reconstructions (clean surfaces) and reactive chemisorption pathways.

This work has been done in close collaboration with P. H. Citrin, whose contribution is here greatly acknowledged. Support from the staff of the Stanford Synchrotron Radiation Laboratory is also very much appreciated.

REFERENCES

1. F. Comin, J. E. Rowe, and P. H. Citrin, Phys. Rev. Lett. *56,* 2402 (1983).

2. K. L. I. Kobayashi, S. Sugaki, A. Ishizaka, Y. Shiraki, H. Daimon, and Y. Murata, Phys. Rev. B*25*, 1377 (1982).

3. R. T. Tung, J. M. Gibson, and J. M. Poate, Phys. Rev. Lett *50,* 429 (1983).

4. The interference spectra is very sensitive to the bond distances. See, for example: J. Mimault, A. Fontaine, P. Lagarde, D. Raux, A. Sadoc, and D. Spanjaard, J. Phys. F*11*, 1311 (1981).

5. F. Evangelisti, M. G. Proietti, A. Balzarotti, F. Comin, L. Incoccia, and S. Mobilio, Solid State Comm. *37* 413 (1981).

6. F. Comin, L. Incoccia, P. Lagarde, G. Rossi, and P. H. Citrin, (to be published).

X-Ray Absorption Spectroscopy and Scattering for Surfaces

P. Rabe

Institut für Experimentalphysik der Universität Kiel
D-2300 Kiel, Fed. Rep. of Germany

1. Introduction

Various sample systems are collected under the generic term surface. The technical surface, i.e. catalysts, distinguishes itself by an exceptionally large area and, EXAFS studies, can be investigated in general by comparable simple transmission experiments. Large surface areas as well are characteristic for pseudo surfaces like intercalated layered crystals. As an example we mention here graphite, which can be produced with surface areas of the order of $1m^2/cm^3$. Again these systems can be investigated in transmission experiments so that the demands concerning the intensity of X-ray sources are moderate. Essentially greater demands on the source are made from clean surfaces or surfaces with low coverages of absorbates. A coverage of a single monolayer on a single crystal surface represents an area density of $\sim 10^{15}$ atoms/cm^2 and is (compared to a bulk experiment) equivalent to a dilution of 10^{-5} to 10^{-6}. Obviously it is trivial to state that X-ray sources as intense as possible are required in surface structural analysis.

Improvements of the experimental possibilities in the future to investigate two dimensional structures can be expected from three directions:
a) Wigglers and undulators at dedicated synchrotron radiation sources provide a higher intensity compared to X-ray sources available in the past.
b) New experimental techniques have been developed or are in the process of development, where the X-rays are absorbed in superficial regions rather than uniformly in the bulk.. c) Through a combination of different experimental techniques (X-ray absorption, X-ray scattering) the structural properties are illuminated from different points(local order, correlation or long range order). In the following, guided by selected examples it shall be demonstrated how these different aspects have on the experimental possibilities in the past and possibly will affect the development in the future.

2. Improved X-ray sources

Before answering the question of the highest possible intensity one has to look for the relevant parameter of the source which describes the finite signal at the experiment. For absorption experiments at synchrotron radiation sources this parameter generally is the spectral flux defined by the unit (number of photons/ sec $*$ mrad horizontal $*$ 0.1% band width).

In special cases e.g. with small samples which cannot utilize an X-ray beam with large cross section, the relevant parameter is the spectral brilliance defined by (flux/mrad vertical $*$ mm^2 source area).

To estimate the progress which can be expected in future we shall reproduce some numbers which have been calculated for the planned European Synchrotron Radiation Facility (ESRF)(1). The present layout of the ESRF provides four different types of sources: bending magnets, two types of multipole wigglers, and undulators. Whereas the flux from a bending magnet at a wavelength

243

of $\lambda=1.5$ A amounts to the order of 10^{12} photons/sec $*$ mrad $*$ 0.1% $\Delta\lambda/\lambda$ the flux from the multipole wigglers is increased to the order of 10^{14} photons/sec $*$ mrad $*$ 0.1% $\Delta\lambda/\lambda$. These numbers are mainly limited by the current stored and probably cannot be increased significantly. Different is the development of the spectral brilliance. By decreasing the beam size, a drastic increase of the brilliance is to be expected at the ESRF. At a wavelength of 1.5 A one expects 1.6×10^{15} photons/sec $*$ mrad^2 $*$ 0.1% $\Delta\lambda/\lambda$ from a bending magnet. The corresponding numbers for the multipole wiggler and the undulator are 1.6×10^{16} and 1.8×10^{17}. These numbers have to be compared with 10^{13} photons/sec $*$ mrad^2 $*$ mm^2 $*$ 0.1% typically provided from bending magnets of existing sources.

Which consequences can be drawn from these perspectives for the investigation of e.g. adsorbates?
The arrangement of an Auger yield SEXAFS experiment illustrates that the spectral brilliance contributes essentially to the fraction of Auger electrons collected in the electron energy analyzer. With present synchrotron radiation sources adsorbates have been investigated down to about 1/3 of a monolayer. A typical example is shown in fig. 1(2). In this experiment a Si (111) surface was saturation covered with Cl. The EXAFS spectrum has been measured by monitoring the Cl $KL_{2,3}L_{2,3}(^1D)$ Auger electrons.

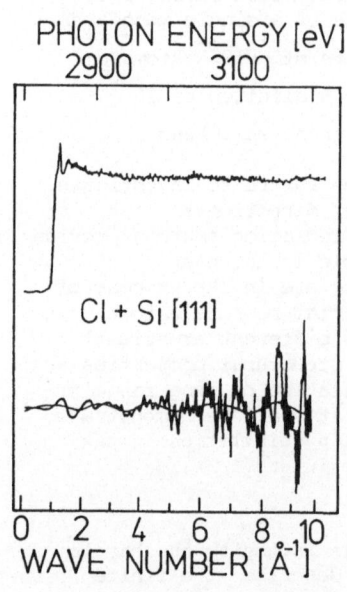

PHOTON ENERGY [eV]

Cl + Si [111]

WAVE NUMBER [Å^{-1}]

Fig.1. Top: Auger yield spectrum at the Cl K-edge measured with polarization vector parallel to the surface. Bottom: Background - subtracted SEXAFS and filtered data (Smooth line)(from ref.2)

It should be emphasized that these spectra show the state of the art for systems of such strong dilution. Nevertheless the significant fraction of statistical noise in the spectra calls for X-ray sources with improved flux and brilliance. A small contribution of statistical noise leads to large uncertainties in the determination of bond lengths from X-ray spectra as has been shown by Mobilio and Incoccia (3). For an EXAFS spectrum of a fictive molecule Fe N_6C_8 (6 N atoms at 2.05A and 8 C atoms at 2.78A) covered with less than 1% noise they calculate uncertainties of the order of 0.1A for nearest neighbour distances. At a noise level of more than 0.4% the second coordination sphere cannot be resolved. This means for the present

SEXAFS experiments that second coordination spheres around absorbed
atoms cannot be resolved for coverages below 1 monolayer. An increasing brilliance at synchrotron radiation sources will essentially lead to improved
accuracies in the determination of bond lengths. In conclusion, a prediction
about the possibilities at dedicated synchrotron radiation sources seems to
be trivial: either with a comparable signal-to noise level adsorbates with
lower coverages are accessible or EXAFS spectra with significantly improved
signal-to-noise and thereby more accurate bond lengths can be measured.
Nevertheless this statement is of fundamental importance for surface science.On
the one hand at coverage of one monolayer electronic interactions between
nearest neighbours of the adsorbate are by no means negligible. At lower coverages surfaces will be accessible where the adsorbed atoms are isolated i.e.
without mutial interaction. On the other hand with improved S/N-levels informations about coordination shells beyond the nearest neighbours can be obtained
which possibly lead to solutions of questions like surface reconstruction.
In the experiments discussed so far the significant fraction of the primary
photons is unused. Typical penetration depths of the primary photons amount
to 10-100 μm compared to the thickness of a surface layer of 0.2-0.3 nm. This
means that only a fraction of 3×10^{-5} - 2×10^{-6} of the incoming photons are
absorbed in the adsorbate. This petty amount is the crux of the SEXAFS experiments and a waste of the expensive synchrotron radiation.

3. Enhanced Absorption

The losses of photons in an absorption experiment can be reduced if the photons
hit the sample surface under grazing incidence. It is well known that with
decreasing glancing angles the penetration of photons decreases and reaches
typical values c 2 nm in the range of total reflection.
Apparently only superficial regions contribute to the absorption in that case,
which means that a reflectivity experiment uses the available photons more
effectively.
Two possible variants, comparable to the transmission EXAFS and Auger yield
EXAFS can be performed:
a) The reflectivity yields a signal which arises from the coherent scattering
in forward direction by all atoms present within the penetration depth of the
sample(4,5). Because the reflectivity depends in a rather complicated form on
both optical constants the observed finestructure in general will differ from
the EXAFS in the absorption coefficient. The price which has to be paid
for the combination of high efficiency and surface sensitivity is a comparable
large effort in the evaluation of the experimental data. However if the transmission SEXAFS experiments reach their bounds according to limited photon
fluxes,this technique may extend surface experiments to even larger dilution.

The grazing incidence of X-rays is combined with an increase of the effective
sample size i.e. the projection of the beam on the surface. For such
experiments the increase of the spectral brilliance will be of essential significance. An example will elucidate this:The effective vertical size of the
electron beam in the W multipole wiggler at the ESRF is expected to be 0.1 mm.
A one-to-one image of this source on the sample would lead to a projected
sample length of about 15 mm for a glancing angle of 0.4^{0}, a typical value
for the critical angle of total reflection. The corresponding parameters at
DORIS would be 1.2 mm and 225 mm. Samples of this size seem to be unacceptable
for routine surface experiments.
Making use of the reflectivity experiments structural analysis at boundaries
between two phases (solid-solid, solid-liquid, liquid-liquid) are possible
which defy present EXAFS experiments. If the electronic densities of the two
phases are different,the glancing angle can be chosen so that the X-rays
penetrate the first and are reflected from the second medium,thus probing the
interface region.

b) A further discrimination of certain atomic species, of course, is also possible in reflectivity experiments by monitoring electrons or fluorescence photons which leave the surface.
Because the escape depth of electrons is comparable to the penetration depth of the X-rays the electron yield will be increased significantly compared to a transmission SEXAFS experiment.

4. Position of adsorbates relative to the bulk and correlation (long range order)

The parameters provided by EXAFS experiments characterize the 2-dimensional structure only in an incomplete way. The long range order is hidden in the near edge structure. Although significant progress has been made in multiple scattering calculation it seems to be hopeless to derive parameters like correlation lengths from X-ray absorption spectra. For this reason alternative surface structure analyses based on the absorption of X-rays under special conditions or X-ray Bragg scattering in superficial regions should be performed in the future to obtain complementary information about the surface structure. Two of these experiments are briefly described in the following.
An experiment where the dependence of the X-ray absorption on selected spacial regions rather than the spectral dependence is explored has recently been introduced by Golovchenko and coworkers (6,7). The experimental arrangement is schematically shown in fig.2.

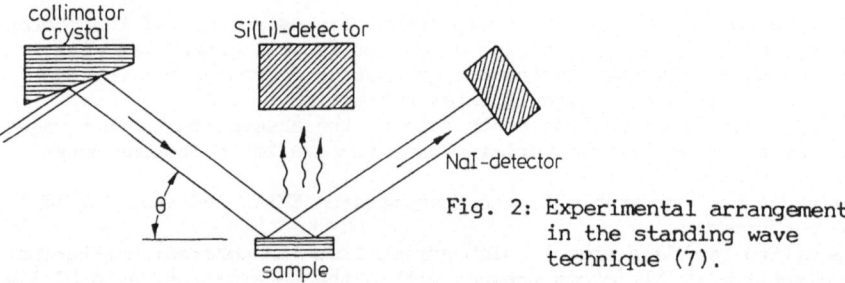

Fig. 2: Experimental arrangement in the standing wave technique (7).

A parallel beam of monochromatic X-rays which is generated by an asymmetric cut crystal hits the single crystal the surface of which is to be investigated. If the angle of incidence complies with the Bragg conditions, the incoming and the reflected beam interfere and form a standing wave field. In the Bragg case shown above (incoming and reflected beam are on the same side with respect to the entrance surface) the nodes of the wave field are parallel to the (hkl)-Fourier component of the charge density distribution. For weakly absorbing crystals the intensity distribution of the standing wave electric field varies sinusoidally with the period of the netplane distance and is projected from the surface by up to 100 nm. Total (Bragg) reflection occurs within a finite angular range. At the small angle boundary the nodes lie on the netplanes. Changing the Bragg-angle over the range of total reflection causes a phase shift in the reflected beam of π so that at the large angle boundary of total reflection the nodes of the wave field lies half way between the net planes. The absorption leads to an energy flow into the crystal which is smaller for the small angle than for the other borderline case. This spacial dependence of the local intensity and therefore the absorption of the electromagnetic field is used to determine distances of adsorbates normal to the reflecting net planes of the crystal. If the adsorbed atoms are in correlated positions with the bulk crystal the X-ray fluorescence yield will follow the X-ray field intensity when the crystal is rocked through the range of total reflection fig.3.

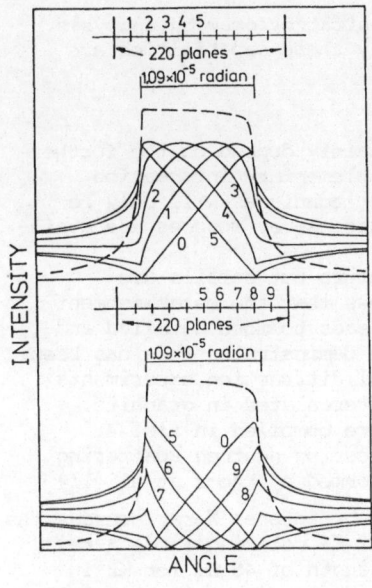

Fig. 3: —— X-Ray field intensities
on the 220 planes of a Si
crystal (0) and at indicated
positions (1-9) between the
planes. - - - reflectivity
for a perfectly collimated
beam at a photon energy of
17.5 keV (from ref. 7)

It should be noted that the positions of adsorbates are measured with respect
to the net planes of the bulk crystal which generate the standing wave field.
Distances of the adsorbates to the first surface layer of the crystal may be
different from those determined by the standing wave technique if e.g. the
surface of the crystal is reconstructed.

A modified version of this experiment using the Laue case (incoming and reflected beam are on opposite sides of the entrance surface) has recently been
presented by Materlik et al (8). In this geometry distances with respect to
netplanes perpendicular to the surface are determined.
A disadvantage of this method is the extremely low intensity. For a Si (220)
crystal, covered with 1/4 monolayer of Br the ratio of Br yield relative to
the primary flux amounts to 3×10^{-8} in the Laue case. A bending magnet unfocused
beam line of DORIS 0.3 Br K_α photons per second and mm^2 detector area are
observed. Compared to a SEXAFS experiment, these low counting rates are partly
compensated by the fact that relatively few data points are necessary to obtain
the structural information. Using a detector of 30 mm^2 in the above-cited experiment, the total counting time amounted to 160 min. A more improved X-ray
source like a multipole wiggler obviously offers the possibility to perform
such an experiment within a couple of minutes.
One point should be stressed at the end of this section:None of the experiments
discussed so far provides information about correlation lengths within the
two dimensional layer. This important parameter can only be determined by a
surface X-ray diffraction experiment. The idea to enhance the surface sensitivity compared to a bulk scattering experiment is basically the same as in the
EXAFS experiment in total reflection. The glancing angle under which the
photons hit the sample is chosen close to the critical angle of total
reflection. The angular dependence of the Bragg reflected X-rays directly
provide the reciprocal lattice vectors of the two dimensional structure. The
widths of the Bragg peaks yield the correlation length in the surface layer.
Such an experiment on a monolayer of Pb on Cu (110) surfaces has recently been
performed by Marra et al (9) at the focused wiggler beam line at SSRL. Remar-
cable high counting rates of 50000cps for the first order reflection and a

247

high signal-to-noise ratio of 500 are reported. A straightforeward analysis
resulted in correlation lengths of 100nm. Obviously these capabilities are
essentially based on a high flux X-ray source.

5. Conclusions

The future in structural surface science will certainly depend on the further
development and use of technique which provide complementary information
about 2-dimensional systems. However most of these techniques will only be
viable if more and more high flux and high brilliance X-ray sources are available for a broader surface science community.
Numerous new results have to be expected in the future but details are
difficult to forecast. However, a backward look shows that the establishment
of more and more advanced experimental equipment leads to more detailed and
in many cases unexpected results. An example which demonstrates this has been
compiled by J. Als-Nielsen (10). Since 1974 several diffraction experiments
have been performed on 2-dimensional structures intercalated in graphite.
The diffraction profiles of the (1,0) reflection are compiled in fig. 4.
From the width of the profile obtained in the pioneering neutron scattering
experiment on N_2 (a strong neutron scatterer)performed by Kjems et al (11)

a correlation length of 10nm was inferred. Using more intense X-ray sources the
correlation lengths increased with time. Using a 12 KW rotating anode X-ray
generator, Horn et al (12) obtained a correlation length of 45 nm for Kr in
graphite. A subsequent experiment carried out at a focused beam line at SSRL
by Moncton et al (13)led to a value of 200 nm.

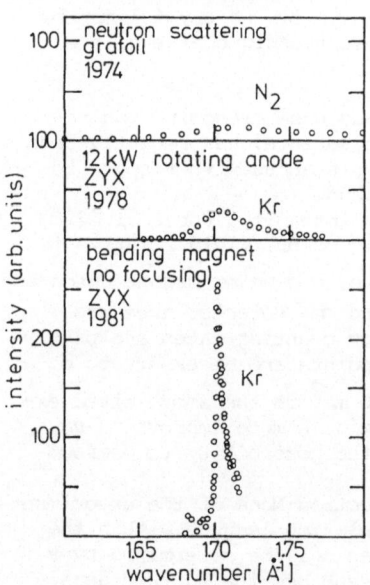

Fig.4: Scattering profiles of
2-dimensional N_2 and Kr
physisorbed on graphite
(compiled by J.Als-Nielsen,
ref. 10)

One can be sure that in 10years from now one can find various papers which
end up with the conclusious of Moncton et al (13)'... that use of synchrotron
source has enabled us to improve our resolution by a factor of 20 and elucidate
physics,which was entirely unexpected'.
However 'it is very dangerous to try to predict the future, and the above
possibilities may turn out to be the least exciting developments. Time will
tell' (14).

References
1. D.J. Thompson, Nucl. Instr. Methods 208, 1 (1983)
2. P.H. Citrin, J.E. Rowe, and P. Eisenberger, Phys. Rev. B28, 2299 (1983)
3. S. Mobilio and L. Incoccia, in 'EXAFS and Near Edge Structure',
   eds. A. Biancon, L. Incoccia, and S. Stipcich,
   Springer Series in Chemical Physics 27 (Springer, Berlin (1983), p. 87
4. G. Martens and P. Rabe, Phys. stat. sol. (a) 58, 415 (1980)
5. G. Martens and P. Rabe, I. Phys. C14, 1523 (1981)
6. P.L. Cowan, J.A. Golovchenko, and M.F. Robbins,
   Phys. Rev. Letters 44, 16801 (1980)
7. M.J. Bedzyk, W.M. Gibbons, and J.A. Golovchenko, I. Vac. Sci. Technol.
   20, 634 (1982)
8. G. Materlik, A. Frahm, and M.J. Bedzyk, Phys. Rev. Letters 52, 441 (1984)
9. W.C. Marra, P.H. Fuoss, and P.E. Eisenberger, Phys. Rev. Letters
   49, 1169 (1982)
10. I. Als-Nielsen, ESRP internol report
11. J.K. Kjems, L. Passell, H. Taub, and J.G. Dash,
    Phys. Rev. Letters 32, 724 (1974)
12. P.M. Horn, R.J. Birgenau, P. Heiney, and E.M. Hammonds,
    Phys. Rev. Letters 41, 961 (1973)
13. D.E. Moncton, P.W. Stephens, R.J. Birgenau, P.M. Horn, and G.S. Brown
    Phys. Rev. Letters 46, 1533 (1981)
14. E.A. Stern and S.M. Heald, in 'Handbook of Synchrotron Radiation',
    Vol. 1B, ed. E.-E. Koch (North-Holland, Amsterdam 1983), Chapter 10.

Near Edge X-Ray Absorption Fine Structure Studies of Chemisorbed Molecules

Francesco Sette

AT & T Bell Laboratories, 600 Mountain Avenue, Murray Hill, NJ 07974, USA

Joachim Stöhr

Corporate Research Science Laboratories,
Exxon Research and Engineering Company
Annandale, NJ 08801, USA

1. INTRODUCTION

The K-edge photoabsorption spectrum of low Z atoms (B,C,N,O,F) in molecules chemisorbed on transition metal surfaces is dominated by strong intramolecular resonances in the energy region within ~40eV from the excitation threshold. We will refer to this part of the spectrum as the Near Edge X-ray Absorption Fine Structure (NEXAFS). For molecules composed of low Z atoms (B,C,N,O,F), two kinds of resonances are observed which are generally referred to as π and σ resonances on the basis of the symmetry of the final states involved in the excitation. Such near edge structures arise from scattering processes within the *intra*molecular potential, which is determined by the atomic cores and the valence charge distribution of the molecule, and are to a large extent decoupled from the extramolecular environment [1]. This fact renders the NEXAFS spectrum of a chemisorbed molecule a powerful local probe for monitoring the modifications of the intramolecular bond induced by the reaction with the metal surface. In spite of the fact that a quantitative interpretation of the NEXAFS spectrum requires detailed calculations where localization and correlation effects are taken into proper account [1,2], we will show that important geometric and electronic structure information, like the orientation of the molecule with respect to the surface, bond-length changes upon chemisorption and molecular orbital rehybridization, can be directly linked with general properties of the near edge molecular resonances.

2. NEXAFS SPECTRA OF GAS PHASE AND CHEMISORBED MOLECULES

In Fig. 1 we report the O-K edge NEXAFS spectrum of gas phase CO [3] and of CO chemisorbed on Cu(100) [4] for two different angles of incidence of the synchrotron radiation with respect to the Cu surface. The spectra for the gas and chemisorbed phases look very similar. All of them show first, a sharp structure, which is a π resonance, and second, a broad structure in the continuum, which is a σ-resonance. The main effect of the chemisorption process is revealed through a broadening of the π resonance and the disappearing of the small Rydberg states observed in the gas phase spectrum around 540eV. Note that in the chemisorbed phase the spectrum is still *totally* dominated by intramolecular excitations without new structures induced by the metal. This confirms the local nature of such excitations. Such observations apply to any molecule composed of low Z atoms. Another striking feature of Fig. 1 is the polarization dependence of the resonances and, in particular, the reversed behavior of the π and σ excitations. This polarization dependence arises from the fact that the σ-shape resonance involves final states which are symmetric with respect to a reflection plane containing the molecular bond axis, while the π resonance involves final states which are antisymmetric with respect to this plane. Therefore, because the K-edge photoabsorption spectrum is governed by the dipole selection rule, the σ-shape resonance is maximized when the electric field E of the radiation is parallel to the bond axis while the π resonance is suppressed and vice versa [5]. Using the polarized nature of synchrotron radiation and measuring the intensity of the π and σ resonances as a function of the direction of E with respect to the surface, it is possible to accurately determine the orientation of the molecule. The results obtained for CO/Cu(100), shown in Fig. 1, indicate that CO stands up on the Cu surface [4].

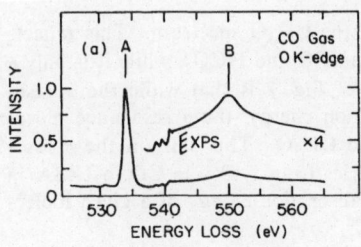

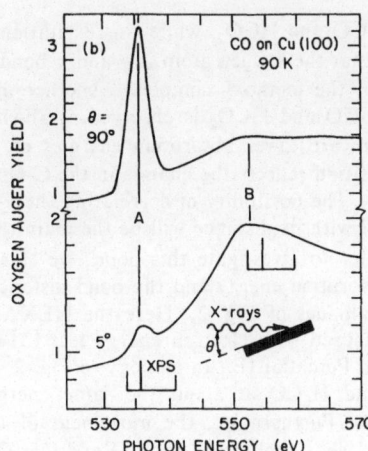

Fig. 1. O K-edge NEXAFS spectra of CO;
a) Electron Energy Loss results
for gas phase CO [3]; b) photoabsorption results for CO/Cu(100)
with the Electric vector of the
radiation parallel (θ=90°) and
almost normal (θ=5°) to the
surface [4].

While a σ-shape resonance is always observed for molecules with bonds between two or more low Z atoms (with the exception of hydrogen which has a vanishingly small backscattering ampli-tude), the π resonance is observed only if the bond configuration contains π bonding and antibond-ing orbitals (i.e., double and triple bonds). The study of the broadening and the presence of a π resonance in a given molecule can therefore be used to monitor chemical reactions with the metal and rehybridization of the molecular bonds. These concepts are illustrated in Fig. 2 where we show the O-K edge NEXAFS spectra [4] of CO, HCO_2 and CH_3O chemisorbed on Cu(100) with the polarization chosen in order to maximize the σ-shape resonance. We observe a π resonance

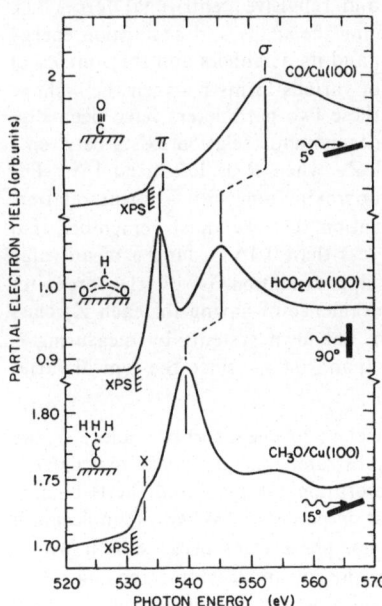

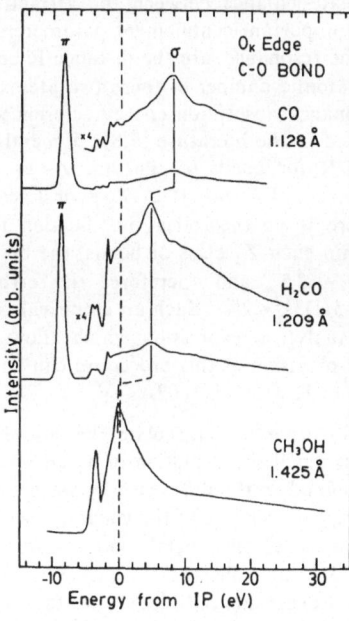

Fig. 2. O K-edge NEXAFS spectra of
CO, HCO_2 and CH_3O on Cu(100)
[4]. The angle of incidence
is chosen in order to maximize the σ-shape resonance.

Fig. 3. Electron Energy Loss
spectra at the O-K-edge of
gas phase CO, H_2CO and
CH_3OH [3]. The energy
scale is referenced to the
Ionization Potential (IP).

only for CO and HCO_2, while this excitation is not present in the CH_3O spectrum. This reflects the fact that the oxygen atom is σ and π bonded to the C atom in CO and HCO_2, while it is only σ bonded in the methoxy complex. Another interesting feature of Fig. 2 is that while the π resonance in CO and HCO_2 is observed at about the same excitation energy, the σ resonance moves 13.5eV toward lower absorption energies on going from CO to CH_3O. This shift in the σ resonance position reflects the change in the C-O distance which varies from 1.13Å in CO to 1.43Å in methoxy. The possibility of correlating the σ-shape resonance absorption energy of a given molecular bond with its distance will be the main topic of this article.

In order to investigate this point, we start with gas phase molecules where the σ-shape resonance absorption energy and the bond distances are accurately known. In Fig. 3 we show the gas phase analogues of Fig. 2. Here the NEXAFS spectra of CO, H_2CO, and CH_3OH at the O-K edge measured with electron energy loss [3] are reported with the energy scale referenced to the Ionization Potential IP. In analogy to Fig. 2, we observe that while the π resonance is present only in CO and H_2CO at about the same energy, the σ-shape resonance is observed for all three molecules. Furthermore, the movement of the σ-shape resonance between CO and CH_3OH is similar to what is observed for CO and CH_3O on Cu. Results obtained for the C-K edge of these molecules and for other groups of molecules consisting of different atoms show a similar correlation between σ-resonance position and bondlength [6].

3. *CORRELATION OF σ SHAPE RESONANCE AND MOLECULAR BOND-LENGTH*

The σ-shape resonance can be described in a molecular orbital picture as a transition to a $(1s^{-1},\sigma^*)$ state which may be located below or above the Ionization Potential. In the multiple scattering theory, such a resonance can be visualized as a scattering process where certain angular momentum components of the photoelectron wave are resonantly trapped by a barrier created by a delicate balance between the attractive Coulomb potential and repulsive centrifugal forces [1]. The important confinement parameters, which directly determine the shape and absorption energy of the resonance, are the distance R between the excited atom and its neighbors and the sum Z_T of the atomic number of these two atoms. In fact, correlations of various forms between the σ-shape resonance absorption energy referenced to IP, δ and one of these two parameters have been suggested in the literature [7,8]. Recently we have shown that a monotomic relation exists between δ and R for bonds of constant Z_T as well as between δ and Z_T when R is left fixed [6]. The behavior of δ with R at Z_T=const. can be fitted to a good approximation with a linear relation. There is no theoretical justification for such a linear correlation, but we must remember that within each Z_T class of bonds, the bond distances change by less than 0.3Å around a mean value of $\sim$1.3Å, and therefore, the error in using a linear approximation is of the order of $(0.15/1.3)^2<2\%$. Such an empirical correlation offers the convenience of having for each Z_T class an analytical expression, which allows to predict distances in unknown systems by measuring δ. The precision of this procedure can be estimated to be better than ±0.05Å since the typical variation of R with δ is 0.02Å/eV.

In order to extrapolate the gas phase results for a given class to chemisorbed molecules, we must recall the definition of δ and how δ is modified by the substrate. For gas phase molecules, δ is defined as the difference between the σ-shape resonance absorption energy ϵ^g and the 1s binding energy E_B^g relative to the vacuum level (E_V), i.e., the 1s ionization potential. When the molecule is chemisorbed on a metal, both ϵ^g and E_B^g can differ from their gas phase value because both involve a deep core hole and the presence of the metal may modify the intramolecular screening. It is found experimentally [5,9] that for weakly chemisorbed molecules, where distance changes are not expected, the σ resonance energy ϵ^s is the same as in the gas phase ($\epsilon^s\approx\epsilon^g$). Examples are CO, N_2, NO and hydrocarbons with C-C single bonds when chemisorbed on noble or quasi-noble metal surfaces. This experimental observation is supported by the theoretical fact [1,8] that the σ-shape resonance is localized in the intramolecular region and decoupled from the surrounding chemical environment. On the other hand, the 1s binding energy relative to the vacuum level, measured by photoemission, is in general different between gas phase and chemisorbed molecules because the photoelectron leaves the molecule and the metal electrons can efficiently screen the core hole. Such considerations lead to the working hypothesis that in order to extrapolate the δ vs R rule from gas phase to chemisorbed molecules, we must correct for changes in the 1s binding energies. Typically binding energies E_B^s for chemisorbed molecules are measured relative to the Fermi level, E_F, which differ by the work function ϕ from those referred to the vacuum level, E_v. If we denote by δ_s the σ resonance position for the chemisorbed molecule relative to E_F and assume $\epsilon^s = \epsilon^g$, we obtain the following relation between σ resonance positions and 1s binding energies:

$$\delta_s = \delta_g + (E_B^g - E_B^s)$$

The easiest case of transferring the correlation of δ_g and R for gas phase molecules to an appropriate one between δ_s and R for chemisorbed molecules is obtained when $E_B^g - E_B^s = $ const. Indeed, this is the case for hydrocarbons [9]. Because in these molecules the C-C bond is nonpolar, the C 1s binding energies in the gas phase and on the surface are remarkably constant. Gas phase values are typically $E_B^g = 290.8 \pm 0.4$ eV. On Pt(111) the E_B^s values are also nearly constant with a mean value of 283.8 ± 0.4 eV. Thus, for chemisorbed hydrocarbons on Pt(111) we find the simple correlation, $\delta_s = \delta_g + 7$ eV valid for any value of R. Note that most of the 7eV difference in E_B^g and E_B^s is accounted for by the work function ($\phi = 6.2$ eV for clean Pt). The above results lead to the plot in Fig. 4. We have used the correlation of δ_s and R in Fig. 4 to determine R for a variety of hydrocarbons on Pt(111) [9]. Our studies included pseudo-diatomic molecules like triple-bonded C_2H_2, double-bonded C_2H_4 and the single-bonded ring structure C_6H_{12} as well as C_7H_8, which is a ring structure containing both double and single C-C bonds. δ_s was found to be 9.0 ± 0.5 eV in C_6H_{12} and 9.5 ± 1 eV and 16.5 ± 2 eV for the single and double bonds of C_7H_8. Using the calibration graph of Fig. 4, we can estimate 1.51 ± 0.03 Å in C_6H_{12}, and 1.37 ± 0.04 for the double bond and 1.50 ± 0.03 Å for the single bond in C_7H_8. These values are close to the respective gas phase distances in agreement with the fact that the intramolecular bonding of ring structures is extremely stable. On the contrary, we find $\delta_s = 12 \pm 1$ eV in C_2D_2 and 10 ± 1 eV in C_2H_4, which correspond to C-C distances of 1.45 ± 0.03 Å and 1.49 ± 0.03 Å, respectively. The dramatic C-C bond stretch observed in these two molecules, which are both chemisorbed flat on Pt(111), amounts to 0.15 Å in C_2H_4 and 0.25 Å in C_2H_2 with respect to the gas phase. These results are corroborated both by earlier photoemission and vibrational electron energy loss studies as well as by the analysis of the residual π resonance in the NEXAFS spectra of these two molecules which show a strong interaction of the π states with the metal surface [10].

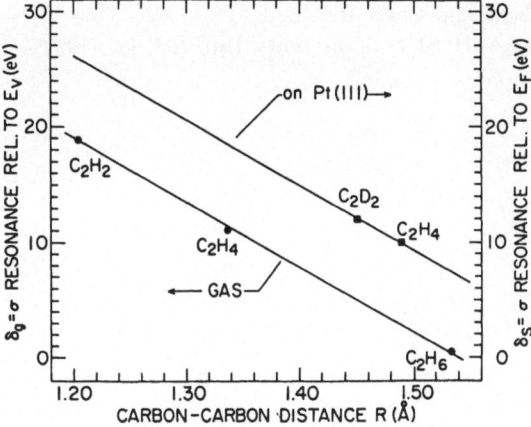

Fig. 4   Correlation between σ-shape resonance absorption energy referenced to the Vacuum level, δ_g (Fermi level, δ_s), and bond distance of C-C bonds in gas phase molecules [molecules chemisorbed on Pt(111)].

4. CONCLUSIONS

In conclusion, the fact that the NEXAFS spectrum of a chemisorbed molecule is dominated by intramolecular resonances, strongly decoupled from the metal surface and the surrounding chemical environment, makes this kind of spectroscopy a valuable local probe to study the intramolecular bonding and bondlength of the chemisorbed molecule. In particular, the correlation between the σ-shape resonance absorption energy and the bond distance, which can be determined empirically from the experimental gas phase results, is extremely useful from the practical point of view. It is hoped that these empirical rules will also find a more satisfying theoretical foundation in the near future.

REFERENCES

1. J.L. Dehmer and D. Dill: *Electron-Molecule and Photon-Molecule Collisions*, T.N. Resigno, B.V. McKoy and B. Schneider (eds.) (Plenum, NY, 1979) p. 225; J.L. Dehmer, D. Dill and A.C. Parr: "Photoionization Dynamics of Small Molecules", in *Photophysics and Photochemistry in the Vacuum Ultraviolet*, S.P. McGlynn, G. Findley and R. Huebner, eds., (D. Reidel, Dordrecht, Holland, 1983).

2. F.W. Kutzler, C.R. Natoli, D.K. Misemer, S. Doniach and K.O. Hodgson: J. Chem. Phys. *73*, 3274 (1980); F.W. Kutzler, K.O. Hodgson and S. Doniach, Phys. Rev. A*26*, 3020 (1982).

3. For a review see A.P. Hitchcock: J. Electron Spectrosc. *25*, 245 (1982).

4. J. Stöhr, J.L. Gland, W. Eberhardt, D. Outka, R.J. Madix, F. Sette, R.J. Koestner and W. Doebler: Phys. Rev. Lett. *51*, 2414 (1983).

5. J. Stöhr, R. Jaeger: Phys. Rev. B*26*, 4111 (1982).

6. F. Sette, J. Stöhr and A.P. Hitchcock: submitted to J. Chem. Phys.; and A.P. Hitchcock, F. Sette and J. Stöhr: these proceedings.

7. A.E. Orel, T.N. Resigno, B.V. McKoy and P.W. Langhoff: J. Chem. Phys. *72*, 1265 (1980); L.E. Machado, E.P. Leal, G. Csanak, B.V. McKoy and P.W. Langhoff: J. Electron Spectrosc. *25*, 1 (1982); T. Gustafsson and H.J. Levinson: Chem. Phys. Lett. *78*, 28 (1981); A. Bianconi: Proceedings of the 1st Int. Conf. on EXAFS and XANES, Frascati, Italy, 1982, Springer Series on Chem. Phys., Vol. 27 (Springer-Verlag, NY, 1983), p. 118; A. Bianconi, M. Dell'Ariccia, A. Gargano and C.R. Natoli, ibid, p. 57; A.P. Hitchcock and C.E. Brion: J. Phys. B*14*, 4399 (1981).

8. R. Natoli: Proceedings of the 1st Int. Conf. on EXAFS and XANES, Frascati, Italy, 1982, Springer Series on Chem. Phys., Vol. 27 (1983), p. 43.

9. J. Stöhr, F. Sette and A.L. Johnson: submitted to Phys. Rev. Lett.

10. R.J. Koestner, J. Stöhr, J.L. Gland and J.A. Horsley, Chem. Phys. Lett. *105*, 332 (1984).

The Interactions of Oxygen and Hydrogen with Cleaved InP(110)

P.J. Love, R.A. Rosenberg, Victor Rehn, and P.R. LaRoe

Michelson Laboratory, Naval Weapons Center, China Lake, CA 93555, USA

C.C. Parks

IBM Corporation, General Technology Division, P.O. Box 390, South Road
Poughkeepsie, NY 12602, USA

The origin of trapping and recombination centers on semiconductor
surfaces is of great importance in the development of high performance
electronic devices. Chemical impurities, imperfections, or adsorbates can
provide electronic states deleterious to electronic performance. In this
paper, we report results on the interactions of oxygen and hydrogen with
vacuum-cleaved InP(110) surfaces. InP has become an important material for
microwave and fast-switching devices, as well as optoelectronic devices
[1]. However, unlike GaAs, relatively little work has been reported on the
surface-electronic structure of InP.

Photon-stimulated ion desorption (PSID) and total photoelectron yield
(PEY) spectra have been obtained near the In(L) and P(K) edges on freshly
cleaved (110) faces of InP before and after exposure to 10^4 L ($1L = 10^{-6}$
Torr-sec) of electron-beam-excited hydrogen (H_2*). PEY spectra were also
recorded near the $In(M_{4,5})$ and $P(L_{2,3})$ edges. The effects of oxidation
by exposure to 5×10^9 L O_2 were examined near the $P(L_{2,3})$ edge. The
experiments were performed on beam lines I-1 and III-3 at the Stanford
Synchrotron Radiation Laboratory. All spectra reported here have been
normalized to the incident photon intensity.

Figure 1 shows the total PEY and H^+-PSID spectra near the In(L) edges
for the hydrogen-exposed surface. No background adjustments have been

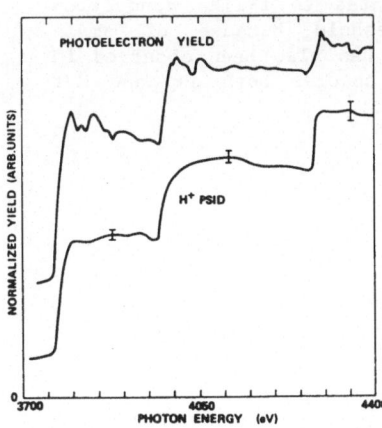

Fig. 1. Total PEY and H^+-PSID
of cleaved InP(110) (dosed with
10^4 L H_2*) near the In(L) edges.

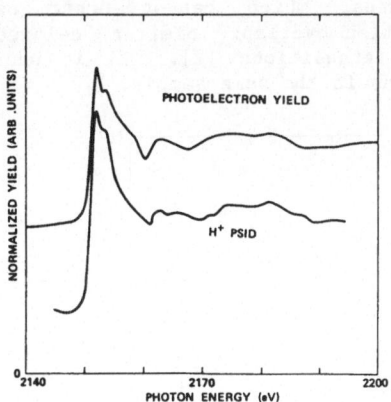

Fig. 2. Total PEY and H^+-PSID
of cleaved InP(110) (dosed with
10^4 L H_2*) near the P(K) edge.

made — both spectra have been plotted with respect to the same zero reference. In each spectrum, the three edge jumps occur with about the same
proportion to the pre-edge background (the PSID edge jumps are somewhat
larger relative to the pre-edge background than the PEY edge jumps).
Figure 2 shows similar data for the P(K) edge. Again, no background
adjustments have been made. Here, the PSID edge jump-to-background ratio
is roughly three times larger than for the PEY edge, but clearly still
within the same order of magnitude. Utilizing the total PEY signal as a
measure of the photoabsorption of a near-surface layer of the InP, it can
be used to normalize the PSID edge jumps as follows: the ratio of the PEY
edge jump to pre-edge background is a measure of the probability of core-
hole production relative to all other photoabsorption mechanisms.
Similarly, the edge jump-to-background ratio of the PSID spectrum represents the relative probability of H^+ ion desorption which occurs as a
result of core-hole production. Therefore, comparing the edge jump-to-
background ratios for PSID and total PEY for In and P core excitations is
equivalent to comparing the H^+ yield per core excitation at these two
sites. Assuming that, at sites containing the appropriate core hole, the
probability of H^+ emission is about the same at In-H sites as it is at
P-H sites, then we must conclude that the populations of the In-H and P-H
bonds on the surface are of the same order of magnitude.

It should be noted that no photodesorbed H^+ was observed on the
freshly cleaved InP and that the total PEY spectra were essentially
unaffected by the H_2^* exposure. It follows that the near-edge structure
observed in the total PEY spectra is representative of the freshly cleaved
surface, while the PSID near-edge structure is representative of the
hydrogenated surface.

Figure 3 shows the total PEY spectrum near the In($M_{4,5}$) edges of a
freshly cleaved InP(110) surface. The two small features, also shown
enlarged fivefold from a high resolution spectrum, are separated by
7.4 ± 0.3 eV, the spin-orbit splitting of the In($M_{4,5}$) initial state.
Their energies (448.9 and 456.3 eV) lie about 5 eV above the free-atom
energies and probably represent dipole-allowed transitions between the
In(3d) core states and p-like conduction band states. The shapes of these
peaks suggest the presence of a pair of unresolved final states separated
by about 2 eV. The strongly rising PEY signal above 470 eV is due to
dipole-allowed transitions from In(3d) core states to f-like conduction
band states, which cannot occur near threshold because of small
wave-function overlap. Similar "delayed onset" has also been observed in
In(4d→4f) transitions [2]. It is unusual to observe both d→p and d→f
transitions in the same sample.

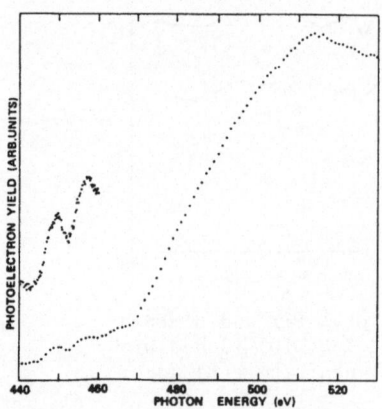

Fig. 3. Total PEY of cleaved
InP(110) near the In($M_{4,5}$)
edges.

The effects of oxidation on the total PEY spectrum near the $P(L_{2,3})$ edge are illustrated in Fig. 4. Power-law background curves fitted to the PEY spectra between 125 and 128.2 eV have been subtracted from the data represented in Fig. 4. Both spectra show two pairs of peaks, separated by the $P(L_{2,3})$ spin-orbit energy of 0.84 ± 0.02 eV. This spin-orbit energy is slightly smaller than the 0.99 eV customarily used [3], but the excellent signal to noise and reproducibility of the present data lead us to consider this the more accurate value. The binding energies of the $P(L_3)$ and $P(L_2)$ states of the freshly cleaved surface agree within experimental error with those obtained previously [2]. For the oxidized surface, these peaks are shifted slightly upward (+0.08 ± 0.03 eV) and narrowed. These changes may reflect the lessened lattice strain near the oxidized surface. Because the total PEY electrons emanate from a range of depths below the surface of the crystal, the lattice strain of surface reconstruction results in a broadening of the PEY near-edge features. Oxidation is known to reduce the surface lattice strain, and Fig. 4 shows, in fact, that the $P(L_{2,3})$ near-edge doublet sharpens upon oxidation.

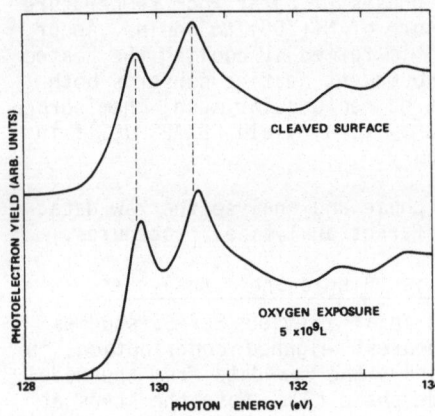

Fig. 4. Total PEY of cleaved InP(110) near the $P(L_{2,3})$ edges before and after dosing with 5×10^9 L O_2.

1. J.B. Boos, H.B. Dietrich, T.H. Weng, K.J. Sleger, S.C. Binari, and R.L. Henry, IEEE Electron Device Lett. EDL-3, 256 (1982).
2. R.A. Rosenberg, P.R. LaRoe, Victor Rehn, G.M. Loubriel, and G.M. Thornton, Phys. Rev. B. 28, 6083 (1983).
3. T.A. Carlson, Photoelectron and Auger Spectroscopy (Plenum Press, New York, 1975), p. 342.

A SEXAFS Study of Iodine on Ni{100}:
The Surface Iodide Phase

Robert G. Jones[1], S. Ainsworth, M.D. Crapper, C. Somerton[2], and
D.P. Woodruff
Physics Department, University of Warwick, Coventry CV4 7AL, United Kingdom

R.S. Brooks, J.C. Campuzano[3], David A. King, and G.M. Lamble
Donnan Laboratories, University of Liverpool,
Liverpool L69 3BX, United Kingdom

M. Prutton
Physics Department, University of York, Heslington
York YO1 5DD, United Kingdom

1. Ni{100}/I - Surface Phases

This adsorption system shows several surface phases [1,2] at room temperature.
A c(2x2) chemisorbed phase is formed by exposure of Ni{100} to iodine vapour
at room temperature, while a "surface iodide" is formed by cooling the heated
sample to room temperature in the vapour. Subsequent heating converts both
structures to a range of variable sized, centred rectangular mesh, chemisorbed
phases. These systems have been investigated by total yield SEXAFS utilising
SERC's Synchrotron Radiation Source at Daresbury.

Here we concentrate on the surface iodide phase and analyse the raw data,
independently at two laboratories, by two different analytical procedures.

2. Fourier-filtered Nearest Neighbour Analysis (Single Shell Analysis)

This is the general analytical procedure used in all previous SEXAFS studies
[3] and concentrates on a comparison of the nearest neighbour contributions in
the unknown and model compounds. Figure 1a shows the raw data from the model
compound, bulk NiI_2(B), and the surface iodide phase (S). Note the steps at
the I L_3 and L_2 edges. The high energy fine structure function (EXAFS > 70 eV)
is extracted and Fourier transformed, Figure 1b. The Fourier transforms are
backtransformed with a "window" around the main (Ni-I nearest neighbour) peak
at R ≈ 2.2 Å yielding a phase difference which gives the bond length change,
-0.05±0.01 Å. The table gives the results of six separate runs.

Table. Bond length changes, surface phase minus bulk NiI_2 value

Run Number	Single Shell Analysis Ni-I change (Å)	Multi Shell Analysis Ni-I change (Å)	I-I change (Å)
794	-0.050	-0.036	-0.039
796	-0.060	-0.041	-0.034
797	-0.050	-0.046	-0.072
798	-0.045	-0.017	-0.055
823	-0.035	-0.025	-0.042
824	-0.055	-0.049	-0.048
Average	-0.05±0.01	-0.036±0.01	-0.048±0.01

1 Chemistry Department, University of Nottingham, Nottingham NG7 2RD, United
  Kingdom
2 British Aerospace, Dynamics Group, Bristol Division, Filton, Bristol
  BS12 7QW, United Kingdom
3 Cavendish Laboratory, University of Cambridge, Madingley Road, Cambridge
  CB3 OHE, United Kingdom

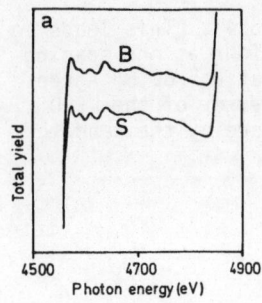 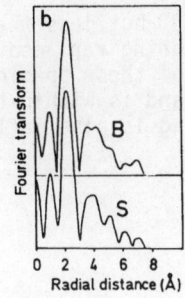

Fig. 1(a) Raw experimental data for the bulk NiI_2(B) and the surface iodide phase (S), 1(b) The Fourier transforms of $\chi(k)k$

3. Multi-shell Theory Comparison

The alternative approach uses a suit of programs [4] at Daresbury and performs a model calculation of the EXAFS for a specified multi-shell structure. Scattering phase shifts are calculated from muffin-tin potentials and optimised to give the best fit to the model, bulk NiI_2, data of known structure. The structure is then adjusted to give the best fit to the, unknown, surface iodide.

Fourier prefiltering is applied over a larger data range than before. The backtransform uses a window from 1.0 to 4.5 Å which only has the effect of removing noise. Figure 2 shows the fits (dashed line) of the two data sets. The table shows the changes in the first two shells (Ni-I and I-I) found. The spectra are dominated by the effects of these two shells, although five shells were included in the analysis.

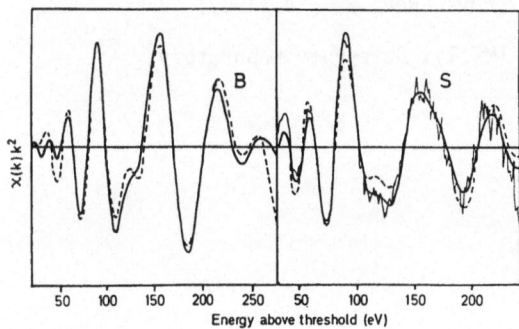

Fig. 2. Fits of the theory (dashed lines) to the filtered experimental data (full lines) for the same bulk (B) and surface (S) iodide spectra shown in Fig.1. In the latter case, the unfiltered data is also shown

4. Conclusions

Despite the obvious importance of the I-I shell in all the data, the agreement between the two methods for the first Ni-I shell contraction (0.04±0.02 Å) is excellent.

The bond length changes can be reconciled with previous LEED data [2]. These showed that the hexagonal single sandwich (I-Ni-I) surface iodide structure is slightly distorted. The iodine mesh is related to the Ni{100} substrate mesh by

$$\begin{pmatrix} 1-\cot\theta & -1-\cot\theta \\ 2\cot\theta & 2\cot\theta \end{pmatrix}$$

with θ the included angle greater than 60° but less than 63.4°. This leads to the original six I-I nearest neighbours being replaced by four at one spacing and two at another. A weighted average of these spacings at θ = 62.5° agrees with the contraction measured by SEXAFS and is within the error of the LEED result (θ = 61.5°±1.5°). The Ni-I spacing is also influenced by the sandwich layer spacing not investigated by LEED.

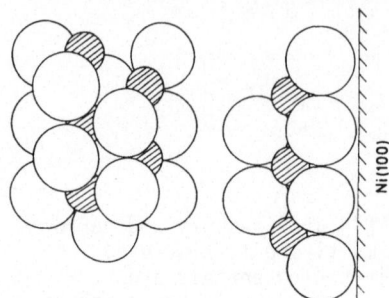

Fig. 3. Plan and sectional view of the sur-face iodide structure

References

1. R.G. Jones and D.P. Woodruff, Vacuum 31 (1981) 411.

2. R.G. Jones, C.F. McConville and D.P. Woodruff, Surface Sci. 127 (1983) 424.

3. e.g. P.H. Citrin, P. Eienberger and R.C. Hewitt, Phys. Rev. Letters 41 (1978) 309.
    S. Brennan, J. Stöhr and R. Jaeger, Phys. Rev. B. 24 (1981) 4871.

4. S.J. Gurman, Technical Memo DL(SCI/TM21T), Daresbury Laboratory, Warrington (1980).

Fluorescence Detection of Totally Reflected EXAFS (FREXAFS) at Interfaces

Edward A. Stern, Edward Keller, Olivier Petitpierre, and Charles E. Bouldin
Department of Physics, FM-15, University of Washington, Seattle, WA 98195, USA
Steve M. Heald and John Tranquada
Brookhaven National Laboratory, Upton, NY 11973, USA

A new detection scheme for measuring the EXAFS signal at interfaces is described. By totally reflecting the x rays at the
interface and detecting the signal by fluorescence, an improvement of the signal-to-noise ratio by three orders of magnitude
compared with electron detection is attained for a 1.5 monolayer
of Au atoms.

Surface EXAFS (SEXAFS) has had much utility in determining the structure of
adsorbed atoms and surface atoms at the vacuum-solid interface [1,2]. In
principle, SEXAFS can determine the same information as bulk EXAFS, i.e.,
coordination number, types of surrounding atoms, average bond distances
and directions, and the disorder about the average distance. However, in
practice, because of the experimental limitations of the standard method of
SEXAFS detection [2], usually only average bond distances and directions are
determined. The experimental limitations usually lead to low counting rates
and a large background to signal.

Standard SEXAFS employs electron detection to turn EXAFS into a surface-
sensitive technique [2]. The x rays that are used to excite the atoms penetrate deep into the sample so that the absorption signal is dominated by the
bulk. By detecting only the electrons that can escape from the surface, the
EXAFS signal becomes surface-sensitive, but at the expense of a loss in total
signal, since only a very small fraction of the x rays contribute to the
detected signal.

If the x rays were to penetrate only a few atomic layers below the surface,
the EXAFS signal would inherently be surface sensitive. In addition, the
background would be correspondingly reduced. As is well known [3], x rays
can totally reflect from surfaces if they are incident below a critical
glancing angle $\alpha_c = (4\pi n e^2/m\omega^2)^{\frac{1}{2}}$ in c.g.s. units, where n is the number of
electrons per cubic centimeter, e is the electron charge, m is the electron
mass, and ω is the x-ray radial frequency. Typically, $\alpha_c \simeq 5~mr$ at hard
x-ray energies. In this total reflection mode the x rays penetrate about
30 Å and are thus inherently surface sensitive. In addition, the construc-
tive coherence between the incoming and reflected x rays increases the x-ray
intensity (and thus the EXAFS signal) by an additional factor of 4.

Experiments of this type [4] have been performed in which the reflected
intensity was monitored to detect the EXAFS. The disadvantage of this method
of detection is that the reflected intensity is only weakly modulated by
the EXAFS from a monolayer on the surface, and one has to detect a small
signal on a large background. For example, in the case that we measured
(discussed below), the edge step is only about 3% of the total reflected
intensity and the EXAFS is about 10^{-2} times smaller. If one desires to
measure the EXAFS to 3% accuracy, it would be necessary to measure the

reflected signal to an accuracy of 10^{-5} and eliminate all glitches in the background to 3×10^{-4}, a formidable task not attainable at present. For these reasons the totally reflected mode of SEXAFS has not been pursued.

The solution to the detection of a small signal on a large background is well known in bulk EXAFS. By using fluorescence detection with an appropriately designed detector it is possible to eliminate the background substantially relative to the signal and make the detection of the signal feasible. We have used the fluorescence detection mode, which we denote by FREXAFS [5], to measure a coverage of 2.6×10^{15} atoms/cm^2 of Au (~1.5 monolayers) on a glass substrate. As described below we find a fluorescent signal at the L_3-edge of 10^7 photons/sec and a background of only half that amount. The total acquisition time of a spectrum of the quality of a bulk sample was less than 30 minutes.

The experimental setup used is shown in Fig. 1. The incident x-ray beam was limited to a cross section of 0.1×10 mm^2 by a slit. Ionization chambers were used to monitor the incident (I_0), the fluorescence (I_f), and the reflected (I_r) intensities. Measurements were made at CHESS during parasitic running and at SSRL during dedicated time. The sample was oriented so that measurements were made with the x-ray polarization both perpendicular and parallel to the surface. Measurements were made in the ambient atmosphere.

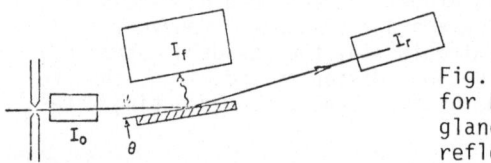

Fig. 1. Experimental arrangement for FREXAFS. X rays incident at a glancing angle of θ are totally reflected from the sample

The EXAFS signal $\chi(k)$ is shown for the surface layer in Fig. 2(a) with the x-ray polarization normal to the surface and an Au foil in Fig. 2(b), both measured at room temperature. In this case the surface layer actually has less noise than the bulk film! There is no significant polarization dependence of the surface layer EXAFS as the polarization is varied from perpendicular to parallel. The bulk and surface layer data of Fig. 2 can be compared with one another.

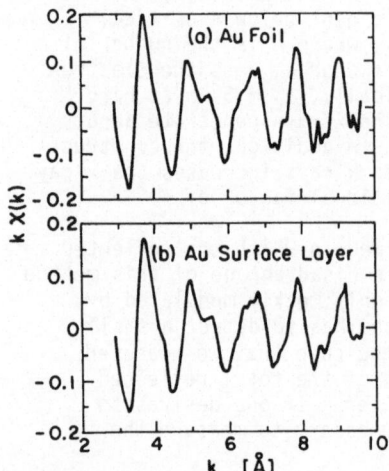

Fig. 2. EXAFS signal $k\chi(k)$ for (a)bulk and (b) surface layer gold atoms measured at room temperature. The x-ray polarization is normal to the surface in (b)

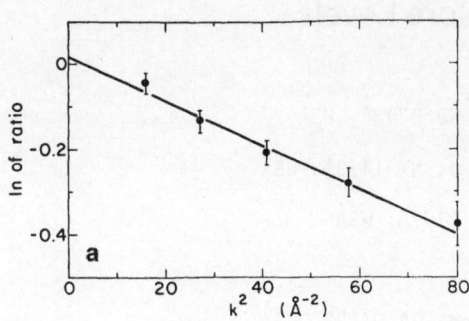 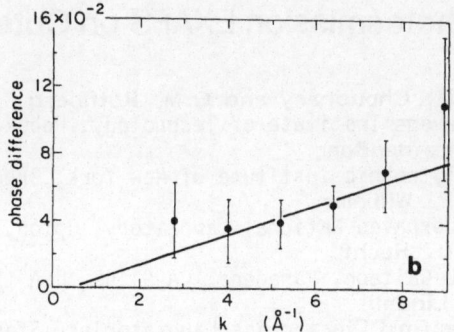

Fig. 3. Ln of ratio (a) and phase difference (b) for the surface and bulk data of the first coordination shell contribution of Fig.2

Analysis of the plot in Fig. 3(a) indicates by its intercept an equal number of first neighbors in the surface layer and the bulk to within 2% and by its slope an increase in the mean square disorder σ^2 of the surface layer relative to the bulk Au at room temperature of 0.0025 ± 0.0002 Å^2. The plot in Fig. 3(b) indicates by its slope that the first-neighbor distances between the bulk and surface layer samples are the same within about 0.004 Å. The lack of polarization dependence indicates that the Au is not wetting the glass surface but "balls up" into small three-dimensional particles. This result is confirmed by an electron microscope study done at Brookhaven Laboratory.

Our measurements indicate that FREXAFS is a powerful technique for SEXAFS. Besides about a 10^3 increase in the rate of collecting data compared with the standard SEXAFS electron detection method for heavier atomic number (Z) surface atoms, FREXAFS has additional advantages. The signal compares well with bulk measurements and can be used to obtain the full EXAFS information. The penetrating nature of the fluorescence signal allows interfaces to be probed and surfaces to be studied in an ambient atmosphere. As the Z of the surface atom decreases, the efficiency of emission of fluorescence radiation decreases so that at about oxygen the expected fluorescence signal becomes equal to that of the electron detection technique. However, the background may well be less, and measurements in that region should be undertaken to assess how useful FREXAFS will be for low Z surface atoms.

This work was supported in part by the DOE Materials Sciences Division under contracts DE-AS05-80-ER10742 and DE-AC02-76CH00016, and by the National Science Foundation under grant DMR80-22221.

We thank CHESS and SSRL staffs for their generous help. SSRL is supported by the DOE Office of Basic Energy Sciences and by the NIH, Biotechnology Resources Program, Division of Research Resources. CHESS is supported by the National Science Foundation.

1. S.M. Heald and E.A. Stern: Phys. Rev. B 17, 4069 (1978)
2. For a review of surface EXAFS spectroscopy using electron detection, see
    J. Stöhr, R. Jaeger, and S. Brennen: Surf. Sci. 117, 503 (1982)
3. L.G. Parrott: Phys. Rev. 95, 359 (1954)
4. R. Barchewitz, M. Cremonese-Visicato, and G. Onori: J. Phys. C 11, 4439
    (1978); R. Fox and S.J. Gurman, J. Phys. C 13, L249 (1980); and
    G. Martens and P. Rabe: Phys. Status Solidi A 58, 415 (1980)
5. S.M. Heald, E. Keller, and E.A. Stern: Phys. Lett. A (in press)

Photoemission EXAFS on Outer Core Levels

K.M. Choudhary and G.M. Rothberg
Stevens Institute of Technology, Hoboken, NJ 07030, USA
M.L. denBoer
Polytechnic Institute of New York, Brooklyn, NY 11201, USA
G.P. Williams
Brookhaven National Laboratory, Upton, NY 11973, USA
M.H. Hecht
JPL Caltech, Pasadena, CA 91109, USA
I. Lindau
Stanford Electronics Laboratories, Stanford, CA 94305, USA

The first definitive measurements of EXAFS by monitoring the direct photoelectron emission as a function of photon energy have been made using the Mn 3p and F 2s core levels in evaporated films of MnF_2 [1,2]. Good agreement is found with bulk transmission EXAFS obtained with the Mn 1s level. This development is important because: (1) Almost all elements have suitable core levels of low binding energy. (2) Since photopeaks from adjacent core levels do not cross as the photon energy is varied, the limitation imposed on other EXAFS techniques by this interference is removed. Auger electrons, which have fixed kinetic energies and so can cross photopeaks, will generally have low energies if they originate from the outer core levels and be unlikely to appear in the EXAFS region of kinetic energies, i.e., in excess of about 50 eV. (3) VUV light is used, which greatly extends the useable range of photon energies. (4) More than one element in a solid may be studied with the same monochromator.

In this work all diffraction effects except the backscattering that gives rise to the EXAFS [3] were eliminated by using a polycrystalline evaporated thin film and by making use of the 2π azimuthal acceptance angle of a cylindrical mirror analyzer. Previously MARGARITONDO and coworkers [4] observed EXAFS-like oscillations in photopeaks as a function of photon energy when a CMA was used; however, their data did not extend over a sufficient photon energy range to prove conclusively that EXAFS was involved.

The experiments were carried out at the National Synchrotron Light Source; preliminary work was also done at the Stanford Synchrotron Radiation Laboratory. Data were obtained with VUV light and using constant initial state spectroscopy (CIS) to measure photopeak intensity I as a function of photon energy over the range 90 to 280 eV in 1 eV steps. The CMA was operated in the fixed retarding ratio mode [5]. Photon flux I_0 was monitored by measuring the total electron yield from an 80 percent transmission grid coated in situ with lithium fluoride. The ratio I/I_0 is used for the EXAFS analysis. At a monochromator resolution of about 1 eV and at the maximum of the output at 175 eV the photon flux was approximately 1.5×10^{10} photons/S at a typical average storage ring current of 50 mA. One scan took less than ten minutes; 3 to 5 scans were sufficient for adequate statistical accuracy as the photopeaks used here have relatively good signal-to-background ratios of 2 to 3. In this work the ^7P component of the Mn 3p multiplet was used.

Figure 1 shows the CIS spectra of (a) Mn 3p and (b) F 2s. The ratio I/I_0 in arbitrary units is shown as a function of photon energy in eV. The EXAFS function $\chi(k)$ was obtained by subtracting fitted splines approximating the background, which is a more complicated function of energy than

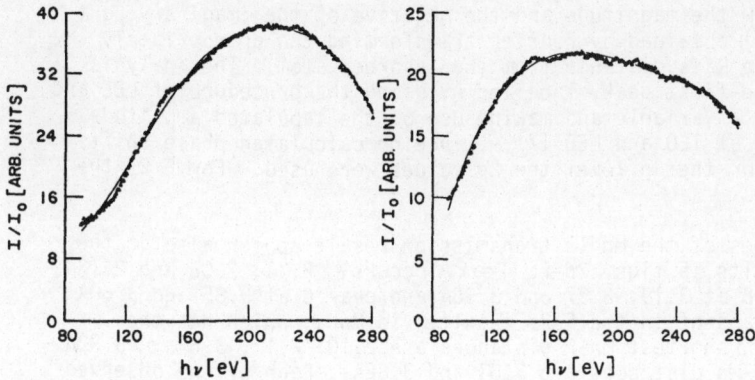

Fig. 1. CIS spectra of (a) Mn 3p and (b) F 2s. Solid line is background.

in the case of transmission EXAFS. Three cubic splines fitted to the full
data range seem to describe the background best, and these fits are shown
as solid lines in the figure. No attempt was made to obtain the absolute
amplitude of $\chi(k)$.

Figure 2 shows the results of EXAFS analyses of the photoemission data
and of transmission data obtained using the Mn 1s absorption edge in the
photon energy range 6400 to 7100 eV. The transmission measurements were
carried out at SSRL. In the transmission case the pre-edge absorption was
removed before fitting the background. Figs. 2a-c show $\chi(k)$ weighted by k^3
for the 1s, 3p and 2s cases, respectively. For these plots the parameter
E_0, which is the electron energy at which the kinetic energy is zero, was
arbitrarily taken to be 6537 eV, the binding energy of the Mn 1s level, and
48 and 28 eV, respectively the binding energies of the Mn 3p and F 2s levels
relative to the valence band edge.

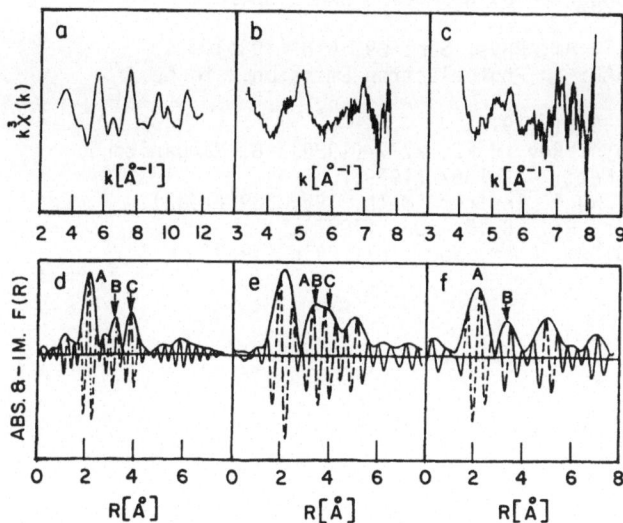

Fig. 2. EXAFS function $k^3\chi(k)$ vs electron wavevector $k[\mathring{A}^{-1}]$ and Fourier
transform F(R) vs. $R[\mathring{A}]$. Solid line Abs F(R), dashed line -Im F(R). (a,d)
Mn 1s. (b,e) Mn 3p. (c,f) F 2s.

Figures 2d-f show the magnitude and the negative of the imaginary part
of the function F(R) obtained by Fourier transforming the appropriately
weighted $\chi(k)$, where R is distance from the absorber atom. The analysis
was optimized on the first peak, labelled A, using the procedure of LEE and
BENI [6] with E_0 as a variable and making use of the tabulated amplitude
and phase functions of TEO and LEE [7]. Since no calculated phase shifts
are available yet for the 3p level the 2p values were used. For F 2s the
1s values were used.

The major features of the Mn 1s transmission result appear also in the
Mn 3p and F 2s results in figs. 2d-f. Peak A occurs at 2.11, 2.06 and 2.13Å,
respectively, peak B at 3.19, 3.37 and 3.36Å and peak C at 3.85 and 3.60Å
and is essentially absent in the F 2s result. In MnF_2, which has the
rutile structure, the shortest Mn-F distances are 2.10, 2.13, 3.52 and 3.92Å
and the shortest Mn-Mn distances are 3.31 and 3.82Å. Each of the observed
peaks arises from two closely spaced atomic shells. Agreement between the
three measurements and with known distances is highly satisfactory; agreement for peaks B and C can be improved by optimizing the analysis to those
peaks. The photoemission results give wider peaks because the k range is
shorter. The absence of peak C in F 2s provides further evidence that
EXAFS is being observed. For an F absorber peak C is due to four Mn atoms
at 3.92Å. For an Mn absorber the peak is due to eight Mn atoms at 3.82Å
and four F atoms at 3.92Å. The peak should thus be considerably smaller in
the former case.

The Lee and Beni procedure seeks to align a peak in Abs F(R) with a peak
in -Im F(R). This is achieved for peak A in the Mn 1s transmission case
with E_0 of 6538 eV. The alignment in the Mn 3p and F 2s cases, with E_0
values of 58 and 22 eV, is not quite as good. If $\pi/2$ is arbitrarily subtracted from the Teo and Lee Mn 2p absorber phase shifts and added to the
F 1s absorber phase shifts, alignment is achieved in each case with substantially no change in E_0 or peak positions.

This work was supported by DOE contract DE-AC02-8/ER10908 and grant
DE-FG02-84ER45091. MLdB is supported by NSF grant DMR 8306426.

1. G. M. Rothberg et al.: Bull. Am. Phys. Soc. 29, 516 (1984).
2. G. M. Rothberg et al.: "EXAFS in Photoelectron Emission," to be
published.
3. P.A. Lee: Phys. Rev. B 13, 5261 (1976).
4. G. Margaritondo et al.: Phys. Rev. B 22, 2777 (1980): G. Margaritondo
and N. G. Stoffel: Phys. Rev. Lett. 42, 1567 (1979).
5. M. H. Hecht and I. Lindau: Nucl. Instrum. Meth. 195, 339 (1982).
6. P. A. Lee and G. Beni: Phys. Rev. B 15, 2862 (1977).
7. B. K. Teo and P. A. Lee: J. Am. Chem. Soc. 101, 2816 (1979).

Part VI
Amorphous Materials and Glasses

The EXAFS of Disordered Systems and the Cumulant Expansion

Grant Bunker

Institute for Structural and Functional Studies, University City Science Center, 3401 Market Street, Room 320, Philadelphia, PA 19104, USA

In the 15 or so years since SAYERS, STERN and LYTLE [1-4] demonstrated the utility of EXAFS spectroscopy for local structure determination, the technique has been successfully applied to a wide variety of materials. EXAFS is especially appropriate for studying liquids and amorphous solids because long range order need not be present in the sample. As the technique has matured, workers have become more aware of its strengths and limitations. Foremost among the latter are the "lack of information at low k" and the complications ("non-gaussian disorder") that occur when large disorder is present[5]. The main points of this paper are that such limitations are not as severe as has been supposed, and that practical schemes have been developed for analysis of moderately disordered systems.

Accessibility of Low-k Information

Well above a K-absorption edge, the oscillatory part $\chi(k)$ of the spherically averaged absorption coefficient can be expressed as [2,6,7]

$$\chi(k) = \frac{\mu-\mu_0}{\mu_0} = \sum_i \frac{f_i(\pi,k)}{kr_i^2} S_0^2 (k) e^{-2r_i/\lambda(k)} \sin[2kr_i+2\alpha(k)+\beta_i(k)] \qquad (1)$$

where $f_i(\pi,k)$ and $\beta_i(k)$ are respectively the amplitude and phase of backscattering from neighboring atom i, $\alpha(k)$ is the central atom p-wave phase shift, and $S_0^2(k)$ and $\exp(-2r/\lambda)$ represent inelastic losses from multielectron excitations.

For the purpose of data analysis, it is convenient to rewrite (1) as

$$\chi(k) = \sum_i \frac{B_i(k)}{r_i^2} e^{-2r_i/\lambda(k)} \sin[2kr_i + \delta_i(k)] \qquad (2)$$

where $B_i(k)=f_i(\pi,k)S_0^2(k)/k$ and $\delta_i(k)=2\alpha(k)+ \beta_i(k)$. The "electronic" quantities are absorbed in $B(k)$, $\delta(k)$ and $\lambda(k)$. It is known that (1) and (2) break down close to the edge for several reasons: the curvature of the photoelectron wavefront at the backscattering atom cannot be neglected, the asymptotic form of the Hankel function cannot be used, and multiple scattering of the photoelectron becomes more pronounced. The role of many-electron excitations also is not well understood near the threshold except in simple model systems. The first two approximations ("small atom" and "short wavelength" approximations) in fact are not very good even over the lower end of the EXAFS range $k>3\text{Å}^{-1}$ [6,8], but the errors tend to cancel if the standard is well chosen, or curved waves are used in simulations.

HARTREE et al [9], LEE and PENDRY [6], MULLER and SCHAICH [10] (and probably others) have derived more general equations than (1) that do not make the small atom and short wavelength approximations, but their expressions involve complicated sums over partial wave phase shifts, and are rather awkward to use for analysis of experimental data. If we are willing to make the small atom approximation, but not the short wavelength approximation, then it can be shown [11] that (2) generalizes to :

$$\chi(k) = \sum_i \frac{B_i(k)}{r_i^2}[1+(1/kr_i)^2]e^{-2r_i/\lambda(k)} \sin[2kr_i + \delta_i(k) + 2Tan^{-1}(1/kr_i)]. \quad (3)$$

The correction terms are only substantial at very low values of k; the phase correction for k>3 imitates a shift in energy threshold E_0. Even with these correction terms, substantial errors are incurred from the small atom approximation. At present I know of no simpler way to deal with the small atom approximation than to explicitly calculate the sum over phase shifts from theoretical calculations, or in experimental work, to use standards that have bond lengths very close to the unknown's. The criteria for a "good" standard are more stringent near the edge than in the EXAFS region [12], in part because of the small-atom approximation.

Another reason that the low-k information has been thought to be inaccessible is the occurrence of multiple scattering [6,13-15]. It is known that multiple scattering (MS) is especially important between atoms that are nearly colinear with the central atom, because the intervening atom focuses the photoelectron wave. This forward scattering effect ("type 1 MS") is important over the whole energy range. Experimental evidence concerning the importance of large angle scattering has been scarce, however. Recent work of G. BUNKER and E. STERN [11,16] indicates that the large angle ("type 2") MS to SS amplitude ratio rapidly decays to 1/e within 1 Ryd past the edge in the experimental K-absorption spectra of tetrahedrally coordinated Mn in $KMnO_4$.

The first shell MS and SS are distinguishable because the MS signal oscillates more rapidly [6]. In $KMnO_4$ (for which the crystal structure is known [17]) the Fourier transform peaks from the shortest MS paths do not overlap the SS peaks from distant shells, so it is possible to isolate them by Fourier filtering. Because the first shell tetrahedron of oxygen atoms is very rigid, the first shell SS and MS show very little temperature dependence. On the other hand, the SS from atoms outside the first shell shows a strong temperature dependence because those atoms are not covalently linked to the central MnO_4 tetrahedron. The lack of temperature dependence of the MS signal is strong evidence that it arises from the first shell scattering.

We have also performed numerical simulations [11] of the XANES of MnO_4 tetrahedral clusters, using both the single scattering formalism of MULLER and SCHAICH [10] and the full multiple scattering formalism of DURHAM, PENDRY and HODGES [14]. The potentials were constructed by superimposing self-consistent atomic charge densities and chemically shifting the muffin tin potentials to eliminate discontinuities. The spectra are compared in fig. 1, which shows the region immediately above the edge. As predicted by theory [6], the MS spectrum exhibits rapid oscillatory structure in addition to the slower oscillations from single scattering. Fourier transformation vs k of the difference spectrum [11] yields peaks that correspond to the MS path lengths (divided by 2) and whose positions and sizes scale with the cluster size, in complete accord with the theory.

It is possible to compare the experimental MS vs SS contributions in a similar manner, using Fourier filtering methods. Much of the fine structure in the experimental $KMnO_4$ edge is due to scattering from atoms beyond the first shell but these signals were eliminated by Fourier filtering. Figure 2 compares the experimental first shell SS signal with the first shell SS plus low order MS signals, as in figure 1. Despite the crudity of the potential construction, the qualitative similarity between the calculated and experimental

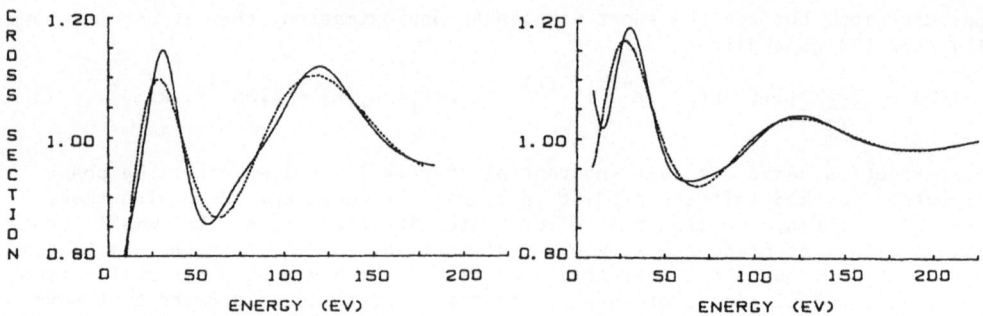

Fig. 1 MnO_4^- theoretical multiple scattering (solid) vs single scattering (dashed) normalized absorption coefficient.

Fig. 2 MnO_4^- experimental multiple scattering (solid) vs single scattering (dashed) normalized absorption coefficient.

spectra is striking, which further supports our claim of having experimentally isolated the MS signals.

Theoretically [6] the nth order scattering scales as $a^{-(n+1)}$ where a is the length scale of the cluster of atoms; e.g. single scattering scales as $1/a^2$, double scattering scales as $1/a^3$ etc. We therefore expect MS to be relatively less important for tetrahedral clusters of bond length larger than in in $KMnO_4$ (1.63Å)[17] and with similar atoms.

Furthermore, we expect type 2 MS to be smaller in randomly oriented (e.g. polycrystalline) samples containing octahedral sites, because the contributions from the shortest type 2 MS paths spherically average to zero [18]. Type 1 MS (forward scattering) can be appreciable in this case, however, as is well known [19].

It is often said that, since the photoelectron mean free path becomes large at low energy [20], the very long MS paths must make an important contribution. It is important to realize however that in EXAFS the finite core-hole lifetime limits the range of the photoelectron, which suppresses the divergence of $\lambda(k)$ near threshold [11]. This effect only occurs in x-ray absorption, not in photoemission or LEED. At k=0, in atomic units $\lambda = \eta_o^{-1/2}$, where η_o is the core hole width. For a core level width of 1 ev (as for Fe), attains a constant value of 2.8Å at k=0, and only becomes longer than several Angstroms at high energies. Thus for central atoms of sufficiently large atomic number, only relatively short MS paths contribute.

Below the edge both types (1 and 2) of MS are important. It is well known that certain features below the edge (e.g. 3d transitions) depend on the coordination symmetry. Such a dependence on bond angles can only arise from MS [21]. This is true even if the orbital is regarded as a bound state; the wavefunction in that case decays exponentially with distance, which explains the strong dependence of the 3d transition strength on bond length that is observed experimentally [22,11].

MULLER and SCHAICH [10] have proposed that their SS theory can account for the XANES and EXAFS, whereas the Daresbury group [14] claim that MS is important. Our results indicate that the Daresbury group is correct within about 1 Rydberg of the threshold and below, but Muller and Schaich are correct above that point, provided that type 1 MS is included with the single scattering. This suggests that an EXAFS-like analysis can be extended to lower energies than has been thought justified, if the small-atom approximation can be dealt with in some way. It also gives hope that the information at low k in disordered materials can be used for structure determination, which would eliminate (or minimize) a major perceived limitation of EXAFS spectroscopy. Further testing of this hypothesis is clearly needed, however.

Cumulant Expansion

The following discussion is after BUNKER [23]. In an EXAFS experiment, one always simultaneously measures the absorption of a large number of atoms. Equations 1 and 2 must therefore be averaged over an ensemble of photoabsorbing atoms whose environments may differ because of thermal motion and inhomogeneities. The finite Fourier transform range in data analysis makes it necessary to group together neighboring atoms whose distances cannot be resolved from each other. These shells of atoms often correspond to coordination shells. In the following we confine our attention to a single shell of atoms of the same atomic number.

Following SAYERS, STERN and LYTLE [2-4] we write the average of (2) for a single shell of identical atoms as:

$$\chi(k) = NB(k) \int \frac{p(r)}{r^2} e^{-2\gamma r} \sin[2kr + \delta(k)]\, dr \tag{4}$$

where $Np(r)dr$ is the probability of finding an atom between r and $r+dr$, and $\gamma = \lambda^{-1}$. This can be trivially rewritten as

$$\chi(k) = NB(k) \mathrm{Im}[e^{i\delta(k)} \int P(r;\gamma)e^{2ikr}\, dr] \tag{5}$$

where we have defined the "effective distribution" $P(r,\gamma) = p(r)e^{-2\gamma r}/r^2$. REHR [24] has shown that (5) can be expanded as

$$\chi(k) = NB(k) \mathrm{Im}[e^{i\delta(k)} \exp(\sum_{n=0}^{\infty} \frac{(2ik)^n}{n!} C_n(\gamma))] \tag{6}$$

where the C_n are the cumulants [25] of the effective distribution. The EXAFS amplitude $A(k)$ and phase $\phi(k)$, which satisfy $\chi = A \sin\phi$, are:

$$\ln[A(k)/NB(k)] = \frac{(-1)^n}{(2n)!} (2k)^{2n} C_{2n} \tag{7}$$

and

$$\phi(k) - \delta(k) = \frac{(-1)^n}{(2n+1)!} (2k)^{2n+1} C_{2n+1} \tag{8}$$

With the cumulant formalism it is straightforward also to include the effects of the k-dependent mean free path, and to relate the cumulants of the real and effective distributions to their power moments [23].

In most current EXAFS analysis, only the first term of the phase expansion and the first two terms of the amplitude expansion are kept, and the cumulants of the effective and the real distributions are assumed to be the same. These approximations break down when the distribution is very disordered, however. This fact has prompted some workers to claim that EXAFS gives incorrect distances for disordered systems, when in fact the analysis procedure, not EXAFS per se, is to blame. It should be stressed that it is quite incorrect to use the Fourier transform peak position as a measure of the distance unless the disorder is small (k <1). The peak position reflects the average slope of the EXAFS phase, not just the linear term which describes the distance. Substantial errors can occur if the higher order cumulants are neglected, k-space

271

analysis methods are most suitable, because the curvature of the phase is obvious if it is present. Very often the spectra can be accurately described by simply including one more term in the phase (k^3) and amplitude (k^4) expansions. This minor complication is an advantage in many cases, because the extra terms contain additional experimental information. Using this approach and a straightforward extension of the classical ratio method [4,23], TRANQUADA and INGALLS [26,27] have studied anharmonicity in solids and BOULDIN and STERN [28,29] have studied disordered adsorbed films. BOULDIN [30] has also used the known behavior of the phase and amplitude at low k (expressed by (7) and (8)) to formally invert (4) and determine the effective distribution.

To summarize, recent work has shown that moderately disordered systems can be successfully analyzed using Rehr's cumulant expansion approach, and experimental results and numerical simulations indicate that EXAFS-like analysis methods may be applicable to the low k region, rendering EXAFS an even more useful probe of the structure of disordered systems.

References

1. D.E. Sayers, F.W. Lytle, and E.A. Stern, Advances in X-Ray Analysis, 13, Plenum Press 248 (1970)
2. E.A. Stern, Phys. Rev. B, 10, 3027 (1974)
3. F.W. Lytle, D.E. Sayers, and E.A. Stern, Phys. Rev. B, 11, 4825 (1975)
4. D.E. Sayers, F.W. Lytle, and E.A. Stern, Phys. Rev. B, 11, 4836 (1975)
5. P. Eisenberger and G.S. Brown, Sol. St. Comm., 29, 481 (1979)
6. P.A. Lee and J.B. Pendry, Phys. Rev. B, 11, 2795 (1975)
7. C.A. Ashley and S. Doniach, Phys. Rev. B, 11, 1279 (1975)
8. R.F. Pettifer, P.W. Mcmillan, and S.J. Gurman, The Structure of Non-Crystalline Materials, P.H. Gaskell ed., p63., Taylor and Francis (1977)
9. D.E. Hartree, R. de L. Kronig, and H. Petersen, Physica (Utrecht) 1, 895 (1934)
10. J.E. Muller and W.L. Schaich, Phys. Rev. B, 27, 6489 (1983)
11. G.B. Bunker, University of Washington dissertation, (1984)
12. E.A. Stern, B. Bunker and S.M. Heald, in EXAFS Spectroscopy: techniques and Applications, B.K. Teo and D.C. Joy eds., Plenum N.Y. (1981)
13. W.L. Schaich, Phys. Rev. B, 8, 4028 (1973)
14. P.J. Durham, J.B. Pendry, and C.H. Hodges, Comp. Phys. Comm 25, 193 (1982)
15. C.R. Natoli, D.K. Misemer, S. Doniach and F.W. Kutzler, Phys. Rev. A, 22, 1104 (1980)
16. G. Bunker and E.A. Stern, Phys. Rev. Lett., 52, 1990 (1984)
17. G.J. Palenik, Inorg. Chem. 6., 503 (1967)
18. J.J. Boland, S.E. Crane, and J.D. Baldeschwieler, J. Chem. Phys., 77, 142 (1982)
19. G. Tolkiehn, P. Rabe, and A. Werner, Daresbury Laboratory Report DL/SCI/R17 (unpublished)
20. C.J. Powell, Surf. Sci., 44, 29 (1974)
21. J.B. Pendry, Comments. Sol. St. Phys., 10, 219 (1983)
22. J. Wong, F.W. Lytle, R.P. Messmer, and D.H. Maylotte, Submitted to Phys. Rev. B, (1984)
23. G.B. Bunker, Nucl. Inst. Meth., 207, 437 (1983)
24. J.J. Rehr, in Extended X-Ray Absorption Fine Structure, R.W. Joyner ed., Plenum (to be published)
25. R. Kubo, J. Phys. Soc. Japan, 17, 100 (1962)
26. J.M. Tranquada and R. Ingalls Phys. Rev. B, 28 3520 (1983)
27. J.M. Tranquada, University of Washington dissertation (1983)
28. C.E. Bouldin and E.A. Stern, Phys. Rev. B, 25, 3462 (1982)
29. C.E. Bouldin, University of Washington dissertation (1984)
30. C.E. Bouldin, this volume

EXAFS Studies of Amorphous Semiconductors

C.E. Bouldin* and E.A. Stern

Department of Physics, FM-15, University of Washington
Seattle, WA 98195, USA

EXAFS has shown itself to be a useful tool for the study of amorphous materials [1,2]. The well-known advantages of EXAFS compared with diffraction and scattering methods are that EXAFS requires no long-range order and is atom specific, greatly simplifying analysis of multicomponent systems[3]. It is also well known that EXAFS is primarily sensitive to sharp features in the radial distribution function[4] (r.d.f.) and that this often limits the radial distance that EXAFS can probe in an amorphous system since distant coordination shells tend to be quite broad. However, the sensitivity of EXAFS to sharp features can be useful even in the study of amorphous systems, and there are some prospects for extending EXAFS studies to systems with high disorder ($\sigma^2 \gtrsim 0.02$ Å^2). This paper presents a study of the amorphous-to-crystalline transition in amorphous germanium (a-Ge) in which the sensitivity of EXAFS to sharp features was crucial to the discovery of a minority (~20%) of small (~30 Å) microcrystals in the a-Ge. Finally, an assessment is given of the prospects for using EXAFS to study systems with large disorder.

The a-Ge which was studied was prepared by sputtering Ge onto Kapton films which were held at a fixed temperature, T_s, between 175°C and 330°C. Details are given elsewhere [5]; here it is sufficient to note that the structural properties of the material are controlled by varying the substrate temperature. A measure of any increase in the ordering of the a-Ge with increasing T_s is provided by the variation in $\Delta\theta$, the average angle between first-shell bonds. In the crystal, $\Delta\theta = 0$, except for thermal motion, while in a-Ge $\Delta\theta_{rms} \simeq 10°$. This has a dramatic effect on σ_2^2, as shown in the Fourier transforms in Fig. 1. The values of σ_2^2 can be used to determine the changes occurring in $\Delta\theta$ and hence to monitor the onset of crystallization.

An important issue is whether the ordering associated with crystallization occurs homogeneously or begins with heterogeneous nucleation. This is a difficult question to resolve with other structural probes since diffraction is limited by finite-size broadening, broad background from the a-Ge which is not nucleated into microcrystals, and a weak signal since only the nucleation sites contribute to the lines. X-ray scattering is useful for looking at broad features in the r.d.f., but it can easily miss the presence of a small, sharp feature embedded in a much broader structure since it has lower spatial resolution than EXAFS. This is because the wave number q is defined through the product qr in scattering, while in EXAFS it is defined by $2kr$ so that $q = 2k$. Since scattering data stop at $q \simeq 15$ Å^{-1} this is equivalent to $k_{max} = 7.5$ Å^{-1} in EXAFS. EXAFS data typically extend to $k_{max} \simeq 13$ Å^{-1}, so that EXAFS has significantly higher spatial resolution. Since small changes

---

(*) Address as of Oct. 1, 1984: U.S. Dept. of Commerce, National Bureau of Standards, Washington, DC 20234.

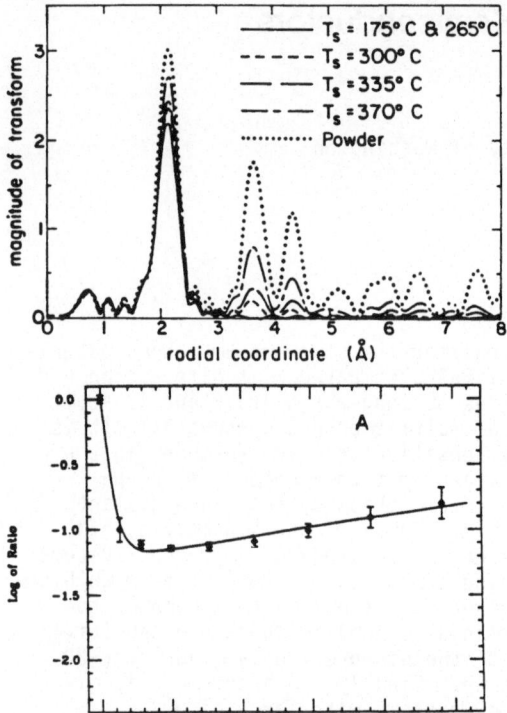

Fig. 1. Fourier transforms of the $\chi(k)$ of crystalline Ge and a-Ge with various values of T_s

Fig. 2. Log ratio of second-shell a-Ge (T_s=300°C) measured at 80 K, taken with crystalline Ge measured at 300 K. The solid line is fit of the two-phase model described in Text

in the *sharp* part of the r.d.f. are expected to reveal details of the crystallization process, EXAFS is well suited to this problem.

Using the ratio method to analyze the second shell in the data for T_s = 300°C in Fig. 1 gives the result shown in Fig. 2. If the linear part of the ratio at $k \gtrsim 25$ Å² is extrapolated to $k = 0$ we find that apparently $N_2^{Am} \simeq 0.2\ N_2^{Cr}$. However, an analysis of the first shell shows that $N_1^{Am} = N_1^{Cr}$ to within 1%. This means that $N_2^{Am} = N_2^{Cr}$ to similar accuracy because a second neighbor to a given center atom is a first neighbor to a first neighbor of the center atom. This inferred value of N_2^{Am} does not determine the *distribution* of the second shell atoms, but it does show that $N_2^{Am} = 12$. This means that the $k = 0$ extrapolation of the ratio must pass through the origin and that the upturn in the data at low k is real. The solid line shown is a fit to the data of

$$\log_e [\alpha \exp -2k^2\Delta\sigma_a^2 + (1 - \alpha)\exp(-2k^2\Delta\sigma_{\mu c}^2)] , \qquad (1)$$

where α is the fraction of the a-Ge with disorder $\Delta\sigma_a^2 = \sigma_a^2 - \sigma_{cr}^2$ and the rest has disorder $\Delta\sigma_{\mu c}^2 = \sigma_{\mu c}^2 - \sigma_{cr}^2$. From Table 1 it is clear that $\Delta\sigma_a^2$ is characteristic of the disorder found in a-Ge, while $\sigma_{\mu c}^2$ is much closer to σ_{cr}^2. This leads to an interpretation of our results in terms of a two-phase model in which the Ge is in two different forms in different structures in different parts of the solid. Since the σ^2's are so different it seems clear that the ordered parts are the nucleation sites at which crystallization begins. We also determine how $\sigma_{\mu c}^2$, σ_a^2, and α vary with T_s. As T_s increases we see that $\alpha \rightarrow 0$ and $\sigma_{\mu c} \rightarrow \sigma_{cr}^2$. These results all indicate that we are observing the amorphous-to-crystalline transition as T_s is increased. The value $\sigma_{\mu c}^2$ is always slightly larger than σ_{cr}^2 and this must be due to the fact that the microcrystals are present inside a matrix of a-Ge. The requirement that the microcrystals join continuously with the surrounding a-Ge will introduce strains into the microcrystals and increase the disorder in them. Since we observe a substantial change in $\sigma_{\mu c}^2$ with T_s the strain must propagate through a substantial fraction of the microcrystals. Assuming an exponential decay of the strain, we conclude the microcrystals start out ~30 Å in size, consistent with the fact that no x-ray diffraction peaks were observed in the samples with the two largest $\sigma_{\mu c}^2$'s.

Table 1. Results of fitting second-shell $\log_e$ ratios of 80 K a-Ge with 50 K crystalline Ge with two-phase model, and $\Delta\phi(k)$ with a linear fit. For T_s below 250°C no second-shell signal was detected.

α	$\Delta\sigma_{\mu c}^2$ [$\text{Å}^2 \times 10^{-3}$]	$\Delta\sigma_a^2$ [Å^2]	R_2 [Å]	T_s [°C]
0.75(5)	5.55(10)	> 0.07	-0.02(1)	250*
0.69(3)	6.00(10)	> 0.07	-0.01(1)	250*
0.68(6)	6.10(10)	> 0.07	-0.01(1)	250*
0.77(3)	6.30(10)	> 0.07	-0.01(1)	300
0.73(1)	5.60(15)	> 0.07	-0.02(1)	300
0.56(5)	4.15(10)	> 0.07	-0.01(1)	335
0.20(10)	2.00(10)	> 0.07	-0.01(1)	370

*The T_s of this sample is uncertain due to poor thermal contact.

To investigate the structure of a-Ge beyond the first coordination shell before any crystallization begins is a challenging problem because $\sigma^2 \gtrsim$ 0.07 Å^2 for distant shells. A general approach to high disorder is provided by the cumulant moment expansion [6]. This approach can be used to determine a single-shell r.d.f. for a disordered material. Single-shell $\chi(k)$ can be written as

$$\chi(k) = N\ B(k) \int_0^{\infty} \frac{\rho(r)\exp(-2r/\lambda)}{r^2}\ \sin[2kr + \delta(k)]dr \ , \tag{2}$$

and by canceling $B(k)$ and $\delta(k)$ using a suitable standard this can be reduced to

$$P'(k) = \frac{N_u}{N_s} \int \rho'(r)\sin(2kr)dr \ , \tag{3}$$

where u and s subscripts refer to the standard and the unknown and $\rho'(r)$ is defined as the difference $\rho_u(r) - \rho_s(r)$. This is easily inverted to give $\rho'(r)$ by

$$\rho'(r) = \frac{N_s}{N_u} \int_{kmin}^{\infty} P'(k)\sin(2kr)dr \ , \tag{4}$$

and hence $\rho(r)$, since $\rho_s(r)$ is known. Ordinarily, $kmin \simeq 3\ \text{Å}^{-1}$ so that this inversion is almost worthless [7]. However, the low k information can be restored by using the cumulant expansion to parameterize the log ratio and phase difference. The cumulant moment expansion shows generally that these functions have the form

$$\log_e \frac{A_u(k)}{A_s(k)} = a_0 + a_2 k^2 + a_4 k^4 + \ldots \tag{5a}$$

$$\Delta\phi(k) = a_3 k^3 + a_5 k^5 + \ldots , \tag{5b}$$

where the a's represent the cumulant moments and the coefficients in the expansion. The coefficient a_0 depends on the coordination number in the unknown and is *always* determined by some extrapolation of the EXAFS to $k = 0$. If N_u is known, as in the case of a-Ge, then a_0 is determined and the missing information for $0 < k \lesssim 3 \ \text{Å}^{-1}$ can be interpolated with the cumulants. This means that the inversion in Eq. (4) can be carried out since the information is now available from $k = 0$ to k_{max}.

To test this procedure, $\chi(k)$ data were generated for a model r.d.f. which is highly disordered and has been shown to be useful for interpreting EXAFS data from liquid metals [8]. The functional form of this model is

$$\rho(r) = \begin{array}{ll} (r - r_0) \exp[-B(r - r_0)] & r \geq r_0 \\ 0 & r < r_0 \end{array} , \tag{6}$$

and with $B = 5.14 \ \text{Å}^{-1}$ the first cumulants are $C_2 = 0.1136 \ \text{Å}^{-2}$ and $C_3 = 0.0442 \ \text{Å}^{-3}$, showing that the model is both highly disordered and non-Gaussian. Figure 3 shows the model and the results of reproducing the model from the EXAFS $\chi(k)$ corresponding to the r.d.f. The agreement between the original r.d.f. and what is recovered from the $\chi(k)$ data is good, except for the extreme part of the tail at high r.

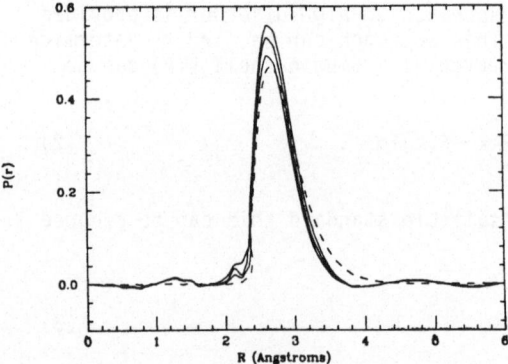

Fig. 3. $\rho(r)$ recovered from $\chi(k)$ by analysis described in text. Data are used only over range $k > 3 \ \text{Å}^{-1}$. Original $\rho(r)$ is dashed line; three solid lines ±10% uncertainty in N_u.

This model system shows that, in principal, $\rho(r)$ can be recovered from EXAFS data. A problem for amorphous semiconductors which is usually not discussed comes from the fact that the first shell remains highly ordered compared with the large disorder in the second shell. This means that the leakage of the first-shell signal into the second-shell r-space bandpass can be a serious problem, since the leakage can be as large as the second-shell signal. In systems in which the disorder in the first and second shells is not so radically disparate, or in analyzing a highly disordered first shell, this is less of a problem.

Amorphous systems have another property which is encouraging for the study of large disorder--the existence of a good standard through the crystalline counterpart of the material. This opens up prospects for using the $\chi(k)$ data down to $k \sim 1 \text{ Å}^{-1}$. Since the amorphous unknown and the crystalline standard often have the same *average* distances, the small-atom corrections (which are only distance dependent) cancel out [9]. In addition, it has been shown recently that mulitple scattering effects at low k are much smaller than had been thought [10], giving further confidence in the use of data at low k. These facts, combined with the existence of good standards, indicate that there is still room for further progress in EXAFS studies of amorphous materials.

This work was supported by the National Science Foundation, grant no. DMR80-22221 (Stern), and by NASA grant NAGW199 (Bouldin). We thank the staff of SSRL for their help. The SSRL is supported by the DOE Office of Basic Energy Sciences, and by the NIH, Biotechnology Resources Program, Division of Research Resources. Discussions with G. Bunker are gratefully acknowleged.

1. F. Evangelisti, M.G. Paoietti, A. Balzarotti, F. Comin, L. Incoccia, and S. Mobilio: Solid State Commun. 37, 413 (1981)
2. E.A. Stern, C.E. Bouldin, B. von Roedern, and J. Azoulay: Phys. Rev. B 27, 6557 (1983)
3. S.H. Hunter: Stanford Synchrotron Radiation Project Report No. 77/04, 1974.
4. P. Eisenberger and G.S. Brown: Solid State Commun. 29, 481 (1979)
5. C.E. Bouldin, E.A. Stern, B. von Roedern, and J. Azoulay: Phys. Rev. B (accepted for publication)
6. G.B. Bunker: Nucl. Inst. Meth. 207, 437 (1982)
7. G.S. Brown: in Synchrotron Radiation Research. ed. H. Winick and S. Doniach (Plenum Press, New York)
8. E.D. Crozier and A.J. Seary: Can. J. Phys. 58, 3027 (1979)
9. G.B. Bunker: Ph.D. thesis, University of Washington, 1984
10. G.B. Bunker and E.A. Stern: Phys. Rev. Lett. 52, 1990 (1984)

EXAFS Measurements of Hydrogenated Amorphous Germanium

C.E. Bouldin* and E.A. Stern

Department of Physics, FM-15, University of Washington
Seattle, WA 98195, USA

Hydrogenated amorphous semiconductors (a-Si:H and a-Ge:H) are proving to be technically useful in the production of low-cost solar cells. Hydrogenation of a-Si and a-Ge is a useful means of saturating dangling bonds which are present in the pure amorphous materials. After hydrogenation, it is possible to dope the amorphous materials to produce p-n junctions which are similar to those constructed from crystalline material [1]. It is not known exactly where the H binds in the a-Ge or a-Si, but this information could clearly be useful in enhancing the production of devices from amorphous semiconductors. In this study we report the results of EXAFS measurements on a-Ge and a-Ge:H with H concentrations of 4.9 and 6.0 at%.

As has been noted previously [2,3], a-Ge shows only a first-shell EXAFS signal due to the large (~10°) disorder in the bond angle. Our results show that a-Ge also has a first-shell coordination which is, at most, 1% less than the crystal. The addition of H does not change the value of the first-shell coordination or cause the appearance of any other peaks in the Fourier transform. We do find a *decrease* in σ_1^2 as H is added, as shown in Table I.

Table 1. Results for first-shell EXAFS

Sample	$\Delta N/N$	ΔR [Å]	$\Delta\sigma^2$ [Å²]	$\Delta\sigma^2$ (300–80)
Amorphous Ge	-0.005(15)[a]	0.003(2)[a]	0.00163(10)[a]	0.00146(10)
4.9% hyd. Ge	-0.010(15)[a]	0.000(2)[a]	0.00140(10)[a]	0.00160(10)
	-0.010(10)[b]	0.003(2)[b]	-0.00033(10)[b]	
6.0% hyd. Ge	-0.015(15)[a]	0.002(2)[a]	0.00110(10)[a]	0.00145(8)
	-0.010(10)[b]	0.008(2)[a]	-0.00066(10)[b]	
	0.007(10)[c]	-0.003(3)[c]	-0.00028(8)[c]	

[a]Relative to crystalline Ge
[b]Relative to unhydrogenated a-Ge
[c]Relative to 4.9% hydrogenated a-Ge

The interesting result of these measurements is that σ_1^2 is decreased by the addition of H and that no other structural changes are observed. Since σ_1^2 changes without any simultaneous change in N, we can rule out substitutional H since this would lower the Ge-Ge coordination number. We believe that our results indicate that the H coats the surfaces of internal voids in the a-Ge and attaches to dangling bonds on the surface of the voids. It is known that a-Ge has a lower density than crystalline Ge and so voids must be

---

(*) Address as of Oct. 1, 1984: U.S. Dept. of Commerce, National Bureau of Standards, Washington, DC 20234.

present, since the first-shell bond lengths are the same [4]. By combining the known density for a-Ge with our measured $\Delta N/N$ values, we estimate that the voids are ~30 Å in size.

This work was supported by National Science Foundation grant DMR80-22221 (E.A. Stern) and NASA grant NAGW199 (C.E. Bouldin).

References

1. W. Paul, A.J. Lewis, G.A.N. Connell, and T.D. Moustakas: Solid State Commun. 20, 969 (1976)
2. E.D. Crozier: in EXAFS Spectroscopy (Plenum, New York 1981), pp. 89-103
3. E.A. Stern, C.E. Bouldin, B. von Roedern, and J. Azoulay: Phys. Rev. B 27, 6557 (1983)
4. R.J. Timkin, W. Paul, and G.A.N. Connell: Adv. Phys. 22, 581 (1973)

Structural Investigations of Amorphous Iron-Germanium Alloys by EXAFS

R.D. Lorentz

Department of Applied Physics, Stanford University, Stanford, CA 94305, USA

S.S. Laderman and A.I. Bienenstock

Department of Materials Science and Engineering, Stanford University
Stanford, CA 94305, USA

We present the amplitudes and the interatomic distance results returned by one shell fitting of EXAFS data obtained at both the iron and germanium K absorption edges from a-Fe_xGe_{1-x} alloys, with $0.0 \leq x \leq 0.72$. The first shell bond lengths determined for a-$FeGe_2$ are found to be about 0.15 Angstroms shorter than corresponding distances in c-$FeGe_2$. This difference may be due to a true decrease in the average first neighbor bond length in the amorphous material or may result from an asymmetric broadening of this shell relative to that in the crystal.

I. Introduction

The work reported here is the first part of a structural study of a-Fe_xGe_{1-x} alloys ($0.0 \leq x \leq 0.72$) utilizing EXAFS, anomalous x-ray scattering, and small angle x-ray scattering. Over this composition range, the system changes from a tetrahedrally coordinated semiconductor to a close packed amorphous metal, and displays a semiconductor-metal transition [1] as well as magnetic moment formation [2]. Our aim is to understand this system's structure vs. composition and the relation of this structure to the observed electrical and magnetic properties.

II. Experiment

Amorphous Fe_xGe_{1-x} samples have been prepared as thin films by sputtering from pure Fe and Ge targets onto Kapton substrates. Transmission EXAFS data have been collected at SSRL at both the Ge and Fe K absorption edges for all amorphous samples and c-$FeGe_2$ at liquid nitrogen temperature.

III. Results and Discussion

A one coordination shell model with a variable edge energy Eo was fit to this EXAFS data. Calculated phase shifts and backscattering amplitudes were used [3]. For the amorphous samples, both model and data were windowed from k=4 to k=15 inverse Angstroms, but not Fourier filtered. Fits to data from both edges converged to physically reasonable parameters with either Fe or Ge assumed as first shell neighbors. The distances returned were approximately 0.04 Å longer when Fe neighbors were assumed than with Ge as neighbors.

This method returned a Ge-Ge distance of 2.44 Å in a-Ge, which compares well with the c-Ge value, 2.45 Å. The Ge-first shell distances are essentially constant in each of three composition ranges. The Fe-first shell distances can also be grouped into three ranges, though these distances increase with x in the lowest iron content range, and are constant for the other two. When comparing results from both edges, the Ge-Fe and Fe-Ge distances are seen to be equal, within experimental error, over the entire range 33% to 72% Fe. For 33-49% Fe, these bond lengths are equal to the sum of the Fe and Ge covalent radii, 2.39 Å. The Fe-Ge distances found in the 5-18% Fe alloys increase from 2.35 Å to 2.38 Å, and are shorter than any found in a Fe-Ge crystal compound. The Ge-first shell distances are constant in this range, but decrease by 0.09 Å on going from 30% to 33% Fe. No change in Fe to first shell distance occurs at 33% Fe. The distances about Fe and Ge both increase by about 0.05 Å from 49% to 65% Fe, and may indicate an expansion in interatomic distances as the iron content is increased in the high x alloys.

Table 1. Distances in Angstroms returned by fits assuming one species of neighbors about a central atom vs. sample iron concentration

Central Atom	Neighbor Atoms	Distance returned by fitting in composition region:			
		0 - 18%Fe	18 - 30%Fe	33 - 49%Fe	65 - 72%Fe
Ge	Ge	<------ 2.45+0.01 ------>		2.36+0.01	2.41
Ge	Fe	<------ 2.49+0.01 ------>		2.40+0.01	2.45
Fe	Ge	2.35-2.38	<------ 2.39+0.01 ------>		2.44+0.01
Fe	Fe	2.39-2.42	<------ 2.42+0.01 ------>		2.48+0.01

The distances determined for the 33 at% Fe (a-FeGe$_2$) sample are in striking contrast to those of c-FeGe$_2$. The filtered first shell Ge EXAFS of c-FeGe$_2$ gives a Ge-Fe distance of 2.55 Å, close to the known value of 2.56 Å. However, the Ge edge EXAFS of a-FeGe$_2$ indicates a much shorter Ge-Fe distance of 2.40 Å. The Fe edge data from c-FeGe$_2$ assuming Ge neighbors gives 2.55 Å, close to both the average distance in the first shell, 2.54 Å, and to the Ge neighbor distance. In the amorphous material, the Fe-Ge distance found is shorter, 2.38 Å.

Table 2. Comparison of actual c-FeGe$_2$ distances, those determined for the crystal with this EXAFS data, and those determined for a-FeGe$_2$

Central Atom	Actual # of neighbors and distance in c-FeGe$_2$	Assumed neighbor	Distances returned by fitting for	
			c-FeGe$_2$	a-FeGe$_2$
Ge	1 Ge at 2.59	Ge	2.51	2.36
Ge	4 Fe's at 2.56	Fe	2.55	2.40
Fe	8 Ge's at 2.56	Ge	2.55	2.38
Fe	2 Fe's at 2.48	Fe	2.58	2.42

The Fe-Ge and Ge-Fe average bond lengths found for a-FeGe$_2$ may be about 0.15 Å shorter than the corresponding distances in the crystal. Since these distances are very close to the covalent distance, bonding in the amorphous phase may be more covalent than in the crystal. These distances have analog in a crystal other than c-FeGe$_2$. They are about the closest neighbor distance, 2.39 Å, and are about 0.04 Å less than the average distance in the low temperature polymorph of c-FeGe. These same distances are found for the 33-49% Fe alloys. Their structure may be more similar to c-FeGe (low T) than to c-FeGe$_2$, and the reduction in bond length seen at 33% Fe may be due to a switch from primarily Ge-Ge bonding below this iron concentration to Ge-Fe covalent bonding above it.

The trends in derived distances appear to be correlated with trends in the EXAFS amplitudes. The Ge edge EXAFS amplitudes vs. iron concentration show a minimum at 30% Fe (Fig.1). This is close to where the Ge to first

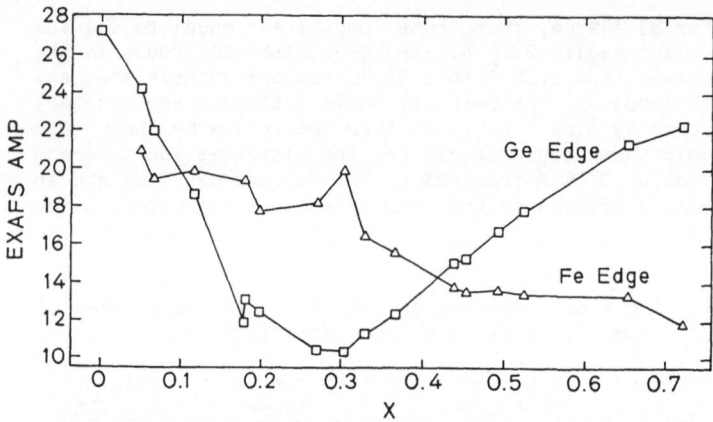

Fig. 1. EXAFS amplitudes vs. iron concentration

shell distances seem to suddenly decrease. The Fe edge amplitudes instead decrease with increasing iron content. No sudden Fe-first shell distance drop was seen across the composition range either.

These distance and amplitude trends are difficult to understand when using EXAFS alone. A reduction in Ge edge amplitudes could be produced by well ordered Ge-Ge and Ge-Fe bonds. However, a reduction could also result from a high degree of disorder in these bonds. This possibility influences the interpretation of the distance reduction seen at 33% Fe. The EXAFS may be returning the distance to the leading edge of a disorder-broadened distribution of first shell distances [4]. We hope to resolve this uncertainty with Differential Anomalous Scattering (DAS) experiments performed on this sample. DAS provides similar information on the average environment about a particular species as EXAFS, but doesn't suffer from this sensitivity to the leading edge of a distribution of distances [5]. The DAS data have been collected and are being analyzed.

IV. Conclusion

The most striking feature of the possibilities for the interatomic distances reported here is the dramatic reduction in bond lengths in a-FeGe$_2$, where the distances are about 0.15 Angstroms shorter than in the crystal. It remains to be seen from further work on this sample, especially DAS, if this is a real reduction in first shell average distance or is instead due to disorder in the local environments of a-FeGe$_2$ and to the problems encountered with EXAFS on disordered materials.

V. Acknowledgements

The authors wish to thank C.L. Tsai for providing the c-FeGe$_2$ sample. This work was performed at SSRL which is supported by the Department of Energy, Office of Basic Energy Sciences; and the National Institutes of Health, Biotechnology Resource Program, Division of Research Resources, and was supported by the NSF-MRL program through the Center for Materials Research at Stanford University.

VI. References

1. H. Daver, O. Massenet, and B. Chakraverty, Proc. Fifth Int. Conf. Amorphous and Liquid Semiconductors, p.1053, Garmish (1973).
2. P. Mangin, M. Piecuch, G. Marchal, and C. Janot, J.Phys.F 8, 2085 (1978).
3. B.-K. Teo and P.A. Lee, J. Am. Chem. Soc. 101, 2815, (1979).
4. P. Eisenberger and G.S. Brown, Sol. St. Comm. 29, 481, (1979).
5. J. Kortright, W. Warburton, and A. Bienenstock, in "EXAFS and Near Edge Structure", ed. by A. Bianconi, L. Incoccia, and S. Stipcich (Springer-Verlag, 1982), p.362.

Short Range Order In a-Ge$_x$Si$_{1-x}$:H Alloys

L. Incoccia* and S. Mobilio

PULS-INFN Laboratori Nazionali di Frascati, I-00044 Frascati (Roma), Italy

M.G. Proietti, P. Fiorini, and F. Evangelisti

Dipartimento di Fisica, Università "La Sapienza", I-00185 Roma, Italy

1 Introduction

The possibility of "tailoring" optical and electronic properties makes the study of the semiconducting binary alloys of great interest. In particular the hydrogenated amorphous silicon-germanium alloys (a-Ge$_x$Si$_{1-x}$:H) have been recently extensively studied and applied in efficient photovoltaic devices [1]. It is the aim of the present work to perform a structural characterization of these alloys by means of EXAFS spectroscopy.

2 Experimental

The samples were grown in a R.F. capacitively coupled glow discharge apparatus by a mixture of SiH$_4$ and GeH$_4$. The gas composition r=GeH$_4$ /(SiH$_4$ +GeH$_4$) was varied in the range 0-0.93. The deposition temperature was 250 °C, but a few samples were also deposited at 190 °C. The chemical composition x was determined by plasma emission spectroscopy at CISE Laboratories, Segrate (MI). The X ray absorption spectra at the Ge K-edge were taken at room temperature at the Frascati Synchrotron Radiation facility.

3 Results and Discussion

In Fig. 1 we show some EXAFS spectra of the samples studied. It is apparent that the lineshapes of the two extreme cases (spectra (a) and (f) corresponding to x=0.07, and x=1 respectively) are quite different: spectrum (a) having a monotonically decreasing envelope function , spectrum (f) a nearly gaussian envelope function with a maximum at 7 Å . This dissimilar k-dependence results from the different backscattering function of the atoms which form the first coordination shell in the two cases: Ge in spectrum (f) and predominantly Si in spectrum (a). With decreasing Ge concentration the remaining spectra of Fig.1 exhibit a progressive evolution toward a higher backscattering amplitude at low k and a decrease of the maximum at 7 Å. This behavior confirms a smooth variation of the first shell composition from pure Ge to pure Si.

Therefore at intermediate composition the spectra result from a contribution of Ge-Ge and Ge-Si pairs whose bond lengths and relative weight should be determined. To this end we performed a complete k-space analysis by Fourier filtering the experimental data. The EXAFS spectra of the samples with the two extreme compositions x=1 and x=0.07 are the obvious representative of Ge-Ge and Ge-Si pair, respectively. These two samples match all the conditions required to be good models for investigating the alloys with intermediate compositions.

---

* Istituto di Struttura della Materia del CNR, Frascati (Roma), Italy

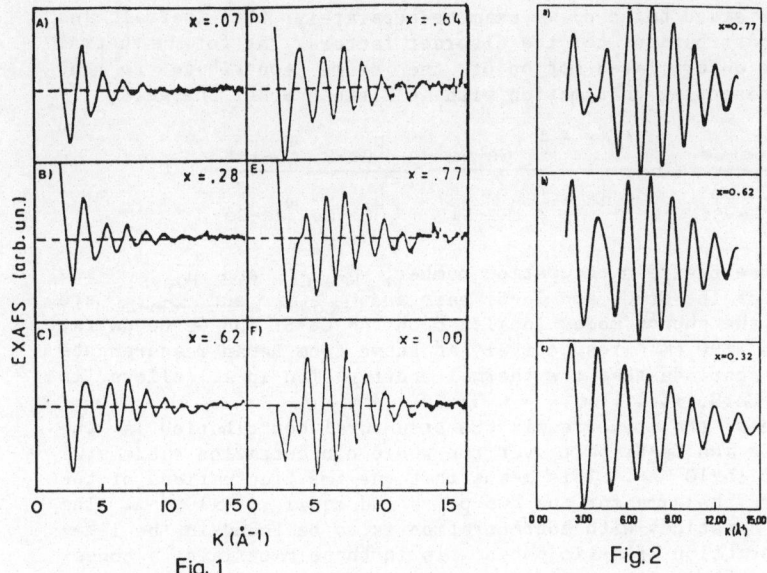

Fig. 1

Fig. 2

Figure 1 Experimental χ (k) at different Ge concentrations.

Figure 2 Comparison of the experimental Fourier filtered k χ (k) with the result of the fit (dashed line).

A fitting function was built in k-space to be compared to the Fourier filtered kχ (k) by using the phases and amplitudes derived from these model samples, as reported in Ref.2. The only assumption we made was that the average coordinaton number around the Ge atoms does not vary as a function of the concentration. This hypothesis is reasonable since in both the two extreme cases considered (x=1 and x=0.07) the coordination number was equal to four. Therefore we performed a two shells fit by keeping fixed the total coordination number around Ge atoms, but varying the relative concen-tration of the two components, their σ^2 factors and R values. In Fig.2 we show the excellent quality of the fits obtained, and in Table 1 we report the numerical results. We want to stress here that:

a) the bond lenghts are equal to 2.45 Å for the Ge-Ge pair and to 2.38 Å for the Ge-Si pair and both remain constant over the whole concentration range;

b) the relative concentrations of the Ge-Ge and Ge-Si pairs are in excel-lent agreement with the results of the chemical analysis (compare co-lumns one and two);

c) σ^2 values obtained are so small that the disorder factor can be consi-dered constant over the whole range of concentration and equal to the value of the two models, namely $\sigma^2=0.5 \times 10^{-2}$ Å.

Our results on distances are the first direct determination of first-nearest neighbour bond length in Ge-Si:H amorphous alloys. They con-firm previous results of a diffraction study of $Ge_x Sn_{1-x}$ at x=0.5 and x=0.75 [3] and suggest that a concentration-independent bond distance is probably a general feature of the disordered alloys.

The excellent agreement found between the relative number of Ge-Ge and Ge-Si pair in the first coodination shell and the overall chemical composi-ton of the alloy demonstrates the presence of a large compositional disord-er superimposed to the topological one. This conclusion is in agreement with results obtained by the analysis of Raman data [4].

285

In order to understand point c) we examine separately the thermal and the structural contribution to the disorder factors. As for the thermal component, only the uncorrelated motion of the atoms contribute to the EXAFS . By representing this motion with an Eistein model one gets:

$$\frac{\Delta\sigma^2_{Ge-Ge}}{\Delta\sigma^2_{Ge-Si}} = \frac{<u^2>_{Ge-Ge}}{<u^2>_{Ge-Si}} = \frac{\left(<n> + \frac{1}{2}\right)_{Ge-Ge}}{\left(<n> + \frac{1}{2}\right)_{Ge-Si}} \cdot \frac{\mu_{Ge-Si}\,\omega_{Ge-Si}}{\mu_{Ge-Ge}\,\omega_{Ge-Ge}} \sim 1$$

where $<n>$ is the Bose-Einstein occupation number, μ_{Ge-Ge} and μ_{Ge-Si} are the reduced masses of the Ge-Ge and Ge-Si pair and ω_{Ge-Si} and ω_{Ge-Ge} are the frequencies of the phonon modes localized on the Ge-Si and Ge-Ge pairs, equal to 400 and 290 cm^{-1} respectively as shown from Raman measurements [5]. Therefore we conclude that the thermal contribution in all alloys is equal to that of a-Ge:H, i.e. $\sigma^2 = 0.33 \times 10^{-2}$ $\mathring{A}^2$.

As a consequence of the above result the structural contribution is the same for the Ge-Ge and Ge-Si pair over the whole concentration range studied and equal to $0.16 \times 10^{-2} \mathring{A}^2$. This means that the rms fluctuations of the bond distances are the same for the two pairs and equal to 4×10^{-2} $\mathring{A}$. The explanation of its constancy with concentration is to be found in the likeness of the interaction elastic potentials in these materials, a consequence of their very close electronic structure. This closeness is reflected in the nearly equal values of the Keating potentials in pure Ge and Si.

Table 1 Numerical values of the parameters obtained from the fitting procedure.

% Ge	%Ge by EXAFS	R_{Ge-Ge} ($\mathring{A}$)	$\Delta\sigma^2_{Ge-Ge}$ ($10^{-3} \mathring{A}^2$)	R_{Ge-Si} ($\mathring{A}$)	$\Delta\sigma^2_{Ge-Si}$ ($10^{-3} \mathring{A}^2$)
28	26	2.45	−0.21	2.37	−0.03
32	30	2.45	0.42	2.37	−0.20
62	62	2.45	−0.05	2.37	0.17
64	65	2.45	−0.73	2.38	0.74
77	89	2.45	−0.11	2.38	2.7
89	100	2.45	−0.09	--	--

1 Y.Yukimoto: Jarect Vol.6 "Amorphous Semiconductor Technology and Devices", Ed. Y.Hamakawa, North Holland Publ. Co. 1983.
2 L.Incoccia, S.Mobilio, M.G.Proietti, P.Fiorini, C.Giovannella and F.Evangelisti: to be published.
3 R.J.Temkin, G.A.N.Connel and W.Paul: Solid State Commun. 11, 1591 (1978).
4 B.K.Agrawal: Solid State Commun. 37, 271 (1981).
5 S.Minomura, K.Tsugi, M.Wakagi, T.Ishidate, K.Inoue and M.Shibuya: J. Non-Cryst. Solids 59-60, 541 (1983).

EXAFS Study of $(GaSb)_{1-x}(Ge_2)_x$

Fred Ellis and Edward A. Stern

Department of Physics, FM-15, University of Washington, Seattle, WA 98195, USA

Linda Romano

Metallurgy Department, University of Illinois, Urbana, IL 61801, USA

The nature of the metastable compounds $(GaAs)_{1-x}(Ge_2)_x$ and $(GaSb)_{1-x}(Ge_2)_x$ has been the subject of recent speculation. Phase separation which would normally occur in thermal equilibrium is inhibited during the sputtering process by low-energy ion bombardment of the growing film [1]. Both single-crystal GaAs and Corning 7059 glass have been used as substrates producing single-crystal and preferentially oriented polycrystalline films, respectively. Optical absorption and x-ray diffraction measurements have led NEWMAN et al. [2] to adopt a zincblende-diamond order-disorder transition model for the structure of the film as a function of concentration x. This model predicts a direct band gap consistent with their optical results, and the superlattice lines associated with the zincblende structure were also observed to decay in a consistent manner.

EXAFS provides a direct probe in the case of $(GaSb)_{1-x}(Ge_2)_x$ for the zincblende-diamond transition, since the first coordination shell for all edges involved should vary in a simple way with concentration and signify either order or disorder. In particular, if $\overline{N}_{Sb}$ is the average number of Sb near-neighbors, for the Ga center atom we have

$$\overline{N}_{Sb} = 4(1 - x) \tag{1}$$

in the ordered case (zincblende) and

$$\overline{N}_{Sb} = 2(1 - x) \tag{2}$$

for the completely disordered state (diamond). The Ge center atom follows (2) for both the ordered and the disordered state. It is clear that a determination of $\overline{N}_{Sb}$ will show the degree of ordering (Ga center) and also provide a check on the uniformity of Ge mixing assumed so far (Ge center).

1. Experiment

$(GaSb)_{1-x}(Ge_2)_x$ was chosen for the EXAFS work because it was expected that the drastically different k dependence of the Sb back-scattering amplitude compared with Ga or Ge would allow a unique determination of $\overline{N}_{Sb}$ needed for comparison of (1) and (2). The EXAFS for the different concentrations was measured in the standard abosrption mode on beamline 4-I at SSRL. To keep the background absorption as small as possible the films were specially grown on Corning #00 glass slides. Both sides of the glass were used, and each sample consisted of two slides, typically giving a total sample thickness of 20 μm and glass thickness of 150 μm. Only the Ga and Ge edges were measured, both at a low temperature (50 K) so that effects due to a thermal Debye-Waller factor could be ignored.

287

2. Results and Discussion

The Ge concentrations determined from the Ga and Ge edge steps were $x = 0.25$, 0.29, 0.38, 0.46, and 0.58. These values sandwich the expected transition concentration of $x = 0.35$. The EXAFS signals were Fourier-filtered in the usual way to isolate the first coordination shell EXAFS data. Pure GaSb and Ge were also processed in the same way. The pure χ data can then be corrected by theory for central atom and back-scattering atom changes between Ga and Ge to generate a complete set of "pseudo χ" standards. These changes involve $\Delta Z = 1$, so they are fairly small. The first-shell data of the mixtures can then be fit to a superposition of the appropriate standards with numbers and distances allowed to be either independent or dependent parameters.

As a first trial, based on the simplest expectations of random Ge substitution, we require that the atoms be fully coordinated and that the number of Ge near-neighbors be fixed at $4x$. We also require that the Ga-Sb and Ge-Ge distances be fixed at the pure values. The results of this fit for the number of Sb neighbors are shown in Fig. 1. The circles are for the Ga centers and the plus signs are for the Ge centers. Equations (1) and (2) are plotted as the solid lines for comparison.

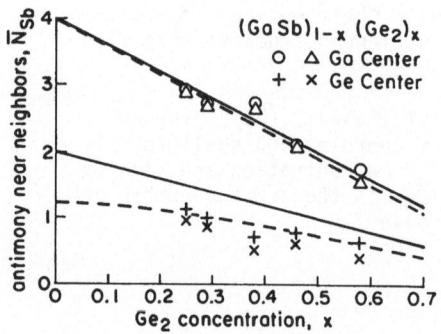

Fig. 1. Fitting results for N_{Sb} for Ga centers (circles and triangles) and Ge centers (plusses and crosses)

The first observation to make is that the number of Sb atoms around the Ga fall along the ordered line. This is not consistent with the completely random diamond structure at any concentration measured. The result for the Ge edge is disturbing because it implies that the Ge are not randomly distributed and the simple relationship observed for the Ga data should not have been seen. In addition, the fits were not as good as the Ga centers, and the Ga-Ge distance for both edges did not agree. This forced us to remove further constraints on the fitting. We allowed the total number of near neighbors to deviate from four. This had little effect on the Ga fits, but it greatly improved the Ge. Results are shown in Fig. 1 as triangles and crosses. The improvement in the Ge fits required the addition of vacancies. The Ga atoms remained fully coordinated. The atom-atom distances also agreed between both edges.

A model which describes the above results better than an ordered state with random substitution is one in which the Ge atoms preferentially substitute Sb with vacancies in some of the Ga sites neighboring the Ge. This is the simplest model that would allow fewer Sb around the Ge without substantially decreasing the amount around the Ga. The dashed line in Fig. 1 shows the prediction of this model with the measured vacancy fraction as input.

We then checked the behavior of the superlattice diffraction lines in our samples. We could see those lines only in a 10% Ge sample not used in our EXAFS measurements. For all of our low concentration samples, therefore, we found that the superlattice line decreases much faster than the $(1 - x)^2$ expected for Ge dilution alone.

3. Conclusions

We have shown that short-range order about the Ga atoms is 100%; i.e., the number of Sb near-neighbors is the maximum value expected for all concentrations measured. In addition, the Ge appears to preferentially substitute Sb atoms and has vacancies associated with it. We believe our diffraction results are inconsistent with an order-disorder transition. The short-range order and the lack of long-range order at low concentrations can be understood in terms of a model which excludes Ga-Ga and Sb-Sb first neighbors but permits the atoms to be otherwise randomly distributed. Such a model is also consistent with the band gap results [3]. It is possible that these results are dependent on the substrate used.

This research was supported by National Science Foundation grant PCM82-04234. We gratefully acknowledge the assistance of the staff of SSRL, at which these experiments were conducted. SSRL is supported by the DOE Office of Basic Energy Sciences, and by the NIH, Biotechnology Resources Program, Division of Research Resources.

References

1. J.E. Greene, S.A. Barnett, K.C. Cadien, and M.A. Ray: J. Crys. Growth 56 389 (1982)
2. K.E. Newman and J.D. Dow: Phys. Rev. B 27, 1744 (1983)
3. H. Holloway and L.C. Davis, unpublished

Measurements of the Fe Impurity Site in Si_3N_4 Using EXAFS

C.E. Bouldin* and E.A. Stern

Department of Physics, FM-15, University of Washington, Seattle, WA 98195, USA

Silicon nitride is a ceramic that is promising as a structural material which retains its strength at high temperatures. Possible applications include high-efficiency turbines and other heat engines [1]. The routine use of this material is inhibited due to remaining uncertainty about the variables that control the consistent production of defect-free, fully nitrided products. Of particular concern is the presence of ~0.5 wt% metallic impurities in the Si powder used to produce Si_3N_4. The most important of these is Fe, since the powder is typically worked with iron milling tools during the processing [2]. Small quantities of Si_3N_4 have been produced from "semiconductor grade" Si, but this is expensive. In a practical sense, a small Fe impurity is always present, so that it is important to understand the effect of this impurity on the Si_3N_4.

EXAFS measurements were made on six samples from various stages in the production of Si_3N_4. Three of these contained only the "native" Fe impurity (which is probably introduced by iron milling tools) with ~0.5 wt% Fe, and the other three had 1.5 wt% Fe_2O_3 added. Additional Fe is sometimes added, since this has been empirically observed to speed the reaction and produce more fully nitrided Si_3N_4.

The EXAFS measurements indicate that the Fe is converted to $FeSi_2$ in all cases before the silicon is nitrided. This occurs because the silicon is exposed to temperatures of from ~1000°C to 1200°C before nitridation starts. According to the Fe-Si phase diagram, the stable composition at low Fe concentrations and these temperatures is $FeSi_2$; this is directly confirmed by the EXAFS measurements. This is an important result because it means that for low Fe concentrations (~1-3 wt%) any source of Fe impurity should affect the reaction in the same way, since the Fe is converted to $FeSi_2$ before nitridation begins. It is important, however, that the Fe is initially well dispersed since large local Fe concentrations can change the connectivity of the Si_3N_4 by causing voids. At an atomic level, the Fe is always in $FeSi_2$ and affects only the reaction kinetics.

After nitridation is ~95% completed we find that the $FeSi_2$ structure has changed only slightly, with slight increases in the Fe-Fe and Fe-Si σ^2's. We interpret this to mean that the $FeSi_2$ is now present in some form with a large surface-to-volume ratio, possibly in the form of a thin film coating the particle surfaces.

This work was supported by the National Science Foundation, grant no. DMR80-22221 (Stern), and by the National Aeronautic and Space Administration, grant no. NAGW199 (Bouldin).

1. J.J. Burke, A.E. Gorum, and R.N. Katz (editors): Ceramics for High-Performance Applications (Brook Hill, Chestnut Hill, Massachusetts, 1974)
2. A. Atkinson, A.J. Moulson, and E.W. Roberts: J. Am. Ceram. Soc. 59, 285 (1976)

(*) Address as of Oct. 1, 1984: U.S. Dept. of Commerce, National Bureau of Standards, Washington, DC 20234.

EXAFS and Edge Studies of Transition Elements in Silicate Glasses

J. Petiau[1], G. Calas, T. Dumas[2], and A.M. Heron[1]

Laboratoire de Minêralogie-Cristallographie (LA CNRS 09), Universitês Paris 6 et 7, 4 place Jussieu, F-75230 Paris Cedex 05, and

LURE, Orsay, France

1 INTRODUCTION

The atomic selectivity of X-ray absorption spectroscopy is an essential property for the study of the elements present as minor components in silicate glasses. Besides the atomic organization in the first shell (studied by EXAFS and edge structure), a partial insight into the middle range order is given by the contribution of the second atomic shell to EXAFS. In this way we have investigated the structure around most of the 3d elements in various silicate and borosilicate glasses. The transition elements (including rare-earths and actinides) are also of particular interest because they have important effects on the properties of those glasses, even at low concentrations. One of the important questions concerns the relation between the local structure around
nucleating elements – as Zr and Ti – and the processes of nucleation and crystallization. More recently we have initiated studies on the local environment and chemical state of uranium, thorium and 4d-elements in borosilicate glasses in an attempt to gain information concerning their properties in glasses proposed for the storage of nuclear waste.

2 INFORMATION FROM THE EDGE

It first concerns the oxidation number of the absorbing element and its inference from the energy position of the edges. At high energies (>12 keV) when the resolving power of the monochromator is small and the core-level width is large, that does not raise major questions as the main part of the edge is nearly featureless in crystalline compounds such as glasses. For uranium -containing borosilicate glass, the U-LIII edge is compared in Figure 1 to the edge of uranium in UO2 and uranyl -nitrate. Its position clearly confirms that the oxidation state is predominantly six. The study of the M-edges has also been made in the same compounds. The MIV and MV edges exhibit large resonance peaks associated to 3d--5f transitions, which prove a very large density of empty f-states. The resolving power of the monochromator as well as the core level width seem to favour studies at M rather than at L edges. However the energy shifts are very small for M-edges because the transitions to 5f final states are essentially atomic-like, as was recently discussed in uranium intermetallic compounds (1).

For 3d elements the features, pronounced in crystalline compounds, are blurred in glasses probably because various symmetries are mixed. This often makes questionable a determination of oxidation states by comparing edge position in glasses and crystals. As a consequence, the position of the edge does not really informs quantitatively but give useful

---

(1) and Ecole Normale Supérieure, MONTROUGE, FRANCE.
(2) and Saint-Gobain Recherche, AUBERVILLIERS, FRANCE.

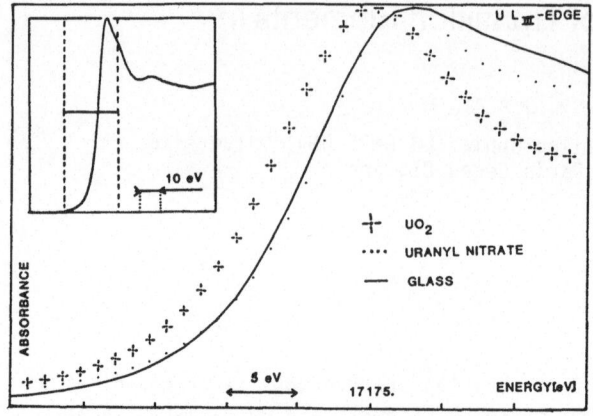

Fig.1 Comparison of oxidation states of uranium in :
- a borosilicate glass
  - UO2 (+4)˜
  - uranyl nitrate (+6).

indications as it was shown in Fe-containing glasses synthesized under various oxidation conditions (2). The position of the prepeak gives more reliable information when the final state is mainly determined by the absorbing atom (small mixing of its 3d-states with the ligand p-states). Such a systematic shift is indeed observed in Fe-containing glasses and crystalline compounds and found to be about 1.3eV between Fe(II) and Fe(III) (3). However the position of the pre-edge does not give reliable information in cases of strong covalency, and especially in complexes corresponding to extreme oxidation states.

Another information concerns the identification of the coordination numbers. Several studies have been published (3,4,5,6) concerning the evolution from small features in nearly perfect octahedral symmetry to the large peaks which characterize the tetrahedral symmetry and which are especially intense for totally empty d subshells (Ti(IV),V(V)..). As for us we have studied the prepeaks of iron edges in various crystalline compounds and silicate glasses (3). They may be separated into Lorentzian components which are interpreted as deriving from crystal-field split 3d states. The percentage of 4-fold versus 6-fold coordinated iron has thus been found to be about 15% in a Fe(II)- containing silicate glass, when Fe(III) is predominantly in tetrahedral symmetry in disilicate glass. The K-edge of vanadium was studied by Bianconi et al.(5) in phosphate glasses and more recently by Dumas (6) in lithium silicate glass. The situation is complicated by the large asymmetry of the oxygen coordination shell: the presence of oxygen atoms at short distances induces a shift of the peak towards lower energy and this covalency effect competes with the formal valence state effect. In Figure 2, K-vanadium edges are compared in a lithium silicate glass and in crystalline compounds. The valence state 5 is accounted for by the intensity of the peak, higher than in any crystalline compound, when its position at low energy is related to the presence of oxygen atoms at short distances which has been confirmed by EXAFS analysis (6). In the case of cobalt and nickel in sodium borate glasses, the increase of the percentage of 4 fold-coordinated ions with alkalinity has been followed by comparison of the prepeak amplitude (7).

3 EXAFS OF THE FIRST OXYGEN SHELL

It is difficult to get accurate backscattering amplitudes and phases for oxygen. The first problem is general for light backscattering elements: the use of theoretical values is not suitable because the chemical and geometrical situations play an important role and large window effects are unavoidable in experimental determination. The second difficulty is

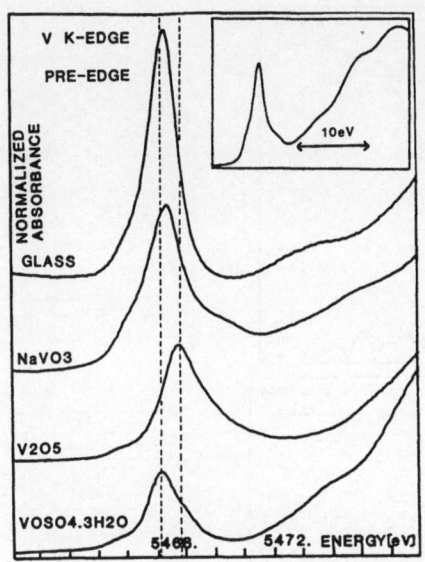

REFERENCE COMPOUND	VALENCE STATE	V-O DISTANCES (Å)	ENERGY OF "PRE-PEAK" MAXIMUM
V_2O_5	+ 5	1.59 , 1.78, 1.87 (x2), 2.02	5467.9
$NaVO_3$	+ 5	1.63 , 1.65 (x2) 1.80	5467.4
$VOSO_4 \cdot 3H_2O$	+ 4	1.56 , 2.02 (x2) 2.05 , 2.08	5467.1

Fig.2 Vanadium K-preedge in a lithium silicate glass and in crystals

related to the choice of crystalline reference compounds for some absorbing elements as zirconium, because all the known crystalline references show broadly distributed M-O distances. Beyond these problems, the effect of structural disorder in the glass takes gradually more importance with increasing coordination number and when complexes give rise to distinct subshells.

Transition elements of the 3d-series which use to be 6-fold as well as 4-fold coordinated in crystalline compounds show a tendency to be in the lower coordination state in oxide glasses. This fact has been shown in many cases by edge studies and has always been confirmed by analysis of the oxygen shell contribution to EXAFS spectra. The coordination number N can be directly deduced from amplitude measurements or indirectly from distance determinations. Actually these two data give complementary information, the eventual discrepancy observed having to be discussed in terms of distance distribution. For example in a silicate glass of cordierite composition (49% SiO_2, 29%Al_2O_3, 9%MgO, 9%TiO_2 and 4%ZnO), Ti and Zn have been found to be present in 4-fold coordination by distance measurements, and edge study in the case of titanium, although the coordination value is found significantly smaller for zinc (Table 1).

TABLE 1

M	M-O (Å)	N	σ (Å)	M-Si(Al,Mg)(Å)	M-O-Si
Zn	1.97±0.02	3.1±0.5	0.06±0.02	3.25	129°
Ti	1.82±0.02	3.8±0.5	0.08±0.02	2.93	119°
Zr	2.09±0.02	4.6±0.5	0.11±0.02	3.32	126°
Si	1.62	4		3.14	152°
Ge	1.74	4		3.14	130°

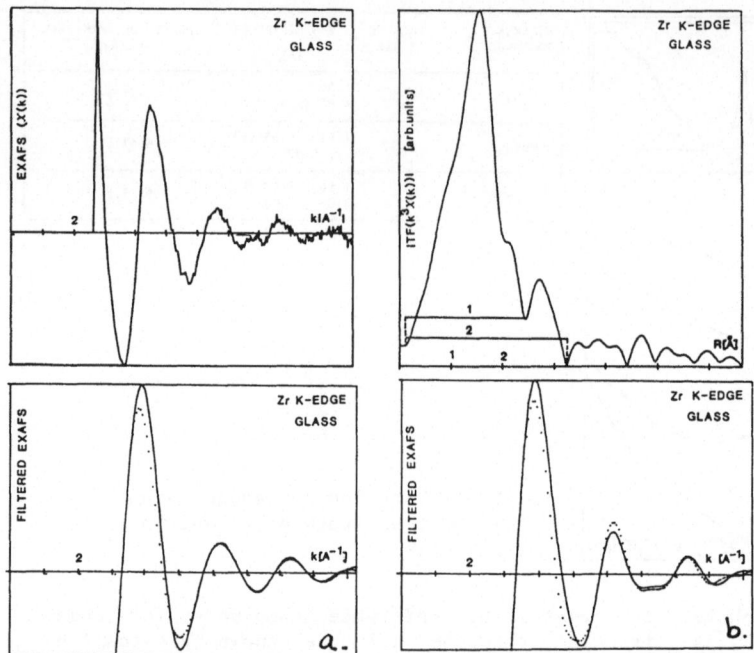

Fig.3 EXAFS analysis at Zr K-edge in a borosilicate glass (exp. signal = dotted curve, simul. = solid curve) for a- 0 shell, b- 0 shell + Si shell.

For elements having a high coordination number in crystals (6 to 9) there are less reasons for it to have an unique value in a glass; information given by EXAFS about that question is only partial. A typical example is zirconium that we studied in a silicate of cordierite composition and in a borosilicate glass (Fig. 3). The organization around zirconium is similar in these two glasses ; for the second glass, we used different structural assumptions for calculation of spectra to be compared with the experimental Fourier filtered signal and, in each case, we made an optimization of the structural parameters. With the hypothesis of one shell with a gaussian distance distribution, we found 5.7 oxygen neighbours at a distance of 2.08 Å. As in zircon (ZrSiO4) the coordination shell is made of 4 oxygens at 2.13 Å and four at 2.27 Å, we tried a fit with two subshells. This solution improves clearly the quality of the fit in the zircon when it has only a small effect on the factor of quality for the glass ; in his latter case the optimization was obtained for a splitting of 0.10A between the two subshells and a total occupation of 6 atoms. An exponential distribution of distances of the form used by De Crescenzi et al. (8) was also tried ; the best fit is obtained for an exponential width of 0.10-0.15Å but its quality was again not significantly better than with the one shell assumption. Consequently it can be said that zirconium is surrounded by about six atoms within a shell of 0.10-0.15Å broadness. It is not possible to eliminate the presence of some more oxygens apart from this main part of the distribution, though a strong argument again this possibility is the fact that the mean Zr-0 dstance for the six measured neighbours is consistent with Zr in 6-fold coordination.

The last type of situation is the presence of elements in extreme oxidation numbers which give rise to complexes with oxygen atoms. Two

294

examples have been studied. For Vanadium, EXAFS supported the results of edge measurements in letting out the presence of two oxygen subshells at 1.62Å(3 atoms) and 1.81Å (2 atoms). In this case, the fitting procedure clearly put in evidence a two sub-shells configuration. In the borosilicate glass under study, uranium is proved to have an oxidation number equal to six by the L-edge position, as it was said before, and optical absorption has shown the presence of $UO2^{++}$ group. To these results, EXAFS added information about U-O distances ; within the uranyl-group, the U-O distance is identical to that found in uranyl silicate and accurately determined by EXAFS (1.77Å). The distance to further oxygen atoms is found shorter than in crystalline compounds : 4.7 atoms are counted at a mean distance of 2.18Å instead of 5.4 at 2.45Å in uranyl nitrate. The actual situation in uranyl nitrate is 6 oxygen atoms between 2.38 and 2.50Å   In uranyl silicate, which gives a more realistic comparison but that we have not studied by EXAFS, there are five oxygen atoms in the range 2.22-2.45Å. A tendency to compacity of the coordination shell in glasses is thus found also in these types of situation.

4 EXAFS OF THE SECOND ATOMIC SHELL

Its investigation gives a partial but interesting view of middle range order, especially when comparisons are made at different stages of an evolution or for different concentrations.Firstly it must be pointed out that, in some cases, not any contribution of a second shell comes from the experimental data. It is the situation for Mn, Co, Fe and Ni in most of silicate and borate glasses studied (apart from a broad range of non-zero amplitude in radial distributions generally found in borate glasses) and it could be chosen as a criterion of modifier role. When a usable information is obtained in the spectra, the appropriate way to describe the result is in term of bond angle at the oxygen site which informs about the linking of the elemental polyhedra. The second neighbours shell is likely to be made of the major components of the glass, that is silicon in most cases ; when aluminium or magnesium are also present at high concentrations, they could not be recognized from silicon. On the contrary it was possible, for example in nucleation studies, to use differences in backscattering amplitudes and phase functions for deciding whether the element under study is dispersed in the glassy matrix or directly bound to transition elements polyhedra.

For zinc and titanium in silicate glasses of cordierite composition, a second shell could be Fourier filtered, giving the results reported in Table 1 where the germanium and silicon cases in $GeO2$ and $SiO2$ are also reported for comparison with typical formers environments. As an example figure 4 shows a radial distribution around zinc where the second shell is directly visible. The number of atoms found by analysis of this second shell is always small (2 atoms) so that it is only an indication relative to the central part of a broad distribution of distances.

In case of zirconium, for the two glasses studied, the second shell contribution was found too close to the oxygen one to be analyzed separately .
Therefore , the Fourier filtering was carried out for the two shells taken together and the fitting procedure used for these two shells (Fig.3). The result is shown in Table 1; as expected because of the small contribution of the second shell, the position and the number of atoms determined for the oxygen shell remain nearly unchanged when the second contribution is added.

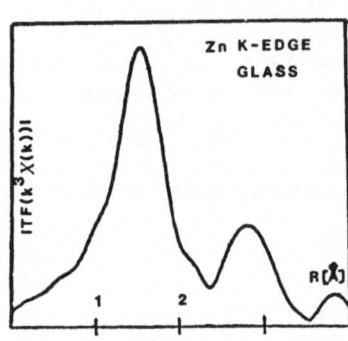

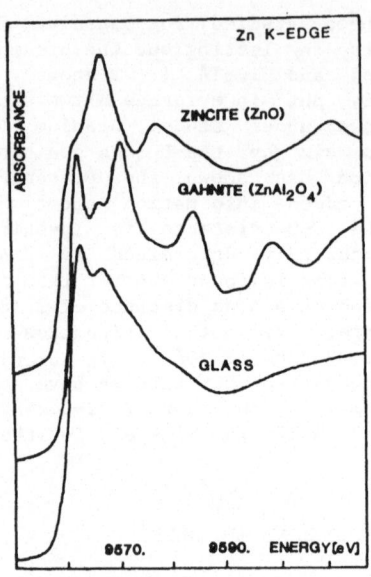

Fig.4 RDF for Zn in silicate
glass (first shell = oxygen
atoms, second
        shell = silicon atoms)

Fig.5 Comparison of XANES in a
glass and in crystalline
compounds.

As a last remark concerning middle range order, it must be pointed out
that the XANES energy range (from the absorption maximum to 40eV above)
is nearly structureless in all glasses. This fact is clearly related to
the small degree of order beyond the first coordination shell which does
not allow multiple scattering contribution. An example is shown in Fig. 5
for the case of zinc in a cordierite glass.

1   J.L. Lawrence, M.L. den Boer and R.D. Parks, Phys. rev.B, 29 , 568,
    (1984).
2   G.Calas and J.Petiau, Bull. Minéral. 106 , 33 (1983)
3   G.Calas and J.Petiau, Solid State Comm., 48 , 625 (1983).
4   R.B.Greegor, F.W.Lytle, D.R.Sandstrom, J.Wong and P.Schultz,
        J. Non Cryst. Sol. 55 , 27 (1983).
5   A.Bianconi, A.Giovanelli, I.Davoli, S.Stizza, L.Palladino, O.Growski
        and L.Murawski, Solid State Comm., 42 , 547 (1982).
6   T.Dumas, Thesis (Univ. de Paris, 1984).
7   J.Petiau and G.Calas, Jour. de Phys., C9 , 47 (1982).
8   M.de Crescenzi, A.Balzarotti, F.Comin, L.Incoccia, S.Mobilio and
        N.Motta, Solid State Comm. 37 , 921 (1981).
9   C.Lapeyre, J.Petiau and G.Calas in "The structure of Non-Crystalline
        Materials", Ed. P.H.Gaskell et al. (1982).
10 G.S.Knapp, B.W.Veal, D.J.Lam, A.P.Paulikas and H.K.Pan,
        Mat. Letters 2 , 259 (1984).

The Environments of Modifiers in Oxide Glasses

G.N. Greaves

Science and Engineering Research Council, Daresbury Laboratory
Daresbury, Warrington WA4 4AD, United Kingdom

N. Binsted and C.M.B. Henderson

Department of Geology, The University, Manchester M13, 9PL, United Kingdom

EXAFS studies of network modifiers in oxide glasses have made an important
contribution to our understanding of the structure of glass. In this paper
work on the environments of Na and Ca is reviewed. The environment of Fe,
which is intermediate between a modifier and a network former, is also discussed. In each case the local structure of the metal in the glass is contrasted with its environment in the crystalline state.

1 Introduction

Modifying atoms like Na and Ca are the cations least strongly bonded to the
oxygen sub-lattice in glasses. The more strongly bound network-forming atoms
like Si or Al have a rigid local structure by comparison. Curiously modifiers
in glasses usually exhibit a distinct EXAFS spectrum indicative of a discrete
environment [1] – an environment that often alters with the stoichiometry and
physical properties of the glass. On the other hand the network-forming unit
(e.g. the SiO_4 tetrahedron) seldom changes to the same extent. For this
reason the atomic environments of modifiers like Na or Ca in oxide glasses are
generally more sensitive to the conformation of the network than those of the
network-forming cations themselves. The atomic environments of transition
metals, on the other hand, are often intermediate between those of modifiers
and network formers. Whilst Fe or Ti for instance are ionically bonded in
oxide glasses they can sometimes be tetrahedrally coordinated like the network
forming atoms [10].

EXAFS provides a particularly valuable technique for studying the local
structure of metal sites in glasses. By isolating the pair distribution
functions of the various cations, the correlations involving ionically bonded
modifiers and intermediates can be distinguished from those associated with
the network. In this paper we will consider the K-edge EXAFS of Na (1080 eV),
Ca (4039 eV) and Fe (7113 eV) and the local structure which is revealed.
Measurements of the Na K-edge require soft X-rays and use of thin blown films.
Details of EXAFS techniques at these energies have been given elsewhere [2].

Although the EXAFS spectra of modifiers often carry some information about
cation-cation separations, analysis to date has concentrated on the cation-
oxygen pair distribution functions. These are particularly interesting as
they are not always symmetrical and are sometimes structured. In this context
the use of a fast curved wave least squares program [3] is essential as the
extent of modifier EXAFS spectra are limited by the relatively weak backscattering from the surrounding oxygens. The importance of model compounds
must also be stressed as these enable the accuracy of the phase shifts calculated for atoms in an ionic lattice to be optimised. Fig. 1 shows an example.
The weighted K-edge EXAFS of Ca in CaO and its Fourier transform are compared
to the LEE and PENDRY [4] curved wave theory (dotted line). Agreement between
radial geometry from EXAFS analysis and the known structure is encouraging
[5].

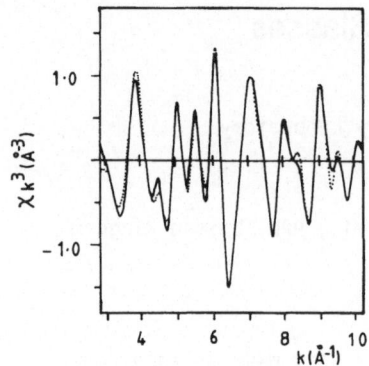

Fig. 1 k^3 weighted EXAFS of the Ca K-edge in CaO. Dotted line is theoretical fit using curved wave approximation with multiple scattering corrections [4] and the known structure

2 Sodium Environments

Turning first to the crystalline state, in binary silicates like Na_2SiO_3 and $Na_2Si_2O_5$ the Na environment consists of a trigonal biprism of 5 O's [6]. Although the O shell is distorted,the static displacement is less than the room temperature thermal disorder. However in ternary silicates like Na_2CaSiO_4 or $Na_2Mg_2Si_6O_{15}$ the Na environment is very irregular with near neighbour O's both at bonding (2.3 Å) and non-bonding (~3Å) distances. The same structural syndrome has been observed for glasses from analysing the Na K-edge EXAFS of binary and ternary systems [1, 7]. Results for $Na_2Si_2O_5$ and $Na_2CaSi_5O_{12}$ are reproduced in Fig.2. Note the diminished fine structure in the ternary glass. Detailed analysis reveals the O shell is split for this glass [7]. For $Na_2Si_2O_5$, on the other hand, the first coordination sphere is well-defined with a mean spread in Na-O distances (σ) of 0.07Å.

The Na K-edge EXAFS of $Na_2B_2O_7$ glass has also been measured [7]. In this case the O shell is broader (σ = 0.18Å) but interestingly there is a strong second shell at 3.8Å comprising Na's. No Na-B separations are detected, presumably because of the limited back-scattering from B.

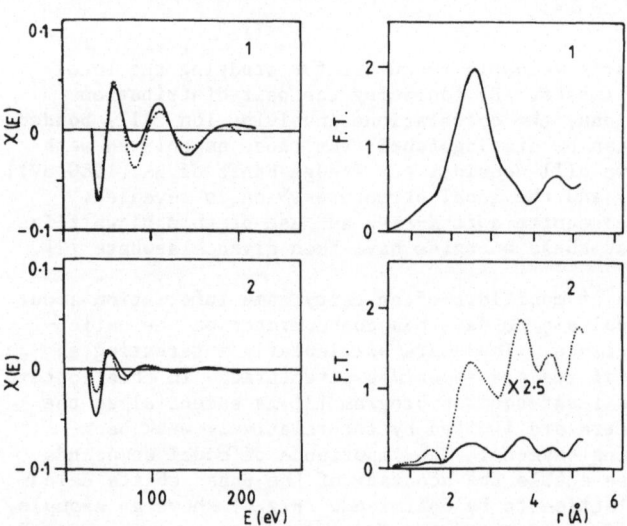

Fig.2 EXAFS of the Na K-edge in $Na_2Si_2O_5$ (1) and $Na_2CaSi_5O_{12}$ (2) glasses together with curved wave fits. Fourier transforms are shown alongside

3 Calcium Environments

Compared to Na, the O shells of Ca in crystalline silicates and also in phosphates are extremely distorted. Whilst the local symmetry is crudely octahedral, – the first coordination sphere will often comprise 8 or 9 O's – with the distribution peaking at ~2.4Å. The mean static displacement in Ca-O distances typically lies in the range 0.1 to 0.15Å [5, 8]. By contrast the relative mean phonon displacement in Ca-O distances in CaO (Fig.1) is 0.09Å. Because the static disorder is significantly greater than the thermal disorder this can be deconvoluted in the EXAFS analysis to yield the Ca-O pair distribution function. We have done this by replacing the O shell with single atom subshells and least squares fitting the Fourier filtered EXAFS. A variety of crystalline silicates [5] and phosphates [9] have been analysed in this way. Although the wavevector range is limited ($3Å^{-1} < k < 10Å^{-1}$), the EXAFS pair distribution functions compare favourably with those obtained from crystallography data.

The pair distribution functions for two mineral glasses – $CaAl_2Si_2O_8$ (Anorthite) and $CaMgSi_2O_6$ (Diopside) – are given in Fig.3. Each subshell has been gaussian broadened by $\sigma = 0.09Å$. It is evident from Fig.3 that the Ca sites in these glasses are substantially distorted. The mean displacement in O distances is around 0.2Å. Similar conclusions were drawn earlier for Wollastonite glass ($CaSiO_3$) [8]. Note, however, that the pair distribution functions shown in Fig.3 are not symmetrical. The distributions for the crystalline silicates are shown by the dotted lines in Fig.3. Where the feldspar (Anorthite) has similar Ca sites to its glass, this is not true for the pyroxene (Diopside). In the latter case the whole pair distribution function for the mineral is better ordered and falls at shorter Ca-O distances than the glass.

There is also an interesting correlation between density and the mean Ca-O distance: Anorthite and its glass have similar densities ($2.7gm\ cm^{-3}$) whereas the density of Diopside and its glass are quite different (3.2 and $2.8gm\ cm^{-3}$

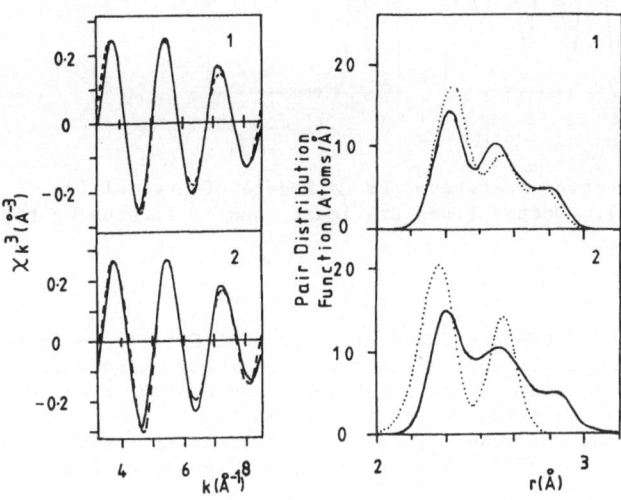

Fig.3 k^3 weighted EXAFS of Ca K-edges in $CaAl_2Si_2O_8$ (1)and $CaMgSi_2O_6$ (2) glasses. The dashed lines are the least squares fits for the pair distribution functions shown alongside. Results for the crystalline materials are indicated by the dotted lines

respectively). It is clear from Fig.3 that differences in density between one silicate and the next are governed by the various modifier environments.

Pair distribution functions for modifiers in amorphous biogenic phosphates have also been investigated [9]. In intracellular granules in Helix Aspersa the Ca environment has similarities with $Ca_2P_2O_7$. The local structure of introduced Mn on the other hand is better ordered.

4 Iron Environments

The ionic radii of Fe(II) and Fe(III) are 0.77Å and 0.65Å respectively and are substantially shorter than either Na or Ca which are both close to 1Å. Fe-O bond lengths accordingly are around 2Å and as such are "intermediate" between the bond length for Na or Ca at 2.4Å and those for Si at 1.6Å or Al at to 1.7Å. The Fe-O bonds are nevertheless ionic but the coordination may be octahedral or tetrahedral depending on the structural chemistry. This is manifest in the crystalline state [6]. The octahedral sites, however, are seldom symmetrical which is the reason for considerable detail in the XANES. Fig.4 shows the XANES measured for $NaFeSi_2O_6$ (Aegerine) for instance.

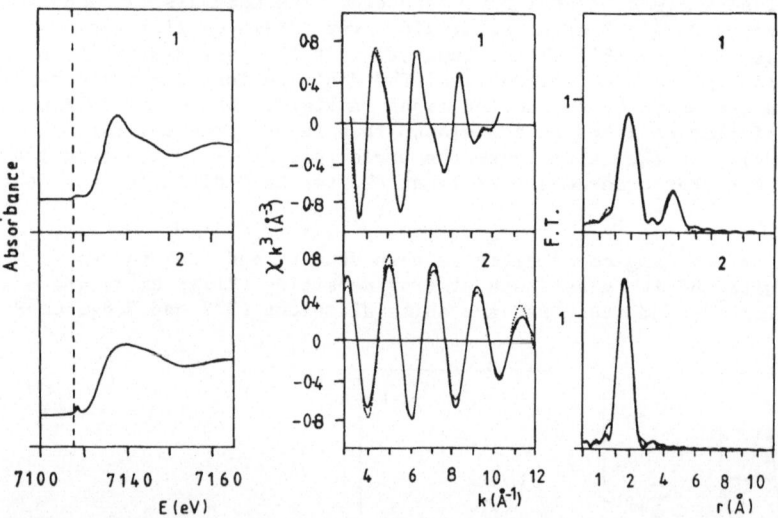

Fig.4 XANES, EXAFS and its Fourier transform for Ca K-edge of crystalline $NaFeSi_2O_6$ (1)and its glass (2). Dotted lines are least squares fits using the curved wave approximation.

We have recently studied the XANES and EXAFS of a range of Fe(III) containing minerals and rapidly quenched glasses in order to follow any changes in the intermediate character of Fe between the crystalline and amorphous state [5]. Results for Aegerine and its glass are contrasted in Fig.4. Compared to the crystals, Aegerine glass and the other glasses have less XANES detail, indicating either a larger variety of Fe sites or perhaps sites which are more symmetrical.

The glasses, however, are also characterised by a prominent pre-edge feature. This relates to transitions to empty states in the 3d band. Dipole transitions will be allowed if there is some hybridisation of the orbitals

300

from this band. CALAS and PETIAU [10] have demonstrated this is the case for tetrahedrally coordinated Fe in minerals. Tetrahedral coordination is accompanied by a shorter Fe-O bond length which should be detectible in the EXAFS. We found this to be the case for all the glasses. It can be clearly seen in Fig.4 where the Fe-O peak in Aegerine glass (at 1.86Å) is noticeably shorter than in the crystal (2.03Å). Moreover the relative mean displacement in Fe-O distances is also smaller in the glass (0.06Å) than in the crystal (0.11Å) which is compatible with the more symmetric tetrahedral site. Qualitatively the same picture emerges for other Fe(III) containing glasses [5]. In all cases Fe sites are best described as 'intermediate' in character – being ionically bonded but tetrahedrally coordinated.

5 Conclusion

A proper knowledge of the environments of modifiers in glasses is a prerequisite for constructing models for the structure of glass [1]. For these to be realistic they must at least account for :
   (a) the huge changes in viscosity that are brought about by varying the amount and balance of modifiers present
   (b) the high electrical conductivity of modifying cations like Na^+ or Ag^+.
EXAFS remains the key technique for tackling these problems.

Acknowledgments

We acknowledge the joint financial support of the SERC and NERC for much of this work. We are also grateful to the director and staff of Daresbury Laboratory for providing facilities for EXAFS measurements and analysis.

References

1 G.N. Greaves, A. Fontaine, P. Lagarde, D. Raoux and S.J. Gurman: Nature 293, 611 (1981); G.N. Greaves: La Recherche 137, 1184 (1982)
2 G.N. Greaves and D. Raoux: The Structure of Non-Crystalline Materials 1982 (Taylor & Francis, London 1983) p.55
3 S.J. Gurman, I. Ross and N. Binsted: J. Phys. C 17, 143 (1984)
4 P.A. Lee and J.B. Pendry: Phys. Rev B11, 2795 (1975)
5 N. Binsted, G.N. Greaves and C.M.B. Henderson: submitted to Cont. Mineral. Petrol; N. Binsted, G.N. Greaves and C.M.B. Henderson: Progr. Expt. Petrol (NERC) 10 (in press)
6 R.N.G. Wyckoff: Crystal Structures (Wiley, New York 1964); See also references in [1] and [7]
7 G.N. Greaves: EXAFS for Inorganic Systems (Daresbury Laboratory Research Report 17) p.115
8 R.G. Geere, P.H. Gaskell, G.N. Greaves, J. Greengrass and N. Binsted: EXAFS and Near Edge Structure (Springer-Verlag, Berlin Heidelberg 1983) p.256
9 G.N. Greaves, K. Simkiss and M. Taylor: Biochem J., Part 3, 221, 855 (1984)
10 G. Calas and J. Petiau: Bull. Minéral 106, 33 (1983)

Comparative Examination by EXAFS and WAXS of GeO$_2$- SiO$_2$ Glasses

R.B. Greegor and F.W. Lytle
The Boeing Company, P.O. Box 3999 2T-05, Seattle, WA 98124, USA
J. Kortright and A. Fischer-Colbrie
Department of Material Sciences and Engineering, Stanford, CA 94305, USA
P.C. Schultz
Corning Glass Works, Corning, NY 14830, USA

1. Introduction

Germania-silica glasses have been used in the manufacture of low attenuation
optical waveguides for communication [1]. Various measurements have been
conducted on this type of glass including IR spectra [2], refractive indices
and IR spectra [3], linear expansion coefficients and viscosities [4], composition, density and refractive index [5] and UV absorption spectra [6].
The measurements reported here extend the observations to structural determination by EXAFS and WAXS.

2. Experimental Techniques

The glasses examined were prepared by a vapor-phase oxidation process (fully
described elsewhere [5]) using vapor mixtures of GeCl$_4$ and SiCl$_4$. Briefly, in
this process a porous preform of pure GeO$_2$-SiO$_2$ glass is prepared by depos-
iting the "soot" reaction products (from the hydrolysis/oxidation of GeCl$_4$/
SiCl$_4$ vapor mixtures in a methane/oxygen flame) onto a bait rod rotating in
air. This porous preform is then sintered to a high-quality bubble-free
glass in a He-atmosphere electric furnace for 1 h at 1000° C to 1450°C depending on GeO$_2$ content. After sintering, the glasses are air-cooled to
ambient in ~0.5 h without annealing. Glass composition is controlled by
varying the SiCl$_4$ and GeCl$_4$ vapor flows.

The EXAFS/WAXS experiments on the above fabricated glasses were performed
at the Stanford Synchrotron Radiation Laboratory (SSRL). At SSRL side sta-
tions of wiggler lines were used, with the synchrotron beam at a current of
approximately 60 mA and an energy of 3 GeV. A Si(220) double crystal monochromator was used approximately 50% detuned. The EXAFS data was gathered
at room temperature using a fluorescent X-ray detector [7]. For the WAXS
measurements the incident intensity was measured by radiation air-scattered
through 90° into a NaI scintillation detector with an energy dispersive Ge
detector for the WAXS beam. The data were gathered close to the Ge K-edge
and at 20 KeV.

3. XANES/EXAFS and WAXS Results

The Ge K-edge of both the 24 wt% and 9.8 wt% GeO$_2$/SiO$_2$ were shifted by 5.7
eV with respect to the Ge polycrystalline standard. The XANES signatures
of both GeO$_2$/SiO$_2$ samples were also similar. There was a greater detail of
structure in the XANES for the GeO$_2$ standard which may be due to multiple
scattering from the unique atomic array in GeO$_2$ rather than the more complex
mixture expected in glassy GeO$_2$/SiO$_2$. On the basis of the XANES observations, little difference was evident in the electronic structure or the near
neighbor geometry of Ge atoms as the wt% GeO$_2$ was changed from 9.8 wt% to
24 wt%.

The EXAFS Fourier transforms for polycrystalline hexagonal GeO_2, and the 9.8 wt% and 24 wt% GeO_2 in SiO_2 glasses are shown in Fig. 1 These transforms were produced by including the Ge-O phase shift in the transform kernel so that the first major Ge-O peak appears at the correct position in R-space. The position of the near neighbor oxygen peaks were at 1.72 ± .02 Å for both the 9.8 wt% and 24 wt% GeO_2/SiO_2 samples. Least squares curve fitting to the inverse transform spectra of the near neighbor peaks was also in agreement with these values. The hexagonal GeO_2 standard from which the Ge-O phase shift was derived had a much more intense second neighbor Ge-Ge peak than did the glass samples. This suggests that the second neighbors in the glass are predominantly Si (lower EXAFS scattering amplitude) and/or with a greater degree of disorder than the second neighbor Ge atoms in the GeO_2 standard.

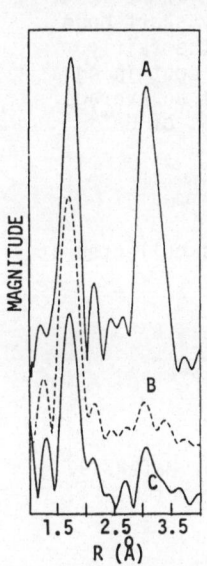

Figure 1. Fourier Transforms of Ge K-edge EXAFS of A) GeO_2, B) 24 wt% GeO_2/SiO_2, and C) 9.8 wt% GeO_2/SiO_2

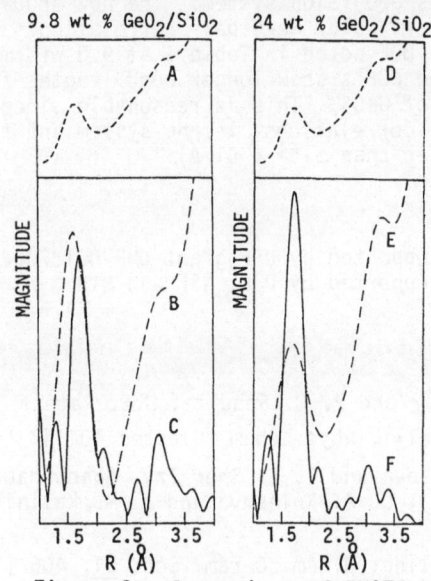

Figure 2. Comparison of EXAFS and WAXS Results; A, D) RDF (20 KeV); B, E) DDF; C, F) Fourier Transforms of EXAFS

The WAXS data was analyzed using software developed by Fuoss (8). The results are given in Fig. 2 and Table I. Both the radial distribution function (RDF) and differential distribution function (DDF) are shown. The DDF was obtained from two WAXS patterns taken at -200 and -8 eV relative to the Ge K-edge. For the RDF a slightly longer bond length is found for the first neighbor peak in the 24 wt% glass than in the 9.8 wt% glass, in contrast to the EXAFS results. This is due to the averaging over all Ge-O, Si-O pair correlations by the RDF whereas the EXAFS sees only the Ge-O distance. The apparent short bond for the 9.8 wt% DDF is probably due to fairly noisy data.

TABLE I. First Neighbor Bond Lengths, $\overset{o}{A}$

SAMPLE	EXAFS	RDF	DDF
9.8 wt% GeO_2/SiO_2	1.72 ± .02	1.64 ± .03	1.62 ± .03
24 wt% GeO_2/SiO_2	1.72 ± .02	1.68 ± .03	1.68 ± .03
GeO_2 (Hexagonal)	1.74 [9]	SiO_2 (Glass)	1.62 [10]

4. Discussion

These results support a model in which Ge atoms substitute for Si in glassy SiO_2. This is expected since hexagonal GeO_2 is isomorphous with α-quartz SiO_2 (in both of these structures Si and Ge cations are tetrahedrally coordinated by oxygen). Over the range of GeO_2 concentration investigated 9.8 - 24 wt% GeO_2 in SiO_2 (2 - 5 atomic % Ge) the results support a random solution with no clustering. The second neighbor, Ge-Si (or Ge-Ge) peak in the EXAFS would be sensitive to clustering of GeO_2. The fact that this peak does not grow with GeO_2 concentration indicates that clusters do not form.

The near agreement of bond distances from EXAFS and WAXS measurements are expected in this GeO_2/SiO_2 system. The DDF should be more like EXAFS in isolating the particular Ge-O pair correlation. The apparently short bond distance in the DDF noted in Table I at 9.8 wt% may be due to the fairly noisy data. The RDF's show longer bond lengths for the sample containing a greater amount of GeO_2. This is reasonable since the RDF's are an average of all the pair correlations in the system and the ionic radius of Ge^{+4} (.53 Å) is larger than Si^{+4} (.41 Å).

Acknowledgements

This work was supported by NSF grant DMR-8013706. The data was collected at SSRL which is supported by DOE, NSF and NIH.

References

1. R. D. Maurer and P. C. Schultz, U.S. Patent 3,884 (1975).

2. H. F. Borrelli, Phys. Chem. Glasses 10, 42 (1969).

3. U. A. Kolesova and E. S. Sher Izv. Akad. Nauk SSSR. Neorg. Mater. 9, 1018 (1973); U. A. Kolesova and A. M. Kalinima Fiz. i Khim. Stekla 1, 70 (1975).

4. E. F. Riebling, J. Am. Ceram. Soc. 51, 406 (1968).

5. Y. Y. Huang, A. Sarkar and P. C. Schultz, J. Non. Cryst. Sol. 27, 29 (1978).

6. P. C. Schultz, Paper Presented at XI Congress on Glass, Prague, Czechoslovakia (1977).

7. F. W. Lytle, R. B. Greegor, D. R. Sandstrom, E. C. Marques, J. Wong, C. L. Spiro, G. P. Huffman and F. E. Huggins, "Measurement of Soft X-ray Absorption Spectra With a Fluorescent Ion Chamber Detector", Nucl. Inst. and Meth.(In Print).

8. P. H. Fuoss, Ph.D. Thesis, Stanford University (1980).

9. G. S. Smith and P. B. Isaacs, Acta Cryst. 17, 842 (1964).

10. R. L. Mozzi and B. E. Warren, J. Appl. Cryst., 2, 164 (1969).

EXAFS Studies of Sodium Silicate Glasses Containing Dissolved Actinides*

G.S. Knapp, B.W. Veal, A.P. Paulikas, A.W. Mitchell, D.J. Lam, and
T.E. Klippert

Materials Science and Technology Division, Argonne National Laboratory
9700 South Cass Avenue, Argonne, IL 60439, USA

Sodium silicate glasses containing dissolved Th, U, Np, and Pu have
been studied using the EXAFS technique. Th^{4+}, U^{4+}, Np^{4+}, and Pu^{4+} ions in
the silicate glasses are 8-fold coordinated to oxygen neighbors. The
higher valent U^{6+} and Np^{5+} ions have complex local symmetries. The U^{6+}
ions appear in a uranyl configuration with 2 oxygen atoms at 1.85Å and 4
at 2.25Å from the U ion. The Np^{5+} local symmetry is more complex and
difficult to determine uniquely. The U^{6+} glasses show substantial
clustering of the uranium atoms. A structural model, with nearly planar
uranyl sheets sandwiched between alkali and silica layers, is used to
explain the U^{6+} EXAFS data. This model allows us to understand why U^{6+}
ions are much more soluble in the glasses than the actinide 4^+ ions.

The bonding properties of actinide elements in silicate glasses are of
current interest because such glasses are being considered as storage
media for radioactive wastes. Solubility measurements show that the
bonding depends strongly on valence and type of the actinide ions. For
example, when Th, U, Np, or Pu are in the 4+ state the solubility is
quite low (less than 5%).[1] However U^{6+} and Np^{5+} exhibit much greater
solubilities, being 30% in the case of U^{6+} and 10% for Np^{5+}. We have made
EXAFS measurements on a number of sodium disilicate and trisilicate
glasses containing various actinides in different charge states, and have
found that U^{6+} and Np^{5+} bond quite differently than 4+ ions.

The EXAFS measurements were made using our laboratory EXAFS facility.
[1,2] Both Si(620) and Si(840) bent crystals were used. Resolution
varied depending on the absorption edge studied but was always better than
17 volts. Deconvolution procedures were used to minimize the effects of
the low resolution. All measurements reported here were made at
temperatures between 79 and 84K and all samples were powders encapsulated
in polyethylene. The glasses were made by methods described earlier, [3]
except the 4+ glasses which were made in a reducing atmosphere. The data
were analyzed using the standard EXAFS equation, with the phase shifts and
amplitudes determined from the corresponding dioxides. [4]

In an earlier work we made a detailed study of some U^{6+} glasses. [3]
In Fig. 1 we show $k\chi(k)$ versus k for a series of these U^{6+} glasses. Note
the large amplitude oscillations at high k which vary as a function of
composition. From this composition, amplitude, phase, and temperature

---

*Work supported by the U. S. Department of Energy.

[1]Here we define the % concentration as 100x, where x is defined from the
formula $(0.25Na_2O \cdot 0.75SiO_2)_{1-x}(AnO)_x$ and An stands for an actinide.

dependance (heavy atoms have low Einstein temperatures), we concluded that these oscillations resulted from uranium containing clusters. The uranium bonds in a uranyl configuration (a linear UO_2^{2+} ion) with two oxygen atoms at 1.85 A and four at 2.25 A, and as many as four uranium atoms at approximately 3.3 A. The large U backscattering intensity suggests extensive U clustering. By making a simple planar model (see below) we concluded that the clusters, on the average, must contain at least six uranium atoms.

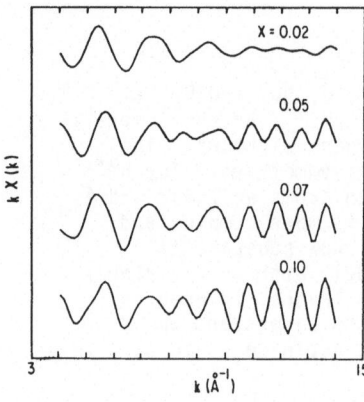

Fig. 1. Experimental EXAFS spectra for a series of sodium trisilicate glasses contain U^{6+} ions at various concentrations, measured at approximately 80K

The glasses containing the 4+ ions have a very different local environment. In Fig. 2(a) and 2(b) we show the radial distribution functions of PuO_2 and a 1.5% Pu glass. Note that the oxygen shell peak is virtually identical to that of the PuO_2. Since the radial distribution function is a sensitive function of both the numbers of neighbors and their local symmetry, this means that the local environment is nearly identical out to the first near neighbor. Results of nonlinear least-squares fitting confirm this, the only difference being that the Pu-O distance is slightly smaller in the glass. The other 4+ glasses are very similar. There are 8±1 oxygen near-neighbor atoms at a distance which is slightly smaller than the distance in the corresponding dioxides.

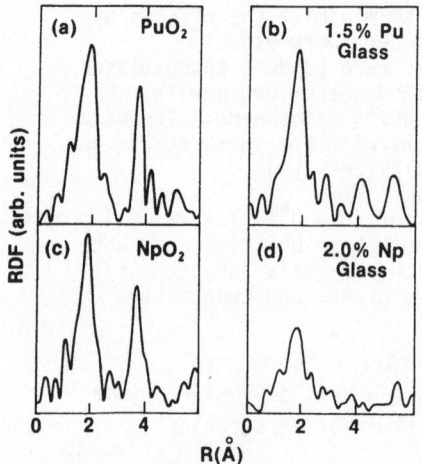

Fig. 2. The radial distribution functions (at ~ 80K) of (a) PuO_2, and (b) a Pu^{4+} glass with x = 0.015, (c) NpO_2, and (d) a Np^{5+} glass with x = 0.02

The Np^{5+} glasses are more complex. In Fig. 2(c) and 2(d) we show the radial distribution functions of NpO_2 and a 2% Np^{5+} glass. Note the striking differences in the first neighbor peaks. Nonlinear least-squares fits with one-shell and two-shell models have been attempted for several of these glasses but no satisfactory fits have been achieved. The average distance is much smaller than that of NpO_2 and the number of neighbors is smaller.

How are we to understand these results and how do we understand the differences in solubilities? First consider the U^{6+} glasses. The local symmetry is that of the uranyls, which in their crystalline forms can be considered as layered compounds in which the U and O are in layers, separated by weakly bonding alkali ions. Crystalline sodium silicates are also layered compounds with corner-bonded double silicate layers separated by sodium ions. We previously proposed a model in which the uranyl clusters are almost planar, between silica layers. The 4+ ions, on the other hand, are 8-fold coordinated in a locally cubic symmetry (in analogy with the CaF_2 structure of the actinide dioxides). In order to have eight near-neighbor oxygen atoms coordinated with the actinide ions, the Na ions have to be displaced and the silicate network has to be highly distorted and strained. This is why the solubility is low. The Np^{5+} glasses may be more similar to the U^{6+} glasses. However the results are more complex and more work is necessary for full understanding.

References
1. G.S. Knapp and P. Georgopoulos, A.I.P. Conf. Proc., Ed. E.A. Stern, No. 64, 7, 1980.
2. G.S. Knapp, H.K. Pan, P. Georgopoulos, and T.E. Klippert, EXAFS and Near Edge Structure Ed. A. Bianconi, L. Incoccia, S. Stipcich, Springer-Verlag, Berlin, 402, 1983.
3. G.S. Knapp, B.W. Veal, D.J. Lam, A.P. Paulikas, and H.K. Pan, Materials Letters 2 (1984) p. 253.
4. For a good review of the EXAFS technique, see P.A. Lee, P.H. Citrin, P. Eisenberger, and B.M. Kincaid, Rev. Mod. Phys. 53 (1981) 769.

Local Coordination Environment of Na in a Series of Silica-Rich Glasses and Selected Minerals Within the Na$_2$O-Al$_2$O$_3$- SiO$_2$ System

David A. McKeown and Gordon E. Brown, Jr.

Department of Geology, Stanford University, Stanford, CA 94305, USA

Glenn A. Waychunas

Center for Materials Research, Stanford University, Stanford, CA 94305, USA

1. INTRODUCTION

Photoelectron total yield Na EXAFS spectra were gathered on the JUMBO mono-chromator line at SSRL to characterize the Na environment for a collection of Na$_2$O-Al$_2$O$_3$-SiO$_2$ minerals and glasses. Included in the collection are crystalline compounds albite (NaAlSi$_3$O$_8$), and nepheline (KNa$_3$(AlSiO$_4$)$_4$), as well as glasses at these compositions, Na$_2$Si$_2$O$_5$ glass, and a series of 75 mole % silica glasses ranging in R (=Al/Na) from 0.03 to 1.61.

The purpose of this study is to determine if the information extracted from Na EXAFS data can accurately depict the known Na environment of the crystalline samples, and to characterize the unknown Na environment in the glasses. Information on the aluminate and silicate tetrahedral network of the glasses in this study was obtained from Raman spectroscopy [1], and of albite and nepheline glasses from X-ray scattering experiments [2]. Since little knowledge of the Na network modifying environment was gained from Raman and X-ray scattering, Na EXAFS provides complementary informat-ion on glass structure.

The 75 mole % silica glass series is of particular interest, since phys-ical properties of melts and glasses at these compositions have a minimum or maximum value as R approaches unity [3]. From Raman, Al EXAFS, and Al XANES data on the glass series, it appears that Al remains as a network former throughout the series [1,4], and therefore, is not primarily respon-sible for the observed changes in physical properties. The possible effects of sodium's local coordination on physical properties will be evaluated in the present study.

2. EXPERIMENTAL

The EXAFS spectra were collected over an energy range from 1050eV to near 1500eV, as shown in Figure 1., using beryl (0001) monochromator crystals under dedicated conditions at SSRL (3.0GeV and 80mA). The quality of the data for the crystalline samples is consistently poorer than the data for the glasses. EXAFS data of Na$_2$Si$_2$O$_5$ glass, three out of five of the glasses in the series, nepheline glass, and nepheline were of sufficient quality for analysis. The analysis of the EXAFS data used an E$_0$ value of 1057eV near the edge maximum, and a phase shift function that was modified slight-ly from Teo and Lee tabulated values for Na as the absorber and oxygen as the backscatterer [5].

3. RESULTS AND DISCUSSION

The only model compound analyzed was nepheline, where the back-transformed data was fit by a theoretical function using the average Na-O distance (r)

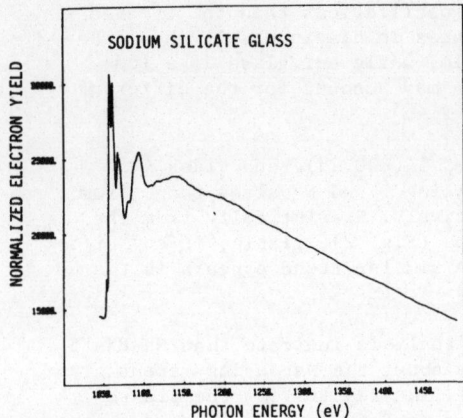

SODIUM SILICATE GLASS

NORMALIZED ELECTRON YIELD

PHOTON ENERGY (eV)

Figure 1. Na EXAFS data of $Na_2Si_2O_5$ glass, containing the Na K-edge and EXAFS regions.

and coordination number (N) for the entire known Na coordination in nepheline [6]. The refinement lowered the r and N values to 2.56Å and near 5 respectively. A Debye-Waller factor of 0.035Å was calculated from positional and thermal vibration disorder of the closest five oxygens about Na in nepheline; this value was constrained during the subsequent refinements. The refinement fit improved yielding the final values of r=2.57Å and N=5.5. Applying the same scaling factor and phase shift function used in the nepheline refinement to the glasses, r, N, and the Debye-Waller term for each glass refinement were varied. The nepheline glass data refined to an r of 2.62Å and an N of 5.1. These results appear to follow the trends found for a high temperature crystal structure refinement of nepheline [7], where the first five oxygens about Na increase in average distance from 2.55Å (at 25°C) to 2.65Å (at 908°C). The average Na-O distance observed for nepheline glass from EXAFS analysis is similar to that observed for crystalline nepheline at high temperature.

Our results for $Na_2Si_2O_5$ glass (r=2.61Å and N=6.4) differ from r and N values reported by Greaves [8]. Our $\chi(k)$ data for this glass appears to

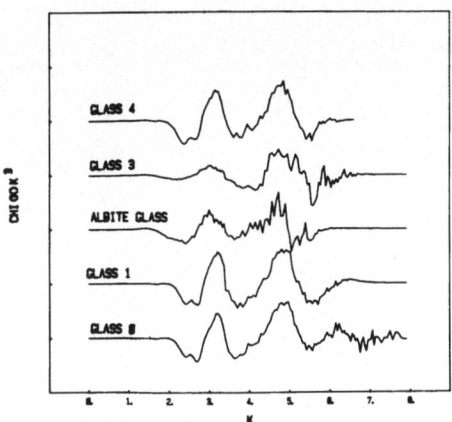

GLASS 4

GLASS 3

ALBITE GLASS

GLASS 1

GLASS 0

CHI (K) K³

K

Figure 2. k^3 weighted background subtracted data for the 75 mole % silica glass series.

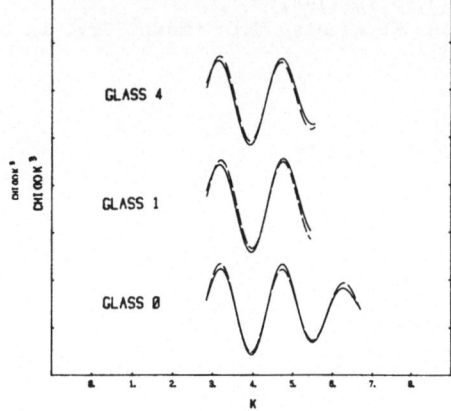

GLASS 4

GLASS 1

GLASS 0

CHI (K) K³

K

Figure 3. Fitted back-transformed data for the analyzed glasses in the 75 mole % silica glass series.

309

have a slightly longer wavelength of EXAFS oscillations than the Greaves data, which is consistent with the differences in distances. It should be noted that the Greaves glass was a thin film, while our glass is a 1cm thick fragment; such differences in samples may account for the different results.

Na EXAFS data for glass 0 (R=0.03), glass 1 (R=0.28), and glass 4 (R=1.61) were fitted (Fig. 3) and the resulting r and N values range from 2.60Å to 2.61Å and from 7.6 to 5.7 respectively. Statistically from the fitting, and from the appearance of the data (Fig. 2), little, if any, difference in the Na environment is apparent. A similar trend appears in the Na XANES for the glass series [9].

Overall, the results from the nepheline analysis indicate that Na EXAFS appears to sense the inner shell of oxygens about the Na, using recent crystal structure refinement data on nepheline. The results for the glasses indicate little change in the Na envrionment as R is varied; changes in the Na environment cannot be accountable for changes in physical properties in the glass series.

4.ACKNOWLEDGEMENTS

Funding for this study was supported in part by NSF Grant (Brown-Waychunas) and NSF-CMR. We also want to thank M. Rowan and P. Pianetta of SSRL for their assistance on the JUMBO line.

5.REFERENCES

1. D.A. McKeown, F.L. Galeener, G.E. Brown, Jr., J. Non-crystalline Solids to be published (1984).
2. M. Taylor, G.E. Brown, Jr., Geochim. Cosmochim Acta, 43, 1467 (1979).
3. K. Hunold, R.Bruckner, Glastechn. Ber., 53, 149, (1980).
4. G.E. Brown, Jr., F.D. Dikmen, G.A. Waychunas, SSRL Report No. 1983/01, Proposal No. 741.
5. B.K. Teo, P.A. Lee, J. Amer. Chem. Soc., 101, 2815, (1979).
6. W.B. Simmons, D.R. Peacor, Am. Min., 57, 1711, (1972).
7. N. Foreman, D.R. Peacor, Zeits. Krist., 132, 45, (1970).
8. G.N. Greaves, A. Fontaine, P. Lagarde, D.Raoux, S.J. Gurman, Nature, 293, 611, (1981).
9. G.A. Waychunas, G.E. Brown, Jr., in this volume.

Structural Organization Around Nucleating Elements (Ti, Zr) and Zn During Crystalline Nucleation Process in Silico-Aluminate Glasses

T. Dumas[1] and J. Petiau[2]

Laboratoire de Minéralogie-Cristallographie (LA CNRS 09), Universités Paris 6 et 7, 4, place Jussieu, F-75230 Paris Cedex 05, and

LURE, Orsay, France

At the nucleation stage preceding crystalline growth in oxide glasses, some elements, when present at several percent concentration, have a large influence on the nucleation rate and even on the first obtained crystalline phases. The precise role of nucleating elements can only be understood by investigating the very first stages of the nucleation process. Previous studies have mostly been done by methods which cannot give direct structural information about each atomic species (SAXS,SANS,TEM,...).

We present our results on two silicate glasses near the cordierite composition (2MgO,2Al2O3,5SiO2): the nucleating element is titanium in the first one, zirconium in the second one which contains also a few percent zinc. These glasses are the "parent glasses" of a class of well-known glass-ceramics. EXAFS at Zr,Zn and Ti K-edges was carried out in conjunction with transmission electron microscopy in order to relate the two scales of investigation as much as possible [1]; these two methods were applied after successive heat-treatments at increasing temperatures within a range of 40-50°C around the glass transition temperature (Tg).

Changes of the titanium coordination number with the heat-treatment are followed by the edge structure study at Ti K-edge; in the parent-glass, titanium is predominantly four-fold coordinated and part of titanium atoms become progressively six-fold coordinated as heat-treatments are performed. The result for the parent-glass is consistent with the X-ray absorption study of Greegor et al. [2] on the coordination of titanium in SiO2-TiO2 glasses. The Ti-O distance and the number of atoms in the oxygen shell, given by EXAFS at Ti K-edge, are 1.82 Å and 4. for the parent glass and become higher (1.86 Å and 5.) as heat-treatment is performed. The second neighbours are silicon at 2.90 Å in the parent glass and, as heat-treatment is performed, this shell is progressively replaced by a narrower silicon (and/or magnesium and/or aluminium) shell at 3.05 Å. This proves that titanium is progressively embedded in a crystalline phase. The structural parameters of the second shell are incompatible with rutile (TiO2) and geikielite (MgTiO3) structures. In contrast, this environment of titanium is consistent with that of Al2TiO5 (pseudobrookite structure), which has also been identified by neutron diffraction in the same sample [3].

The first crystalline phase detected by electron microscopy and X-ray diffraction is stuffed B-quartz crystals which have grown in a dendritic shape and hinder the observation of Al2TiO5 crystals by electron

---

(1) and Saint-Gobain Recherche, AUBERVILLIERS, FRANCE
(2) and Ecole Normale Supérieure, MONTROUGE, FRANCE

microscopy. If the same glass is first submitted to a heat-treatment below the glass transition temperature (Tg) prior to the same crystallization heat-treatment, the growth of B-quartz crystals does not occur, though the environment of titanium has the same evolution from a vitreous environment to a crystalline one. In that case, electron microscopy can give information about the Al2TiO5 phase which appears as crystallites of a mean size equal to 60 Å, regularly spaced in the material. The absence of growth for the B-quartz phase in this last case allows to infer that the presence of Al2TiO5 crystallites is not the only factor controlling the growth of B-quartz crystals and that the structural relaxation below Tg plays also an important role.

The changes in the environment of titanium are progressive and no intermediate stage has been put in evidence between the vitreous phase and the Al2TiO5 crystalline phase. On the contrary, for the glass containing zirconium and zinc, zirconium environment has an evolution characterized by such an intermediate stage, as it is shown in Fig. 1.

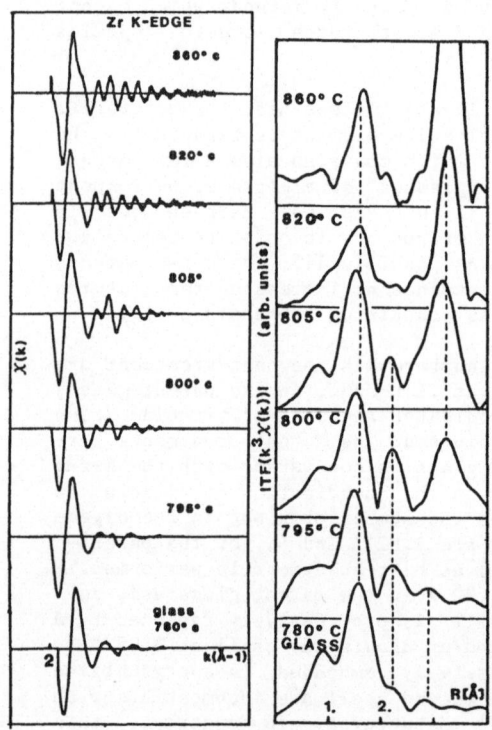

Figure 1 : EXAFS at Zr K-edge different isochronous (2h) heat-treatments and corresponding radial distribution functions

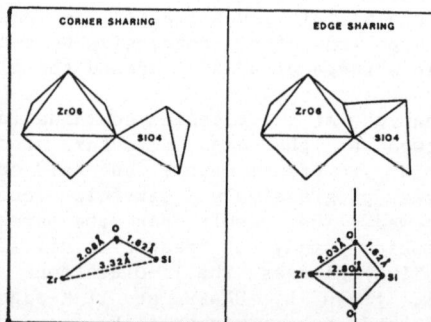

Figure 2 : Local geometry in the cases of edge and corner sharing between a ZrO6 octahedra and a SiO4 tetrahedra. The distances Zr-O and Zr-Si are measured by EXAFS

The evolution of the microstructure has been described elsewhere [1,4,5] and is summed up here. The parent glass is phase-separated, one vitreous phase being dispersed in the Zr-rich other one, as droplets of a mean size equal to 500 Å. For isochronous heat-treatments between 780 and 820 °C, electron microscopy reveals the presence of crystallites of tetragonal zirconia (ZrO2). For the last heat-treatment (860 °C), stuffed B-quartz crystals are observed by electron microscopy and X-ray diffraction. The environment of zirconium, as determined by EXAFS, corresponds to a six-fold coordination and a small contribution of a silicon shell, in the original vitreous phase, whereas it consists of an eight-fold oxygen coordination and of a narrow zirconium shell in zirconia. During the first heat-treatments, zirconium atoms have necessarily diffused through the vitreous phase to give rise to zirconia crystallites. Therefore, this diffusion process corresponds to the specific environment of zirconium put in evidence in the intermediate heat-treatments (Fig. 1). The second atomic shell, which then appears, has been analysed with the two assumptions: oxygen and silicon. The obtained Zr-O distance (2.89 Å) is too large for an oxygen of the zirconium coordination polyhedra. In contrast, the measured Zr-Si distance (2.77 Å) is too small for a silicon at the center of a SiO4 tetrahedra sharing one corner with the coordination polyhedra of zirconium (fig. 2) but can be interpreted as the Zr-Si distance in the case of an edge sharing of the polyhedra. The narrow distribution of Zr-Si distances can be accounted for by the large electronic repulsion between the cations Zr4+ and Si4+ which does not allow the rotation around the oxygen axis in that case (Fig. 2). During its diffusion path towards zirconia, zirconium,which has the largest ionic radius of all the cations of the glass, has to fit the holes of the vitreous network and this seems to result in the edge-sharing between its coordination polyhedra and SiO4 tetrahedral units.

In the same glass, the environment of zinc changes only for heat-treatments at temperature higher than 860°C, when zirconium has entirely diffused into zirconia crystallites and when B-quartz crystals have grown. An intermediate stage characterized by Zr-Si distance of 2.70 Å is put in evidence. This site is the expected one for zinc in the "channels" of the B-quartz structure [6]. After heat-treatments at higher temperatures, zinc is embedded in a spinel phase having the composition (Mg,Zn)Al2O4.

It must be stressed that B-quartz crystals have grown only in the vitreous phase through which zirconium has diffused and that the site occupied by zirconium during the diffusion process is similar to the octahedral site in the "channels" of the B-quartz structure. The nucleant role of zirconium does not consist probably in the formation of crystallites of zirconia by itself but in inducing a structural rearrangement of the vitreous network during its diffusion.

1   T.Dumas, Thesis, Université P. et M. Curie, Paris (1984)
2   R.B.Gregor,F.W.Lytle,D.R.Sandstrom,J.Wong and P.Schultz, J. Non-cryst. Sol., 55 27-43 (1983)
3   A.F.Wright, unpublished result
4   A.J. Stryjak and P.W.McMillan, J. Mat. Sci., 13 1275-1281, 1794-1804, 1805-1808 (1978)
5   T.Dumas, A.Ramos, M.Gandais and J.Petiau, accepted to J. Mat. Sci. Lett.
6   H.Schulz, W.Hofman and G.Muchow, Z. Kristallogr., 116 61-82 (1961)

EXAFS on Silver Borate Glasses

P. Fornasini and G. Dalba
Dipartimento di Fisica, Università di Trento, I-38050 Povo, Italy
F. Rocca
IRST and Centro di fisica degli stati aggregati e di impianto ionico del CNR
I-38050 Povo-Trento, Italy
E. Bernieri and E. Burattini
INFN, PWA group, I-Frascati, Italy

1. Introduction

Fast ion transport properties in crystalline and amorphous solids have been widely studied in the last years. In particular, new glasses have been produced, exhibiting a conductivity due to alkali or silver ions much higher than the corresponding crystalline compounds.

The great variety of materials sharing these properties has stimulated the search for specific structural characteristics or at least for the simultaneous occurrence of conditions favouring the ionic transport, like the presence of a great number of interconnected low-energy sites and the open structure of the material. It is thus interesting to understand the short and mid-range structures of the fast ion conducting glasses to build up a microscopic picture of each material.

The present work concerns glasses of the type $(AgI)_x(Ag_2O \, n \, B_2O_3)_{1-x}$ where

$$n = [B_2O_3]/[Ag_2O] \qquad \qquad x = [AgI]/([Ag_2O] + [AgI])$$

These glasses can be produced in a wide range of compositions and are thus well suited for a detailed study of the dependence of their microscopic properties both on the composition of the binary matrix $Ag_2O \, n \, B_2O_3$ and on the AgI content.

Previous measurements have shown that, when the Ag_2O content is progressively increased, the coordination of boron is modified and various structural groups grow up following the model of Krogh-Moe.

The insertion of AgI into the binary matrix increases the number of low-energy sites and favours their interconnection, inducing a remarkable increase of the ionic conductivity [1].

Due to its ability to select the atomic species, EXAFS can provide a determinant help in understanding the surroundings of the mobile ion Ag^+. However, the high photon energy of the K edge of silver on the one hand, the short useful k range at the L_3 edges of silver and iodine on the other, require a careful comparative investigation of all the available experimental data.

The EXAFS measurements have been performed at room temperature with synchrotron radiation at the Wiggler facility in Frascati. The Si(111) monochromator allowed to maximize the photon flux at the higher energies, to the detriment of the resolving power. New measurements are planned using a Si(220) monochromator to achieve a higher resolving power.

2. Binary matrix Ag_2O n B_2O_3

The edges L_3 and K of silver have been studied in the binary matrix. The EXAFS at the L_3 edge is too short to give accurate quantitative information. It has been possible, however, to remark that both amplitude and phase of the fundamental oscillation don't change in a sensible way when n is varied (n=2,3,4), indicating a substantial identity of mean distances and coordination numbers of the first shell. A qualitative analysis of second shell effects on the EXAFS signal suggests that a higher degree of order is present beyond the first shell in the glasses with higher B_2O_3 content [2].

Preliminary measurements at the Ag K edge allowed to obtain quantitative information concerning the first shell. A good fit of the experimental data has been obtained for Ag-O distances 2.27±0.05 Å (glass n=4) and 2.31±0.05 Å (glass n=6) utilizing the phase-shifts from the standard compound Ag_2O. These distances, remarkably higher than in crystalline Ag_2O (2.044 Å), are comparable to the shortest Ag-O distance (2.23 Å) measured in the crystalline compound Ag_2O 4 B_2O_3 by X-ray diffraction [3].

3. Ternary glasses $(AgI)_x$ $(Ag_2O$ n $B_2O_3)_{1-x}$

For the ternary glasses the EXAFS analysis is possible both at the K edge of silver and at the L_3 edge of iodine. Crystalline β-AgI has been used as standard compound.

The k range at the L_3 edge of iodine is quite short (2.3-8 Å^{-1}); the Fourier transforms F(R) are however meaningful (Fig. 1).

The iodine ions in β-AgI are surrounded by 4 silver ions at a distance of 2.81 Å and 12 iodine ions at 4.58 Å. In the Fourier transform of β-AgI the

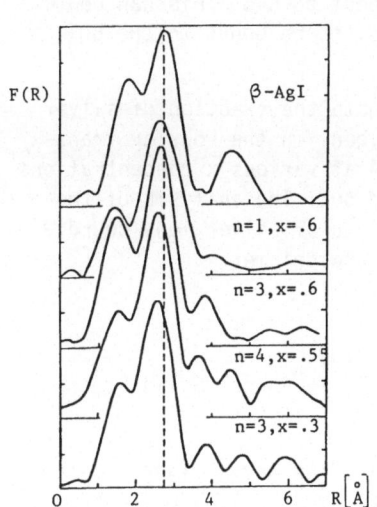

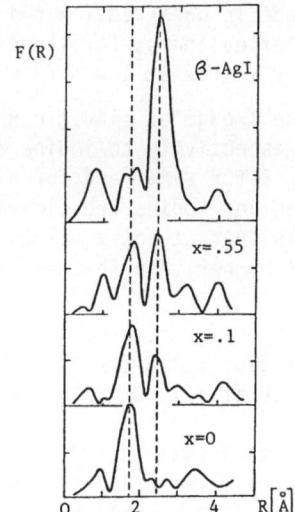

Fig.1. EXAFS at the I L_3 edge. Fourier transforms for β-AgI and ternary glasses in order of decreasing AgI molar content.

Fig.2. EXAFS at the Ag K edge for β-AgI and ternary glasses with n=4.

first two coordination shells are clearly distinguishable. The satellite peak at 1.8 Å is remarkably strong, due to the resonant behaviour of the backscattering amplitude of silver between 3 and 8 $Å^{-1}$ (Fig.3). The amplitude of the backtransformed signal exhibits a deep minimum at $k=5.8$ $Å^{-1}$; this shape is remarkably different from that calculated by Teo and Lee; it is, on the contrary, consistent with both the amplitudes measured for cadmium [4] and that calculated for tellurium by a spherical wave theory [5].

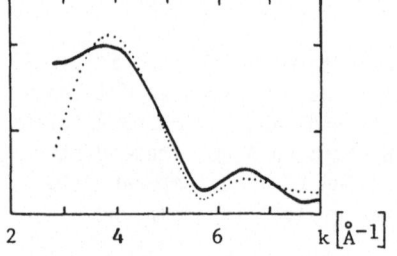

Fig.3. EXAFS at the I L_3 edge. Amplitude of the backtransformed signal for β-AgI (dotted line) and for the glass n=3, x=0.3 (continuous line).

The Fourier transforms of the ternary glasses exhibit a first-shell structure very similar to that of β-AgI. The position of the peak at 2.8 Å slightly shifts towards smaller values of R when the molar fraction of AgI decreases. A more complete data analysis will clarify whether this effect is due to a true shrinkage of the I-Ag distance or to a progressive lack of symmetry of the first coordination shell of iodine.

A comparative analysis of the radial distribution function F(R) (Fig.1) and of the amplitudes of the backtransformed signal (Fig.3) shows a remarkable similarity of the surroundings of iodine between the glasses n=3 and β-AgI. The similarity is weaker for the glasses n=1 and n=4. In particular a reduction of the signal amplitude is observed for n=1 with respect to n=3. This can be related to a greater easiness, for the silver ions, to be bound to the BO_4^- units of the binary matrix at low n values.

The EXAFS at the K edge of silver can discriminate the fraction of silver ions coordinated respectively to iodine and to oxygen. In the Fourier transforms shown in Fig.2 for β-AgI and for glasses n=4 at various x concentrations the peaks of oxygen and iodine are clearly singled out. The analysis of the peak height reveals that at low x values the fraction of silver ions coordinated to iodine is higher than that expected from stoichiometry.

References

1. G. Chiodelli et al.: Solid St. Ionics 8, 311 (1983)
2. E. Bernieri, E. Burattini, G. Dalba, P. Fornasini, F. Rocca: Solid St.Comm. 48, 421 (1983)
3. J. Krogh-Moe: Acta Cryst. 18, 77 (1965)
4. A. Balzarotti et al.: Preprint Lab. Naz. Frascati LNF 84/23 (1984)
5. R.F. Pettifer: in "Inner shell and X-ray physics of atoms and solids", Plenum Press (1980), p.653

EXAFS and XANES Study of Ti Coordination in TiO_2 Containing Glasses Upon Thermal Treatment

M. Emili, L. Incoccia, and S. Mobilio
Laboratori Nazionali di Frascati, I-00044 Frascati, Italy
M. Guglielmi
Università di Padova, I-Padova, Italy
G. Fagherazzi
Università di Venezia, I-Venezia, Italy

1. Introduction

In this work we report an EXAFS-XANES determination of the Ti coordination in TiO_2-SiO_2 sol-gel glasses for three different percentages of TiO_2 (4.5, 10 and 19 wt% TiO_2) for thermal treatment of the initial gel up to temperatures which correspond to the glass phase and glass-ceramic phase. Previous studies have been made on TiO_2-SiO_2 glasses prepared by melt quenching or flame hydrolysis. According to EVANS /1/ titanium atoms substitute silicon in the SiO_2 network, occupying only T_d coordinated sites up to 11.5 wt% TiO_2 (stable region). For higher Ti content up to 16 wt% TiO_2, a fraction of TiO_2 precipitate in the form of O_h coordinated Ti (metastable region). For concentration greater than 16 wt% TiO_2 a phase separation-crystallization occur (crystallization region). Recently GREEGOR /2/ et al., concluded that the O_h coordination is always present in this glass also in the stable region. The three different percentages of TiO_2 studied have been chosen in order to clarify which coordination Ti assumes in the sol-gel glasses.

2. Experimental

Sample preparation is described elsewhere /3/. X-ray absorption spectra at the Ti K edge were collected at the bending magnet X-ray beam line of the Frascati Synchrotron Radiation facility. The radiation was monochromatized by a Si(111) channel-cut crystal.

3. Results and Discussion

Fourier transforms of the spectra (Fig. 1a, 1b) do not show any significant structure above the first coordination shell with the exception of the samples 19 wt% TiO_2 heated at 1000 and at 1200 °C, that show peaks up the 4th coordination shell. Note that 19 wt% TiO_2 heated at 1200 °C is very close to the anatase spectrum, while 19 wt% TiO_2 heated at 1000 °C shows the same peaks but at distances significantly shorter. Also an XRD analysis performed on this sample showed peaks shifted to lower 2θ angles with respect to those of the 19 wt% TiO_2 heated at 1200 °C sample. Moreover some other broad peaks appear at 1000 °C which probably belong to an unknown crystalline phase containing Ti. Table I reports the first shell distances, the coordination numbers and the D.W. factors obtained from a k-space analysis using anatase as model compound. For all non-crystalline samples, the distances range from 1.80 A up to 1.86 A, too short to be compatible with an O_h coordination of Ti. The coordination numbers obtained clearly show that the fraction of O_h sites for Ti is negligible in the glassy state. It is very interesting to note that the effect of the thermal treatement is a progressive decrease of the $\Delta\sigma_G^2$, indicating that one goes from a higly distorted T_d unit at low temperature, when the material is an essicated gel, to a more ordered T_d unit in the more densified glasses. To verify in a more quantitative way the possible presence of a small quantity of sixth fold coordinated Ti^{++} ions, a fit has been performed on the inverse Fourier transform of the first peak of the spectra, using both a single and double shell model. An excellent agreement is obtained with a single shell model with an average coordination number of 4 (Fig. 2a, 2b).

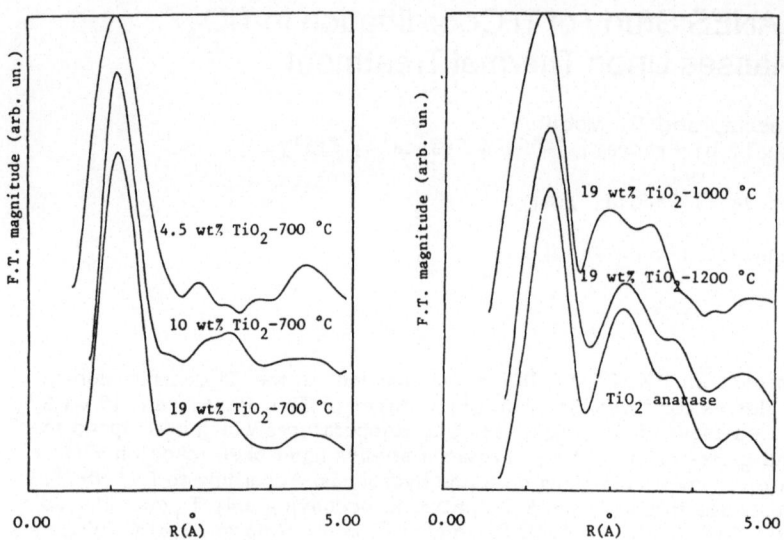

Fig. 1 - Fourier Transform of some EXAFS spectra.

TABLE I - Ti-O R bond distances, coordination numbers N and Debye-Waller factors obtained from the k-space analysis.

Gel-glass composition (wt% TiO$_2$)		250 °C	430 °C	700 °C	1000 °C	1200 °C
4.5	R(k)	n.d.	n.d.	1.84	1.81	1.81
	N	n.d.	n.d.	4.6	4.4	4.1
	$\Delta\sigma^2$	n.d.	n.d.	-0.23×10^{-2}	-0.33×10^{-2}	-0.41×10^{-2}
10	R(k)	1.84	1.84	1.83	1.81	1.81
	N	3.6	3.4	3.8	4.3	3.7
	$\Delta\sigma^2$	0.10×10^{-2}	0.13×10^{-2}	-0.14×10^{-2}	-0.22×10^{-2}	-0.29×10^{-2}
19	R(k)	1.86	1.85	1.84	n.d.	1.93
	N	3.6	3.6	3.8	n.d.	5.9
	$\Delta\sigma^2$	0.45×10^{-2}	0.26×10^{-2}	0.09×10^{-2}	n.d.	0.6×10^{-2}

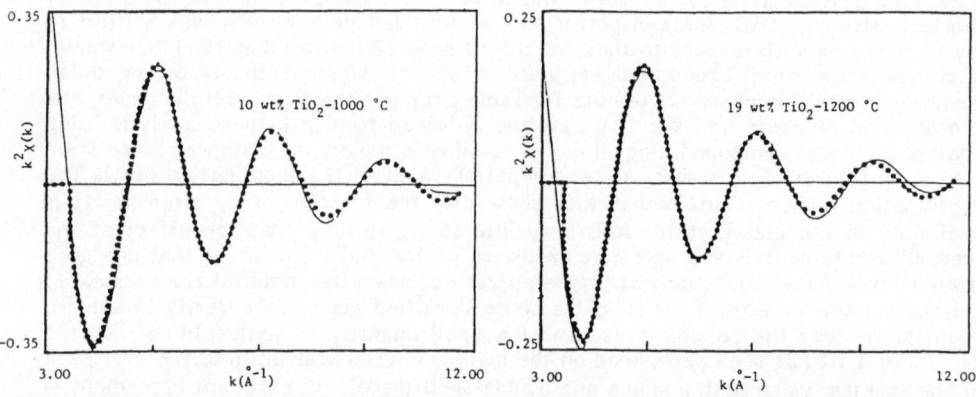

Fig. 2 - Examples of the quality of the fits obtained by using a single shell approach.

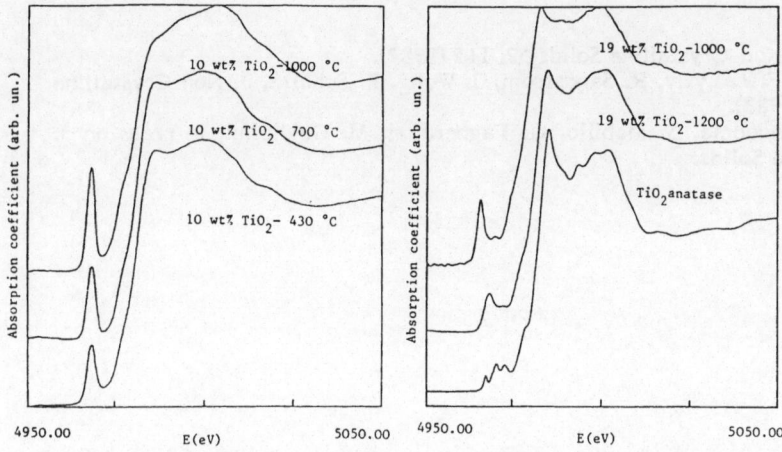

Fig. 3 - Ti K edge absorption coefficient of same of the samples.

In Fig. 3 the near-edge spectra of Ti K edge of some of the samples studied together with TiO_2 anatase are shown. As known, the 1s-3d transition that gives rise to the pre-edge peak, is dipole allowed only when the point group of Ti sites has no inversion symmetry: for this reason its intensity has been related to the ratio of T_d Ti-sites vs O_h Ti-sites in these glasses.

Table II shows the intensity of this peak normalized to the atomic absorption coefficient. Its magnitude increases with increasing temperature and decreases with increasing Ti content. The low intensity for the sample 19 wt% TiO_2 heated at 1200 °C together with the similarity of the general features above the edge again clearly show the crystallization of this sample towards anatase.

TABLE II - Intensity of the pre-edge peak normalized to the single atom.

Gel-glasses Composition (wt% TiO_2)	Heat Treatments				
	250 °C 3h	430 °C 3h	700 °C 3h	1000 °C 3h	1200 °C 3h
4.5	0.38	0.32	0.38	0.60	0.71
10	0.30	0.29	0.32	0.55	0.54
19	0.28	0.32	0.28	0.36	0.19

4. Conclusions

We conclude that the Ti ions occupy only four-fold (T_d) sites at any concentration up to the temperature where an "intermediate" ordered phase comes out. Thus, for our sol-gel glasses a complete dissolution of TiO_2 in the SiO_2 network occurs, at least up to 10 wt% TiO_2: we think that this is due to the high homogeneity of the sol-gel glasses. The thermal treatment causes a progressive ordering of the T_d units and the shortening of the bond length resulting in a more ordered and compact structure. The intensity variations of the 1s-3d pre-edge peak are due to the shortening and to the ordering of the tetrahedral units: the increase of the overlap integrals between Ti 3d and O 2p states results in an increased oscillator strength of the transition.

The crystallization occur only for the 19 wt% TiO_2 glass towards anatase through an unknown intermediate phase.

References

/1/ D.L. Evans, J. Non-Crystalline Solids 52, 115 (1982).
/2/ R.B. Greegor, F.W. Lytly, R. Sandstrom, J. Wong, P. Schultz, J. Non-Crystalline Solids 55, 27 (1983).
/3/ M. Emili, L. Incoccia, S. Mobilio, G. Fagherazzi, M. Guglielmi, in press on J. Non-Crystalline Solids.

Local Order of Metallic Glasses by EXAFS and X-Ray Scattering: CuY Alloys

D. Raoux and A.M. Flank

LURE, Bât. 209C, Faculté des Sciences, F-91405 Orsay, France

Metallic glasses of the metal-metalloid type have a wide range of applications based on their magnetic and mechanical properties. Their potential applications have stimulated a lot of fundamental studies these past ten years, many of them being devoted to structural investigation. Recently Exafs has become one of the most popular techniques in this field because of its ability to probe the local environment around each of the components of the glassy metallic alloy, and because measurements are very easy. For these reasons more amorphous metallic alloys have been studied these past few years by Exafs than by either neutron or X-ray scattering methods. References can be found in review papers (1,2).

However, for disordered systems, Exafs suffers from a very severe limitation, since it is usually not analyzed at low k values, ie k < 6 to 8 Å^{-1}. This makes EXAFS rather insensitive to broad features in the radial distribution functions and very often results in underestimated coordination numbers and too short distance measurements. These limitations have been discussed by various authors who suggested the use of asymmetrical atomic distributions to interpret the difference between Exafs and scattering measurements (3,4,5) for metal-metalloid alloys.
They all assume that X-ray scattering gives the average distance of the atomic distribution and that Exafs gives a measurement of its steepest part. Frahm et al., Cargill et al. (6) made model calculations showing that such asymmetrical distance distributions should also be apparent in X-ray or neutron scattering RDF if they are performed up to sufficiently large k value. There has been recently some experimental evidence in neutron scattering data (7,8) for such asymmetrical distributions of distances. The increasing consistency between Exafs and scattering data for metallic glasses will hopefully be improved by extending Exafs analysis to low k values (2 Å^{-1}) by using the curved wave approximation.

However, we like to show in this paper that even if these requirements are not fullfilled, it is often quite possible to get a detailed and unambiguous description of the local structure of a metallic glass by analyzing simultaneously both Exafs and scattering data. Doing this, we use the specific strengths of each technique to overcome their limitations. We take as an example the specific case of the Cu_2Y alloy which is part of a systematic study of Cu_xY_{1-x} alloys with x ranging from 0.33 to 0.9 (9). Results concerning the dilute Cu_9Y alloy have been published elsewhere (10).

I. Experiments and data processing

The CuY alloys have been prepared by sputtering.

The Exafs spectra have been measured at 20°K using the Exafs I station at LURE. A Si(220) channel-cut monochromator has been used for Cu edge measurements and a Si(400) one for yttrium edge measurements. They have been analyzed using the plane wave approximation for k>5 Å^{-1}.

Since the backscattering factors from Teo and Lee (11) are not accurate enough in the low energy range, specially for a heavy scatterer like Y, it is better to use experimentally determined amplitudes and phase shifts. They have been measured from elemental Cu and Y foils because the intermetallic crystalline phases Cu_5Y and Cu_2Y have too complex local structures to yield unambiguuous determinations. The phase-shift for heterogenous pairs like Cu-Y or Y-Cu have been obtained using a semi-empirical procedure previously used for crystalline (12) or amorphous (13) metals.

$$\varphi_{A-B} = \varphi_{B-B}^{(exp)} + \{2\delta_1'(A) - 2\delta_1'(B)\} \quad (theory)$$

A being the central atom, B the scatterer. $\varphi_{(exp)}$ stands for experimentally derived phase shift ; theoretical values for $2\delta_1'$ are taken in (11). With such an empirical procedure, we have to use the same E_0 values for the amorphous sample and for the Cu and Y references.

X-ray scattering data have been measured at room temperature using an X-ray tube with silver target. Data have been analyzed up to 12 Å^{-1} using standard treatments allowing one to calculate the radial distribution function.

II. Results

The first peak of the radial distribution function derived from X-ray scattering (RDF), fig. 1, exhibits a main peak at 2.85 Å due to Y-Cu correlations, a shoulder around 2.5 Å related to Cu-Cu nearest neighbours and a broad tail towards larger distances d > 3 Å which is evidence for a large structural and/or chemical disorder. On the contrary, the

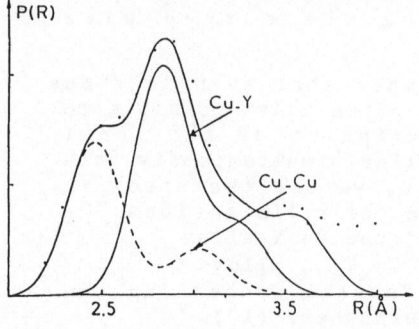

Fig.1. Radial distribution function (...) experimental results (———) final reconstruction using values given in Table I

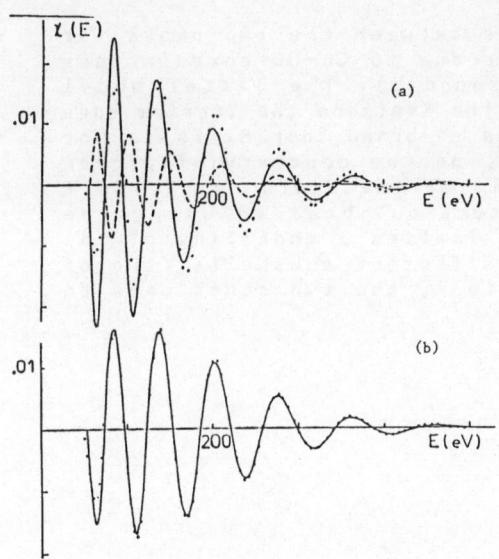

Fig.2. (...) experimental spectrum on yttrium edge (a) (——) calculated EXAFS with one shell of Cu neighbours at 2.84 Å (---) calculated EXAFS with one shell of Cu neighbours at 3.2 Å (b) final reconstruction using a two shell model

Fourier transforms of the Exafs on Cu and Y edges show a single peak.
The phase of the Exafs on the Yttrium edge, Fig. 2a, is rather well fitted by assuming a single Y-Cu distance at
2.84 Å. However the agreement is very poor for the amplitude, specially at low energies. It is very significantly improved by considering a second shell of Cu atoms at 3.20 Å with coordination numbers given in table I (cf Fig. 2b). However, because of the correlation between the coordination number N and disorder parameter σ, this second subshell is not uniquely defined : N can be raised up to 6 by increasing σ to 0.15 Å.

This ambiguity can be removed by fitting the RDF. Because of the large weight of Cu-Y correlations in X-ray scattering, N has to be lower than three. The reconstruction of the RDF with two Cu-Y subshells (parameters are given in Table I)

TABLE I

Interatomic distances R(Å), coordination numbers (N) and mean square relative displacement σ(Å). For comparison, σ =0.06 Å for elemental Cu and σ = 0.07 Å for yttrium. For scattering, results are broadened by the narrow available k range.

			R	N($\pm$0.5)	σ	R	N($\pm$0.5)	σ
Cu–Cu	Exafs	:	2.47 $\pm$ 0.02	4	0.11	3.05 $\pm$ 0.05	2	0.11
	Scattering	:	2.47 $\pm$ 0.05	4.5	0.15	3.03	2	0.15
Cu–Y	Exafs	:	2.83 $\pm$ 0.03	3.3	0.13	3.03 $\pm$ 0.05	1.5	0.13
	Scattering	:	2.84 $\pm$ 0.05	3.3	0.15	3.20	1.3	0.15
Y–Y	Scattering	:	> 3.58	> 5	0.15			

still yields a too weak amplitude between the two peaks, at
about 3 Å, which could be either due to Cu-Cu correlations
(N=2) or Y-Cu ones (N=1.5 Cu around Y). The latter model
spoils the quality of the fit of the Exafs on the Yttrium edge
(unless the atomic distribution is so broad that Exafs is not
sensitive to it). The forner one, on the contrary, improves
very much the quality of the fit of the Exafs on the Cu K
edge. Fig. 3 shows the reconstruction obtained using the
parameters given in table I. It implies a modelling of the
local order around Cu atoms by 4 different subshells, two of
them being due to Cu-Cu correlations, the two other ones to
Cu-Y correlations.

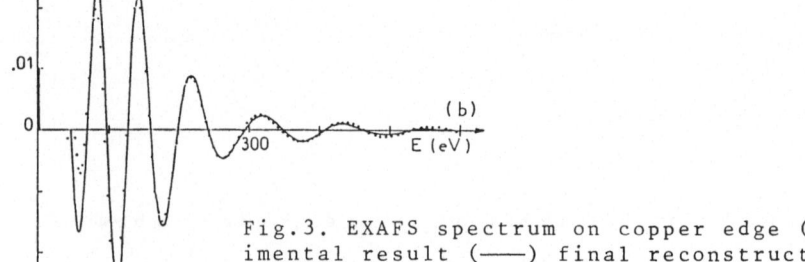

Fig.3. EXAFS spectrum on copper edge (...) experimental result (———) final reconstruction using
values given in Table I

We point out that such a model is necessary to interpret
the scattering data as well as the Exafs ones, and that using
only Exafs we could not have removed some ambiguities. We have
also investigated in the same way other CuY alloys (9,10)
namely CuY, CuY_2 and Cu_9Y and found similar results.
However, in one case (CuY_2 alloy) we were left with two
different models which could fit all the data. We are now
analyzing X- ray anomalous scattering data which hopefully will
remove the ambiguity. This shows that one really needs to
accumulate as many different experiments as possible when
working with so disordered systems.

We also point out that no detailed information could be
obtained for Y-Y correlations which are expected to occur at
larger distances (d>3.5 Å) and are too broadly distributed to
contribute significantly to EXAFS.
A more extensive report of this work will be published
elsewhere,which shows that throughout the composition range
Cu_2Y to CuY_2 some kind of local unit is preserved
around copper atoms in the glassy state. It links each copper
atom to 3 to 4 Yttrium neighbours at a distance 2.85±0.02 Å
which is shorter than in the crystalline Cu_2Y phase (2.97
Å), and to about 2 other yttrium atoms further away at
3.24 ± 0.07 Å. This local unit induces some degree of chemical
order, which increases with the concentration in the late
transition metal, the local structure of the glassy materials
being rather different from that of the crystalline phases. A
similar conclusion has also been found for Ni_2Zr by Frahm
et al. (13). Finally let us notice that such a modelling of
metallic glasses by rather well-defined subshells has also

been evidenced by Exafs studies for CuZr alloys (14,15) and for Ni_2B (16), so that it might be rather general for concentrated metallic alloys.

REFERENCES

1. D. Raoux, A.M. Flank, A. Sadoc, in Exafs and Near Edge Structure, eds Bianconi, Incoccia, Stipcich (Springer Verlag Berlin) 232, (1983)
2. G.S. Cargill III, J. Non Cryst. Solids 61, 261 (1984)
3. R. Haensel, P. Rabe, G. Tolkiehn, A. Werner, in Liquid and Amorphous Metals, eds Lüscher and Coufal (Sitjhoff and Noordhoff, Alphen aan den Rijn, The Netherlands) 459 (1980)
4. M. de Crescenzi, A. Balzarotti, F. Comin, L. Incoccia, S. Mobilio, N. Motta, Solid State Commun. 37, 921 (1981) 5. P. Lagarde, J. Rivory, G. Vlaic, J. Non Cryst. Solids 57, 275 (1983)
6. R. Frahm, R. Haensel, P. Rabe in : Exafs and Near Edge Structure, eds Bianconi, Incoccia, Stipcich (Springer Verlag Berlin) 107 (1983) ; G.S. Cargill III, W. Weber, R.F. Boehm, ibidem p 277
7. P. Lamparter, W. Sperl, S. Steeb, J. Bletry, Z. Naturforschung 37 a, 1223 (1982)
8. T. Fukunaga, N. Watanabe, K. Suzuki, J. Non Cryst. Solids 61, 343 (1984)
9. A.M. Flank, Thèse d'état, Poitiers 1983 unpublished
10. A.M. Flank, D. Raoux, A. Naudon, J.F. Sadoc, J. Non Cryst. Solids 61, 445 (1984)
11. B.K. Teo, P. Lee, J. AM. Chem. Soc. 101, 2815 (1979)
12. D. Raoux, A. Fontaine, P. Lagarde, A. Sadoc, Phys. Rev. B 24, 5547 (1981)
13. R. Frahm, R. Haensel, P. Rabe, J. Phys. F : Met. Phys. 14, 1029 (1984) ; ibid. 1333
14. A. Sadoc, A.M. Flank, D. Raoux, P. Lagarde, J. Non Cryst. Solids 50, 331 (1982)
15. A. Sadoc, Y. Calvairac, A. Quivy, M. Harmelin, A.M. Flank to be published in J. Non Cryst. Solids
16. J. Wong, H. Liebermann, Phys. Rev. B 29, 651 (1984)

EXAFS of Glassy Metallic Alloys: Amorphous and Crystalline $Mg_{70} Zn_{30}$

A. Sadoc

Laboratoire de Physique des Solides, Bâtiment 510 and LURE, Bâtiment 209 C Université de Paris-Sud, F-91405 Orsay Cêdex, France

R. Krishnan and P. Rougier

Laboratoire de Magnêtisme, CNRS, F-92195 Meudon Cêdex, France

We report here an EXAFS study above the Zn edge of amorphous and crystalline $Mg_{70} Zn_{30}$ alloys. An amorphous sample of near the eutectic composition, $Mg_{72} Zn_{28}$, crystallizes into a single-phase of the orthorhombic crystal $Mg_{51} Zn_{20}$ of known structure [1]. This provides an excellent opportunity for a direct structural comparison of the amorphous and crystalline states of a metal-metal system.

The amorphous $Mg_{70} Zn_{30}$ alloy was prepared by single roller techniques. The crystalline specimen was obtained by heating an amorphous sample to 380 K for 7 h. The EXAFS and XANES spectra were taken at the synchrotron radiation facility LURE (DCI) in Orsay. They were measured using channel cut monochromators, Si 220 one for the EXAFS measurements and Si 400 for the XANES (fig. 1) which needs a better energy resolution. The EXAFS data were recorded at 30 K.

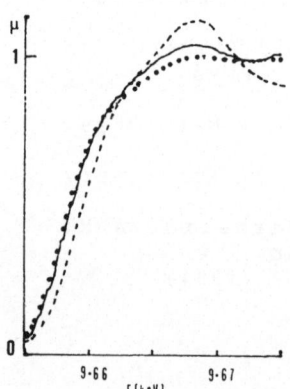

Fig.1. Normalized Zn K-edge XANES spectra in the pure metal (---), crystalline (___) and amorphous (...) alloys

The XANES results are compared in fig. 1 for a zinc foil (0.9 μ m thick) and for the crystalline and amorphous alloys. The position of the first inflection point is the same for both alloys, but is shifted by ~ 1 eV to lower energies with respect to the pure Zn metal. This is an indication of some modification of the electronic structures of the atoms.

As the EXAFS are found similar for both states of the alloy (fig. 2), there is a strong analogy between the local environments in crystalline and amorphous phases. The rather complicated neighbourhood of a Zn atom is illustrated in fig.

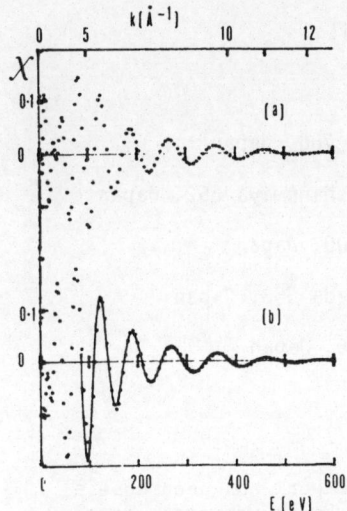

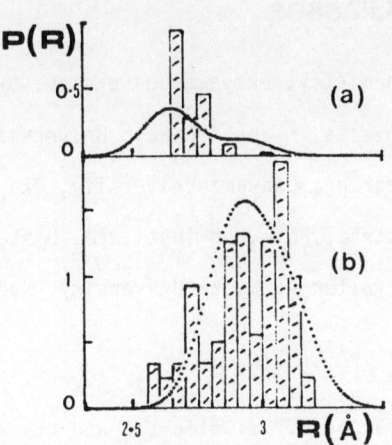

Fig.2. Filtered EXAFS spectra (points) of crystalline (a) and amorphous (b) $Mg_{70}Zn_{30}$. The simulation (solid line) was done using a combination of subshells as described in the text

Fig.3. Histograms of crystalline $Mg_{70}Zn_{30}$ alloy for (a) Zn-Zn and (b) Zn-Mg pairs compared with the radial distributions obtained by EXAFS

3 by the histograms of the radial distribution of the Zn-Zn and Zn-Mg pairs calculated from the crystallographic data of HIGASHI et al [1]. As these data were obtained at room temperature, a 0.05 Å contraction has been taken into account for the histograms.

In order to test the subshell modelling [2-4], in the case of a known structure, the EXAFS were fitted using the theoretical amplitudes and phase shifts of TEO and LEE [5] and taking the zero of the kinetic energy at the maximum of the absorption edge, like for copper and nickel alloys [2]. A good simulation of the EXAFS can be obtained with two gaussian subshells for the Zn-Zn pairs as well as for Zn-Mg (1.5 Zn at 2.65 Å and 0.5 Zn at 2.95 Å, σ = .106 Å, 6 Mg at 2.90 Å and 4 Mg at 3.10 Å, σ = .116 Å) so that the average coordination numbers are kept (1.95 Zn and 9.75 Mg atoms). The distribution of the Zn-Zn pairs reproduces only partly the histogram of fig. 3a. But this could be due to the high relative uncertainty on these less numerous pairs, with ΔN = 0.25 to 0.5 for each subshell. As for the Zn-Mg pairs, the agreement is excellent (fig. 3b). This gives us great confidence in this type of modelling and on the approach based on the analogy between local environments in crystalline and amorphous phases.

1- I. Higashi, N. Shiotani, M. Uda, T. Mizoguchi, H. Katoh, J. Sol. St. Chem. 36, 225, (1981).
2- A. Sadoc, D. Raoux, P. Lagarde, A. Fontaine, J. Non cryst. Sol. 50, 331, (1982).
3- A. Sadoc, J. Non Cryst. Sol. 61, 403, (1984).
4- A. Sadoc, Y. Calvayrac, A. Quivy, M. Harmelin, A. M. Flank, J. Non Cryst. Sol. 65, 109 (1984).
5- B. K. Teo, P. A. Lee, J. Am. Chem. Soc. 101, 2815, (1979).

EXAFS Studies of the Local Environment in Metal-Metal Glasses

H. Maeda
Department of Chemistry, Okayama University, Okayama 700, Japan
H. Terauchi
Department of Physics, Kwansei-Gakuin University, Nishinomiya 662, Japan
M. Hida
School of Engineering, Okayama University, Okayama 700, Japan
N. Kamijo
Government Industrial Research Institute, Osaka , Ikeda 563, Japan
K. Osamura
Department of Metallurgy, Kyoto University, Kyoto 606, Japan

1. Introduction

In recent years, a lot of experimental and theoretical work has been done a
on amorphous materials[1-4]. From this work, it has become apparent that
the nearest neighbor distances in the amorphous state are 0.2 to 0.3 Å less,
and the coordination numbers are significantly lower than those in the corresponding crystals. Several models have been reported for structures of
amorphous states. The properties of amorphous materials are discussed in
terms of the heterogeneity of the structure.

A Fe – Zr amorphous alloy is one example of typical metal – metal amorphous
system. In order to elucidate the local structure of amorphous metal – metal
alloy, the EXAFS studies were carried out for as-quenched and annealed samples of $Fe_{90}Zr_{10}$ amorphous alloy. In this work, we evaluate the structural
terms by comparison with a reference sample, i.e. metallic compound Fe_2Zr.

2. Experimental

Glassy samples in shape of ribbon (3.0 mm width and 0.013 mm thickness) were
prepared by the single roller melt-quenching technique under a vacuum of
10^{-3} torr. The samples obtained were confirmed to be in the really amorphous state by observing the X-ray halo diffraction. The annealed sample
was heat-treated at 623 K for 3 ks. The X-ray absorption measurements were
made with synchrotron radiation by used of the EXAFS facilities[5] installed
at the beam line 10-B of 2.5 Gev storage ring of Photon Factory in the
National Laboratory of High Energy Physics (KEK, tsukuba). The energy
resolution is estimated to be about 1 eV in the present study. All data
were taken at room temperature. The data reduction and curve-fitting were
performed as detailed elsewhere[2]. The background-subtracted Zr and Fe
EXAFS spectra, $k^3\chi(k)$ vs. k (in $Å^{-1}$), of metallic compound Fe_2Zr, as-quenched
and annealed samples of $Fe_{90}Zr_{10}$ amorphous alloy are shown in Fig. 1.

3. Results and Discussion

The Fourier transforms of the normalized EXAFS spectra provide the radial
structure functions as shown in Fig. 2. A comparison of Fig. 2(a) and (b)
indicates that the first peak at 2.23 Å is considerably shorter than crystalline distance, 2.60 Å.

In order to determine the structural parameters, a curve-fitting analysis
was performed using Teo and Lee's[6] phase shifts and backscattering amplitudes. To obtain a good solution, the optimum number of parameters is determined based on the data by applying the Akaike's minimum AIC procedure[7].

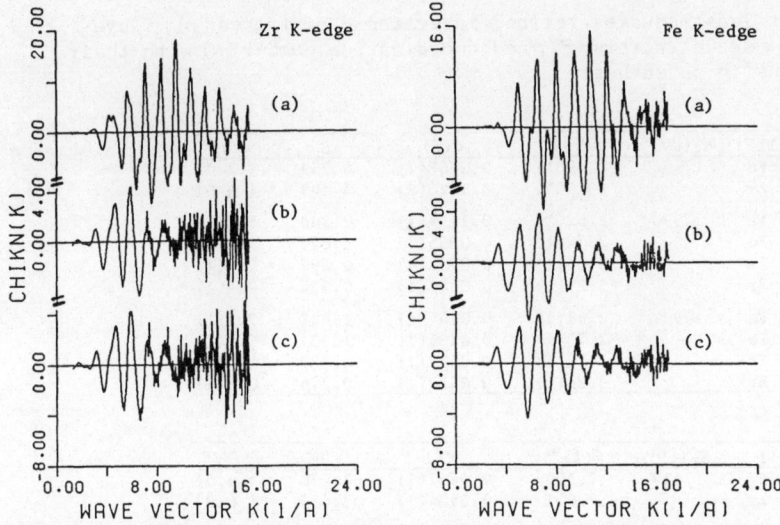

Fig. 1. Normalized $k^3\chi(k)$ versus k plots for Zr and Fe in (a) metallic compound Fe_2Zr, (b) as-quenched $Fe_{90}Zr_{10}$ and (c) annealed $Fe_{90}Zr_{10}$

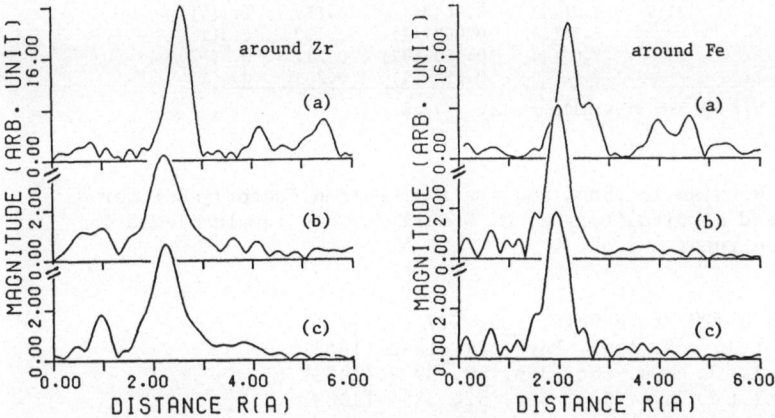

Fig. 2. Radial structure functions around Zr and Fe for (a) metallic compound Fe_2Zr, (b) as-quenched $Fe_{90}Zr_{10}$ and annealed $Fe_{90}Zr_{10}$

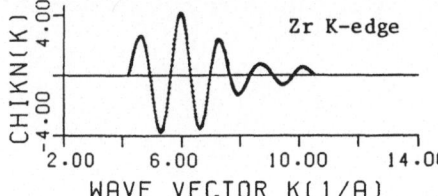

Fig. 3. Fourier filtered zirconium EXAFS spectrum(dotted curve) and the foure shell nonlinear least-squares best fit(solid curve) for as-quenched $Fe_{90}Zr_{10}$(window: $R = 1.34 \sim 3.35$ Å)

From these results, the four-shell model is in best agreement with the Fourier-filtered EXAFS data. This implies that the amorphous structure could be described as an amorphous solid solution with two compositionally fluctuated regions represented by zirconium-dense regions and zirconium-rare regions.

Table 1. Best fit least-squqres refined interatomic distances (R), Debye-Waller factors (σ), scale factors (B), and coordination number (N) with their standard deviations in parentheses

(a) Zr K-edge EXAFS

Sample	Shell	E_0(eV)	R(Å)*	σ(Å)	B	N*
Fe$_2$Zr	Zr – Fe	17992	{2.931}	0.0768(1)	6.259	{12.0}
(crystal)	Zr – Zr		{3.061}	0.1028(2)	4.109	{ 4.0}
Fe$_{90}$Zr$_{10}$	Zr – Fe	17990	2.62(1)	0.1179(1)	2.704	5.2(1)
(as-quenched)	Zr – Fe		2.72(1)	0.0835(1)	1.673	3.2(1)
	Zr – Zr		3.04(1)	0.0755(1)	0.899	0.9(1)
	Zr – Zr		3.29(1)	0.0689(1)	0.325	0.3(1)
Fe$_{90}$Zr$_{10}$	Zr – Fe	17991	2.61(1)	0.0644(1)	1.422	2.7(1)
(annealed)	Zr – Fe		2.73(1)	0.0134(1)	1.131	2.2(1)
	Zr – Zr		3.02(1)	0.0417(1)	0.837	0.8(1)
	Zr – Zr		3.26(1)	0.0573(1)	0.276	0.3(1)

(b) Fe K-edge EXAFS

Sample	Shell	E_0(eV)	R(Å)*	σ(Å)	B	N*
Fe$_2$Zr	Fe – Fe	7117	{2.500}	0.0631(1)	2.430	{ 6.0}
(crystal)	Fe – Zr		{2.931}	0.0934(1)	3.437	{ 6.0}
Fe$_{90}$Zr$_{10}$	Fe – Fe	7116	2.30(1)	0.0849(1)	0.732	1.8(1)
(as-quenched)	Fe – Fe		2.39(1)	0.0633(1)	0.811	2.0(1)
	Fe – Zr		2.64(1)	0.0000(84)	0.098	0.2(1)
	Fe – Zr		2.73(1)	0.1056(1)	0.778	1.4(1)
Fe$_{90}$Zr$_{10}$	Fe – Fe	7119	2.28(1)	0.0618(1)	0.889	2.2(1)
(annealed)	Fe – Fe		2.38(1)	0.0607(1)	1.738	4.3(1)
	Fe – Zr		2.64(1)	0.0001(72)	0.217	0.4(1)
	Fe – Zr		2.76(1)	0.0519(1)	0.205	0.4(1)

*) R and N values in braces are crystallographic values.

Acknowledgements We wish to thank the staff of Photon Factory(KEK) for their assistance and hospitality; Dr. M. Nomura for his invaluable advice and help during the runs.

References

1. G.S.Gargill III: Solid State Phys., 30, 227 (1977)
2. H.Maeda, et al.: Jpn. J. Appl. Phys. 21, 1342 (1982)
3. H.Terauchi, et al.: J. Phys. Soc. Jpn. 50, 3977 (1981)
4. H.Terauchi, et al.: J. Phys. Soc. Jpn. 52, 3454 (1983)
5. H.Oyanagi, T.Matsushita, M.Ito and H.Kuroda: KEK Report, 83-30 (1984)
6. B.K.Teo and P.A.Lee: J. Am. Chem. Soc. 101, 2815 (1979)
7. H.Akaike: IEEE Trans., on Automatic Control, AC-19, 716 (1974)

EXAFS of Amorphous Vanadium Phosphate Oxides: Correlation Between Hopping Conductivity and Local Structure

S. Stizza and I. Davoli
Dipartimento di Matematica e Fisica, Università di Camerino
I-62032 Camerino, Italy
M. Tomellini, A. Marcelli, and A. Bianconi
Dipartimento di Fisica, Università "La Sapienza", I-00185 Roma, Italy
A. Gzowski and L. Murawski
Institute of Physics, Technical University, Gdansk, Poland

1 Introduction

The vanadium phosphate glasses have attracted a large interest as characteristic systems for hopping conductivity [1-5]. The binary $(V_2 O_5)_x (P_2 O_5)_{1-x}$ glasses exhibit semiconducting properties arising from electron transfer between V^{4+} and V^{5+} ions. The presence of V^{4+} ions is due to the loss of oxygen compensated by vanadium ions in the lower oxidation state. Therefore for a glass of fixed composition x mole % $V_2 O_5$ different redox ratio of V^{4+} on the total number of vanadium ions $C=V^{4+}/V$ are possible using different glass preparation procedures. The d.c. conductivity follows an exponential law $\sigma = \sigma_0 \exp(-W_h/KT)$ predicted by the small polaron theory, where the polarization of the lattice around V^{4+} ion follows the diffusion of the electron between vanadium ions in different valence states. W_h is the activation energy, which is approximately equal to half the polaron binding energy [6]. The activation energy is depending on the distance between vanadium sites (V-O-V), on the structure of vanadium coordination shell VO_n and on the arrangement of VO_n poliedra in the glass. A variation of the conductivity and of the activation energy [3] ranging from 0.3 eV to 0.42 eV for glasses with different composition x and redox ratio C has been observed [4]. In spite of the large number of studies on the electronic properties of this class of binary glasses, very little is known on local atomic structure. Moreover V^{4+} is silent to most of spectroscopical methods.

We have undertaken a project on the study of the local structure of vanadate glasses by X-ray absorption methods XANES (X-ray Absorption Near Edge Structure) and EXAFS (Extended X-ray Absorption Fine Structure) [7,8]. We have shown that the mixed valence state (i.e. the presence of both V^{4+} and V^{5+}) of the $(V_2 O_5)_x (P_2 O_5)_{1-x}$ glasses can be measured from the 1 eV energy shift of the sharp core exciton at the absorption threshold of the 1s level of vanadium due to 1s → 3d-derived molecular orbitals localized within the first coordination shell of vanadium ions [8].

Here we report the investigation of the variation of the local structure around vanadium ions by changing the redox ratio C in the 60% and 50% mole $V_2 O_5$ glasses.

2 Experimental

The EXAFS spectra were recorded at the Frascati Synchrotron Radiation Facility using the storage ring Adone operated at 50 mA and 1.5 GeV. Both Si(220) and Si(111) channel cut crystals were used as monochromators. The energy resolution using a 0.5 mm exit slit was less than 0.8 eV at the vanadium K-edge 5469 eV.

3   Results and Discussion

Figure 1 shows the presence of mixed valence state V^{4+}-V^{5+} detected in both the 50% and 60% mole V_2O_5 glasses, by changing the redox ratio C from the splitting of the white line due to the core exciton at the K absorption threshold. The XANES spectra of V_2O_5 and vanadyl-bis-acetylacetonate were used as model compounds where vanadium is five coordinated. The white line is assigned to a bound state formed by a core hole in the 1s level and one excited electron in the unoccupied molecular orbital (derived by mixing of V-3d and oxygen 2p orbitals) deeper below the continuum threshold and localized on the vanadium site by the core hole Coulomb attraction.

The intensity of this core exciton depends on the presence of l=1 orbital components in the localized 3d-derived state because transitions to l=2 components are dipole forbidden. For a fixed geometry the intensity depends on the interatomic distance changing the mixing of vanadium 3d with oxygen 2p orbitals [9]. The weaker intensity of the white line in the

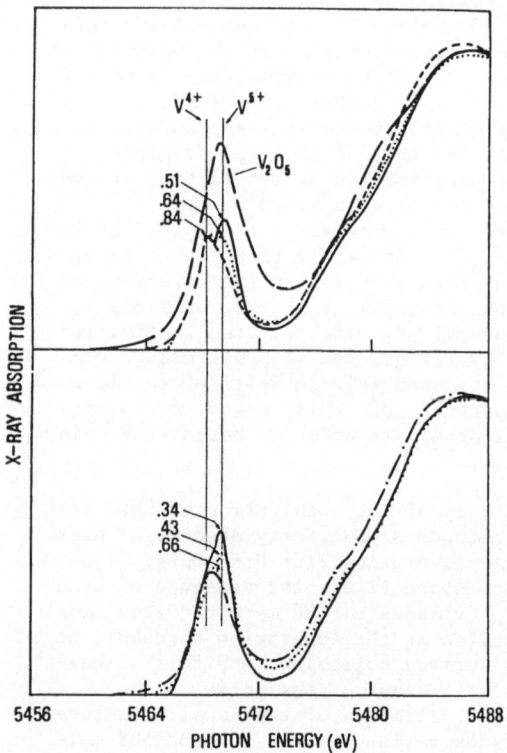

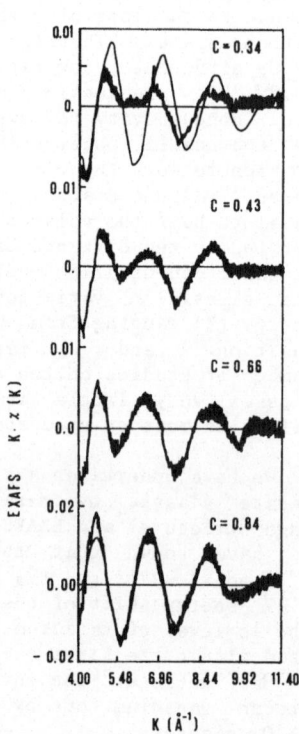

FIG. 1 Vanadium K-XANES of 50% (upper panel) and 60% (lower panel) V_2O_5 posphate glasses at different redox ratio C. The splitting of the core exciton at 5468 eV is due to the mixed valence state i.e. to the presence of both V^{4+} and V^{5+} ions in the same glass.

FIG. 2 EXAFS of 60% mole V_2O_5 glasses (the lowest panel shows the EXAFS of 50% mole V_2O_5) at different redox ratio C, showing the effect of reduction.

glass compared with V_2O_5 indicates a longer average interatomic distance V–O like in the vanadyl-bis-acetylacetonate where there is a single 1.56 Å double bond (where in V_2O_5 is 1.58 Å) and 4 single bond oxygens at 1.97 Å nearly 0.1 A longer than in V_2O_5. The intensity of the core exciton localized on V^{5+} site at ~1 eV higher energy is stronger than that at V^{4+} site as it is shown for the glasses close to the redox ratio C=0.50. This is in agreement with a shorter V–O distance in V^{5+} sites if the coordinaton geometry in V^{4+} and V^{5+} is the same.

Figure 2 shows the EXAFS of 60% mole V_2O_5 glasses at redox ratios C=0.34, 0.43, 0.66 and of the 50% mole V_2O_5 at C=0.84. A characteristic feature of the EXAFS oscillations of these amorphous oxides is that they are very weak, of the order of 1%. The amplitude of the oscillations becomes larger at values of the redox ratio C larger than 0.60. A single shell contribution is the main characteristic of glasses with high redox ratio. In fact the back-Fourier of the first shell, shown as a solid line, of the 50% mole V_2O_5 glass at redox ratio C=0.8 is clearly sufficient to give account of the experimental spectrum. Increasing the V concentration, a destructive interference effect appears at K~5.5 Å in the EXAFS oscillations. This effect is due to a second shell EXAFS oscillation. The comparison between the Fourier transforms of the glasses show that the second shell peak in the radial distribution function at 3.0 Å is increased with reducing c.

The results on the variation of the V–O average interatomic distance as a function of redox ratio C are reported in Fig.3. We have found a dilatation of the V–O average distance of ~0.07+0.02 Å going from c=0.34 to c=0.66. This result clearly shows a large distortion of the local structure of the vanadium sites accompaning the reduction of V^{5+} sites. The variation of the contribution of the second shell with the redox ratio gives information on the formation of a regular network of the microscopic structural units of the glass giving a regular arrangement of atoms in the second shell. This network is modified (broken) and the second shell contribution becomes negligible reducing the oxide i.e. increasing the number of V^{4+} sites. We want here to suggest a possible model of the V_2O_5 60% mole glass local structure in agreement with our data but which deserves further experimental work to be established. We assume that the amorphous binary oxide is formed by sheets of VO_5 square piramids like V_2O_5. The microscopic structural unit of vanadium is formed by four oxygens (close to

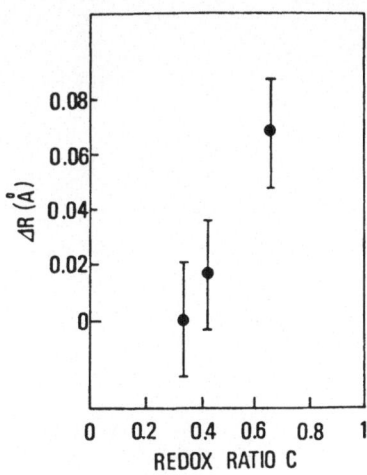

FIG. 3 Variation of V–O interatomic distance (ΔR) with change of redox ratio C in the $(V_2O_5)_{.6}(P_2O_5)_{.4}$ glasses.

333

the sheet plane) and their V-O single bonds length increases by reducing the V^{4}ion which predicts a polaron effect on the sheet surface. The second shell in the radial distribution function of glasses at low redox ratio is mainly assigned to vanadium ions. The presence of the vanadium ions in the second shell of a vanadium site predicts the high probability of electron hopping in the adiabatic regime as it has been found [4]. Our data show that this is the case at low C ratios. Increasing C the sheets are broken and the probability of finding a vanadium ion in the second shell decreases. This structural effect induced by reduction of binary (V_2O_5)-(P_2O_5) glasses can give an explanation for the maximum of conducibility occurring at low redox ratio C=0.2 and not at C=0.5 as is expected from the theory of hopping conductivity.

References

1 N.F.Mott: J. Non Crystalline Solids 1, 1 (1968)
2 M.Sayer and A.Mansingh: Phys. Rev. B6, 4629 (1972)
3 M.Sayer and A.Mansingh: J. Non Crystalline Solids 58, 91 (1983)
4 J.N.Greaves: J. Non Crystalline Solids 11, 427 (1973)
5 L.D.Bogomolova, T.K.Pavlushkina and A.V.Roshchina: J. Non Crystalline Solids 58, 99 (1983)
6 I.G.Austin and N.F.Mott: Advan. Phys. 18, 41 (1969)
7 A.Bianconi, A.Giovannelli, I.Davoli, S.Stizza, L.Palladino, O.Gzowski and L.Murawski: Solid State Commun. 42, 547 (1982)
8 I.Davoli, S.Stizza, M.Benfatto, O.Gzowski, L.Murawski, A.Bianconi: in "EXAFS and Near Edge Structure" Ed. by A.Bianconi et al., Springer Series in Chemical Physics 27, 162 (1983)
9 J.Wong, R.P.Messner, D.H.Maylotte and F.W.Lytle: as in Ref.8 pag.130 (1983)

Part VII **Geology and Geochemistry and High Pressure**

Application of EXAFS and XANES Spectroscopy to Problems in Mineralogy and Geochemistry

G.A. Waychunas

Center for Materials Research, Stanford University, Stanford, CA 94305, USA

G.E. Brown, Jr.

Geology Department, Stanford University, Stanford, CA 94305, USA

1. Introduction

Mineralogy and geochemistry have long used conventional diffraction and spectroscopic methods to structurally characterize natural crystalline phases [1] and silicate glasses [2]. However, the relatively new spectroscopic methods made available by high intensity synchrotron sources have been utilized only recently by a few geochemists and mineralogists [3,4]. In particular, EXAFS and XANES spectroscopies are capable of yielding unique structural information about most elements of geochemical importance (most of the periodic table), whether they occur as trace (0.1 Wt.%), minor (0.1-1.0 Wt.%), or major elements in crystalline or amorphous solids, aqueous electrolyte solutions, or organic matter such as coal. This paper is a brief survey of EXAFS and XANES results we have obtained recently on a variety of earth materials intended to illustrate the utility of these methods in addressing problems of geologic importance. The topics considered include: Structural characterization of local environments of Al and Na in glasses in the system $Na_2O-Al_2O_3-SiO_2$; the study of Fe site partitioning and Fe/Mg clustering in minerals; the characterization of transition metal complexes (Fe and Zn) in aqueous chloride solutions; and the study of structural changes accompanying amorphitization of mineral phases such as zircon ($ZrSiO_4$).

Our EXAFS and XANES work on Al and Na was carried out on SSRL branch line III-3 (JUMBO) using polished glass samples and freshly cleaved crystalline samples under vacuums of at least 10^{-9} torr. Our other studies were accomplished on SSRL beam lines I-5 and VII-3 using both transmission and fluorescence detection techniques.

2. Crystalline and Amorphous Aluminosilicates

A key geochemical problem is the delineation of structure-property-composition relationships of aluminosilicate melts to the extent that physical properties of melts can be structurally understood. As an example, the transport of silicate melts in the earth's upper mantle is critically dependent on the variation of viscosity with temperature (T), pressure (P), and composition (X). Figure 1 shows the well-known variation of viscosity with composition (at constant T and P) for melts in the $Na_2O-Al_2O_3-SiO_2$ system at three fixed SiO_2 concentrations [5]. As Na_2O is replaced by Al_2O_3, viscosity rises steeply to a maximum occurring near Al_2O_3/Na_2O (=R) of 1.0, then drops with further increase in R. The conventional explanation for these variations is incorporation of Al into IV-coordinated network former sites in the melt up to R=1.0, thus increasing tetrahedral polymerization and resistance to flow; Al in excess of Na has been proposed to enter the melt as VI-coordinated cations, which would act as network modifiers and reduce viscosity [6]. In order to shed light on the structural role of Al in glasses and melts within this system, we have carried out Al K-edge XANES and EXAFS using the total electron yield technique and quartz (1010) monochrom-

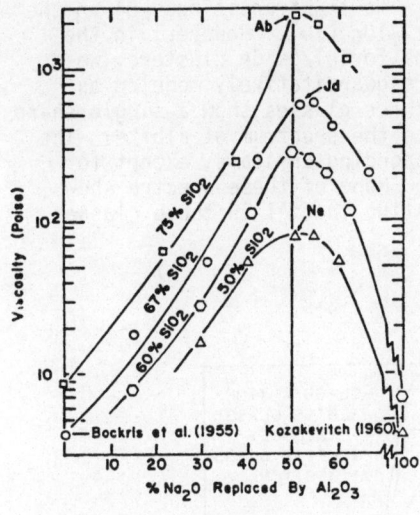

Fig.1. Viscosity of melts in the system $Na_2O-Al_2O_3-SiO_2$

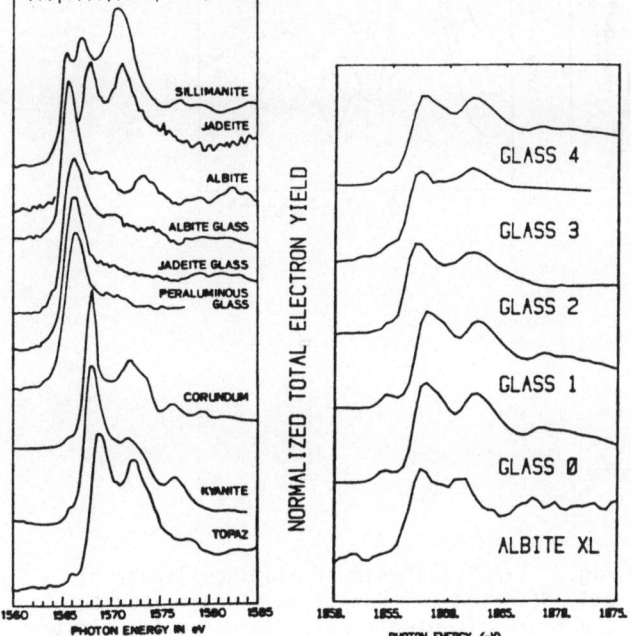

Fig.2. K-edge XANES of Al and Na model compounds and glasses

ator crystals [7]. Crystalline albite ($NaAlSi_3O_8$) served as a model compound for Al^{IV} and exhibits a single, strong white line centered at 1566 eV (Fig.2). Crystalline jadeite ($NaAlSi_3O_6$), Corundum (α-Al_2O_3), kyanite (Al_2SiO_5), and topaz ($Al_2SiO_4F_2$) were used as model compounds for Al^{VI} and exhibit strong edge features at 1568 and 1572 eV. The mineral sillimanite (Al_2SiO_4), with equal amounts of Al^{IV} and Al^{VI} displays all three features in its K-edge spectrum at the same positions as the compounds noted. The white line at 1566 eV in albite and sillimanite is assigned to Al^{IV} and is interpreted as a bound state transition $1a_1$ to t_2 [8,9]. The lines at 1568 and 1572 eV in the XANES of the Al^{VI} compounds and sillimanite are believed to be characteristic of

337

Al^{VI} and have been tentatively assigned to la_{1g} to t_{1u} transitions [8] on the basis of $X\alpha$-scattered wave MO calculations for AlO_6 [10]. However, in the absence of multiple scattered wave calculations for Al-oxide clusters, we cannot rule out that these features can be more quantitatively modeled as shape resonances. Al K-XANES spectra of the three glasses show a single sharp feature at 1566 eV and are generally similar to the spectrum of albite. The glass compositions are identical to the corresponding crystals, except for the peraluminous glass 4 ($Na_{0.75}Al_{1.2}Si_{2.90}O_8$). None of these spectra shows resolvable features at 1568 or 1572 eV, indicating that Al in these glasses is predominantly IV-coordinated by oxygen.

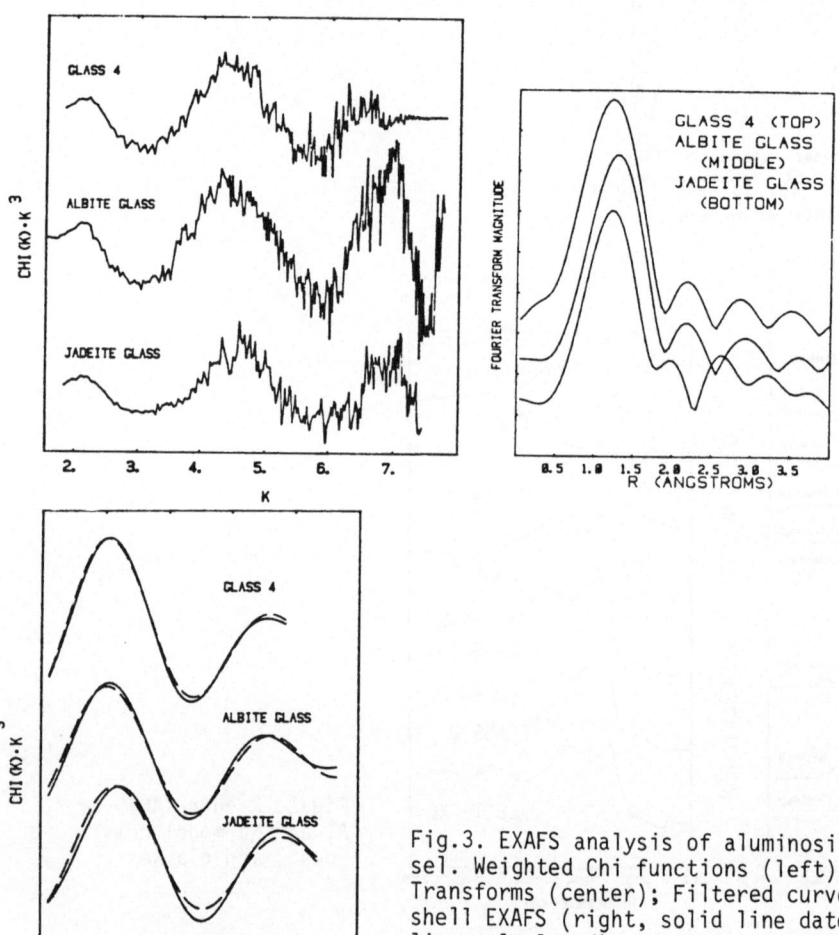

Fig.3. EXAFS analysis of aluminosilicate glassel. Weighted Chi functions (left); Fourier Transforms (center); Filtered curve-fit first shell EXAFS (right, solid line date, dashed line calculated)

This conclusion is supported by analysis of the Al K-edge EXAFS spectra of these glasses [11] shown in figure 3. For jadeite glass the EXAFS fits gave 4.9(1) neighbors at 1.70(1) A with a Debye-Waller factor of 0.005 A^2 and a Chisquare value of 0.127. For albite glass and glass 4 these figures were 1.0(1), 1.74(2) A, 0.031 A^2, 0.023 and 1.3(1), 1.73(1) A, 0.037 A^2, 0.015 respectively. We conclude that no significant changes in Al coordination occur in these melts as Al_2O_3 replaces Na_2O at 75 Wt.% SiO_2. Thus, changes in Al coordination can be ruled out as the primary cause of viscosity changes.

We have also carried out Na K-edge XANES and EXAFS studies on these same glasses to investigate the possible role of Na coordination in controlling physical properties [12]. Figure 2 shows Na XANES of crystalline albite and a series of sodium aluminosilicate glasses (0 to 4) with R values of 0.03, 0.28, 1.0 (albite glass= glass 2), 1.26 and 1.61, respectively. All have 75 Wt.% SiO_2. Little changes in the Na XANES are observed, suggesting that the Na environments in these glasses is approximately constant. This interpretation is supported by results from Na K-edge EXAFS of the same glasses [13]. We conclude that variations in Na local environments in these glasses is small and is not the primary cause of viscosity variations.

3. Fe Site Characterization in Oxides and Silicates

Fe K-edge XANES and EXAFS have been used to characterize Fe sites in silicate and oxide minerals where investigation by alternative methods is difficult. Such work yields information on site geometry and intrasite partitioning, the latter having application to geothermometry. An example of this work is the characterization of the 0.3 Wt.% Fe in a plagioclase feldspar $(Ca_{0.7}Na_{0.3})(Al_{1.7}Si_{2.3})O_8$. The data are shown in Fig. 4 as collected with a fluorescence detection system. The Fourier transform shows mainly one large asymmetric first shell peak at 1.5 A (uncorrected for phase shift). Curve fitting analysis of the back transform of this peak indicates that most of the Fe is Fe^{3+} on the tetrahedral site with bond length 1.85(1) A, with a lesser amount of Fe^{2+} divided between the tetrahedral and large calcium-sodium site (bond lengths 1.94(1) and 2.38(2) A respectively).

Clustering of cations may also be determined. An example is afforded by Fe^{2+} and Mg^{2+} rocksalt structure oxide solid solutions. Fe^{2+} and Mg^{2+} have backscattering phases differing by approximately pi radians (Fig. 5). Hence,

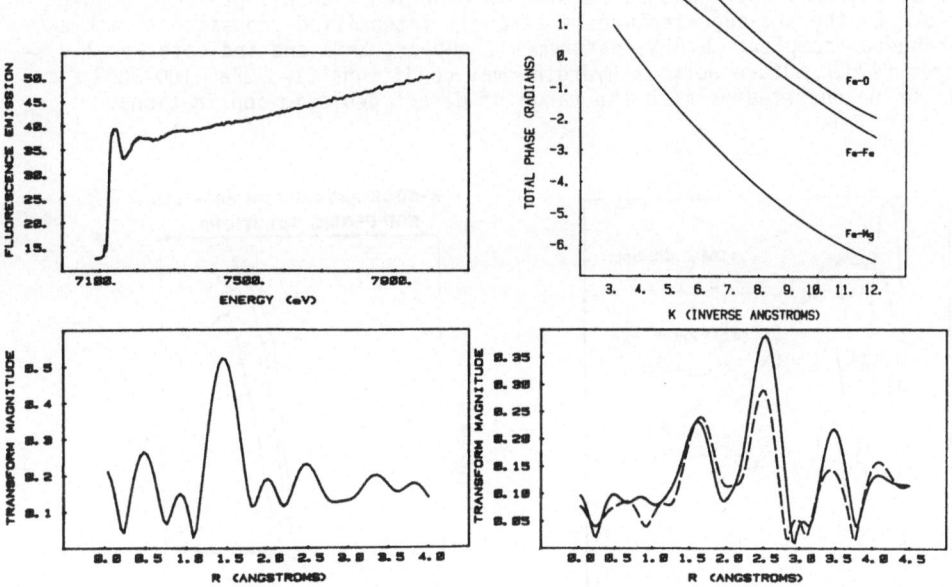

Fig.4. EXAFS spectrum of 0.3 Wt.% Fe in Lake Co., Oregon Plagioclase (top) Fourier transform (bottom)

Fig.5. Backscattered phase for Fe absorber with O, Fe and Mg backscatterers (top); Fourier transforms of magnesiowustite (solid line $Mg_{0.9}Fe_{0.1}$ dashed line $Mg_{0.7}Fe_{0.3}$)

mixing of these cations in a single shell will result in partial destructive interference of the backscattered photoelectric wave at the probe Fe atom, making the observed amplitude of the Fe,Mg shell peaks (at 2.5 and 3.5 A in Fig. 5) sensitive to the compositions of these shells. In Fig. 5 , the Fe,Mg shell peaks are considerably larger in $Mg_{0.9}Fe_{0.1}O$ over $Mg_{0.7}Fe_{0.3}O$ since most second and fourth shell neighbors are Mg. This is consistent with random shell occupation. If clustering were occurring, most of the second and fourth shell neighbors would be Fe, and little variation in the peak amplitudes would be observed. Similar analysis has indicated Ni^{2+} clustering in layer silicates [14], and , in principle, Al/Si ordering is measurable.

4. Zinc and Iron Chloride Complexes in Aqueous Solutions

The aqueous complexes responsible for the transport and precipitation of ore minerals have been little studied due to the low concentrations involved (typically millimolar). EXAFS and XANES are ideal for such studies, if fluorescence detection can be utilized. Here we report on solutions of two basic types, $ZnCl_2$, where a change in complexing occurs at low Zn and Cl molarity, and $Fe^{3+}Cl_3$, where complexing changes at high Cl molarity.

Representative Chi functions for two $ZnCl_2$ solutions and anhydrous solid $ZnCl_2$ are shown in Fig. 6. Curve fitting analysis indicates only 4.7 Cl first neighbors in the 4.5 M solution, but a mixed shell with 5.2 water molecules at 1.0 M. At lower molarity, the number of water molecules appears to increase, to 6.8 at 0.1 M and 7.6 at 0.001 M. Increase in the Cl:Zn ratio tended to decrease the number of water molecules at any $ZnCl_2$ concentration. Raman [15] and Acoustic velocity [16] measurements are consistent with these findings.

The XANES for representative $FeCl_3$ solutions are shown in Fig. 7. At 1.0 M Fe^{3+} the formation of the tetrachloro complex is controlled by chlorinity. At 7.8M Cl, only the hexaaquo complex is observed with nil pre-edge absorption. At 15 M Cl the pre-edge feature is strongly intensified, consistent with a tetrahedral complex. EXAFS measurements support this and indicate Fe-Cl distances [17]. Future work at hydrothermal conditions (1-3 GPa, 100-300° C) will bring our studies into the range of direct geologic applications.

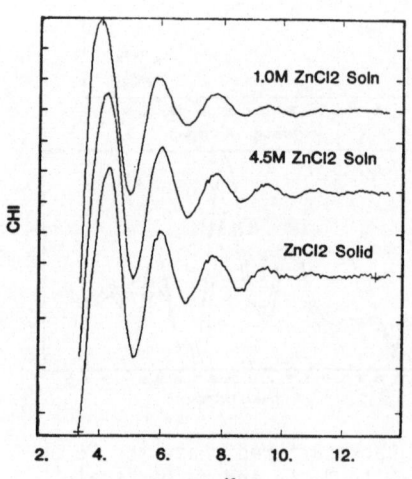

Fig.6. Chi functions of $ZnCl_2$ solutions and $ZnCl_2$ model compound

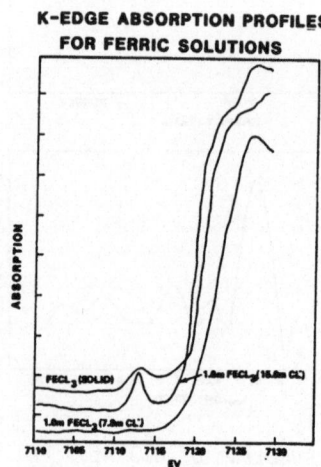

K-EDGE ABSORPTION PROFILES FOR FERRIC SOLUTIONS

Fig.7. K-edge XANES of $FeCl_3$ solutions and $FeCl_3$ model compound

5. Zr Environment in Zircon and Sodium Zirconium Silicate Glass

We have carried out an EXAFS study of the local environment of Zr in a sodium silicate glass (17.3 Na_2O-17.3 ZrO_2-65.3 SiO_2: mole%) and in the mineral zircon ($ZrSiO_4$) [18]. Chi functions are shown in Fig. 8 and the Fourier transforms in Fig. 9. The structure function for zircon shows multiple shells out to 8 A due to Zr-O, Zr-Si and Zr-Zr correlations, whereas that for the glass shows only first shell Zr-O and second shell Zr-Si correlations. Curve fitting of the filtered backtransform of the Zr-O peak in the glass gives 5.3 oxygens at 2.10 A, using zircon as a model compound. The Zr-Si distance is about 3.2 A. The loss of correlations other than first shell Zr-O and second shell Zr-Si in going from crystal to glass Zr silicate is consistent with results for crystalline and metamict Ti-Nb-Ta oxides where Ti EXAFS analysis of the metamict phase showed only first shell Ti-O and second shell Ti-Ti correlations are preserved [19].

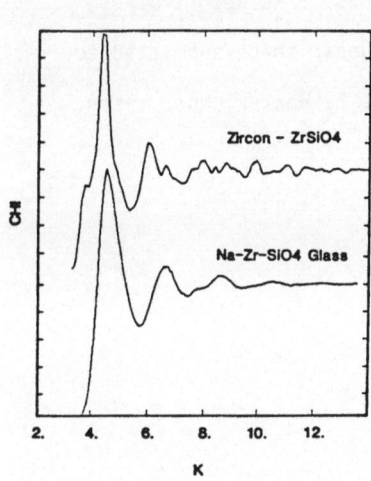

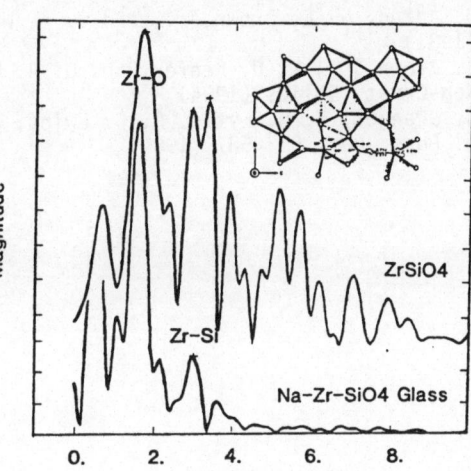

Fig.8. Chi functions of Zircon (top) and Na Zr Silicate glass (bottom)

Fig.9. Fourier transforms of Zircon and Na Zr Silicate glass

Acknowledgements

This work was supported by NSF grant EAR 8016911 and by the NSF-MRL program through the Center for Materials Research at Stanford. We thank D. McKeown for some of the data analysis and M. Rowen and P. Pianetta for help with the JUMBO line at SSRL.

References

1   A. S. Marfunin: Spectroscopy, Luminescence and Radiation Centers in Minerals (Springer-Verlag, New York 1979)
2   G. Calas and J. Petiau: Bull. Mineral. 106, 33 (1983)
3   G. A. Waychunas, M. J. Apted and G. E. Brown,Jr.: Phys. Chem. Mineral. 10, 1 (1983)
4   G. Calas, P. Levitz, J. Petiau, P. Bondot, and G. Loupias: Rev. Phys. Appl. 15, 1161 (1980)
5   K. Hunold and R. Bruckner: Glastechn. Ber. 53, 149 (1980)
6   D. E. Day and G. E. Rindone: J. Amer. Ceram. Soc. 45, 579 (1962)

7   G. E. Brown,Jr., F. D. Dikmen, and G. A. Waychunas: SSRL Report 83/01, VII-146 (1983)
8   G. E. Brown,Jr., G. A. Waychunas and F. D. Dikmen: to be submitted to Sol. State Comm. (1984)
9   J. A. Tossell: J. Amer. Chem. Soc. 97, 4840 (1975)
10 J. A. Tossell: J. Phys. Chem. Solids 36, 1273 (1975)
11 D. A. McKeown, G. A. Waychunas, and G. E. Brown,Jr.: to be submitted to J. Non-Cryst. Solids (1984)
12 D. A. McKeown, G. A. Waychunas, and G. E. Brown,Jr.: to be submitted to J. Non-Cryst. Solids (1984)
13 D. A. McKeown, G. A. Waychunas, and G. E. Brown,Jr.: this volume
14 A. Manceau, G. Calas and J. Petiau: this volume
15 M. P. Fontana, G. Maisano, P. Migliardo, and F. Wanderlingh: J. Chem. Phys. 69, 676 (1978)
16 R. Carplo, M. Mehiche, F. Borsay, C. Petrovic, and E. Yeager: J. Phys. Chem. 86, 4980 (1982)
17 M. J. Apted, G. E. Brown,Jr., and G. A. Waychunas: SSRL Report 81/03, 21 (1981)
18 G. E. Brown,Jr., K. D. Keefer, and G. A. Waychunas: to be submitted to J. Non-Cryst. Solids (1984)
19 R. B. Greegor, F. W. Lytle, R. C. Ewing, and R. F. Haaker: Nuc. Instr. Meth. Phys. Res. B1, 587 (1984)

EXAFS/XANES Studies of Metamict Materials

R.B. Greegor and F.W. Lytle

The Boeing Company, P.O. Box 3999 2T-05, Seattle, WA 98124, USA

R.C. Ewing and R.F. Haaker

Department of Geology, University of New Mexico, Albuquerque, NM 87131, USA

1. Introduction

Radiation damage effects and the kinetics of annealing in crystalline materials are important in the evaluation of the long-term stability of crystalline radioactive waste forms. The selection of a radioactive waste form which
will maintain its chemical and physical integrity for periods of 10^4-10^6 years
requires a thorough understanding of the structural and bonding controls on
radiation damage, as well as the thermal conditions under which the alpha,
alpha-recoil and fission fragment damage will be annealed. One approach in
evaluating these effects is to study the transition from the crystalline to
the metamict state in naturally-occurring minerals which contain uranium,
thorium and their daughter products. Metamict minerals are a special class
of "amorphous" materials which were initially crystalline, but due to damage
caused by alpha particles and their recoil nuclei have become "X-ray diffraction amorphous." Recent simulations of the metamictization process have included a variety of techniques but these simulation experiments generally
have not been validated by comparative studies of the properties of irradiated
materials with those observed in natural metamict minerals of great age. This
paper summarizes the results of EXAFS and XANES spectroscopy on the evaluation
of the coordination geometries of Ti [1] and Ca [2] in selected natural
metamict minerals.

2. Samples and Experimental Technique

The metamict Ti-Nb-Ta oxides of the type formula $A_x B_y O_z$ (A = lanthanides,
Fe^{2+}, Mn, Na, Ca, Th, U, Pb; B = Ti, Nb, Ta, Fe^{3+}) include three structure
types investigated here whose ideal formulas are represented by euxenite,
$YNbTiO_6$, aeschynite, $CeNbTiO_6$ and zirconolite, $CaZrTi_2O_7$. The nomenclature
and crystal chemistry for these minerals are summarized by EWING and
CHAKOUMAKOS [3]. In the high temperature euxenite structure (Pcan), B-site
cations have six-fold coordination; and A-site cations, eight-fold coordination. The B-site octahedra are slightly distorted and connected along two
edges into zig-zag chains. These chains are arranged into two-dimensional
layers that are connected by single layers of A-site cations. Blomstrandine
is a low temperature priorite - aeschynite (Pbnm) structure type, where A and
B site cations have eight and six fold coordination respectively, but B site
octahedra are joined in pairs along edges, and each pair is connected to another pair of B-site octahedra at the apices and thus form zig-zag three-
dimensional chains. Sample R25 (euxenite) was annealed to the high-temperature
structure type (1100°C for 60 h), sample R13 (blomstrandine) to the low-
temperature structure type (510°C for 52 h). For comparison, synthetic standards of the euxenite structure type ($YNbTiO_6$) and aeschynite structure type
($LaNbTiO_6$) were prepared. Zirconolite (C2/c, Z = 8) is a fluorite-related
superstructure [4] similar to pyrochlore. The structure has sheets of
$Ti(Zr)O_6$ octahedra sharing corners in three- and six-membered ring arrange-

343

ments. These sheets are interleaved by planes containing Ca and Zr(Ti) atoms ordered into alternating rows with Ca coordinated by eight oxygens and Zr(Ti) coordinated by seven oxygens. A Ti atom close to the center of a six-membered ring arrangement occupies one of a pair of sites which is five-fold coordinated by O. For comparison both metamict and annealed (1100°C for 2 hours) zircono-lite were examined as well as several Ca standard reference compounds.

The EXAFS/XANES experiments were performed at the Stanford Synchrotron Radiation Laboratory (SSRL). At SSRL the side station of Beam Line VII (a Wiggler line) was used, with the synchrotron beam at a current of approxi-mately 60 mA and an energy of 3 GeV. A Si(220) double crystal monochromator was used for the Ti K-edges and a Si(111) for the Ca K-edges. The angle be-tween the crystal faces was adjusted to detune the diffracted beam by 50% in order to reduce the harmonic content of the beam in the sample chamber. The monochromator was moved in 0.25 eV steps in the vicinity of the K-edge. The data was gathered at room temperature using a fluorescent X-ray detector [5]. Nitrogen was used in the I_0 and I chambers, with He in the sample holder area to enhance transmission of the X-rays and minimize background scattering.

3. Ti K-Edge Results

Evidence of significant geometrical rearrangement due to radiation damage at the Ti site is illustrated in Figs. 1 and 2 which show the Fourier transforms of the $\chi(K)$ of the unannealed, annealed and synthetic standards for R13 and R25. The first major peaks in the transform (indicated by arrows in the figure) are due to first neighbor oxygen atoms around Ti. The peaks at low R values are an artifact of the data analysis and are not due to coordinating atoms. Although the first neighbor oxygen peaks show interesting differences, the most striking feature in the Fourier transforms is the second neighbor peak. For both the R13 and R25 sets of Fourier transforms there is a sub-stantial enhancement of the second neighbors for the standard and annealed samples as compared to the unannealed samples. This suggests that the met-amictization process has the effect of disordering the second neighbors around the Ti (i.e., the metal-metal coordinated sphere).

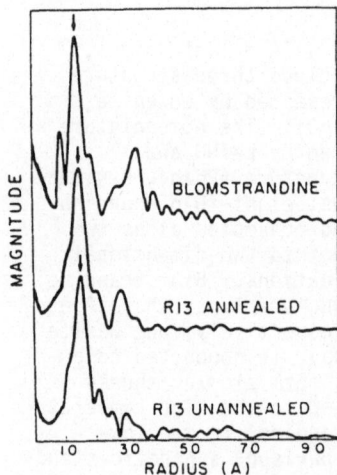

Figure 1. Fourier transforms of Ti K-edge EXAFS of synthetic blomstran-dine, R13 annealed, R13 unannealed

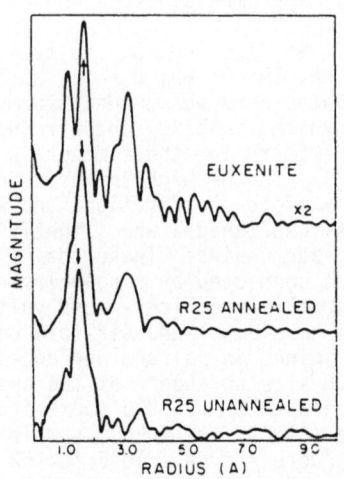

Figure 2. Fourier transforms of the Ti K-edge EXAFS of synthetic euxenite R25 annealed, and R25 unannealed

To investigate the more subtle differences in the first neighbor oxygen peaks, inverse Fourier transforms were performed on these peaks. The resulting $\chi(K)$ functions were then analyzed using iterative nonlinear least-squares curve fitting techniques. At this point in the analysis the least square fitted values are expected to have fairly large errors. This analysis only suggests trends, rather than allowing the assignment of precise values of coordination number, bond length and disorder.

The results suggest that in the metamict state (i.e., prior to annealing) the first neighbors are reduced by 10-20%, with most of the reduction being due to loss of the longer (i.e., > 2.0 Å) bonded oxygens. Also a slight contraction in the average bond length was seen in the metamict samples. The picture that emerges from this annealing study is that the effect of radiation damage at the Ti-site is to disrupt the nearby oxygen atoms increasing the asymmetry, with even greater disruption in the second neighbor metal atoms.

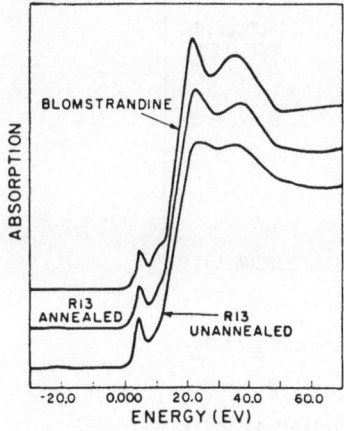

Figure 3. Ti K-edge XANES of synthetic blomstrandine, R13 annealed, and R13 unannealed

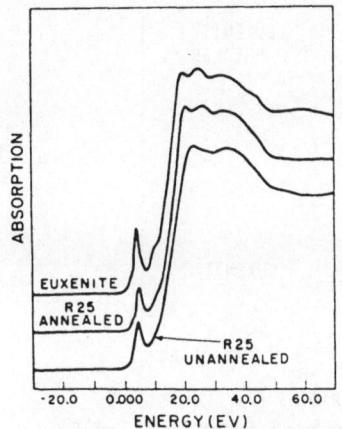

Figure 4. Ti K-edge XANES of synthetic euxenite, R25 annealed, and R25 unannealed

A comparison of the XANES of the unannealed, annealed and synthetic standards is shown in Figs. 3 and 4. Interestingly, the unannealed samples of both R13 and R25 have remarkably similar XANES. After annealing, however, the XANES of R13 and R25 show significant differences. As expected, they no longer resemble each other but instead are similar to the XANES of their crystalline counterparts. The height of the pre-edge feature is indicative of the degree of asymmetry around Ti [6]. In the unannealed R13 and R25 samples the pre-edge height is $\sim$ 15% and $\sim$ 9% higher, respectively, than the corresponding annealed samples. This provides further support that one effect of radiation damage is to increase the degree of asymmetry in the Ti-O polyhedra. In addition to site symmetry, the oxidation state of Ti can also be inferred. Examination of the edge shifts of the annealed and unannealed metamict samples indicates that, within the accuracy of the technique, all of the Ti is in a 4^+ oxidation state.

4. Ca K-Edge Results

The Ti atoms in the standard structure types are located at the center of an octahedron of oxygen atoms with the edges or apexes of the octahedra connected

to form a network whose voids are filled with other atoms such as Ca. The
Fourier transforms of the Ca K-edge EXAFS of the natural (unannealed) and
annealed samples are shown in Figs. 5 and 6. The data for euxenite sample
were not of high quality. From the transforms we see that the natural samples
have lower second coordination shell magnitudes than do their annealed counterparts. The observed change in the second coordination shell around the Ca is
consistent with the observations of the Ti site previously discussed. That
is, the second coordination sphere has undergone substantial rearrangement as
a result of the metamictization process. Our interpretation of this data
is that the change in the higher coordinating atoms is due to increased disorder as these crystalline phases proceed to a more glass-like random network.

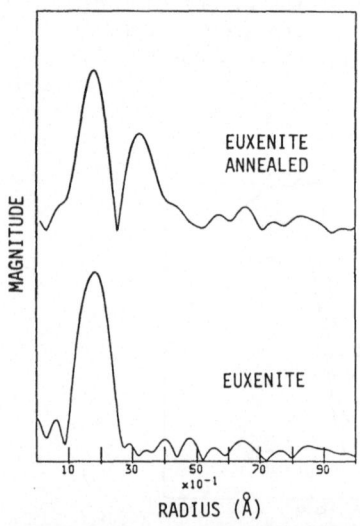 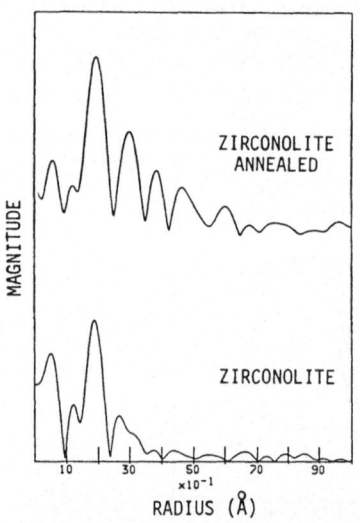

Figure 5. Fourier transforms of Ca
K-edge EXAFS of natural and annealed euxenite

Figure 6. Fourier transforms of Ca
K-edge EXAFS of natural and annealed
zirconolite

Inverse Fourier transforms were performed on the first oxygen coordinating
shells around the Ca for the zirconolite sample. The resultant spectra for
the annealed and natural samples were then compared using the ln of the ratio
technique. This analysis showed that the average 1st shell oxygen coordination number was reduced in the metamict sample so that N(Nat)/N (Ann) = .62
and the disorder $\sigma^2(Nat) - \sigma^2(Ann) = - .001 \text{ Å}^2$.

Using this result as a guide we proceeded to a nonlinear least square fit
of the inverse 1st shell spectra. The initial trial values for the fit to
the annealed data were obtained from the X-ray diffraction studies on zirconolite by GATEHOUSE [4]. The eight Ca-O distances were combined into three subshells (N_1 = 4, N_2 = 1, N_3 = 3) for the first coordinating oxygen shell around
the Ca. The EXAFS phase and scattering amplitude were obtained from a Ca-O
standard. The R values were held fixed, and N and σ allowed to vary so that
the ratio of $N_1:N_2:N_3$ was 4:1:3. The values from the fit to the annealed
spectra were then adjusted so that the sub-shell N values were 75% (using as
a guide the 62% reduction indicated by the ln of the ratio technique) and the
sub-shell R values were 98% (using the peak positions of the Fourier transforms)
of the annealed fitted values. The σ value was the only free variable in the
fitting of the metamict spectra which resulted in a good fit of the data.

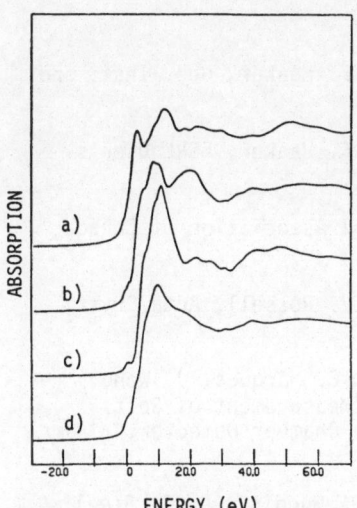

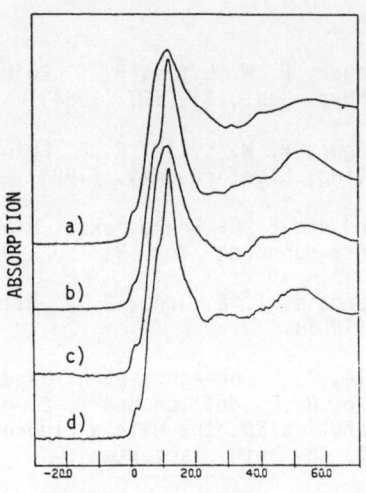

Figure 7. Ca K-edge XANES of standard reference materials; a) CaO,
b) CaCO3, c) gypsum, d) Ca acetate
[6]

Figure 8. Ca K-edge XANES of a)
zirconolite natural, b) zirconolite
annealed, c) euxenite natural and
d) euxenite annealed

Thus we see the fit results of the Ca sites are in qualitative agreement with
the Ti sites. Upon being damaged by alpha-recoil nuclei events the overall
1st shell coordination is reduced with a slight decrease in bond length of
the coordinating oxygen atoms.

The Ca K-edge XANES of several standard materials is shown in Fig. 7 and
the XANES of the natural and annealed euxenite and zirconolite samples in
Fig. 8. The natural metamict specimens show a similar signature which is com-
paratively broadened with a small shoulder on the low energy side of the first
major peak. The annealed samples all show narrower main peaks, the same
shoulder on the low energy side of the major peak and the addition of a step
on the steep increase of the major peak. The general sharpening and finer
resolution of the spectral features was also seen in the Ti edge XANES. The
leading edges of the first major peaks of both natural and annealed samples
were within $\sim$ 1 eV of each other and all were displaced $\sim$ 4 eV higher in
energy as compared to the Ca-O standard.

5. Conclusion

The nearest neighbor environment of Ti and Ca in selected metamict Nb-Ta-Ti
complex oxides have been examined in order to evaluate the effect of alpha-
recoil damage on these structures. Comparison of the EXAFS/XANES data for
metamict samples with data for annealed and crystalline samples suggests
minor changes in the first coordination sphere, Ti-O or Ca-O, but major disruption of the second coordination sphere, for the material in the metamict
state. These data suggest a mechanism for the transition from the crystalline
to the metamict state in which tilting of cation coordination polyhedra is a
possible effect of damage caused by alpha-recoil events.

Acknowledgements

This work was supported by DOE, Division of Materials Sciences, under Contract
Number DE-AC06-83ER45010. The data was collected at SSRL which is supported
by DOE, NSF and NIH.

References

1. R. B. Greegor, F. W. Lytle, R. C. Ewing and R. F. Haaker, Nuc. Inst. and Meth. in Phys. Res., B1, 587 (1984).

2. R. B. Greegor, F. W. Lytle, R. C. Ewing and R. F. Haaker, SSRL User's Group Meeting, Stanford Univ. (1983).

3. R. C. Ewing and B. C. Chakoumakos, Mineralogical Association of Canada Short Course Handbook, Vol. 8, 1982, p. 239.

4. B. M. Gatehouse, I. E. Grey, R. J. Hill and H. J. Rossell, Acta Cryst, B37, 306 (1981).

5. F. W. Lytle, R. B. Greegor, D. R. Sandstrom, E. C. Marques, J. Wong, C. L. Spiro, G. P. Huffman and F. E. Huggins, "Measurement of Soft X-ray Absorption Spectra With a Fluorescent Ion Chamber Detector," Nucl. Inst. and Meth. (accepted).

6. Ca standards courtesy of G. P. Huffman and F. E. Huggins, U. S. Steel Corporation, Monroeville, PA.

EXAFS and Near Edge Absorption in Glasses for Nuclear Waste Storage

M. Antonini
Dipartimento di Fisica and GNSM, Università di Modena, I-41100 Modena, Italy
A. Merlini
Physics Division, Joint Research Centre of the European Communities
Ispra Establishment, I-21020 Ispra, (VA), Italy
F.R. Thornley
Department of Applied Physics, University of Strathclyde
Glasgow G4, 0NG, United Kingdom

Knowledge of the local coordination and of the valence states of fission and actinide metals embedded in a glassy matrix, used for the storage of nuclear waste, is important for assessing the structural stability of the glass and the leaching rate of the metals. Results relative to technetium and uranium in borosilicate glasses are presented and discussed.

1. Introduction

Among the various radionuclides of nuclear waste incorporated in a borosilicate glass, Tc and U are of special interest, technetium is highly volatile in the high valence state (Tc^{+7}) and has high leaching and migration rates in water [1]. Uranium, on the other hand, may represent the behaviour of other toxic or radioactive actinides. At the Frascati Conference, we described some preliminary results about Tc in glasses [2]. Other results relative to hexavalent uranium in glass were recently reported by Knapp et al. [3].

In the following, we present and discuss new EXAFS and Near Edge Absorption data in Tc and U glasses.

2. Results and Discussion

Borosilicate glasses were prepared by mixing and melting at 1200°C a glass powder (58% SiO_2, 16% Al_2O_3 and 20% Na_2O) with a liquid solution of NH_4TcO_4. Two different procedures corresponding to higher and lower reducing conditions, were employed. The glasses are named RX-glass and OX-glass, respectively. The concentration of Tc in all the glass samples was lower than 1% by weight. Reference compounds were: metallic Tc, NH_4TcO_4 and TcO_2. A near edge spectrum only was obtained for TcO_2 owing to difficulties in sample preparation.

Uranium glasses contained U at two different concentrations: 2.5 and 5% by weight. They were prepared by melting the appropriate amounts of UO_2 with the same glassy matrix indicated above. Melting was under reducing conditions (argon atmosphere, graphite crucibles). Reference compounds were: U, UO_2 and U_3O_8. Near Edge Absorption and EXAFS spectra were measured at the LURE DCI Synchrotron Radiation Source (Orsay, France), the energy resolution was about 5 eV.

In Fig. 1 the normalized EXAFS spectra above the K-edge of Tc (21044 eV) in RX- and OX-glasses are compared with computed spectra of the model compounds indicated in the figures. The local structure of these systems, in addition to that of NH_4TcO_4 is shown in Fig. 2. The validity of the computed Tc model was previously tested against experimental results.

In Table 1, interatomic distances measured by X-ray diffraction of metallic Tc and TcO_2 [4,5] are compared with those obtained by Fourier transform of the EXAFS spectra of the glasses. It is concluded that the EXAFS spectrum of the RX-glass can be explained by equimolar concentrations of metallic Tc clusters and of TcO_2, while the presence of TcO_2 only is sufficient to explain most of the spectral features observed in the EXAFS pattern of the OX-glass. In Fig. 3, the results of a similar comparison between U-glasses and UO_2 is shown. The

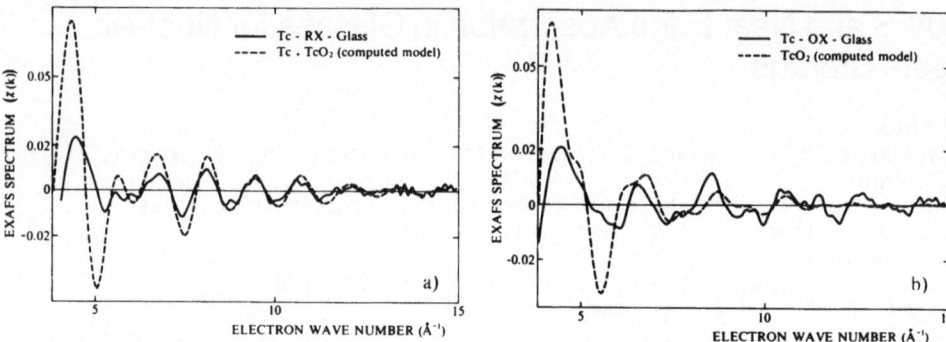

Fig. 1 a) Comparison between the experimental EXAFS spectrum of Tc in RX-glasses with a model compound consisting of equal fractions of metallic Tc (2 shells) and TcO_2 (8 shells);
b) same as a) for OX-glasses. The model compound is TcO_2 (8 shells).

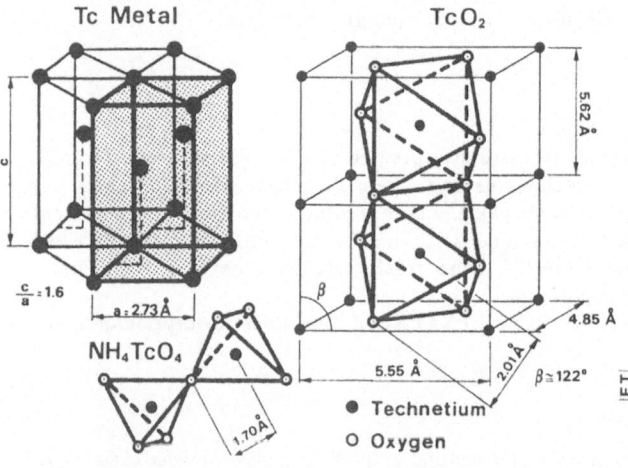

Fig. 2 Local structures around Tc in: Tc metal, TcO_2 and NH_4TcO_4.

Table 1 - Comparison between X-ray and EXAFS interatomic distances in model compounds and in Tc glasses

	Interatomic		Distances (Å)	
Model TcO_2	Compounds* Tc metal	RX-glass**	OX-glass**	
(O) 2.01		1.9 (6)	2.0 (4)	
(Tc) 2.51			2.4 (7)	
	2.73	2.80		
(Tc) 3.11				
(O) 3.21		3.0 (7)		
(O) 3.41			3.4 (5)	
(Tc) 3.69				
(O) 3.69		3.8 (0)		
	3.80	3.8 (5)		

*   from X-ray diffraction data
** from the experimental EXAFS data of the present work

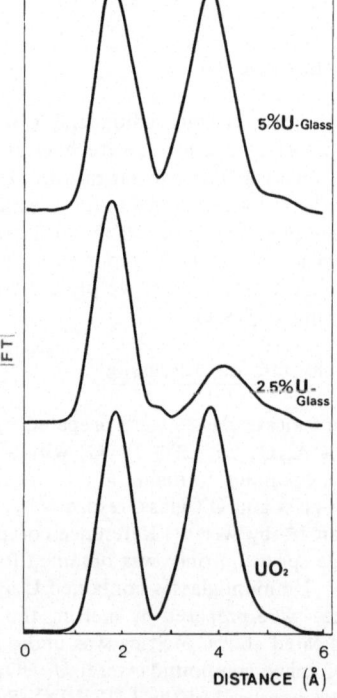

Fig. 3 Comparison among Fourier transforms of EXAFS experimental data obtained from U glasses and UO_2.

EXAFS patterns of UO_2 and U-glasses are dominated by two main frequencies which produce the peaks observable in the correspondent Fourier – transforms.Other frequencies are also present in the EXAFS spectra; their origin is still unclear. The two maxima are at almost the same distance in both glasses and in the reference oxide and can be associated with first oxygen and uranium neighbours located at 2.4 Å and 3.9 Å in the UO_2 crystal structure, respectively. A determination of the uranium atomic phase shifts and backscattering amplitude is under way. We conclude that most of the uranium incorporated in the glass has the same valence and atomic environment as UO_2, although other configurations may also be present.

Chemical shifts of the reference compounds and of the Tc glasses were derived from the maximum of the derivative curve in the near edge region. The plot of these shifts vs the oxidation number of the cation is illustrated in Fig. 4. Assuming a linear relationship between chemical shift and oxidation number, as found in other transition metal compounds [6], an average oxidation number of $\sim$ 1.1 corresponds to Tc in the RX-glass. This value also brings into agreement the results of EXAFS and Near Edge Absorption.

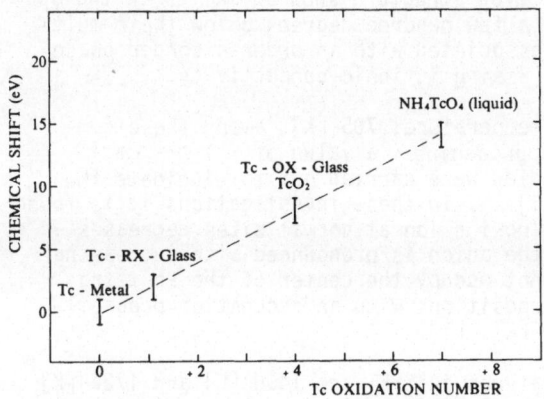

Fig. 4 Graphical representation of chemical shifts measured in Tc glasses and reference compounds.

This work was supported by the programme of Radioactive Waste Management of the Ispra Establishment of the Joint Research Centre of the European Communities.

References

1.   S. Paquette, J.A.K. Reid and E.L.J. Rosinger: "Review of technetium behaviour in relation to nuclear waste disposal", Tech. Rep. TR-25, Atomic Energy of Canada Ltd. (1980).
2.   M. Antonini, C. Caprile, A. Merlini, J. Petiau, F.R. Thornley: "EXAFS and XANES investigation of the coordination of technetium in borosilicate glass", in: Proc. Int. Conf. on EXAFS and Near Edge Structures, A. Bianconi, L. Incoccia, S. Stipcich (Eds.), Springer (1983) 261-264.
3.   G.S. Knapp, B.W. Veal, D.J. Lam, A.P. Paulikas and H.K. Pan, Materials Lett. 2, 253 (1984).
4.   R.C.L. Mooney, Acta Cryst. 1, 161 (1948).
5.   D.B. Rogers, R.D. Shannon, A.W. Sleight and J.L. Gillson, Inorg. Chem. 8, 841 (1969).
6.   P.R. Sarode, S. Ramasesha, W.H. Madhusudan and C.N. Rao, J. Phys. C 12, 2439 (1979).

EXAFS Studies on Fluorites of β-PbF$_2$ and SrF$_2$

Nagao Kamijo
Government Industrial Research Institute, Osaka., Ikeda Osaka 563, Japan
Kichiro Koto and Yoshiaki Ito
Institute of Sci. and Ind. Res., Osaka University, Suita Osaka 565, Japan
Kazuhiro Tanabe and Hikaru Terauchi
Department of Physics, Kwansei-Gakuin University, Nishinomiya 662, Japan
Hironobu Maeda
Dept. of Chem. Okayama Univ., Okayama 700, Japan
Moritaka Hida
School of Eng. Okayama University, Okayama 700, Japan

1. Introduction

Many compounds with calcium fluorite type structure show anomalies in the specific heat at elevated temperatures, a few hundred degrees below their melting temperature. The anomalies are associated with an order-disorder phase transition which is accompanied by a rising in ionic conductivity.

β-PbF$_2$ has the lowest transition temperature, 705 [K], among these compounds and high ionic conductivity approaching a value of $\simeq$ 1 Ω^{-1} cm^{-1}. Single crystal X-ray diffraction studies were carried out to elucidate the mechanism of high ionic conduction [1]. In these investigations it is found that the occupation-probability of fluorine ion at normal sites decreases and anharmonic thermal vibration of the anion is pronounced with increasing temperature. The fluorine ion does not occupy the center of the Pb tetrahedron but displaces into four split positions with an occupation-probability of 1/4 for each position at 800 [K], [1].

The transition and melting temperatures in SrF$_2$ are 1400 [K] and 1723 [K], respectively. The occupation probability of fluorine ion at normal sites decreases (ca. 20%) at high temperature, 1500 [K]. The mean square amplitude of the fluorine ion drastically increases around the transition temperature.

In the present paper we report the nature of disorder of fluorine ion in β-PbF$_2$ as a function of temperature below the transition temperature at 80 [K], 293 [K] and 640 [K], and in SrF$_2$ at 293 [K] and 640 [K] using EXAFS spectroscopy.

2. Experimental

In order to obtain the sample of β-PbF$_2$, powder of α-PbF$_2$ (99.9 wt.%) was heated up to 940 [K] in the nitrogen atmosphere and gradually cooled down to room temperature. It was confirmed to be β-PbF$_2$ by X-ray diffraction method. α-PbF$_2$ was used as the reference sample for curve-fitting analysis. Closed cycle cryocooler (CTI) was used for low temperature measurement and the Lytle type of the furnace was used for high temperature measurement. The sample was inserted between two sheets of thin mica film which was put into the cell fabricated by boron nitride for high temperature measurement.

For SrF$_2$ study crystalline powder (99.9 wt.%) was used.

The X-ray absorption measurements were made with Synchrotron radiation by use of the EXAFS facilities [2] installed at the beam line 10B of the 2.5 Gev storage ring of photon Factory in the National Laboratory for High Energy Physics (KEK. Tsukuba).

Data processing and analysis were applied to the EXAFS spectra in the same manner as in the previous study [3].

3. Results

Figure 1 (a),(b) and (c) show the EXAFS spectra of β-PbF$_2$ (a) at 80 [K], (b) at 293 [K] and (c) at 640 [K] (from now on we will abbreviate these to L-PbF$_2$, R-PbF$_2$ and H-PbF$_2$ respectively).

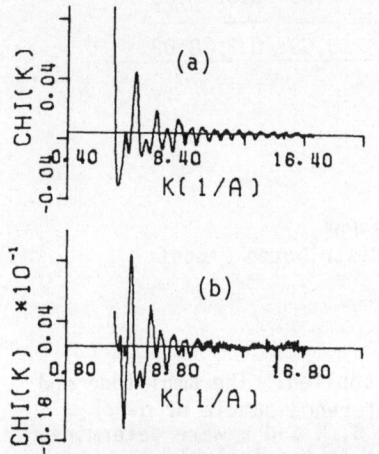

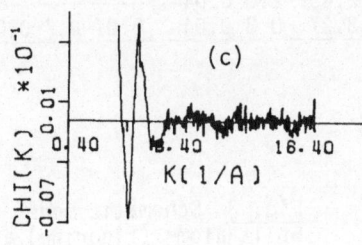

Fig. 1 EXAFS spectra χ(k) of β-PbF$_2$, (a) at 80[K], (b) at 293 [K] and (c) at 640 [K] respectively

Fourier transform (in R-space) of these spectra multiplied by k^3 provides RSFs as shown in Fig. 2, where no phase shift corrections being made. In Fig. 2 "G" indicates the ghost peaks in the Fourier transformation. Arrows indicated by solid and dotted lines in each RSF show the normal fluorine and lead ion sites in ideal fluorite at room temperature. The RSF features in L-, R- and H-PbF$_2$ clearly show that fluorine ion does not occupy the center of normal site but displaces into several separate sites. Considering the previous X-ray diffraction studies [1] and RSFs in this study, we can suggest that these sites will distribute almost along the body diagonals in the direction of Pb tetrahedral faces even at very low temperature, 80 [K]. For

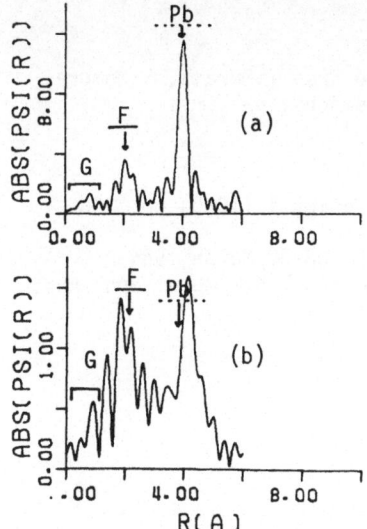

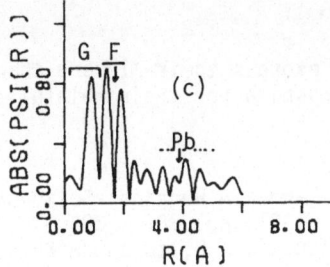

Fig. 2 Fourier transformations of EXAFS spectra multiplied by k^3 of β-PbF$_2$, (a) for L-PbF$_2$ (k = 3.0 ~ 18.0 A^{-1}), (b) for R-PbF$_2$ (k = 3.0 ~ 14.0 A^{-1}) and (c) for H-PbF$_2$ (k = 3.0 ~ 13.0 A^{-1})

353

Table 1 Atomic distances (R), mean square displacement of R: (σ^2), and coordination numbers (N) of β-PbF$_2$

number	schell	80 [K]			293 [K]			640 [K]		
		R(A)	N	σ	R(A)	N	σ	R(A)	N	σ
(1)	Pb-F(1)	2.06	0.2	0.04	2.06	0.3	0.01	2.02	0.2	0.02
(2)	Pb-F(2)	2.53	2.1	0.08	2.52	1.1	0.03	2.48	0.5	0.06
(3)	Pb-F(3)	2.63	2.2	0.06	2.65	1.5	0.04	2.59	0.5	0.06
(4)	Pb-F(4)	2.83	2.0	0.04	──	──	──	──	──	──
(5)	Pb-F(5)	3.27	0.8	0.04	3.07	0.7	0.04	3.00	0.3	0.02

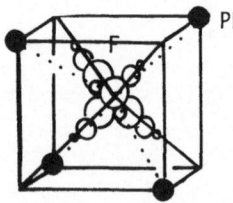

Fig. 3 Schematic model of β-PbF$_2$.
Split atoms (fluorine) are distributed almost along <111>

each RSF of β-PbF$_2$, curve-fitting technique was applied. The amplitude and phase shift parameters were obtained from the reference sample of α-PbF$_2$. Using these parameters the structure parameters, R, N and σ were determined by the least-squares parameter fitting. They are listed in Table 1. From the peak positions in Table 1 a schematic anion disorder model for β-PbF$_2$ is obtained as shown in Fig. 3. We call it "split atom model". The anion occupancy probability ratio of the interstitial sites increases as temperature increases. These anions will play a key role in ion conduction process. It should be noted that the coordination numbers of fluorine ions decreases as temperature increases as in Table 1. This result suggests that many anion defects grow in the crystal as temperature increases.

Similar features of RSFs are observed in SrF$_2$. Process of anion conduction will be similar to β-PbF$_2$.

4. Acknowledgement

The authors wish to express their sincere thanks to Drs. T.Murata, S.Nomura, H.Oyanagi and T.Matsushita for their helpful suggestions.

References

1. K.Koto, H.Schulz and R.A.Huggins: Solid State Ionics 1 (1980) pp. 355 ∿ 365, ibid., 3/4 (1981) pp. 381 ∿ 384
2. H.Oyanagi, T.Matsushita, M.Ito and H.Kuroda: KEK 83-30 March 1984 P
3. H.Maeda, H.Terauchi, K.Tanabe, N.Kamijo, M.Hida and H.Kawamura: Jpn. App. Phys., 31 (1982) pp. 1342 ∿ 1346

EXAFS in Mixed Defect Chalcopyrite $ZnGa_2(S_xSe_{1-x})_4$

P.P. Lottici, C. Razzetti, G. Antonioli, and A. Parisini

Dipartimento di Fisica dell'Università and GNSM-CNR, Via D'Azeglio 85
I-43100 Parma, Italy

1. Introduction

Defect chalcopyrites of the $A^{II}B_2^{III}X_4^{VI}$ family (tetragonal $I\bar{4}$ space group) have a structure derived from that of chalcopyrite by the inclusion of an ordered array of vacancies. Each cation is surrounded by four anions at the corners of a regular tetrahedron whereas each anion has one A atom, two B atoms and a vacancy as first neighbours [1]. The position of the anion in the unit cell is close to the zincblende ideal site, but the anisotropic local structure displaces it by an amount described by three "free" parameters δ_1, δ_2, δ_3, whose determination has been reported only for $CdGa_2S_4$ crystals. In defect chalcopyrites with small tetragonal distorsion (i.e. a c/a ratio close to 2) like $ZnGa_2S_4$ and $ZnGa_2Se_4$, one expects very small δ_i parameters and ion distances close to the zincblende ideal case. The study of the local distortion was the main purpose of this EXAFS work .

The mixed phases obtained by cation or anion cross substitution are of particular interest due to the "tailoring" possibilities in the physical properties [2]. As measured by X-ray diffraction, the lattice constants in the mixed phase $ZnGa_2(S_xSe_{1-x})_4$ vary with composition x in a nearly linear way between the values of the end members. As already found in pseudobinary mixed compounds [3], the NN distances remain approximately constant with x. The second main task of this EXAFS work was to check if this behaviour is presented also by pseudoternary anion mixed compounds.

2. Experimental

The EXAFS measurements at the Zn,Ga and Se K-edges were made in transmission geometry at the PULS facility , Frascati, Italy, at RT and LNT. The materials ($ZnGa_2(S_xSe_{1-x})_4$ x=0,0.25,0.5,0.75, 1) were grown by the I_2-CVD technique [4] at the MASPEC-CNR Institute, Parma, Italy. The crystalline samples were powdered and deposited from a water dispersion on Millipore membranes.

3. EXAFS Data Analysis and Discussion

The experimental data (see Fig.1) have been analyzed by the usual procedure (preedge background subtraction, polinomial fit to the atomic -like background above the edge and Fourier Transform) [5] . The threshold energy E_o has been taken, in all cases, at the inflection point of the absorption coefficient. The contribution of the main peak in the FT was filtered and back-transformed to extract BFT $k\chi(k)$, total phase and amplitude functions.

355

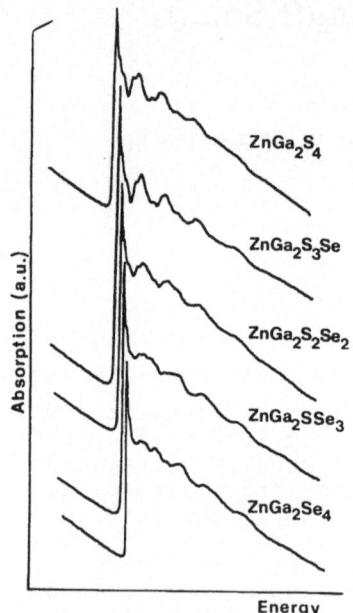

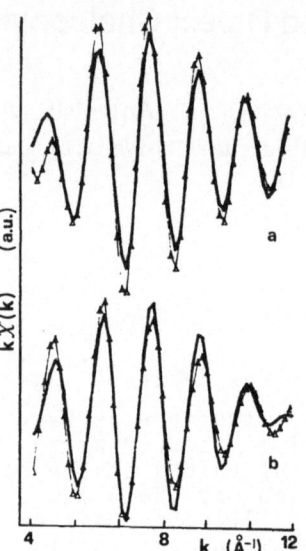

Fig.1 - X-Ray absorption of mixed $ZnGa_2(S_xSe_{1-x})_4$ at the Ga K-edge.

Fig.2 - BFT $k\chi(k)$ (solid line) and two-shells fitting of:
(a) $ZnGa_2SSe_3$, Ga-edge, RT
   $d(Ga-S)=2.23$ Å, $\sigma_2^2=0.007$ Å^2
   $d(Ga-Se)=2.39$ Å, $\sigma^2=0.0065$ Å^2
(b) $ZnGa_2S_3Se$, Se-edge, LNT
   $d(Se-Zn)=2.40$ Å, $\sigma_2^2=0.005$ Å^2
   $d(Se-Ga)=2.36$ Å, $\sigma^2=0.005$ Å^2

Due to the very low tetragonal compression ($ZnGa_2S_4$: a=5.28 Å, c=10.50 Å, c/a=1.989; $ZnGa_2Se_4$: a=5.51 Å, c=11.01 Å, c/a=1.998), the two different distances between gallium atoms and the chalcogen atom cannot be resolved by EXAFS analysis. In the case of a single NN shell ($ZnGa_2S_4$ and $ZnGa_2Se_4$ at the Zn and Ga K-edges) the distances were determined by a linear fitting vs 2k of the difference between the experimental BFT total phase and theoretical Lee and Beni's phases [6] ."Standard" compounds (ZnSe,ZnS, GaS,GaSe) were used as a test and the agreement was good.
In the mixed compounds or at the Se-edge, a least squares fitting of the BFT $k\chi(k)$ by superposition of two gaussian shells was used. Two examples of this fitting procedure are shown in Fig.2.

The distances determined in this way are reported in Fig.3 for all the investigated compositions. The values obtained for cation - Se distances are, within the experimental error 0.02 Å, equal to those determined by the analysis of the EXAFS at the Se K-edge, indicating that our results are correct. In Fig.3 is also reported, for a comparison, the ideal zincblende NN distance $a(x)\sqrt{3}/4$ as calculated from X-Ray measurements.
Our results confirm that the NN distances are conserved in the mixed phase, with a slight tendency to follow a VCA-like behaviour.

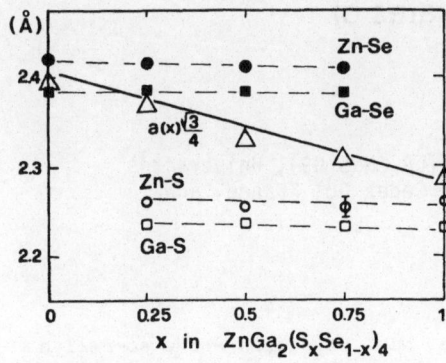

Fig.3 - NN distances
as determined from
EXAFS data at LNT
by the BFT k χ(k)
fitting method.
Triangles represent
ideal zincblende NN
distance as measured
by X-Ray diffraction.
Solid and dashed li-
nes are only visual
aids.

The curve fitting procedure has been used,in the hypothesis
that the chemical composition indicates the actual coordination
numbers of the cations, in spite of a random distribution of
S and Se atoms on the anionic sublattice and then different
cationic environments with different probabilities. Whereas
the NN distances as seen by EXAFS are an average over all configurations, we have assumed the most probable situation.The
Debye-Waller like parameters are not too high (0.008-0.009 Å^2
at RT and 0.006 Å^2 at LNT) and are evidence of a relatively
low structural disorder induced by the anion mixing or by some
cation antisite disorder [2]. Calculations on these subjects
are in progress, as well as a study of the deviation from a
linear $2k^2$ fitting of the log of the ratio of EXAFS amplitudes
at different temperatures, to ascertain possible asymmetries of
the pair distribution functions.

4. References

1 - C.Razzetti,P.P.Lottici and L.Zanotti, Mat.Chem.Phys.(in pr.)
2 - P.P.Lottici and C.Razzetti, J.Mol.Structure 115,133 (1984)
3 - J.C.Mikkelsen and J.B.Boyce, Phys.Rev.B 28,7130 (1983)
4 - C.Paorici,L.Zanotti and G.Zuccalli, J.Cr.Growth 43,705(1978)
5 - P.A.Lee,P.H.Citrin,P.Eisenberger and B.M.Kincaid, Rev.Mod.
    Phys. 53,769 (1981)
6 - B.K.Teo and P.A.Lee, J.of Am.Chem.Soc. 101,2815 (1979)

Cation Ordering in Ni-Mg Phyllosilicates of Geological Interest

A. Manceau, G. Calas, and J. Petiau[1]

Laboratoire de Mineralogie-Cristallographie (LA CNRS 09), Universités Paris 6 et 7, 4 place Jussieu, F-75230 Paris Cedex 05, France, and LURE, Orsay, France

Nickel containing phyllosilicates from New Caledonia - the so-called "garnierites" - consist of a stacking of layers built by tetrahedral and octahedral sheets occupied by various metallic atoms. Ore phyllosilicates belonging to the lizardite-nepouite (serpentine) and the kerolite-pimelite (talc) series have been investigated in order to study the mechanism of Ni-Mg substitution. The local order around Ni has been studied by optical spectroscopy and X-ray absorption spectroscopy using synchrotron radiation DCI/LURE, France. Two types of information are obtained.

I SHORT RANGE ORDER

Absorption K-edges of nickel in the studied clay minerals have been compared with two reference samples, a 0.5M aqueous solution of nickel sulfate and LaNiO3:Ni(III) [1] . Based on the observed dependence of the edge energy on oxidation state and coordination number in Fe-silicates and a comparable trend in Ni-silicates, the position of the inflexion point suggests that Ni is divalent in these hydrous silicates. A splitting of the maximum is observed in these latter phases and has been interpreted as resulting from a site distortion by comparison with Fe-bearing minerals [2,3] . It is to be pointed out that in Ni hexahydrate, where the Ni coordination is a regular octahedron, the K-edge does not exhibit such a splitting. This finding is confirmed by visible-near infrared spectroscopy [4] . Spectra of Ni phyllosilicates show three main absorption bands which are characteristic of the presence of 6-fold Ni(II). Comparison with the Ni hexahydrate spectrum reveals that cubic field degeneracy is removed : divalent Ni atoms are 6-fold coordinated in slightly distorted sites. The VI coordination is also confirmed by EXAFS data relative to the first atomic shell as the mean Ni-O distance is equal to $2.07\text{Å} + 0.01\text{Å}$ for all samples of the two series. Finally, all these results are consistent with the fact that Ni is substituted to Mg into the octahedral sheet of these phyllosilicates.

II MIDDLE RANGE ORDER

The EXAFS relative to the second atomic shell permits discussion of the Mg-Ni cation ordering, due to the backscattering amplitude and phase shift differences among these elements. The magnitude of the Fourier transform relative to minerals belonging to the two series under investigation are superimposed in Fig.1 with decreasing Ni content. The second peak of the Fourier transform has a similar shape and amplitude for all samples of the kerolite-pimelite series although the Ni content ranges from 38.9%(wt.) NiO (Ni/Ni+Mg=0.87) to 7.3%(wt.) NiO (Ni/Ni+Mg=0.23). In contrast, in the lizardite-nepouite series, the functions of partial radial distribution are varying. The amplitude of the peak beyond the first oxygen one continuously diminishes with decreasing Ni content.

---

(1) and Ecole Normale Supérieure, Montrouge, France.

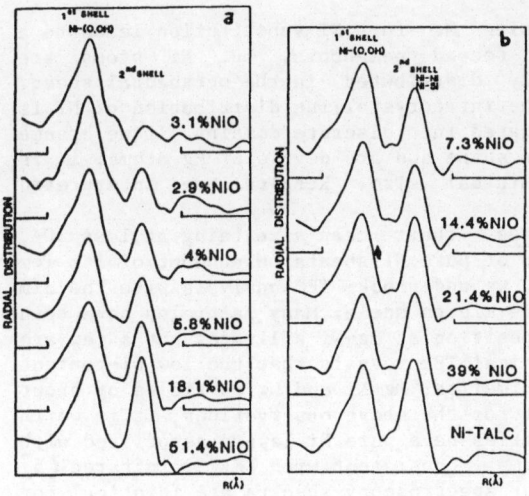

Fig. 1: Comparison of the radial distribution functions.
a) lizardite-nepouite series;
b) kerolite-pimelite series

Curve fitting has been performed on the inverse transform of the second structural peak for all samples, first by assuming that all the atoms of the second shell are Ni atoms, then by assuming only Mg atoms in this shell. The fact that the interatomic distance is well defined (d(cation-cation)=b/3) allows us to overcome the indetermination about the nature of the second nearest neighbors. This is true because the distance determined by EXAFS varies depending on the type of backscattering atom, and inversely it is possible to determine its type by knowledge of the actual distance. A good agreement is found between the experimental and calculated curves (Fig. 2a) by taking nickel as backscattering atom. On the contrary, it is not possible to get a correct fit by considering a second shell only composed of Mg atoms (Fig. 2b), unless the Ni-Mg distance is set at the unrealistic value of 3.13Å instead of 3.05Å-3.06Å.

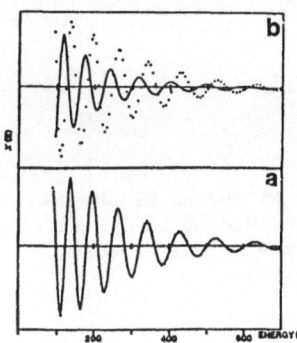

Fig.2: Fourier filtered contribution of the second shell (dotted line) and the corresponding calculated one (solid line). a) kerolite with 7.3% NiO: 5.2Ni and 0.8Mg at 3.07A and 4Si at 3.25A. b) the same sample but calculations are performed by assuming 6Mg neighbors at 3.07A; the fit is clearly unacceptable.

Since nickel is essentially surrounded by heavy atoms, even in phyllosilicates of low Ni content, we have an evidence for the heterogeneous character of the substitution of Ni for Mg. The number of Ni second neighbors vs. Ni content, provided by the fitting procedure, is reported in Fig. 3. In the first series (K-P), the number of second Ni neighbors diminishes slightly with decreasing Ni content but it must be pointed out that this number remains always greater than 5, even for the lowest Ni-containing kerolites. In contrast with the former series, in the lizardite-nepouite series, increasing Mg for Ni substitution leads to a decrease of the number of Ni second neighbors, but Ni atoms are nevertheless always heterogeneously distributed in the octahedral sheet. EXAFS results clearly prove that the intracrystalline distribution of Ni is not random: Ni atoms are segregated into discrete domains. If we assume that these domains have a circular shape and are devoid of Mg atoms, it is possible to calculate their minimal size. Results are interpreted differently depending on the series.

In the kerolite-pimelite series their mean size being at less 30A, EXAFS cannot exclude the existence of pure Ni sheets on account of its low sensitivity in this size range. So, we undertook TEM analysis with the aim of studying the Ni distribution at the micron scale. Many particles have been analyzed. Many of them have a composition of Mg-Ni silicates but a few are pure Ni or Mg silicates. Analytical TEM reveals that the low Ni-content kerolites are in fact mixtures of pimelite, Mg-Ni and Mg particles of about 1 micron or less. By way of support for the above observations, it is worth noting that the minerals in this series have pure Ni layers associated with Mg layers. This distribution scheme is consistent with EXAFS, infrared [5] and optical data, as for this latter spectroscopy spectra are identical for all minerals in the series.

In the lizardite-nepouite series the segregation is more or less important depending on the sample, as the number of Ni second neighbors is varying from 1.5 to 5.2 (Fig. 3). Two Ni distribution patterns are consistent with EXAFS data. (i) Existence of pure Ni clusters whose mean size diminishes with decreasing Ni content. (ii) Existence of Ni enriched areas whose extension, or Ni content, decreases with increasing Mg for Ni substitution. A marked difference in CFSE (Crystal Field Stabilization Energy) values has been observed by optical spectroscopy between nepouite and Mg-Ni lizardite [4], so that we are led to rejection of the first distribution pattern, as the coexistence of pure Ni clusters in lizardite would have given optical spectra similar to that of nepouite.

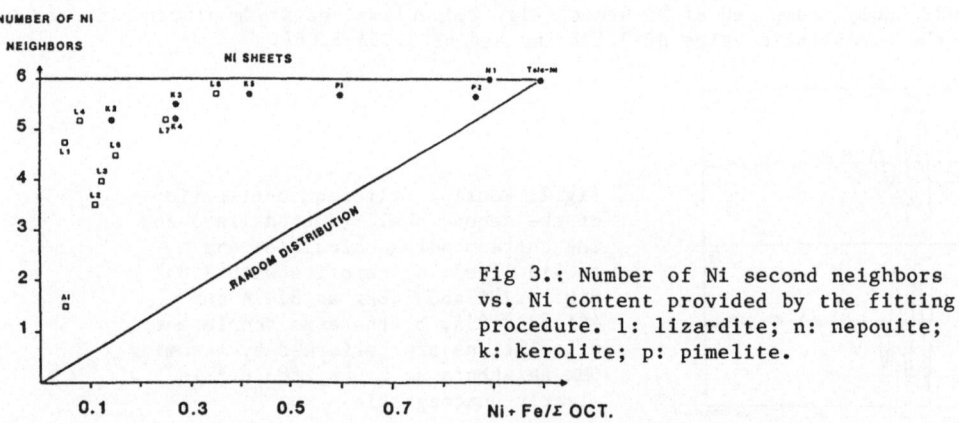

Fig 3.: Number of Ni second neighbors vs. Ni content provided by the fitting procedure. 1: lizardite; n: nepouite; k: kerolite; p: pimelite.

The distinct behavior of these two series as pointed out in this study refers also to distinct geological formation conditions. The lizardite is derived from the transformation of primary serpentines by a Ni enrichment and a subsequent Mg loss. In contrast, the minerals of the kerolite-pimelite series are neoformed products and the existence of distinct Mg- and Ni-octahedral layers can be correlated to this distinct formation process. The Ni-Mg cation ordering seems to be a memory of the formation mechanism of the ore mineral. Finally, EXAFS, when correlated with other spectroscopic data, is very appropriate to study the localization and distribution of 3d transition elements in ore minerals where they are substituted to light elements as Mg and Al on account of their different backscattering factors.

1   A. Manceau, and G. Calas: Am. Miner. Accepted (1984)
2   G. Calas, and J. Petiau: Bull. Miner. <u>106</u> , 33 (1983)
3   G.A. Waychunas, M.J. Apted, and G.E. Brown:Phys.Chem.Min. <u>10</u> ,1(1983)
4   A. Manceau, and G. Calas: Clay Miner. Accepted (1984)
5   P. Gerard, and A. Herbillon: Clays and Clay Miner. <u>31</u> , 143 (1983):

EXAFS and XANES Studies of Trace Elements in Coal

Joe Wong, C.L. Spiro, and D.H. Maylotte

GE Corporate Research and Development, P.O. Box 8, Schenectady, NY 12301, USA

F.W. Lytle and R.B. Greegor

The Boeing Company, P.O. Box 3999 2T-05, Seattle, WA 98124, USA

1. Introduction

Coal contains nearly the entire periodic table as impurity elements. While the common inorganic elements found in coal can be accounted for in the major minerals, the chemical and structural environments of many of the impurity elements (i.e., those with a concentration 1000 ppm or less) cannot be confidently determined by conventional microscopic, spectroscopic and diffraction techniques. Even though these elements are present only as trace quantities, their total effects can be substantial; e.g., a 100 megawatt power plant uses 450 tons/hr of coal which yields 100 lb/hr of a 100 ppm impurity, and they can have serious technological implications in the areas of hot corrosion, catalyst poisoning and pollution. Structural and chemical characterization of the impurities in coal and coal-derived products can make a significant contribution to the chemical processing of coal. At the present time, and with conventional techniques, the state of knowledge about the chemical structure of trace elements in coal is not sufficient for their problems to be dealt with in a rational way [1].

Many of the technological problems associated with the increased use of coal are not due to the organic coal materials but rather to the inorganic elements that are carried along with the organic coal matrix into the coal utilization process. The mineral matter in North American coals averages around 15 wt. percent [2]. The major minerals usually found in coals are quartz, clay, carbonates and sulfides [3]. A typical list of elements found in coal is given in Figure 1 [4]. All of the elements in the first transition row occur in coal. In the sample of thirteen different coals analyzed for Figure 1, all of these elements appeared in every sample.

H ND																	
Li 4-163	Be 0.4-3											B 1-230	C ND	N ND	O ND	F 1-110	
Na 100-1000	Mg 500-3500											Al 3000-23,000	Si 5000-41,000	P 6-310	S 700-10,000	Cl 10-1500	
K 300-6500	Ca 800-6100	Sc 3-30	Ti 200-1800	V 2-77	Cr 26-400	Mn 5-240	Fe 1400-12,000	Co 1-90	Ni 3-60	Cu 3-180	Zn 3-80	Ga 0.3-10	Ge 0.03-1	As 1-10	Se 0.04-0.3	Br 1-23	
Rb 1-150	Sr 17-1000	Y 3-25	Zr 28-300	Nb 5-41	Mo 1-5	Tc ND	Ru <0.1	Rh <0.1	Pd <0.1	Ag <0.01-3	Cd <0.01-0.7	In Standard	Sn 1-47	Sb <0.01-2	Te <0.1-0.4	I <0.1-4	
Cs 0.2-9	Ba 20-1600	La 0.3-29	Hf <0.3-4	Ta <0.1-8	W <0.1-0.4	Re <0.2	Os <0.2	Ir <0.2	Pt <0.3	Au <0.1	Hg <0.3-0.5	Tl <0.1-0.3	Pb 1-36	Bi <0.1-0.2	Po ND	At ND	
Fr ND	Ra ND	Ac ND															

Ce 1-30	Pr 1-8	Nd 4-36	Pm ND	Sm 1-6	Eu <0.1-0.4	Gd <0.1-3	Tb <0.1-2	Dy <0.1-5	Ho <0.1-0.4	Er <0.1-0.4	Tm <0.1	Yb <0.1-0.5	Lu <0.1-0.3
Th <0.1-5	Pa ND	U <0.1-1											

Fig. 1 Concentration Range of Trace Elements in 13 Raw Coals (ppm).

Certainly, sulfur is the most ubiquitous impurity in coal, occurring in both mineral and inorganic phases. Vanadium, titanium and potassium are among the more pernicious trace elements with regard to catalyst poisoning and hot corrosion. In this paper, these elements will be emphasized and described in some detail. Their bonding and local atomic environment in some selected coals are elucidated using a combined XANES and EXAFS technique.

2. Experimental

The ability to obtain soft x-ray absorption spectra down to 2.4keV under non-vacuum conditions with the use of a fluorescent ion chamber detector arrangement has recently been shown by Lytle et al. [5]. This non-vacuum procedure is briefly summarized as follows. Sulfur and potassium K-edge spectra were obtained on beam line VII-3 wiggler side station at SSRL during a dedicated run of SPEAR at an electron energy of 3.0 GeV and an injection current of 75 mA. The synchrotron x-ray beam from SPEAR was monochromatized with double Si(111) crystals and a 1 mm entrance slit, which yielded a resolution of approximately 0.5eV at the sulfur K-edge of 2472.0eV. For Ti and V K-edge XANES and EXAFS at higher energies, double Si (220) crystals were used. Spectra of trace elements in coal and model compounds of K and S were measured by the fluorescence EXAFS technique [6]. This technique monitors the K_α fluorescence intensity, which is proportional to the degree of absorption of the incident beam, and, hence, monitors the x-ray absorption spectrum. A Stern-heald type fluorescence detector [7] was used, the construction of which has been described in detail 5. For sulfur spectra the incident beam was detuned up to 90% to minimize harmonic contents, and an all-helium path from the beryllium window to the sample was used to minimize absorption and scattering by air. Spectral specimens were prepared by packing powdered samples into 6 micron x-ray polypropylene envelopes sufficiently large that the x-ray beam impinged only on the sample.

3. Results and Discussion

A. Sulfur in Coal

Sulfur appears in coal as three major forms: elemental, inorganic, and organic. Least common is elemental sulfur, and is generally associated with weathered material. Inorganic sulfur occurs primarily as undifferentiated pyritic forms (pyrite, marcasite, chalcopyrite, arsenopyrite, galena), and as the sulfate anion, usually with calcium (gypsum) or ferrous iron from weathering. Organic sulfur is that which is covalently bonded to the organic matrix of the coal, and may appear in any of several functional groupings such as mercaptan, thioether, disulfide, thiophene, any of several possible heterocyclic combinations, and possibly in oxidized forms such as sulfonyl or sulfonic acid moieties in weathered coals.

Figure 2 shows the K-edge absorption spectra of sulfur for four model organic sulfur-containing compounds employed in this study. These are thiosalicylic acid, thianaphthene (benzothiophene), bis(4-hydroxyphenyl)disulfide, and thionin. These four compounds contain, respectively, a thiol SH group, S in a five-membered heterocycle, a disulfide linkage and a S-heterocycle bridging two aromatic rings. The zero of energy is taken at the first inflection point of elemental sulfur at 2472.0eV. The spectra are clearly well differentiated in terms of pre-edge features, and the height and shape of various near-edge absorption characteristics. Thus, they provide the opportunity to discriminate among organic-sulfur bearing functional groups in coal [8]. Figure 3 shows the S K-edge spectra of iron pyrite and potassium sulfate. The pyrite spectrum consists of two characteristic absorption peaks at 1.4 and 11.8eV and a shoulder at 8.8eV. K_2SO_4 exhibits an edge shift of 10eV with respect to elemental sulfur, yielding a sharp peak at 11.5eV and a doublet feature just below 20eV. The large edge shift is expected due to the high formal

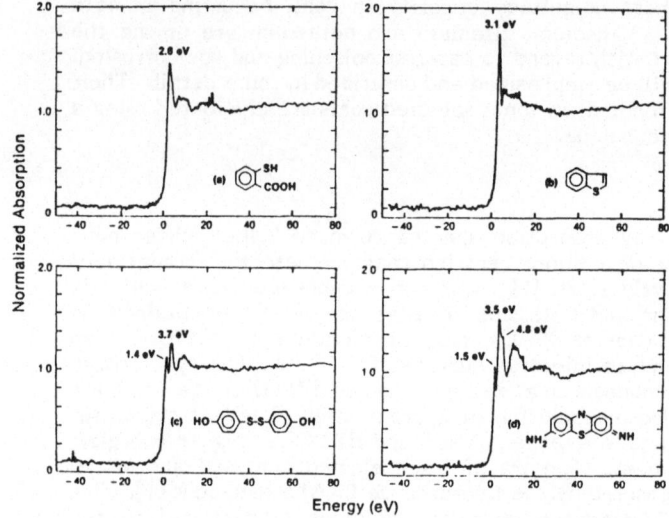

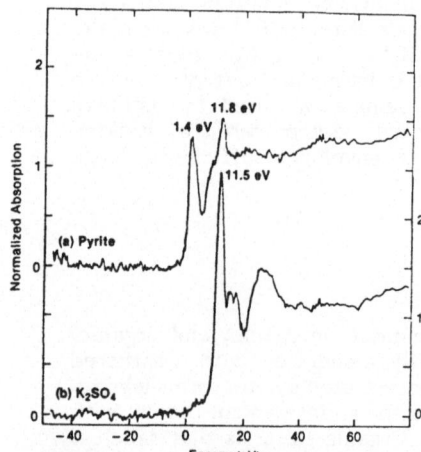

Fig. 2 K-edge Spectra of S in various organic bonding and coordination environments: (a) Terminal Thiol SH group; (b) in 6-member Heterocyclic ring; (c) -S-S- bridge, and (d) in ring bridging. The zero of energy is taken at the first inflection point of Elemental S at 2472.0 eV

Fig. 3 Normalized K-edge Spectra of S in (a) Pyrite and (b) K_2SO_4

oxidation state of sulfur in the sulfate ion. This reflects a higher energy requirement to photoionize the innermost core electron due to an increase in nuclear attraction at the oxidation state of VI.

Figure 4a shows the spectrum obtained on Winifrede seam coal from Lambric, Kentucky, USA. The coal is classified as high volatile A bituminous and was taken as two large lumps from the bottom 10" of channel cut in a freshly exposed seam. The lumps were stored under nitrogen and ground just prior to transport to the synchrotron laboratory. By wet chemical analysis, this sample was determined to have 0.5% pyritic sulfur and 0.94% organic sulfur by weight, or 35% pyritic sulfur and 65% organic sulfur. To elucidate the nature of the organic sulfur species in the coal, a series of XANES simulations were made to match the normalized spectrum of sulfur in coal (Fig. 4a) with a linear combination of the pyrite spectrum (Fig. 3a) and each of the various organic model compounds shown in Fig. 2. Based on the absorption features at 3.0eV and 11.8eV and the shape of the edge in the coal spectrum, the best simulation was obtained by a linear summation of the spectrum (Fig. 2b) representing the 5-membered heterocyclic thianapthene (benzothiophene) and spectrum (Fig. 3a) for iron pyrite. Indeed, when the simulated spectrum utilized

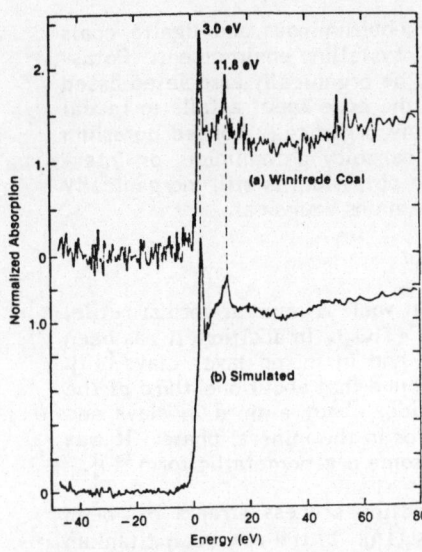

Fig. 4 (a) K-edge Spectrum of S in a Winifrede Coal; (b) a simulated S Spectrum formed by adding .65 the Thianaphthene Spectrum (Fig. 2b) and .35 that of Pyrite (Fig. 3a)

the percentages determined by chemical analysis (35% pyritic, 65% organic) a reasonably quantitative fit is obtained, as shown in Fig. 4b. Heterocycles of this sort have been previously proposed [9] as likely candidates for organic sulfur forms in coal. In that indirect study, mass spectral analysis of esterified products of sodium dichromate treated coal yielded benzo- and dibenzothiophene derivatives.

Simulations using other model compounds yielded spectra which bore little resemblance to the coal spectrum. For example, simulations employing sulfate as the inorganic component yielded incorrect relative intensities for the absorptions at 3.1eV and 11.8eV, together with peaks at 15.0, 17.4, and 27.1eV that are not present in the coal spectrum. Simulations with other organic models and in other ratios yielded peak positions and/or intensities which did not correspond to those observed in the coal spectrum. We therefore conclude that the thiophene unit is the most likely candidate as the main organic sulfur functional group in this particular coal.

K-edge spectra of sulfur in coal has been measured by Hussain et al. [10] using a total electron yield technique under ultrahigh vacuum conditions. In that study, no identification of the sulfur species in coal was made since organic compounds of sulfur, which are usually vacuum incompatible, were not measured.

B. Potassium in Coal

Potassium is another pernicious impurity in coal. The alkali metals are inherently corrosive toward hot metal surfaces, and form low melting phases with iron and sulfate moeities. In addition, potassium fluxes other normally refractory materials which leads to extensive deposits in combustors. These fused deposits can intercept non-fluxing materials, further exacerbating deposition phenomena. In order to remove potassium at the outset, or to somehow intercept potassium prior to the deposition/fusion/corrosion cascade, it has been our premise that an understanding of its inherent chemical environment is a desirable prerequisite.

Analysis of the potassium K-edge XANES by matching near-edge features in coal with model compounds has been effective in determining the identity of indigenous coal impurities [11]. For all older coals tested, potassium is found to occur in a twelve-fold oxygenated site exemplified by illite, a layered clay related to muscovite mica. These include subbituminous, high and low volatile bituminous, cannel, and anthracite coals. Only the youngest subbituminous and lignite coals show deviation, where potassium is found in a non-crystalline environment. Potassium in these lower rank coals has been presumed to be organically associated based on its ion exchange behavior [12]. Examination of the edge spectra fails to reveal similarities between low rank coal potassium and any organically bonded potassium model compounds including carboxylic, phenolic, benzoic, phthalimide, or intercalated groups. It is possible that the exchangeable potassium is still inorganically associated on disordered clay surfaces, though this remains equivocal.

C. Titanium in Coal

Titanium has been reported to occur in coals in various mineral forms: rutile, anatase, brookite, sphene ($CaTiSiO_5$), and ilmenite ($FeTiO_3$). In addition, it has been detected in a vanadium-titanium silicate, in illite and in mixed-layer clays [13]. Semi-quantitative SEM work on a Waynesburg coal found that about one-third of the titanium could be accounted for in titanium dioxide, about a third in clays and accessory minerals, leaving one-third unaccounted for in the mineral phase. It was suggested that the missing third could be present in some organometallic form [14].

The fate of trace Ti in a typical coal liquefaction process stream has been elucidated using high resolution XANES spectroscopy [15]. In the raw coal, titanium exists in at least two chemical and structural environments: anatase (but not rutile) occurs in the sink fraction and the other resembles that in a Ti-organo tyzor compound in the float fraction. Both sites are 6-fold coordinated by oxygens in a distorted octahedral configuration. The organo form was found to persist in both the liquefaction product and the desulfurization catalyst, while the anatase form was found in the coke by-product adjacent to the catalyst.

D. Vanadium in Coal

A number of sink-float separation studies have been done on vanadium-bearing coals [16,17]. Nearly all of these studies point to vanadium having a "high organic affinity" in coals, but some suggest that the element may have both an inorganic and an organic association. In 1936, Triebs reported metal-porphyrins in a number of coals and other organic sediments [18]. He found vanadyl porphyrin a Boghead coal. Since that time, it has been well established that vanadyl porphyrins are common in petroleum oil [19]. No recent studies on vanadyl porphyrins in coal have been found. In an ESR study on bitumens and oils, vanadyl porphyrins were readily identified. No signal was found from a coal [20].

Our EXAFS work [21,22] on vanadium in the Kentucky No. 9 coal shows that the vanadium is in at least two different environments. The vanadium environment in the heavier sink-float fractions has a spectrum that can be identified with that in roscoelite and the vanadium in the lighter fractions (i.e., that part that shows "organic affinity") does not fit the vanadyl porphyrin spectrum, rather it would seem to be V^{++} in an oxygen environment. We do not find a sulphur environment for any of the vanadium. We have also shown that the vanadium environment changes during a liquefaction process and that therefore vanadium may be liberated into the product stream during liquefaction.

4. Future Work

Plans are now underway to extend the measurement to a variety of sulfur-bearing coals of different ranks, as well as to increase our data base for a larger class of relevant inorganic and organic sulfur model compounds. Of particular importance is the study of how sulfur in coal transforms through natural and various thermochemical processes during coal conversion and combustion. Furthermore, by signal averaging of multiply scanned spectra, the signal to noise ratio shown in Figure 4 can much be improved so that quality spectra in the EXAFS region at high energy may be obtained and analyzed to complement the XANES data.

5. Acknowledgment

We are grateful for experimental opportunities at SSRL which is supported by the U.S. Department of Energy.

6. References

1. V. Valkovic: "Trace Elements in Coal", CRC Press, Inc., Boca Raton, Valumes I & II (1983)
2. H.J. Gluskoter: Ad. Chem. Ser. 141, 1 (1975)
3. R.G. Jenkins and P.L. Walker, Jr.: "Analytical Methods for Coal and Coal Products", Vol.II, C. Kerr, Ed., Academic Press, New York, (1978) Chapter 26
4. A.G. Sharkey, T. Kessler and R.A. Friedel: "Trace Elements in Fuel", S.P. Babu, Ed., Adv. in Chemistry, Amer. Chem. Soc. (1975) Vol. 141
5. F.W. Lytle et al.: Nucl. Instrum. Methods (1984) in press
6. J. Jaklevic et al.: Solid State Comm. 23, 679 (1977)
7. E.A. Stern and S. Heald: Rev. Sci. Instrum. 50, 1579 (1979)
8. C.L. Spiro, J. Wong, F.W. Lytle, R.B. Gregor, D.H. Maylotte and S. Lamson: Science (1984) in press
9. R. Hayatsu, R.G. Scott, L.P. Moore and M.H. Studier: Nature 257, 378-380 (1975)
10. Z. Hussain et al.: Nucl. Instrum. Methods 195, 115 (1982)
11. C.L. Spiro, J. Wong, F.W. Lytle, R.B. Gregor, D.H. Maylotte, S. Lamson and B. Glover: in paper on "Nature of Potassium Impurities in Coal". This proceeding.
12. R.W. Bryors, Ed. "Ash Deposits and Corrosion Due to Impurities in Combustion Gases", Hemisphere Publishing, Washington, D.C. (1978)
13. R.B. Finkelman: Ph.D. Thesis, University of Maryland (1980)
14. R.B. Finkelman: Scanning Electron Microscopy, SEM Inc., Vol. 1 (1978) p.143
15. J. Wong, D.H. Maylotte, F.W. Lytle, R.B. Gregor and R.L. St. Peters: Springer Series in Chemical Physics. Vol. 27, p.280 (1983)
16. H.J. Gluskoter: in ref. 4, p.1
17. P. Zubovic: In "Coal Science", R.F. Grould Ed., Adv. In Chem., Amer. Chem. Soc. 55, 221 (1964)
18. A. Triebs: Agnew. Chem. 49, 682 (1936)
19. D.A. Skinner: Ind. Eng. Chem. 44, 1159 (1952)
20. M.B. Hocking and P.I. Premovic: Geochimica et Cosmoschimice Acta. 42, 359 (1978)
21. D.H. Maylotte, J. Wong, R.L. St. Peters, F.W. Lytle and R.B. Greegor: Science, 214, p.554 (1981)
22. J. Wong, D.H. Maylotte, R.L. St. Peters, F.W. Lytle and R.B. Greegor: Process Minerology I: Proc. of the Metallurg. Soc. of AIME (1982) p.335

Nature of Potassium Impurities in Coal

C. Spiro, J. Wong, D.H. Maylotte, S. Lamson, and B. Glover

General Elecrtic Corporate Research and Development Laboratory, P.O. Box 8 Schenectady, NY 12301, USA

F.W. Lytle and R.B. Greegor

Boeing Company, P.O. Box 3999 2T-05, Seattle, WA 98124, USA

1. Introduction

Potassium is a particularly pernicious impurity in coal.[1] The alkali metals are inherently corrosive toward hot metal surfaces. They form low melting phases with iron and sulfate moieties. In addition, potassium fluxes other normally refractory minerals which leads to extensive deposits in combustors. These fused deposits can intercept non-fluxing materials, further exacerbating deposition phenomena and corresponding corrosion. In an engine, fused sub-micron particulates may add to erosion of impact surfaces.

In order to remove potassium at the outset, or to somehow intercept potassium prior to the deposition-fusion-corrosion cascade, a basic understanding of its inherent chemical environment is desired. By following the fate of potassium during processing and utilization, key intermediate phases might be identified and redirected by varying conditions or by adding chemical agents.

Impurities in coal such as potassium are not readily amenable to structural study by conventional microscopic, diffraction, and spectroscopic techniques. These techniques have been applied to major mineral phases in coal[2] with considerable aggregate success. However, for trace elements present in concentrations of 1000 ppm or less in a matrix which is spectroscopically opaque over a wide range of wavelengths, and in non-crystalline or non-unique sites, the results from these measurements are often equivocal. High resolution XANES and EXAFS measurements using intense synchrotron radiation sources provides a novel method for characterizing the bonding and local atomic structure of these trace elements in coal[3,4].

2. Experimental

Potassium K-edge spectra were obtained on beam line VII-3 wiggler side station at SSRL during a dedicated run of SPEAR at an electron energy of 3.0 GeV and an injection current of 75 mA. The beam was monochromatized with double Si(111) crystals and a 1 mm entrance slit, which yielded a resolution of approximately 0.5 eV at the potassium K-edge of 3607.4 eV. Spectra of potassium in coal, coal-derived products and model compounds were measured by the fluorescence EXAFS technique.[5] A Stern-Heald type fluorescence detector was used.[6] Helium and nitrogen were used as detector gases in the incident ion chamber and fluorescence detector respectively. The incident beam was detuned to 50% to minimize harmonic contents. An all-helium path from the Be window to the sample was used to minimize absorption and scattering by air. Spectral specimens were prepared by packing powdered samples into 6 micron x-ray polypropylene envelopes sufficiently large that the x-ray beam impinged only on the sample.[3]

3. Results and Discussion

Figure 1 shows normalized XANES spectra of selected K-bearing minerals. The zero of energy was taken at the first inflection point in the derivative spectrum of pure KCl at 3607.4 eV, which was used to calibrate the spectrometer. Each spectrum

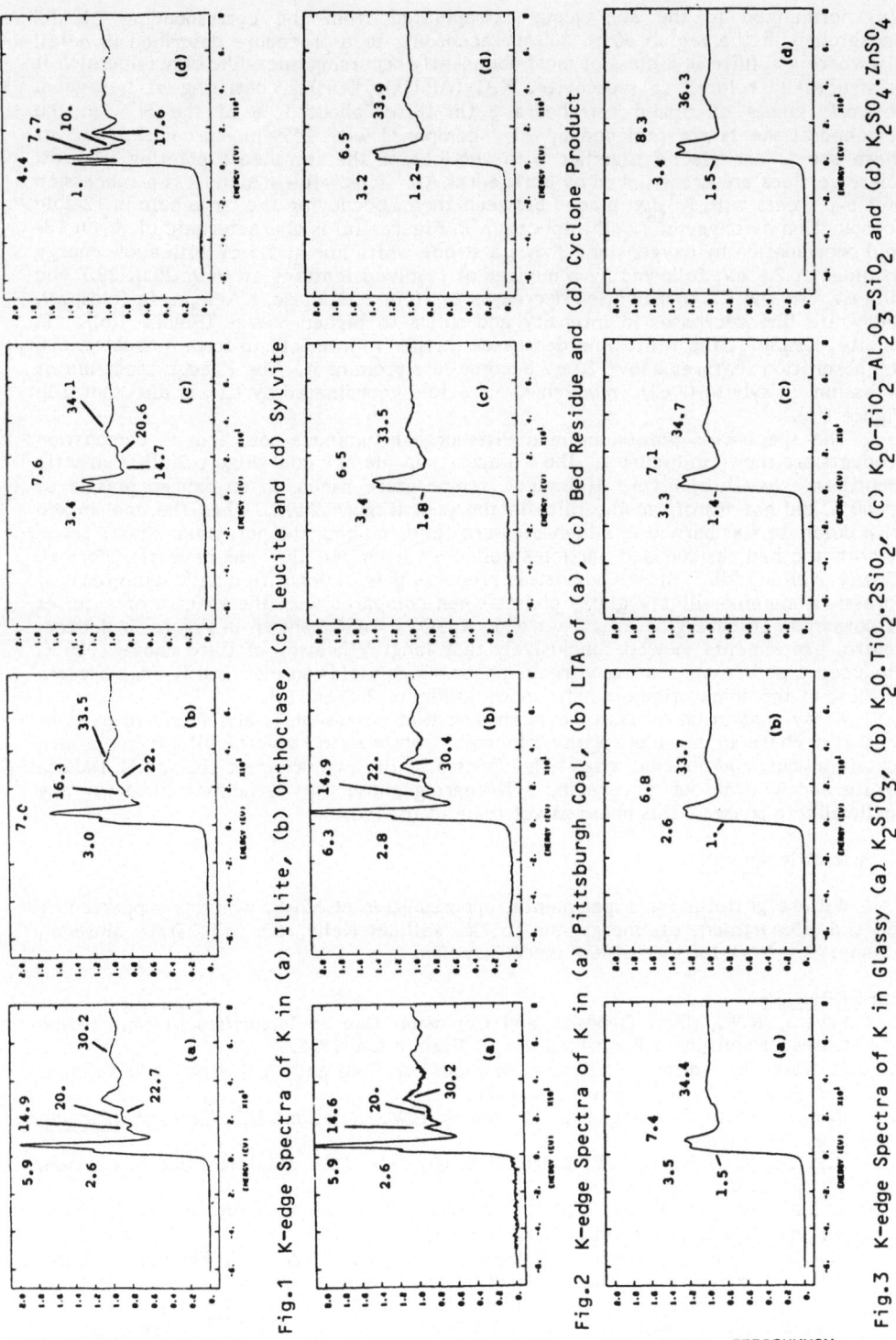

NORMALIZED ABSORPTION

Fig.1 K-edge Spectra of K in (a) Illite, (b) Orthoclase, (c) Leucite and (d) Sylvite

Fig.2 K-edge Spectra of K in (a) Pittsburgh Coal, (b) LTA of (a), (c) Bed Residue and (d) Cyclone Product

Fig.3 K-edge Spectra of K in Glassy (a) K_2SiO_3, (b) $K_2O \cdot TiO_2 \cdot 2SiO_2$, (c) $K_2O-TiO_2-Al_2O_3-SiO_2$ and (d) $K_2SO_4 \cdot ZnSO_4$

was normalized to the step jump extrapolated from the corresponding EXAFS background in the region 50 to 800 eV according to a procedure described in detail elsewhere[7]. Illite is a class of most abundantly occurring mica-like clay minerals. It is structually related to muscovite, KAl $(AlSi_3O_{10})(OH)_2$, consisting of hexagonal network sheets of linked tetrahedra. (In illite, about 15% of the Si^{4+} in the tetrahedral sheets are replaced by Al^{+3} compared with 25% in muscovite.) Two of these sheets are placed together with vertices of the tetrahedra pointing inwards. These vertices are cross-linked by octahedral Al^{3+} ions. The structure is a succession of t-o-t sheets with K^+ ion placed between them, occupying the large hole in 12-fold coordination by oxygens[8]. The spectrum in Figure 1(a) is characteristic of K^+ in 12-fold coordination by oxygens and shows a strong white line at 5.9 eV with a low energy shoulder at 2.6 eV, followed by a number of resolved features at 14.9, 20.0, 22.7 and 30.2 eV. As the K^+ coordination decreases to 10 in orthoclase, $KAlSi_3O_8$ (a feldspar), the white line decreases in intensity and shifts to higher energy (Figure 1(b)). In leucite, $KAlSiO_6$, the white line decreases further in intensity to form a doublet, and the absorption features above 20 eV become less prominent. The K-edge spectrum of potassium in sylvite (KCl), in which K^+ is 6-fold coordinated by Cl^-, is also shown in Figure 1(d).

The spectra of potassium in a Pittsburgh bituminous coal and its combustion products are shown in Figure 2. The K impurity in the raw coal (Figure 2(a)) is directly identifiable as illite, Figure 1(a). Low-temperature ashing in an oxygen plasma at $\sim 100^\circ$C did not transform the illite in the coal (Figure 2(b)). When the coal mixed with dolomite was burned in a high-pressure fluidized bed, the potassium phases found in both the bed residue and particles collected from the flue gas have transformed (Figure 2(c) and (d)). These combustion products may be identified with some form of potassium alumino-silicate glassy phases when compared with the spectra of a series of potassium silicate glasses and vitreous $K_2SO_4 \cdot ZnSO_4$ shown in Figure 3. Indeed, control experiments showed conclusively that lengthy heating of illite above 1000°C produced a potassium-bearing vitreous phase which yields an identical K-edge spectra as those of the combustion products shown in Figure 2(c) and (d).

X-ray absorption measurements showed that potassium is also found to exist as an illitic phase in a wide variety of coals: anthracite, several bituminous coals, subbituminous, and a cannel coal. The spectra of the two youngest coals, a N. Dakota lignite and Wyodak Sub C revealed a K-bearing phase that is neither illite nor any phyllosilicate phases. This phase as yet to be identified.

4. Acknowledgments

We are grateful for experimental opportunities at SSRL, which is supported by the U.S. Department of Energy and to Dr. William Kelly, the N.Y. State Museum, Albany, New York for loan of K-mineral samples.

REFERENCES
1.   Bryers, R.W., "Ash Deposits and Corrosion Due to Impurities in Combustion Gases", Hemisphere Publishing Corp., Washington, 1978.
2.   C. Karr, Jr., editor, "Analytical Methods for Coal and Coal Products", Volumes 1-3, Academic Press, New York, 1978.
3.   D.H. Maylotte, J. Wong, R.L. St. Peters, F.W. Lytle, and R.B. Greegor, Science, 214, 554 (1981).
4.   C.L. Spiro, J. Wong, F.W. Lytle, R.B. Greegor, D.H. Maylotte, and S. Lamson, Science (1984) in press.
5.   J. Jaklevic, J.A. Kirby, M.P. Klein, A.S. Robertson, G.S. Brown and P. Eisenberger, Solid State Comm., 23, 679 (1977).
6.   F.W. Lytle, R.B. Greegor, D.R. Sandstrom, E.C. Marques, J. Wong, C.L. Spiro, G.P. Huffman and F.E. Huggin, Nucl. Instrum. Methods (1984) in press.
7.   J. Wong, F.W. Lytle, R.P. Messmer and D.H. Maylotte, Phys. Rev B (submitted).
8.   W.L. Bragg, "Atomic Structure of Minerals", Cornell University Press, p. 205 (1973).

Investigation of Topics of Interest in Steelmaking Technology by X-Ray Absorption Spectroscopy

G.P. Huffman, F.E. Huggins, L.J. Cuddy, and R.W. Shoenberger

U.S. Steel Corporation, Technical Center, Monroeville, PA 15146, USA

F.W. Lytle and R.B. Greegor

Boeing Company, P.O. Box 3999 2T-05, Seattle, WA 98124, USA

I.   Introduction

The capability of X-ray absorption spectroscopy (XAS) to determine the electronic state and local atomic environment of dilute elements in complex materials is well suited to many problems in steelmaking. Dilute constituents ($\sim$0.01 to 1.0%) often control the bulk properties not only of steels, but also of the raw materials, fuels, and by-products of steelmaking. In this paper, we present two examples of the application of XAS to such problems. In one case, the dilute elements (V and Nb in microalloyed steel) have a beneficial effect, while in the other (K in coke), the effects are harmful.

II. Potassium in Coke

Coke is the carbon-rich fuel used in the production of iron in the blast furnace. It is produced by heating bituminous coal in the absence of air to drive off volatiles, leaving behind a porous, predominantly amorphous, carbon-rich product [1]. Potassium, which is an excellent catalyst for the gasification of carbon [2] causes excessive coke reactivity and poor coke strength in the blast furnace and can also cause serious corrosion and deposition problems.

To investigate the reactions between potassium and coke, a suite of coke samples was saturated with a K_2CO_3 solution and then annealed at 1260°C in argon to simulate conditions in critical regions of the blast furnace. These cokes, the parent coals and untreated cokes, and numerous other coal and coke samples were investigated by fluorescent XAS during a dedicated run at SSRL on beam-line VII-3, using a Si(111) double crystal monochromator, with energy of 3 GeV and beam current $\sim$50 mA.

In all of the coals examined, the dominant K-bearing phase was the clay mineral illite, which exhibits a distinctive XANES spectrum [3]. Both the XANES (Fig. 1a) and Fourier transforms of the EXAFS of untreated cokes (Fig. 2a) indicate that illite is transformed to a predominantly amorphous K-aluminosilicate phase during the coking process. The K-O distance obtained by least squares analysis of the back-transformed O-shell EXAFS signal is 2.96 Å, while the K-Si(Al) distance is approximately 3.9 Å. These distances are reasonable for K that is octahedrally coordinated by oxygen atoms of silica tetrahedra. The XANES of Fig. 1a is similar to those of such phases as leucite ($KAlSi_2O_6$) and K_2SiO_3.

For the K-treated cokes, the XANES (Fig. 1b) and EXAFS (Fig. 2b) are compatible with a mixture of amorphous K-enriched silicate and aluminosilicate phases and an amorphous K-C phase. This conclusion is reached not from the XAS data alone but also from consideration of complementary electron microscopy studies. Attempts to simulate

371

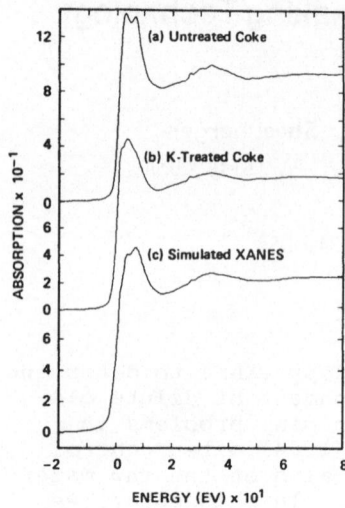

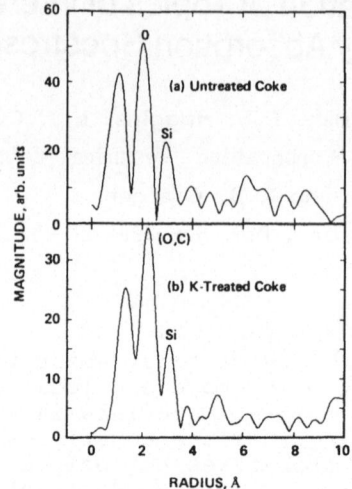

Fig. 1 Typical K XANES of coke.   Fig. 2 Typical Fourier transforms of coke.

the XANES of K-treated cokes by weighted addition of the XANES of standard compounds have been only moderately successful. The simulated XANES of Fig. 1c, obtained by addition of the XANES of K_2SiO_3 (wt. factor = 0.65) and KC_8 (wt. factor = 0.35), is similar to the XANES of Fig. 1b, but leaves much to be desired with regard to the intensities of specific features. This may be because the standard compounds are crystalline, while the K-treated coke is predominantly amorphous. The EXAFS of the main peak of the FT (Fig. 2b) can be analyzed as a two-shell contribution from K atoms with nearest neighbor (nn) O and C atoms. The best fits indicate a nn O to C ratio of roughly 2:1, and K-O and K-C distances of $\sim 3 \overset{\circ}{A}$.

III. Microalloyed Steels
   Microalloying additions (Nb, V, Ti) to low-carbon steels precipitate as carbides, nitrides, and carbonitrides during hot-rolling and tempering. These precipitates interact with dislocations and grain boundaries, producing improved strength and toughness. In order to optimize the processing and composition of these high-strength low-alloy steels, it is essential to know the fractions of the microalloying element present as fine precipitates or in solid solution in the steel. Our investigations of steels containing 0.05 to 0.20% Nb, V, and Ti reacted with C and N by tempering show that XAS is capable of supplying such information.

   In the current paper, only a few typical results are presented. A more detailed report appears elsewhere [4]. Fourier transforms of the EXAFS obtained from tempered samples of a steel containing 0.22%V, 0.29%C, and 0.07%N are shown in Fig. 3. The labels indicate the peaks arising from the first through the sixth Fe nn shells in the ferrite matrix, and the first through the fourth V nn shells and the C/N first nn shell in the carbonitride precipitates. After tempering for six minutes at 640°C, the C/N nn shell peak is just observable, and there is a slight outward shift of the first metallic peak (Fig. 3a). After 60 minutes at 640°C, the peaks from the Fe shells have largely disappeared and the Fourier

transform is dominated by the C/N and V first nn shells in fine,
highly dispersed carbonitride precipitates (Fig. 3b). At this
point, the V-V and V-(C/N) distances are slightly less than those
in bulk VC or VN and the hardness increase produced by tempering
has reached its maximum value. Overtempering, as for the sample
of Fig. 3c, causes all V to be precipitated in the form of large
incoherent V(C/N) precipitates. The Fourier transform is essentially identical to that of bulk VC or VN at this point and the
strength is drastically reduced.

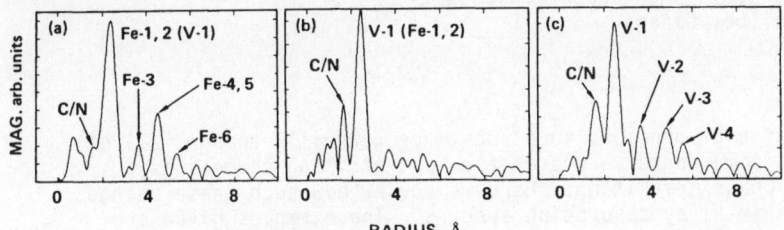

RADIUS, Å

Fig. 3   Fourier transforms of V steel quenched from 1200°C
         and tempered at 640°C for: (a) 6 min; (b) 60 min;
         and (c) 6000 min.

As discussed elsewhere [4], the Fourier transform peak amplitudes, the interatomic distances, and the XANES all vary systematically with tempering time. The XANES spectra exhibit adequate
signal-to-noise ratios for in situ annealing studies. We have
successfully followed the reaction between Nb and C in a HSLA
steel containing 0.07%Nb and 0.27%C during tempering in argon at
590°C [4]. Over a period of approximately one hour, the Nb XANES
evolved from a spectrum characteristic of Nb impurities in a bcc
iron matrix to one characteristic of NbC.

IV.   Conclusions
   In this paper, a brief summary of XAS studies of two important
problems in steelmaking technology has been presented. As these
examples are typical of many problems in the physical and process
metallurgy of steel involving dilute species, it is likely that
XAS will be a valuable technique in steel research.

References

1.   H. Marsh and J. Smith: Chapt. 30, 372-414, in Analytical
     Methods for Coal and Coal Products, II, Acad. Press, 1978.
2.   P. L. Walker, Jr., S. Matsumoto, T. Hanzawa, T. Muira, and
     I. M. K. Ismail, Fuel 62, 140 (1983).
3.   G. P. Huffman, F. E. Huggins, R. B. Greegor, and F. W. Lytle,
     SSRL Report 84/01, IX-31 (1984).
4.   G. P. Huffman, F. E. Huggins, L. J. Cuddy, F. W. Lytle, and
     R. B. Greegor, Scripta Met. 18, 719-724 (1984).

High Pressure X-Ray Absorption Studies of Phase Transitions

J.M. Tranquada* and R. Ingalls

Department of Physics, FM-15, University of Washington
Seattle, WA 98195, USA

E.D. Crozier

Department of Physics, Simon Fraser University
Burnaby, B.C. V5A 1S6, Canada

1. Introduction

The application of high pressure to a substance generally changes all of
its properties to some degree. One of the most striking changes is a
pressure-induced phase transition. Here we review how such phase changes
reveal themselves in x-ray absorption spectra. The examples given are
from our own work at SSRL, the experimental details having been previously
reported [1]. The EXAFS-derived bond compression of a standard are
used to determine the pressure.

2. Pressure Dependence of the EXAFS

We have previously reported the effects of pressure on the XANES of some
materials undergoing phase transitions [2]. Although we shall again
discuss such effects, it is our main purpose here to show the characteristic
behavior of the EXAFS in such cases. We choose to write the EXAFS
interference function $\chi(k)$ in terms of the cumulants $\sigma^{(n)}$ which characterize
the various pair distribution functions [3,4]. For a given shell,

$$\chi(k) = \frac{N}{kR^2} F(k)e^{-2R/\lambda} e^A \sin B \tag{1}$$

where the arguments A and B are given by

$$A = -2\sigma^2 k^2 + \frac{2}{3}\sigma^{(4)}k^4 + \dots \quad \text{and} \tag{2}$$

$$B = 2kR + \phi(k) - \frac{4\sigma^2 k}{R}\left(1 + \frac{R}{\lambda}\right) - \frac{4}{3}\sigma^{(3)}k^3 + \dots \tag{3}$$

As pressure is applied, essentially all of the parameters in Eqs. (1-3)
are expected to vary, with the structural parameters N, R, and σ^2 expected
to exhibit the most significant changes. Using standard EXAFS analysis,
the term

$$\ln\left|\frac{N_a R_b^2}{N_b R_a^2}\right| - \frac{2\Delta R}{\lambda} - 2\Delta\sigma^2 k^2 + \frac{2}{3}\Delta\sigma^{(4)}k^4 + \dots \tag{4}$$

is extracted from the logarithm of the ratio of the amplitude of spectrum a
vs spectrum b. Similarly, the term

$$2k\left[\Delta R - \frac{\Delta\sigma^2}{\overline{R}}\left(1 + \frac{\overline{R}}{\lambda}\right)\right] - \frac{4}{3}\Delta\sigma^{(3)}k^3 + \dots \tag{5}$$

is obtained from the phase difference between two runs. The main point
here is that there is a minor degree of coupling between fits to ΔR and
$\Delta\sigma^2$, vs pressure. Also, because of anharmonicity, one generally must
deal with a nonlinear fit in k^2 or k, respectively.

---

* Present address, Brookhaven National Laboratory, Upton, NY 11973.

The above type of analysis has been made for the Br k-edge spectra in NaBr. Figure 1 shows the transforms of the data for three pressures, where one clearly sees the effects of compression and decreasing thermal disorder. After a back transformation of the nearest neighbor peak, the various parameters in Eqs. (4) and (5) may be determined. We have found it advantageous to first determine the room temperature atmospheric pressure cumulants through comparison with low temperature data, which is assumed to correspond to the harmonic case. The main result from the pressure experiments yields ΔR and $\Delta\sigma^2$ values which appear to be nearly linearly related. For NaBr we express this relation as $\frac{\Delta\sigma^2/\sigma^2}{\Delta R/R}\Big|_{p=0} \simeq 12$. Interpreting these results in terms of an Einstein model would lead to a Grüneisen constant $\gamma_E = \frac{d\ell n\omega_E}{d\ell n\,V}$ of about 1.56 [4].

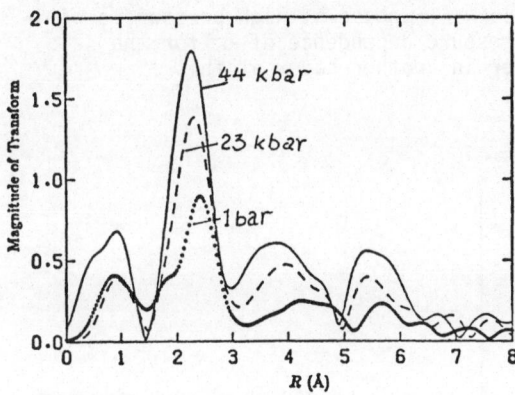

Fig. 1.
Fourier transforms of $k^3\chi(k)$ for NaBr

At atmospheric pressure RbCl has the same B1 (rock-salt) structure as NaBr. However, at about 5.2 [kbar] it transforms to the B2 (cesium chloride) structure. We have followed the EXAFS of RbCl through this phase transition and performed an analysis similar to that for NaBr described above. The main results are shown in Fig. 2. At low pressures $(\frac{\Delta R}{R} = 0)$ the data yield $\frac{\Delta\sigma^2/\sigma^2}{\Delta R/R}\Big|_0 \simeq 16.7$, which would correspond to an Grüneisen parameter of $\gamma_E = 2.78$. This result appears to hold for both phases, that is, both the near neighbor distance and its σ^2 decrease with pressure except at the B1 to B2 transition where they increase slightly. The slight increase in bond length at the transition reflects the fact that the

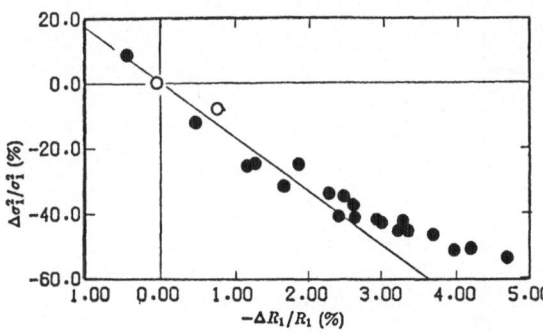

Fig.2. Change in σ^2 vs R for nearest neighbor distance in RbCl. Open(closed) symbols represent the B1 (B2) phase

coordination goes   from 6-fold to 8-fold while the volume itself decreases
by approximately 14%. The accompanying increase in σ^2 signifies a
softening of the appropriate force constants. (The data is also consistent
with N increasing from 6 to 8.)

    We have carried out such an analysis in yet a more complex case, namely
CuBr, which undergoes two phase transitions with pressure: a sharp transition from CuBr III to CuBr V (zincblende to tetragonal) at about 50 [kbar]
and a sluggish transition from CuBr V to CuBr VI (B1 structure) between
60 and 80 [kbar]. The data for both K edges are consistent with a nearest
neighbor coordination of N = 4 for the tetragonal phase, the exact structure
of which is not yet known, and N = 6 for the B1 phase. Again, analyzing
the data in terms of $\Delta\sigma^2$ vs ΔR yields a relation similar to NaBr and RbCl
(Fig. 3). At each transition there appears to be an increase in $\Delta\sigma^2$ and
ΔR followed by a decrease as the pressure increases further. We remark
here that ZnSe also transforms to a rock-salt phase at high pressure; a similar instability is reflected in the pressure dependence of σ^2 for the
second shell. This is discussed further in another paper at this
conference [5].

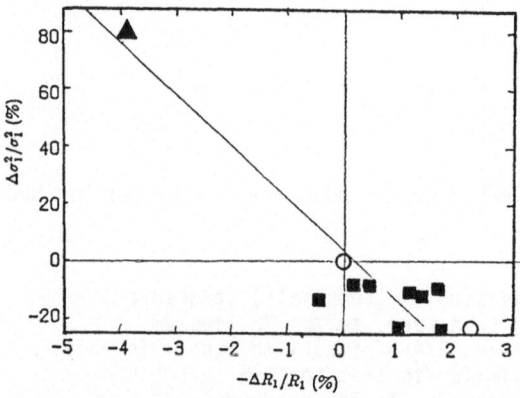

Fig.3. Change in σ^2 vs R for the
nearest neighbor distance in
CuBr III (O), CuBr V (■) and
CuBr VI (▲)

3. Pressure Dependence of the XANES

Figures 4 and 5 show the changes in the XANES with the pressure-induced
changes in crystal structure, for RbCl and CuBr. A definitive detailed
explanation of such behavior is not yet avaliable. We have also been
studying the mixed-valent materials SmS and SmSe which undergo isostructural insulator-metal transitions with pressure. Figure 6 shows the XANES

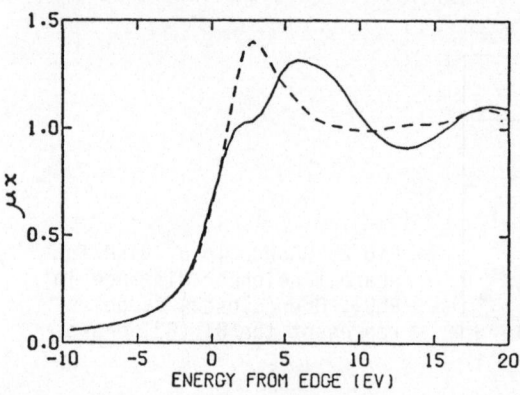

Fig.4. Rb k-edge XANES in the
B1 (solid line) and B2 (dashed
line) phases of RbCl

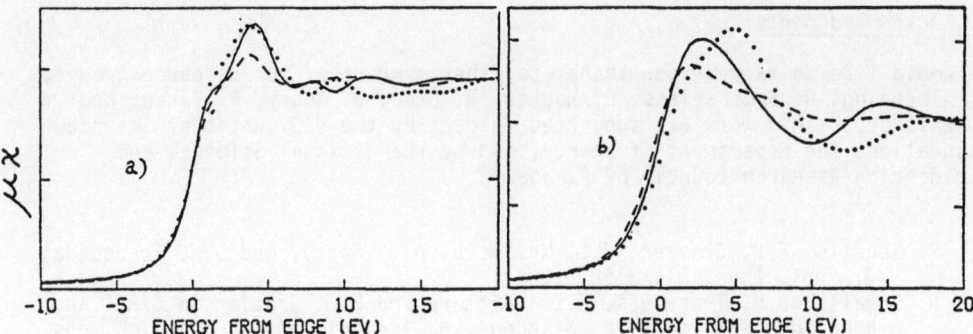

Fig. 5. a) Cu, and b) Br k-edge XANES in the CuBr III (solid line),
CuBr V (dashed line) and CuBr VI (dotted line) phases

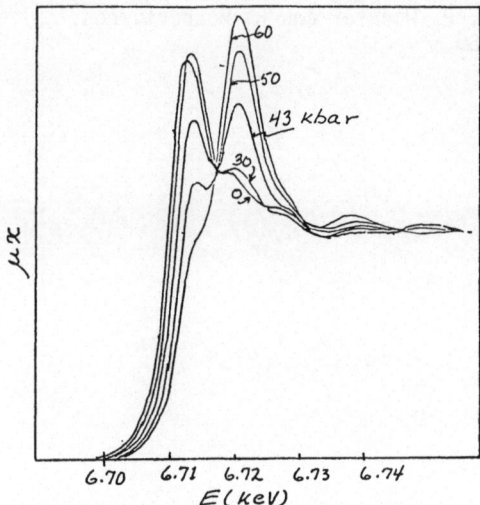

Fig. 6. Sm L3-edge XANES of SmSe at
several pressures

behavior of the Sm L_3 edge in SmSe, for example. Although, to our knowledge,
these spectra have not been explained from first principles, they represent
the change from the divalent $4f^6$ to trivalent $4f^5$ state, in which the photoexcitation threshold is roughly 7 eV higher, accompanied by a volume
collapse.

We also find that if the fractional valence is determined from the relative areas under the two peaks, it does not follow ΔR in a linear manner.
That is after correcting for the usual compressibility, the collapse in ΔR
occurs at a lower pressure than the valence change.

Similarly, and as reported by FRANK et al. [6] for SmS, we find that σ^2
for the nearest neighbor shell appears to increase and then decrease, going
through the transition despite the large decrease in ΔR. This would be in
contrast to the usual Grüneisen relationship described above for the other
systems. It remains to be determined whether such a result is related to
thermal disorder or some type of static structural disorder, possibly
resulting from some type of inhomogeneity.

4. Acknowledgments

We would like to express our thanks to other members of our research groups, N. Alberding, R. Bauchspiess, B. Houser, R. Owen, A. Seary, P. Viren, and J. Whitmore. This work was supported in part by the U.S. National Science Foundation, the Department of Energy, and by the National Sciences and Engineering Research Council of Canada.

1. R. Ingalls, E.D. Crozier, J.E. Whitmore, A.J. Seary, and J.M. Tranquada, J. Appl. Phys. 51, 3158 (1980)
2. R. Ingalls, J.M. Tranquada, J.E. Whitmore and E.D. Crozier in EXAFS and Near Edge Structure, A. Bianconi, L. Incoccia and S. Stipcich, eds. (Springer-Verlag, Berlin, 1983) p. 153
3. J.M. Tranquada and R. Ingalls, Phys. Rev. B 28, 3520 (1983)
4. J.M. Tranquada, Ph.D. Thesis, Univ. of Washington, 1983 (unpublished)
5. J.M. Tranquada and R. Ingalls, this conference.
6. K.H. Frank, G. Kaindl, J. Feldhaus, G. Wortman, W. Krone, G. Materlik, and H. Bach, in Valence Instabilities, P. Wachter and H. Boppart, eds. (North Holland, Amsterdam, 1982) p. 189

High Pressure L$_{III}$-Absorption in Mixed Valent Cerium

Jürgen Röhler

II. Physikalisches Institut, Universität zu Köln, Zülpicherstraße 77
D-5000 Köln 41, Fed. Rep. of Germany

1. Introduction

Intense synchrotron light-sources, high resolution spectrometers and mo-
dern data acquisition techniques brings back old L$_{III}$-absorption spectroscopy into physics and chemistry of rare earth /1,2/. Special interest
in L$_{III}$-absorption arose in physics of valence fluctuating rare earth systems, since L$_{III}$-absorption turned out to be a universally applicable tool
for valence determination /3,4/. Since the discovery of the drastic volume
collapse at the γ-α phase transition in Ce /5/ and Paulings's suggestion of
promotion of one 4f electron in this 15% volume collapse, the degree of 4f
occupation in α-Ce has been the subject of lively controversy.

2. Experimental and Results

L$_{III}$-absorption of Ce was measured under high-pressure up to 120 kbar and
at ambient temperature, using a beryllium gasketed high-pressure cell /6/
and synchrotron radiation of DCI, Orsay (France). Ce (99.95%) was rolled
into sheets of 2μ thickness and sealed with 1μ thick Al-foil in order to
prevent oxidation during preparation of the 200 x 900 μm^2 absorber. Absorption spectra at various pressures are shown in Fig. 1. At ambient pressure
L$_{III}$-absorption of Ce exhibits the significant white line, observed in all
rare earths (La-Lu). Spectra from different batches were identical, and there
was no noticeable influence on annealing. The weak intensity at the high-
energy side of the 0 kbar spectrum is attributed to mixed valent behavior
of γ-Ce. Slight contamination with "tetravalent" CeO$_2$ cannot be excluded.
It is estimated to be within 1%, upon surface inspection.

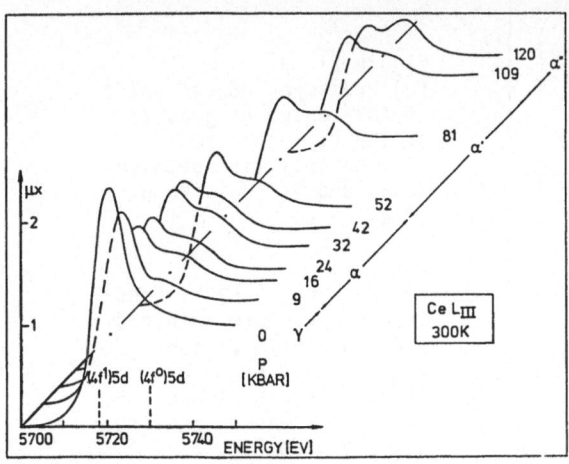

Figure 1:
L$_{III}$-spectra of Ce between 0
and 120 kbar. Pressure is indicated on the right hand
side together with assignment
of Ce high-pressure phases
and phase boundaries at 300 K:
γ-α (8 kbar), α-α' (51 kbar),
α'-α'' (112 kbar).

379

At p>9 kbar the spectra exhibit a pronounced double peaked shape, which by
now is well known from systematic investigations of intermetallic γ-type
and α-type Ce-systems /7/ and further high-pressure experiments at γ-type
intermetallics $CeAl_2$, $CeBe_{13}$ and $CeCu_2Si_2$ /8/. The spectra clearly demonstrate weak mixed valence in γ-Ce and strong mixed valence in α-, α'- and
α''-Ce. This finding is in strong contradiction to Kondo-type theoretical
approaches to the Ce-problem /9/. Tetravalence of α-type Ce is not observed,
rejecting Pauling's suggestion of a full promoted f-electron in the α-phase.
However, partial promotion occurs (≈30%), much less than derived from trivalent and tetravalent metallic radii using Vegard's rule (≈60%) /10/.
The spectrum of roomtemperature, high-pressure α-Ce at 9 kbar is similar
to that from 77K, ambient pressure α-Ce /11/. Although L_{III}-absorption by
now has come into widespread use as a method, yielding 4f occupation numbers
of rare earth mixed valence systems, the interpretation of Ce L_{III}-data
in terms of mixed valence is believed to be falsified by final state effects. In the following we give a detailed description of our data analysis.
We shall prove that L_{III}-absorption in Ce measures the properties of the
mixed valence ground-state. Final state effects due to the creation of a
2p core-hole are not observed. It will turn out that careful data analysis yields quantitative information on 4f-occupation as well as qualitative information on elastic and chemical properties.

3. Valence determination

We are extracting the valence in two independent ways (Fig. 2a and 2b).
The first method involves no assumption on the detailed nature of the photon absorption processes. We consider the pressure-induced shift $\delta(E)$ of
the center of gravity, which we determine by integrating the spectra over a
sufficiently large range (≈50 eV). Dividing $\delta(E)$ by 10 eV gives a number
$\delta v(E)$ which we take as a rough measure of the pressure-induced shift of the
$4f^0$ occupation number v, giving the valence by $3+v$. Here the quantity 10 eV
is the well established shift of $2p$ core-level binding energy in the case
of a configurational change from $4f^1$ to $4f^0$ /12,13/. We find valence $v=3.2$
above 30 kbar. We show below that this method, although ignoring all the
information of the shape of the spectra, already gives nearly the same re-

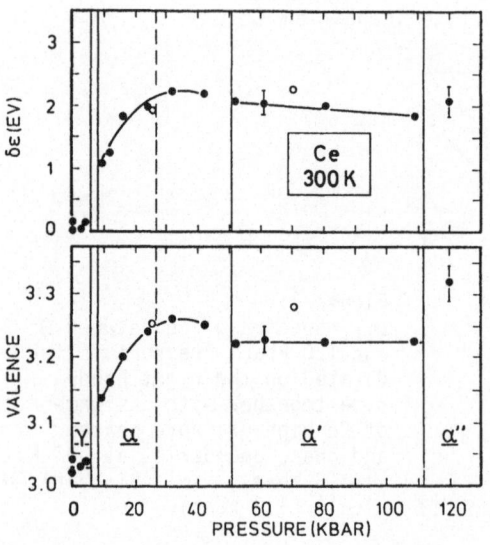

Figure 2:
(a) Pressure-induced shift
of the center of gravity
of the L_{III}-spectrum of
Ce, The splitting between
$4f^1$5d and $4f^0$5d is about
10 eV in Ce

(b) Valence of Ce as function of pressure obtained
from the ratio of the in-
tensities of $4f^1$5d and
$4f^0$5d peaks (pressure increasing ● and decreasing
o). See also Fig. 3

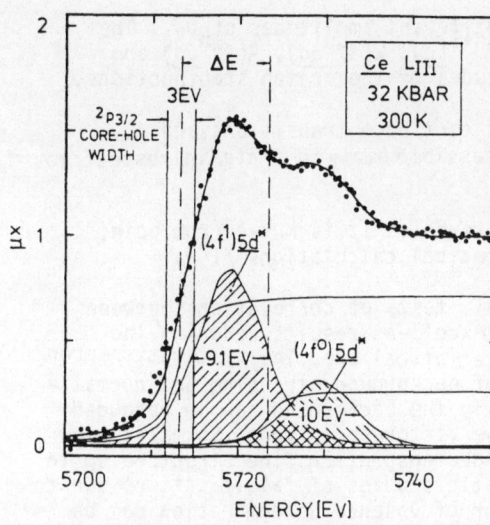

Figure 3:
Analysis of a mixed valent Ce L_{III}-spectrum. Drawn out line, fitting the data points,is a superposition of two distributions of empty 5d-states and correspondent stepfunctions, describing the onset of transitions into continuum states. ΔE assigns the splitting of the two stepfunctions. Both, distributions of empty 5d-states and stepfunctions are folded with a $2p_{3/2}$ core-hole width of 3 eV. FWHM of empty 5d distributions turn out to be different (see text).

sult as extracted in the same straightforward manner as from L_{III} structure of the heavy rare earth elements (Sm-Lu). This means that final state effects have no influence on the ratio of the area of the two peaks.

The second more elaborate method (Fig. 2b) applies the same procedure as commonly accepted in the analysis of L_{III}-structures of Sm-Lu. This method is demonstrated in detail in Fig. 3. The double peaked shape of mixed valent L_{III}-spectra is interpreted to be due to a superposition of two white lines, separated by 10 eV, the shift of 2p core-level binding energy in the case of 4f configurational change. Each white line is due to a dipolar photoabsorption process $2p_{3/2} \rightarrow 4f^1 5d^+$ and $2p_{3/2} \rightarrow 4f^0 5d^+$, respectively. Due to the slow valence fluctuation rate ($\tau \approx 10^{-16} s$), L_{III}-absorption with a characteristic measuring time ($\tau_1 \approx 10^{-16} s$), given by the inverse width of the $2p_{3/2}$ core-hole of 3 eV, clearly resolves the two valence states /14/. The experimentally observed linewidth of one white line is about 6-10 eV, confirming an effective width of the distribution of empty 5d-states in the order of 10 eV. This finding is in agreement with BIS data, optical observations and theoretical calculations, all showing the rather localized nature of 5d wavefunctions in rare earths. Consequently a precise analysis of the observed mixed valent double white line structure should take into consideration the following parameters:

ΔE, describing the shift of $2p_{3/2}$ core-level binding energy in case of configurational change. In Fig. 3 it is determined by the splitting of the two arctan stepfunctions, measured at their inflection points.

D_{5d}, describing the distribution of empty $4f^n 5d$ and $4f^{n+1} 5d$ states. It might be analytically introduced by e.g. a rectangle or semiellipsis. Usual analytical expressions do not fit, since the distribution of empty 5d states reflects coordination, type of chemical bond, bandstructure etc. We extract D_{5d} from the experimental data by a least squares procedure.

$W(4f^{n+1} 5d)$ and $W(4f^n 5d)$,assigning the different widths of D_{5d}. $W(4f^{n+1} 5d)$ is found to be clearly smaller than $W(4f^n 5d)$ due to a higher density of empty 5d states in the $4f^{n+1}$ configuration. This finding is in agreement with L_{III} linewidth variation at the transition from the atomic to solid state /15/.

$A(4f^{n+1}5d)$ and $A(4f^n 5d)$, assigning the different amplitudes of D_{5d}. The area ratio, found from fitting D_{5d}, $W(4f^{n+1}5d)$, $W(4f^n 5d)$, $A(4f^n 5d)$ and $A(4f^{n+1}5d)$ is correlated with the amplitudes of the arctan stepfunctions.

E, describing the energy of the onset of continuum transitions, and δE, describing the position of the first accessible empty 5d state in respect to E. δE is found to be about +1.5 eV.

Γ_L, describing the width of the $2p_{3/2}$ core-hole. It is known from both, fluorescence/Auger experiments and theoretical calculations /14/.

A least squares fit-procedure and careful tests of correlations between the parameters, described above, yield excellent results reducing the error of valence determination to the statistical one. The systematic error is mainly determined by the procedures of background stripping and normalization of the continuum threshold. Usually the background can be regarded as a straight line or approximated by the Victoreen polynom. Normalization of μx is obtained by averaging the extended absorption fine structure up to 200 eV or more. Based on careful systematic studies of falsifications due to wrong normalization, the systematic error of valence determination can be reduced to ±0.02. Finally we wish to point out the importance of thin absorbers. rejection of harmonics and quantitative determination of spectrometer resolution. Due to the strongly asymmetric background even a spectrometer resolution of 3 eV may falsify considerably the intensity ratio of the two lines, depending on which line is dominant /16/. In order to check the possible influence of the spectrometer, we measure during the run the rocking curve of the crystals and deconvolute the spectra with a Gaussian, whose width is taken from the rocking curve /17/.

Fitting the Ce high-pressure data by the procedure described above, we find the pressure dependence of valence, shown in Fig. 2b. It clearly shows a valence shift of about 10% in the γ-α transition and an increases up to 3.25 at 30 kbar, then a clear reversal and a saturation with the onset of the α'-phase. At the α'-α'' phase transition the valence shifts to 3.32. A detailed analysis of the Ce high-pressure valence in terms of valence fluctuations and a discussion of the volume-valence realtion is given by us in /7,18/.

4. Final state effects

L_{III}-absorption is a core-hole spectroscopy. Valence and other quantities extracted from the spectrum are measured in the photoionized state of the absorbing atom. It has been suggested that in L_{III}-absorption of rare earths with a $4f^n$ groundstate, final state configurations $4f^{n+1}$ and $4f^{n-1}$ contribute to the screening of the core-hole, thus producing satellite structures. In valence fluctuating systems these satellite intensities could be located at the energetic positions of the $4f^n$ and $4f^{n+1}$ groundstate. However, up to now there is no theoretical treatment predicting the intensities of satellites in rare earth L_{III}-absorption. From theoretical calculations, treating the same problem in $3d$ XPS/19/, it is expected that the probability for the occurrence of a $4f^2$ satellite (if it exists it would be directly observable 4 eV below the $4f^1 5d$ position) scales with Δ/E. Here Δ denotes the hybridization between 4f- and conduction-electrons and E the shift in core-level binding energy at configurational change. The high-pressure experiment allows one to check this prediction, since high-pressure drastically increases Δ from 0.1 to 0.4 eV at the γ-α transition /6/. A $4f^2$ "shake-down" intensity should occur in the α-type spectra 4eV below the $4f^1 5d$ position. No satellite is observed. In the contrary, the center of gravity of the $4f^1 5d$ distribution

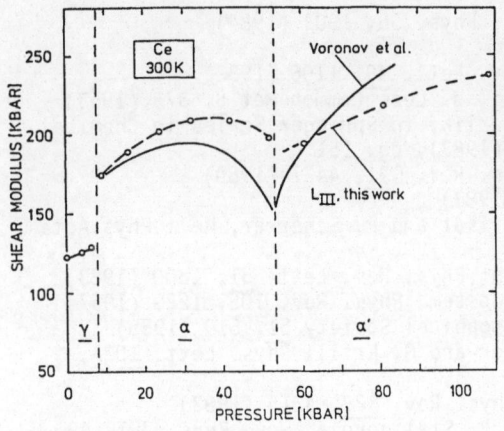

Figure 4:
Shear modulus of Ce under pressure derived from L_{III} linewidths (dashed line), and from ultrasonic experiment of Voronov et al. /20/. Coincidence of the two measurements shows, that pressure dependent L_{III}-linewidth is due to changes in types of ligancy

shifts to higher energies accompanied by a strong line broadening. The position of the continuum threshold remains unaffected! We conclude, that the pressure dependent line-broadening and shifts of center of gravity of 5d -states is due to strong modifications of metallic Ce-Ce binding in the various high-pressure phases. Comparison of high-pressure shear modulus of Voronov et al. /20/ with the shear modulus derived from L_{III}-linewidth is given in Fig. 4. Increase of 4f-5d6s hybridization in α-type Cerium drastically affects the effective density of empty 5d-states (averaged over both configurations), producing strongly directed and anisotropic Ce-Ce bonds. Thus the increase of shear modulus reflects the increase of hybridization, measured via the changes in densities of 5d states. Assuming that the width of empty 5d-states, measured in L_{III}-absorption, gives the 5d bandwidth in the crystal, we can apply the following relation. It is known from physics of transition metals:

$$G = W_{5d} / (4V_{at}) n_d (10-n_d) + G_o$$

$W_{5d} \equiv$ 5d bandwidth; $V_{at} \equiv$ atomic volume; $n_d \equiv$ number of 5d electrons

With G_o = 70 kbar we obtain the dashed line shown in Fig. 4. Although this procedure needs more detailed foundation, it clearly shows that the effects observed at lineshape and linewidth at high-pressure are related to the elastic properties of the crystal. This observation definitely excludes an interpretation of linewidth modifications through final state effects.

Acknowledgements:
I thank D. Wohlleben, J.P. Kappler and G. Krill for valuable discussions. The valuable instrumental support of J. Goulon and R. Cortes at LURE is gratefully acknowledged. This work was supported by SFB 125, Deutsche Forschungsgemeinschaft.

References

/1/ G. Hertz, Z.f.Phys. 3, 19 (1920)
/2/ for an extensive review of rare earth L-absorption data see: Gmelin, Handbuch der anorganischen Chemie, Seltene Erden
/3/ E.E. Vainshtein, S.M. Blokhin and Y.B. Paderno, Sov. Phys. Solid State 6, 2318 (1965)
/4/ H. Launois, M. Rawiso, E. Holland-Moritz, R. Pott and D. Wohlleben, Phys. Rev. Lett. 44, 1271 (1980)
/5/ Lawson, A.W. and T.Y. Tang, Phys. Rev. 76, 301 (1949)
/6/ J. Röhler, J.P. Kappler and G. Krill, Nucl. Instr. Meth. 208, 647 (1983)

/7/ D. Wohlleben and J. Röhler, J. Appl. Phys. 55, 1904 (1984)

/8/ J. Röhler et al., to be published

/9/ J. W. Allen and R. Martin, Phys. Rev. Lett. 49, 1106 (1982)

/10/ K.A. Gschneidner and R. Smoluchowski, J. Less Common Met.5, 374 (1963)

/11/ B. Lengeler, J.E. Müller and G. Materlik, in Springer Series in Chemical Physics 27, Springer-Verlag Berlin (1983), pg. 151

/12/ B. Johansson and N. Mårtensson, Phys. Rev. B21, 4427 (1980)

/13/ J.F. Herbst, Phys. Rev. B28, 4204 (1983)

/14/ H.J. Leisi, J.H. Brunner, C.F. Perdrisat and P. Scherrer, Hel. Phys.Acta 34, 161 (1961)

/15/ G. Materlik, B. Sonntag and M.Tausch, Phys. Rev. Lett. 51, 1300 (1983)

/16/ L.G. Parratt, E.F. Hempstead, E.L. Jossem, Phys. Rev. 105, 1228 (1957)

/17/ F.D. Kahn, Proc. of Cambridge Philosophical Society 51, 519 (1955)

/18/ J. Röhler, D. Wohlleben, J.P. Kappler and G. Krill, Phys. Lett. 103A, 220 (1984)

/19/ O. Gunnarsson and K. Schönhammer, Phys. Rev. B28, 4315 (1983)

/20/ F.F. Voronov, V.A. Goncharova and O.V. Stal'gorova, Sov. Phys. JETP 49, 687 (1979)

High-Pressure Energy Dispersive X-Ray Absorption of EuO up to 300 kbar

Jürgen Röhler and Katharina Keulerz
II. Physikalische Institut, Universität zu Köln, Zülpicherstraße 77
D-5000 Köln 41, Fed. Rep. of Germany
Elizabeth Dartyge, Alain Fontaine, and Alain Jucha
Laboratoire pour l'Utilisation du Rayonnement Electromagnêtique (LURE)
CNRS et Univ. Paris Sud, Bat. 109 C, F-91405 Orsay, France
Dale Sayers
Department of Physics, North Carolina State University
Raleigh, NC 27650, USA

The cubic insulating ferromagnet EuO (T_C = 69 K) is often considered as an ideal Heisenberg ferromagnet and is the subject of extensive investigations, including numerous high-pressure studies /1,2,3/. From the pressure-volume relationship, obtained by high-pressure x-ray diffraction, a pressure induced valence change from Eu^{2+} towards Eu^{3+} at 300 kbar was concluded /2/.

Using the energy-dispersive spectrometer at LURE /4/, we investigated by L_{III} x-ray absorption and high-pressure the mixed valence behaviour of EuO up to 300 kbar. The focussing energy-dispersive spectrometer, equipped with a position sensitive photodiode array (Reticon 1024S, 1024 pixel) allowed one to scan the L_{III}-edge of Eu within 400 msec at 0-50 kbar and within 1900 msec up to 300 kbar. Quasihydrostatic high-pressure was obtained using a beryllium gasketed diamond anvil clamp with a sample size of 50 x 900 μm^2 /5/. Pressure could be varied continously scanning the spectra. A combined translation-rotation stage provided an exact positioning of the sample in the focus of the curved Si 111 crystal. The sample position was optimized for a reproducible smooth background in the edge region. In Fig. 1 typical spectra at pressures from 0-291 kbar are shown, together with a deconvolution of the 291 kbar spectrum. For clarity the spectra at 0,77 and 134 kbar are slightly smoothed, whereas at 291 kbar the original even data points are shown (odd data points are neglected). Compared to results of high-pressure x-ray absorption, obtained with conventional double crystal spectrometers, the signal/noise ratio is increased by a factor of ten. This experiment demonstrates that energy-dispersive x-ray absorption at high-pressure promises to be an excellent tool for the study of dynamical behaviour of high-pressure phases in the range of order 100 msec. Spectra at ambient pressure are compared with those record-

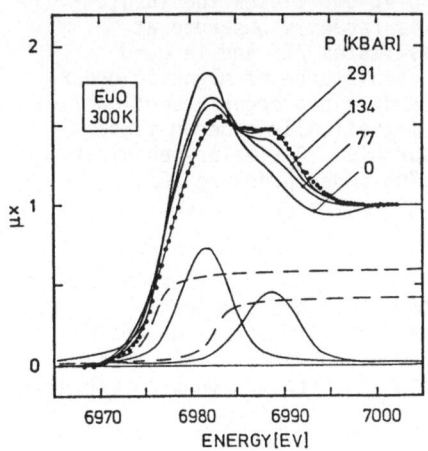

Figure 1:
L_{III}-spectra of EuO at various pressures. Line connecting data points of the 291 kbar spectrum represents a fit which deconvolution is indicated (see text)

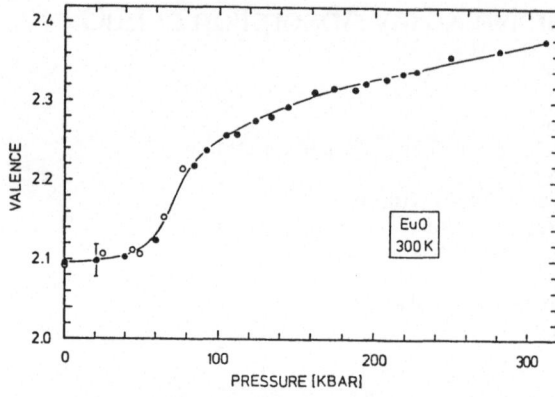

Figure 2:
Valence of EuO under high pressure. Open circles indicate first pressure run, closed circles second pressure run. Data are taken at increasing pressure. Pressure is accurate within 10%

ed at a conventional spectrometer for reasons of energy calibration and comparison of resolution. It appears that due to single Si 111 monochromator in the energy-dispersive spectrometer, energy resolution at the Eu L_{III}-edge is in the order of 3.5 eV, which slightly affects lineshape and amplitude of the "white lines".

Pressurizing EuO, we find EuO to be mixed valence between 0 and 300 kbar, undergoing a S-shaped valence shift from 2.09 to 2.36±0.02 and a critical pressure at about 70 kbar (Fig. 2). Surprisingly EuO, commonly assigned to the stable Eu^{2+} systems, appears to be mixed valent even at ambient pressure. Since the static contributions of trivalent Eu_2O_3 or inhomogenously mixed valence Eu_3O_4 cannot be resolved by L_{III}-absorption, the sample was checked by Moessbauer spectroscopy. We found 6.5% Eu_2O_3; Eu_3O_4 could not be detected. From the absorption edge we derived 15.8% trivalent contribution, which was corrected by the static Eu_2O_3 contribution. Additionally we checked three different samples, which are found to be identical, concerning its "white line" structure. However the near edge region showed slight differences. We derive the valence from deconvoluting the spectra into two lifetime broadened distributions of empty 5d states and correspondent step functions /6/. According to the mixed valence groundstate, we observed two dipolar transitions $2p_{3/2} \rightarrow Xe4f^7 (5d6s)^2$ and $2p_{3/2} \rightarrow Xe4f^6 (5d6s)^3$. Final state effects turn out to be negligible, in agreement with the single-line feature of pure, stable trivalent Eu_2O_3 /7/.

The observed valence shift should be intimately connected with an insulator metal transition since the decrease in volume closes the insulating gap. An onset of metallic reflectivity was recently observed at 100 kbar /9/, in contradiction with earlier estimates /2/ and in good agreement with our results. We presume that coexistence of magnetic ordering and mixed valence in formally Eu^{2+} systems does not occur exceptionally. It seems that an Eu valence of around 2.3 in magnetically ordered mixed valence Eu-systems is energetically most favourable. A similar behaviour was recently observed in antiferromagnetic $EuZn_2$ under pressure /8/.

Acknowledgements:
We thank K. Fischer, KFA Jülich for letting us use his samples, M. Abd El-Meguid for taking the Moessbauer spectra. Two of us (J. Röhler and K. Keulerz) thank D. Wohlleben for valuable discussions. This work was supported by the Deutsche Forschungsgemeinschaft through Sonderforschungsbereich 125.

References

/1/ B.T. Matthias, R.M. Bozorth and J.H. V. Vleck, Phys. Rev. Lett. 7, 160 (1961)
/2/ A. Jayaraman, Phys. Rev. Lett. 29, 1674 (1972)
/3/ U.F. Klein, Thesis TU München (1974)
/4/ A.M. Flank, A. Fontaine, A. Jucha, M. Lemonnier, C. Williams , J. Physique Lettres 43, L315-319 (1982)
/5/ J. Röhler, J.P. Kappler, G. Krill, Nucl. Instr. Meth. 208, 647-650 (1983)
/6/ J. Röhler, this conference
/7/ J. Feldhaus, Thesis FU Berlin (1982) unpublished
/8/ J. Röhler, K. Keulerz, to be published
/9/ H.G. Zimmer, K. Takemura, K. Fischer, K. Syassen, Verhandlungen der DPG (1984) and to be published

An EXAFS Study of the Instability of the Zincblende Structure Under Pressure

J.M. Tranquada

Brookhaven National Laboratory, Upton, NY 11973, USA

R. Ingalls

Department of Physics, University of Washington, Seattle, WA 98195, USA

The tetrahedrally coordinated diamond and zincblende structures are favored by many monatomic and diatomic semiconductors at room temperature and atmospheric pressure. These structures are stabilized by covalent bonding, which provides strong bond-angle-restoring forces. Some years ago, Musgrave and Pople pointed out that if the bond-angle-restoring forces in one of these lattices remain constant under hydrostatic pressure, then there will be a critical pressure at which the lattice becomes unstable and the crystal collapses by shear deformation to a denser tetragonal or orthorhombic structure [1]. In fact, the bond-angle restoring forces in several zincblende structure compounds decrease with pressure; however, for most tetrahedral semiconductors, a 6-fold coordinated phase becomes energetically more favorable long before any lattice instability is observed.

Ionic tetrahedral semiconductors have weak bond-angle-restoring forces and might be expected to exhibit some instability under pressure. Indeed, while ZnSe transforms directly from the zincblende to the NaCl structure at 137 kbar, CuBr and CuCl both pass through intermediate (probably tetragonal) phases before reaching the rocksalt modification [2]. Applying the theory of Musgrave and Pople, elastic constant data for CuCl [3] leads to a predicted instability at a pressure very close to the transition to the intermediate phase observed near 50 kbar. In contrast, the zincblende phase of ZnSe would not become unstable until reaching a pressure roughly twice that at which it transforms to the rocksalt phase [4].

We have used EXAFS to study the high pressure behavior of ZnSe and CuBr up to 47 and 87 kbar, respectively, with special attention given to the phase transitions in CuBr [5,6]. The pressure was generated by a pair of hydraulically driven, opposed anvils made of boron carbide [7]. Each sample, powdered and mixed with epoxy, is held in a gasket together with some RbCl calibrant. By measuring and analyzing Rb K-edge EXAFS from the RbCl and comparing nearest-neighbor distance changes with P-V data, the pressure is determined.

Fourier transforms of the Zn K-edge EXAFS in ZnSe at three different pressures are displayed in Fig. 1. The nearest-neighbor peak at ~2.1 A increases with pressure, while the next-nearest-neighbor peak at ~3.6 A decreases slightly. These changes, which are due to the pressure dependence of the mean square relative displacements, σ^2, have been quantified by applying a ratio method analysis in k-space to the individually Fourier filtered first shell contributions. The results may be summarized in terms of an Einstein model. If the volume dependence of the appropriate Einstein mode is given by a Gruneisen parameter γ_E, then it is easy to show that

$$\Delta\sigma^2/\sigma^2 = 2\gamma_E (\Delta V/V) \quad .$$

For the first shell in ZnSe, we find that $\gamma_E = 1.8 \pm 0.9$, while for the second shell the Gruneisen parameter is -0.8 ± 0.8.

To interpret these observations, it is convenient to consider a simple force constant model which includes just a bond-stretching force constant and a bond-bending force constant. For such a model, one may expect that σ^2 for the first shell is inversely proportional to the bond-stretching force constant and that the second shell σ^2 is inversely proportional to the bond-bending force constant. The EXAFS results then give a direct indication that, under pressure, the bond-stretching and bond-bending force constants increase and decrease, respectively. This is clear evidence of the instability of the zincblende lattice under pressure.

We have obtained EXAFS data on CuBr in the zincblende, intermediate, and rocksalt phases (phases III, V, and VI, respectively). Figure 2 shows Fourier transforms of the Br K-edge EXAFS in the three different phases. As the pressure increases, the net volume decreases at each phase transition, but the nearest-neighbor distance increases. As a consequence, the nearest-neighbor mean square relative displacement increases (and the height of the first shell peak decreases) at each transition.

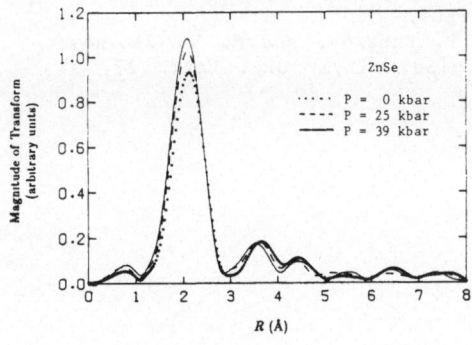 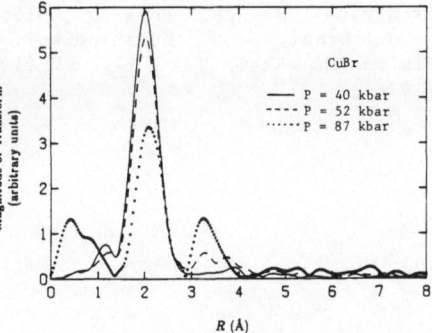

Fig. 1. Magnitude of Fourier transforms of Zn K-edge EXAFS (multiplied by k^2) in ZnSe at three different pressures

Fig. 2. Magnitude of Fourier transforms of Br K-edge EXAFS (multiplied by k^3) in CuBr at three different pressures

A ratio method analysis of the Fourier filtered nearest-neighbor contributions shows that there is no change in the 4-fold coordination in going from zincblende to the intermediate phase, while the data at 87 kbar is consistent with the 6-fold coordination of the NaCl structure. (The high anharmonicity of the highest pressure data makes it difficult to pin down the coordination exactly.) The shells beyond the first are quite small and difficult to analyze except for the second shell at 87 kbar which is quite consistent with the rocksalt structure. The absence of a second shell peak in the zincblende phase indicates that the bond-bending force constant must be quite small in that phase.

Because phases III and V have identical coordinations and the transition occurs over a very narrow pressure region, we believe that the III-V transition in CuBr takes place by shear deformation due to the weakness of the bond-bending forces. The zincblende structure lattice becomes unstable well before the rocksalt phase becomes energetically favorable.

The same explanation should apply to the analogous transition in CuCl and might be related to the anomalous transport properties observed in that compound under pressure [8].

Acknowledgment

This work was supported by the NSF under Grants No. DMR-78-24995 and DMR-77-27489 (in cooperation with DOE).

References

1. M. J. P. Musgrave and J. A. Pople: J. Phys. Chem. Solids 23, 321 (1962).
2. C. W. F. T. Pistorius, Prog. Solid State Chem. 11, 1 (1976).
3. R. C. Hanson, K. Helliwell, and C. Schwab: Phys. Rev. B9, 2649 (1974).
4. B. H. Lee: J. Appl. Phys. 41, 2988 (1970).
5. R. Ingalls, J. M. Tranquada, J. E. Whitmore, E. D. Crozier, and A. J. Seary: in EXAFS Spectroscopy: Techniques and Applications, edited by B. K. Teo and D. C. Joy, (Plenum, 1980), p. 127.
6. R. Ingalls, J. M. Tranquada, J. E. Whitmore, and E. D. Crozier: in Physics of Solids Under High Pressure, edited by J. S. Schilling and R. N. Shelton, (North Holland, 1981), p. 67.
7. R. Ingalls, E. D. Crozier, J. E. Whitmore, A. J. Seary, and J. M. Tranquada: J. Appl. Phys. 51, 3158 (1980).
8. N. B. Brant, S. V. Kuvshinnikov, A. P. Rusakov, and M. V. Semenov: Pis'ma Zh. Eksp. Fiz. 27, 33 (1978) [Sov. Phys. JEPT Lett. 27, 33 (1978)].

Part VIII **Other Applications**

EXAFS and XANES Studies of FeCl₃-Doped Polyacetylenes

Kiyotaka Asakura, Nobuhiro Kosugi, and Haruo Kuroda

Department of Chemistry and Research Center for Spectrochemistry
Faculty of Science, The University of Tokyo, Hongo, Tokyo 113, Japan

Hideki Shirakawa

Institute of Material Science, University of Tsukuba, Sakura-mura
Niihari-gun, Ibaraki 305, Japan

1. Introduction

The simplest conjugated polymer, polyacetylene, shows metallic conductivity
when doped with donors or acceptors [1]. Many compounds are known as effective dopants but their structures in the doped states and interaction with
polyacetylenes have not yet been well understood. When polyacetylene is doped
with $FeCl_3$, the polyacetylene becomes a metallic conductor [2]. We measured EXAFS and XANES spectra of $FeCl_3$-doped polyacetylenes and investigated
the structure of the dopant.

2. Experimental

Chemical doping was carried out by immersing polyacetylene in a nitromethane
solution of $FeCl_3$, and four samples of different doping levels were prepared
by operating the immersing period and the concentration of $FeCl_3$ solutions.
The x-ray absorption spectra were measured, at room temperature and partly
at 80 K, by use of the EXAFS apparatus at Beam Line 10B of Photon Factory in
National Laboratory for High Energy Physics (KEK-PF).

3. Results and Discussion

The Fourier transform of $k^3 \cdot \chi(k)$ of the Fe K-edge EXAFS of $[CH(FeCl_3)_{0.074}]_x$
at 80 K is shown in Fig. 1. Only one prominent peak is found in Fourier
transforms for all the samples, its position being the same irrespective
of the doping levels [3]. This peak can be attributed to the Cl atoms surrounding a Fe atom. The distance and coordination number are obtained by
the curve-fitting analysis as shown in Table 1. The Fe-Cl distance in the
chloro-complexes $[FeCl_x]^{y-}$ is associated with the coordination number x and
valency y [3]. The Fe-Cl distance of 2.19-2.20 Å found in the dopant corresponds to that of $[FeCl_4]^-$.

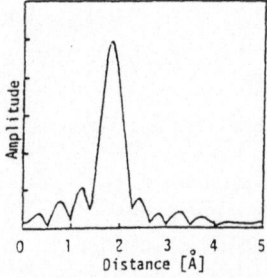

Fig.1. The Fourier transform of $k^3 \cdot \chi(k)$ of the
Fe K-edge EXAFS of $[CH(FeCl_3)_{0.074}]_x$ at 80 K

Table 1. Results of curve-fitting analysis

$[CH(FeCl_3)_y]_x$		Fe-Cl distance[Å]	Coord. No.
y = 0.024		2.20	2.9
y = 0.087	}(at room temp.)	2.19	3.3
y = 0.131		2.19	4.2
y = 0.074	(at 300 K)	2.20	3.6
y = 0.074	(at 80 K)	2.20	3.9

Figure 2 shows a XANES region of the Fe K-edge spectrum of $FeCl_3$-doped polyacetylene together with the spectra of some $[FeCl_x]^{y-}$ complexes. The pre-edge peak which is associated with Fe 1s-3d appears as a relatively strong and single peak; this is characteristic of the high-spin tetrahedral species. Its transition energy of 7104.8 eV is the same as the Fe (III) ones. Furthermore, all the structures observed in the XANES spectrum are similar to those of $[FeCl_4]^-$. From all the above results, we can conclude that the dopant in $FeCl_3$-doped polyacetylene is in the state of $[FeCl_4]^-$ and the doping process is as follows:

$$2\, y\, FeCl_3 + [CH] \rightarrow [CH(FeCl_4)_y] + y\, FeCl_2$$

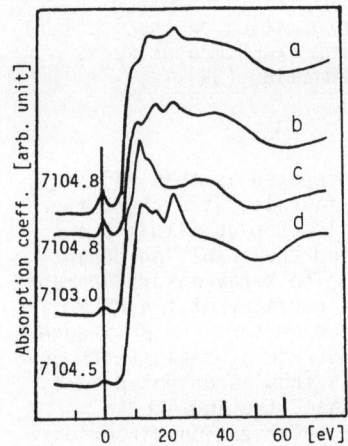

Fig.2. Fe K-edge XANES spectra
(a) $FeCl_3$-doped polyacetylene
(b) $[N(C_2H_5)_4]\cdot[Fe^{III}Cl_4]$
(c) $[N(C_2H_5)_4]_2[Fe^{II}Cl_4]$
(d) $[Co(NH_3)_6]\cdot[Fe^{III}Cl_6]$

References

1. H. Shirakawa, E. J. Louis, A. G. MacDiarmid, C. K. Chiang, and A. J. Heeger: J. Chem. Soc. Chem. Commun., 578 (1977)
2. A. Pron, I. Kulszewicz, D. Billaud, and J. Przyluski: J. Chem. Soc. Chem. Commun., 783 (1981)
3. H. Kuroda, I. Ikemoto, K. Asakura, H. Ishii, H. Shirakawa, T. Kobayashi, H. Oyanagi, and T. Matsushita: Solid State Commun. 46 , 235 (1983)

Polarization-Dependent EXAFS in Br-Doped Polyacetylene

W. Krone, G. Wortmann, K.H. Frank, and G. Kaindl

Institut für Atom- und Festkörperphysik, Freie Universität Berlin
D-1000 Berlin 33

K. Menke and S. Roth

Max-Planck-Institut für Festkörperforschung, Heisenbergstraße 1
D-7000 Stuttgart 80, Fed. Rep. of Germany

1. Introduction

Polyacetylene, $(CH)_x$, is often considered as the simplest prototype of conjugated polymers. It can be synthesized in form of thin films with fibrillar
morphology by Shirakawa's method [1]. Doping with electron acceptors like
halogen molecules has been found to cause an increase in electrical conductivity by many orders of magnitude from the insulating to the metallic regime [2]. While there are detailed models for the crystalline structure of
pristine $(CH)_x$ very little structural information on the microscopic level
is available for its doped forms [3,4,5]. This is due to the poor crystallinity of doped $(CH)_x$, which complicates the use of diffraction methods. It
is the purpose of this paper to show that detailed information on the local
structure of the dopant molecules can be obtained from EXAFS measurements
exploiting the linear polarization of synchrotron radiation (SR).

2. Experimental

Cis-rich polyacetylene samples were conventionally prepared as thin films [1]
consisting of randomly oriented, highly crystalline fibrils. Stretching the
$(CH)_x$ films results in an orientation of these fibrils. Doping with bromine
was performed from gas phase at low Br_2 pressure, and the resulting dopant
concentrations were obtained from the weight uptake. The X-ray absorption absorption measurements were performed in transmission geometry at the ROEMO
beamline at HASYLAB using a channel-cut Ge(220) monochromator. The Br K-edge
EXAFS structure was measured at 77 K from unoriented $(CHBr_y)_x$ foils (with dopant concentrations from y=0.05 to y=0.42) as well as from an oriented foil
(stretching factor 2.4) with y=0.12. In the latter case, the angle α ($\alpha =$
$\sphericalangle$ ($\vec{E},\vec{s}$), where $\vec{s}$ is the stretching vector and $\vec{E}$ the polarization direction
of the SR light) was varied between 0° and 180°.

3. Results

Figure 1 shows the Fourier transforms of those k^2 weighted EXAFS signals which
contain the main information : the spectra from an unoriented sample (y=0.08)
and from the stretch-oriented foil with $\vec{s}$ parallel and perpendicular to $\vec{E}$.
Three dominant peaks labelled P1, P2 and P3 as well as two additional peaks
P4 and P5 are visible. Clear polarization effects are observable for the intensities of peaks P2 and P4.

A detailed data analysis was performed using backscattering and phase shift
functions, in the case of C neighbours theoretical ones from TEO AND LEE [6]
and in the case of Br experimental ones from a Br_2 gas reference measurement.
The first peak P1 in Fig. 1 is dominant in all spectra and is assigned to nn-C
atoms (d_1=1.88(4)Å), since Br-Br distances of this magnitude are impossible.
In the case of the oriented sample the amplitude of P1 exhibits a weak angu-

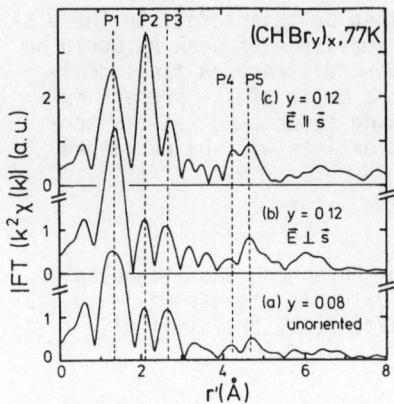

Fig. 1 Fourier transforms of k^2 weighted EXAFS spectra of $(CHBr_y)_x$ at 77 K: (a) y=0.08, unoriented; (b) and (c) y=0.12, oriented; in (b) the polarization vector $\vec{E}$ of the SR light is perpendicular, in (c) parallel to the stretching direction $\vec{s}$.

lar dependence. The second dominant peak P2 can be assigned to nn-Br atoms, since the phase shift corrected distance (d_2=2.52(2)Å) agrees with the Br-Br distance in linear Br_3^- in $(SN)_x$ [7]. The polarization dependence of this peak follows the $\cos^2\alpha$-law expected for linear Br_{2n+1}^- molecules aligned parallel to the stretching direction $\vec{s}$. The presence of these molecules is also in accord with the observation of peak P4, which can be assigned to nnn-Br atoms. While peak P3 can be tentatively assigned to nnn-C atoms (d_3=3.16(5)Å) the origin of peak P5 is not known so far.

Following IKEMOTO et al. [8] the doping range up to y=0.12 is dominated by the intercalation of Br_3^-, while substitution (and addition) reactions take place at higher doping levels. It is therefore tempting to propose model I in Fig. 2 for the local structure at the dopant in $(CHBr_y)_x$. The Br_3^- molecule is assumed to be 1.1 Å above (or below) the $(CH)_x$ plane with the Br atoms in registry with the $(CH)_x$ chains. This model is able to explain both all analyzed distances as well as the polarization effects. The magnitude of the nn-C distance, however, is hard to understand on the basis of van-der-Waals interaction alone. It can be argued that the commensurate intercalation of the Br_3^- molecule as well as the attractive Coulomb interaction leads to the observed reduction of the nn Br-C distance. Model II in Fig. 2 corresponds to the idea of TOKUMOTO et al. [9] which explains P1 and P2 by two different Br species, namely substitutional (or additional) Br atoms and Br_3^- molecules, respectively. Both the observed nn-Br and the nnn-Br distance and its polarization ef-

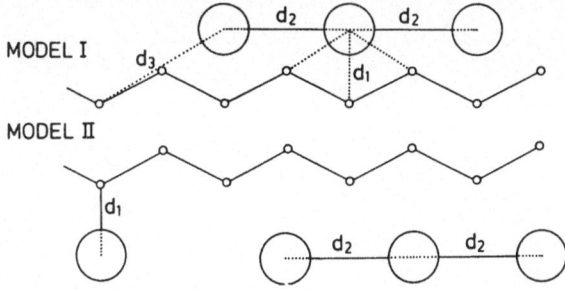

Fig. 2 Schematic view of the two local structure models discussed, assuming trans-$(CH)_x$ in doped regions. Model I: dominant intercalation of Br_3^- molecules at a definite position parallel to the $(CH)_x$ chain. Note that the acceptor molecule lies ∿ 1.1 Å above (below) the $(CH)_x$ plane thereby not interfering with the C-H bonds (not shown here). Model II: intercalated Br_3^- molecules plus substituted Br atoms.

395

fects as well as the nn-C distance can be explained by this model. On the other hand a strong polarization dependence of the amplitude of peak P1 would be expected. In addition, model II gives a wrong nnn-C distance. A final decision between the two models is not possible on the basis of the present experimental results alone. A detailed study of aging effects as well as temperature effects are presently in progress. More details will be published elsewhere [10].

Acknowledgement

This work was supported by the Bundesminister für Forschung und Technologie, contract no. 05 256 KA. The authors acknowledge valuable discussions with Dr. G. Materlik as well as the expert technical assistance by the staff of HASYLAB, Hamburg.

References

1 T. Ito, H. Shirakawa, S. Ikeda: J. Polym. Sci. Polym. Chem. Ed. 12 , 11 (1974).
2 C.K. Chiang et al.: J. Chem. Phys. 69 , 5098 (1978).
3 R.H. Baughman, S.L. Hsu, G.P. Pez, A.J. Signorelli: J. Chem. Phys. 68 , 5405 (1978).
4 C. Riekel, H.W. Hässlin, K. Menke, S. Roth: J. Chem. Phys. 77, 4254 (1982).
5 R.H. Baughman, W.S. Murthy, G.G. Miller, L.W. Shacklette: J. Chem. Phys. 79 , 1065 (1983).
6 B.K. Teo, P.A. Lee: J. Am. Chem. Soc. 101 , 2815 (1979).
7 H. Morawitz et al.: Synthetic Metals 1 , 267 (1979/80).
8 I. Ikemoto et al.: Bull. Chem. Soc. Jpn. 55 , 721 (1982).
9 M. Tokumoto et al.: Solid State Comm. 48 , 861 (1983).
10 W. Krone et al.: to be published.

Polarized X-Ray Absorption Spectroscopy Studies of Halogen-Doped Polyacetylene

H. Oyanagi, M. Tokumoto, and T. Ishiguro
Electrotechnical Laboratory, Umezono, Sakuramura, Niiharigun, Ibaraki, Japan
H. Shirakawa and H. Nemoto
Institute of Materials Science, University of Tsukuba, Ibaraki, Japan
T. Matsushita
Photon Factory, National Laboratory for High Energy Physics, Ibaraki, Japan
H. Kuroda
Department of Chemistry and Research Center for Spectrochemistry
The University of Tokyo, Hongo, Bunkyoku, Tokyo, Japan

1. Introduction

Polyacetylene, $(CH)_x$ is the simplest conjugated system which has been extensively studied because of its unique properties as a conducting polymer. The electrical conductivity of pristine $(CH)_x$ increases by many orders of magnitude upon doping with electron acceptors such as Br_2, I_2, and AsF_5[1]. Transport, optical, and magnetic properties of doped $(CH)_x$ have often been described in terms of "charged solitons" [2], or the spinless defect states as a result of bond alternation domain walls. However, structural studies, important in understanding the doping mechanism and the role of charged solitons in particular, have been severely hindered because of disorder, morphology, and inhomogeneity problems [3]. In our previous paper [4] on the EXAFS studies of Br_2-doped $(CH)_x$, or $(CHBr_y)_x$, the local structure of dopant has been investigated. In this paper, we present the results of polarized x-ray absorption spectra of $(CHBr_y)_x$ and $(CHI_y)_x$ to further elucidate the molecular structure, bonding states, and geometrical arrangements of dopant species.

2. Experimental results and discussion

Polarized x-ray absorption spectra have been measured using synchrotron radiation at Photon Factory [5] on the Br K-edge for $(CHBr_y)_x$ ($0.004<y<0.8$) and I $L_{1,2,3}$-edges for $(CHI_y)_x$ ($0.05<y<0.2$) which were stretch-oriented. A large anisotropy in both the EXAFS and XANES regions between the spectra with the polarization vector $\vec{E}$ parallel ($\vec{E}//c$) and perpendicular ($\vec{E}\perp c$) to the fibril axis has been observed for $(CHBr_y)_x$ ($0.015<y<0.036$) [6]. The Br K-edge $E//c$ oscillations have the envelope which peaks at $k\sim 6$ $[\text{Å}^{-1}]$ and extends to higher-k regions while those of $E\perp c$ oscillations are characterized by a large magnitude in the low-k region which drops off sharply with the increase of k. These results are interpreted as *the direct evidence that polybromine ions exist highly oriented along the fibril axis*. Since the scattering amplitude of bromine has a maximum at a photoelectron wave number k of 6-7 $[\text{Å}^{-1}]$ and extends to a large k of 15 $[\text{Å}^{-1}]$ in contrast with that of low-z element such as carbon which peaks at a small k and falls off sharply with the increase of k, bromine atoms are the dominant scatterer in the parallel direction to the fibril axis whereas carbon atoms contribute to the scattering in the perpendicular direction.

This anisotropy, however, gradually decreases with the increase of bromine concentration. In the heavily doped regime ($y>0.1$), the two spectra with $\vec{E}$ parallel and perpendicular to the fibril axis are similar to the unpolarized EXAFS data [4]. Figure 1 shows the results of Fourier transform of the $E//c$ and $E\perp c$ EXAFS oscillations for *trans*-$(CHBr_y)_x$ ($y=0.036$).

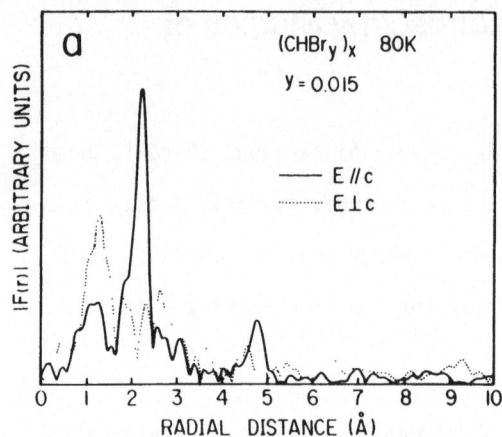

Fig. 1

Results of Fourier
transform for $(CHBr_y)_x$
($y=0.015$) with $\vec{E}$ parallel and perpendicular
to the fibril axis

Prominent peaks located at 2.2 [Å] and 4.8 [Å] in the Fourier transform of
$\vec{E}\|c$ EXAFS oscillations are attributed to the first- and second-nearest Br-Br
spacings. These results rule out the possibilities that bromines exist as
Br^- or Br_2. The average Br-Br distances were determined by a curve fit
analysis using the theoretical amplitude functions and experimental phase
shift extracted from the data for reference compounds, CBr_4 and Br_2. As a
result of the curve fit, the Br-Br distance was determined as 2.55 ± 0.01 [Å]
which is longer than that of Br_2 by 0.27 [Å] due to the charge transfer from
the carbon chain. Dominant bromine species are either Br_3^- or longer Br_{2n+1}^-
chain such as Br_5^-. However, it is difficult to determine the average size
of polybromine ion from the second-nearest neighbor coordination number
which varies between 1.33 and 2 depending on the size of a linear chain due
to the multiple scattering correction.

By employing the Fourier transform using only the low-k region ($2<k<8.5$
[Å^{-1}] to emphasize the scattering by carbon atoms, the peak located at 1.3
[Å] and more distant peaks at 2.6-3.1 [Å] in Fig. 1 are attributed to carbon species. The shortest and longer Br-C distances are determined as 2.0
[Å] and 3.3-3.8 [Å], respectively using the experimental Br-C phase shift
derived from CBr_4 data. This rather short Br-C spacing dominates the ra-
dial distribution of bromine for both $\vec{E}\|c$ and $\vec{E}\perp c$ EXAFS oscillations in the
heavily doped region where the polarization dependence is weak. Taking
the polarization factor into account, the average Br-Br coordination number
in the $\vec{E}\|c$ direction is determined as 0.37 by the curve fit analysis.
Since the coordination would be 1.33 if the dominant bromine species are
Br_3^-, these results imply that less than 27% of bromines are in the form of
polyion. Other bromines are covalently bonded to the polymer chain either
by a substitution or an addition reaction. Longer Br-C spacings located
at 3.3-3.8 [Å] are less dependent on the polarization than the shortest one.
Since these spacings are close to the sum of the Van der Waals radii of
bromine and carbon atoms and the half of the b-axis length in the unit cell
of $trans$-$(CH)_x$, bromine polyions are likely to take the ordered sites intercalated in close-packed $(CH)_x$ chains.

Figure 2 shows the Br K-edge XANES spectra for (a) Br_2 gas, (b) $(CHBr_y)_x$
($y=0.015$) with $\vec{E}$ parallel to the c-axis, (c) $(CHBr_y)_x$ with $\vec{E}$ perpendicular
to the c-axis, and (d) bromobenzene. A sharp spike observed 9-10 [eV] be-
low the threshold, or the continuum limit (13.47 [keV]) is due to the transition from 1s core level to unfilled bound states of p character. In Br_2

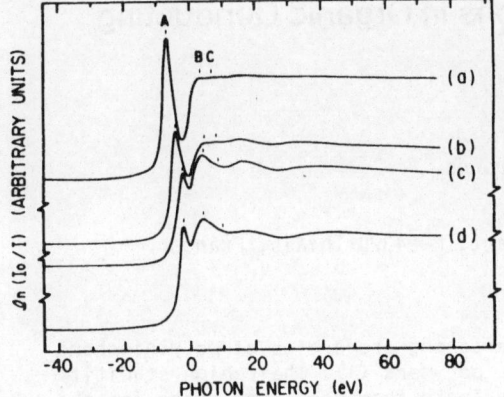

Fig. 2

XANES spectra of (a) Br_2 gas, (b) $(CHBr_y)_x$ with E parallel to the fibril axis, (c) $(CHBr_y)_x$ with E perpendicular to the fibril axis , and (d) bromobenzene

molecule, the lowest unoccupied p-states are $4\sigma^*(p)$ molecular orbital. Isolated halogen ions have the closed shell configuration and therefore would give no sharp transition at this energy. In the E∥c direction, it is known that bromines are mainly in the form of polyion consistent with the EXAFS results. Further, since the overall features of E⊥c XANES spectrum of $(CHBr_y)_x$ are remarkably similar to those of bromobenzene indicating the similarity in the local structure and bonding between them, bromines, which are not in the form of polyion, are covalently bonded with carbons having the sp^2 configuration. These results imply that a substitution reaction takes place in the lightly doped region.

Summarizing the results, we have shown that bromines exist as linear polyions with the average interatomic distance of 2.55 [Å] highly oriented along the fibril axis in *trans*-$(CHBr_y)_x$ with y between 0.015 and 0.036. Other bromines are covalently bonded with carbon chains mostly by a substitution reaction. In the heavily doped regime (y>0.1), the radial distribution of bromine is dominated by the Br-C bonds, which suggests that the fraction of bromines consumed to create polyions decreases with y in this concentration. Based on these results, we propose a new structure model for *trans*-$(CHBr_y)_x$. The key feature of this model structure is *the coex- istence of polybromine ions and bromines covalently bonded with the polymer backbone by a substitution and/or addition reaction.* The anisotropic structure of $(CHI_y)_x$ can also be described by this model. Details of the results of $(CHI_y)_x$ will appear elsewhere.

References

1. H. Shirakawa, E.J. Louis, A.G. MacDiarmid, C.K. Chaing, and A.J. Heeger: J. Chem. Soc. Chem. Commun. 578 (1977)
2. W.P. Su, J.R. Schrieffer, and A.J. Heeger: Phys. Rev. Lett. 42, 1698(1979)
3. R.H. Baughman, S.L. Hsu, L.R. Anderson, G.P. Pez, and A.J. Signorelli: Molecular Metals, NATO Conference Series (Plenum, New York 1979)
4. M. Tokumoto, H. Oyanagi, T. Ishiguro, H. Shirakawa, H. Nemoto, T. Matsushita, M. Ito, and H. Kuroda: Solid State Commun. 48, 861 (1983)
5. H. Oyanagi, T. Matsushita, M. Ito, and H. Kuroda: KEK Report 83-30 (1984)
6. H. Oyanagi, M. Tokumoto, T. Ishiguro, H. Shirakawa, H. Nemoto, T. Matsushita, and H. Kuroda: submitted for publication

EXAFS Study of Copper Inclusions in Organic Conducting Polymers

H. Dexpert and P. Lagarde

LURE, Bât. 209 C, UPS, F-91405 Orsay, France

G. Tourillon

CNRS Photochimie Solaire, 2, rue H. Dunant, F-94320 Thiais, France

We have recently discussed the interesting properties of polythiophene and its derivatives as electroactive polymers (1). Their high stability and their possibility of doping in water make them good candidates for the creating of new catalytic systems. In this way, we have studied the activity and stability of poly 3-methyl thiophene (PMeT) electrochemically loaded with Cu or Ag and Pt aggregates for the proton reduction. These characteristics strongly depend on the location and the size of the metallic clusters and their control is thus obviously important. In this work, we report EXAFS measurements and analysis obtained when either PMeT is electrochemically loaded with copper aggregates or when pulverulent copper is included within this polymer.

PMeT is electrochemically synthetized by the oxidation of the monomer 5.10^{-1}M at +1.35V/SCE (saturated calomel electrode) in a CH_3CN-$N(Bu)_4SO_3CF_3$ 5.10^{-1}M electrolytic medium on a Pt electrode. Two procedures have been used for the copper inclusions :
  - the first one consists of immersing the polymer in a $CuCl_2$ 3M aqueous solution to which a variable cathodic potential is applied,
  - the second one lies in mixing PMeT and pulverulent copper, and then to control the chemical evolutions with time.

1   ELECTROCHEMICAL INCLUSION

Figure 1 is the Fourier transform k^3 weighted obtained for the Cu^{2+} ions in solution and for the polymer polarized at several cathodic potentials (analysis done upon a 700 eV window).
  - for the Cu^{2+} ions in solution (1a), a peak A located at 1.99 Å and characteristic of the Cu-O distance is observed, indicating the complete complexation of the cation by water molecules.
  - when a cathodic potential (-2 V < V < -2.8 V) is applied on the polymer, two different spectra, depending on the $CuCl_2$ concentration, the atmosphere or the electrolysis time are obtained :
  * the first one (1 min. polarization; Fig. 1c) exhibits two peaks at 1.99 Å and 2.9 Å which corresponds to the formation of $Cu_2Cl(OH)_3$ as revealed by X ray diffraction measurements.
  * in the second one (15 sec. polarization; Fig.1b), only one peak at 1.92 Å is detected. This value, characteristic of Cu-O distance, excludes the CuCl formation. It lies between Cu-O bond length in Cu_2O (1.85 Å) and Cu-O (1.95 Å) and could be consistent with the existence of a $Cu^{1+}SO_3CF_3$ link, hypothesis reinforced by the fact that a red complex is obtained when the polymer is treated with a bipyridine solution (a blue complex is formed with Cu^{2+} species).
  - If the potential is cathodically increased (-3 V < V < -3.5 V, Fig.1d), peak A at 1.92 Å is again detected with a new one at 2.51 A characteristic of Cu-Cu distances and therefore of metallic copper aggregates.

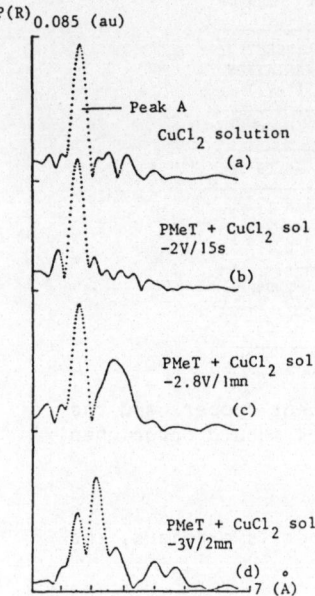

Figure 1 : Evolution of the electrochemical inclusion with voltage and time

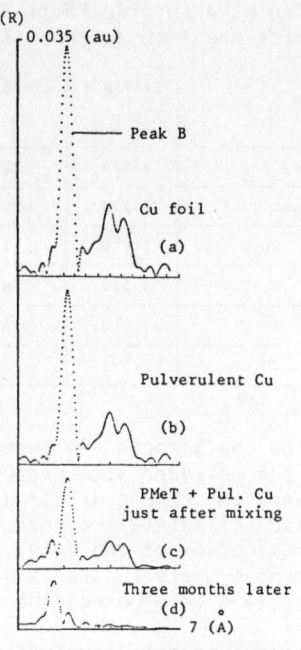

Figure 2 : Pulverulent copper in
PMeT. Evolution with time

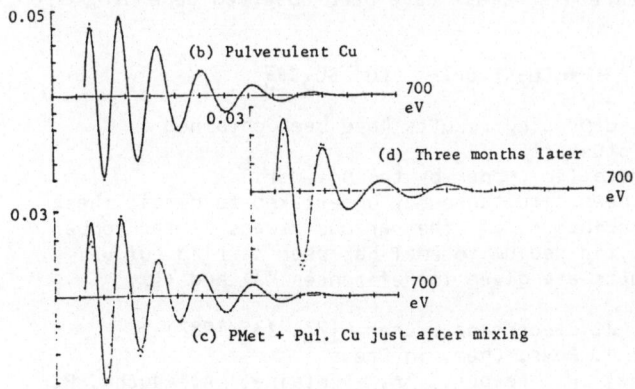

Figure 3 : Fit examples. Experimental (dotted line) and calculated
(full line) spectra

2 PULVERULENT COPPER IN PMeT

The magnitude of the Fourier transform and the Fourier filtered curves
obtained for copper foil, pulverulent copper, PMeT + pulverulent copper
just after mixing and three months later are seen on Fig. 2 and 3.
      - Similar spectra for copper foil and pulverulent copper are obtained
except a smaller neighbours number in the last case : 9 instead of 12
(Table 1).
      - When this pulverulent copper (15% weight) is mixed with PMeT, peak B
at 2.51 Å is well detected with a new one at 1.94 A, characteristic of Cu-
0 bond length. The fit of the Fourier filtered spectra (Table 1) reveals a

Table I : Calculated inverse Fourier transforms. Backscattering amplitudes and phase shifts are from a copper foil and a CuO powder sample

		DISTANCE (A)	ATOMS NUMBER	DEBYE-WALLER VARIATION (au)	ENERGY VARIATION (eV)
Cu foil		2.520	12.00	0.00	0.00
Pulverulent Cu		2.518	9.12	0.02	-0.29
PMeT + Pul.Cu just after mixing	Cu-O	1.944	1.11	0.00	0.11
	Cu-Cu	2.520	4.70	0.00	0.11
PMeT + Pul.Cu three months later	Cu-O	1.940	3.03	0.02	-2.05
	Cu-Cu	2.523	1.05	0.14	-2.05

decrease of Cu neighbours (4 instead of 9 in pulverulent copper) and the presence of one oxygen atom per Cu atom. Two processes should occur when PMeT and pulverulent copper are brought together :
* dissolution of metallic copper in small clusters
* partial oxidation of the metal by PMeT.
- Three months later, almost all the metallic copper is oxidized, the Cu-O peak at 1.94 Å being predominant.

In conclusion, several inclusion processes have been detected by EXAFS measurements when an organic conducting polymer is either cathodically polarized in aqueous $CuCl_2$ or mixed with pulverulent copper:
* In the first case, three species have been observed depending on experimental conditions:

$$Cu^{1+}SO_3CF_3^- , Cu^0 \longleftarrow Cu^{2+} \longrightarrow Cu_2Cl(OH)_3 , Cu^{1+}SO_3CF_3^-$$

* In the second case, two surprising results have been obtained :
    - formation of small Cu clusters,
    - and oxidation of the metallic copper by the polymer.
Kinetic experiments have been simultaneously undertaken to detail these behaviours. An in situ description of the various stages of the copper inclusion from the electrolytic medium to PMeT has been carried out using time-resolved EXAFS. The results are given in references (3) and (4).

1 G. Tourillon, F. Garnier : J. Electroanal. Chem. 173, 135(1982)
2 G. Tourillon, F. Garnier : J. Phys. Chem. in Press
3 G. Tourillon, E. Dartyge, H. Dexpert, A. Fontaine, A. Jucha, P. Lagarde, D.E. Sayers : 3rd International Symposium on Small Particles and Inorganic Clusters, Berlin, July 1984
4 E. Dartyge, A. Fontaine, G. Tourillon : 3rd International EXAFS Conference, Stanford, July 1984

Polarization Use in Anisotropic Elastic Core Effect Determination and Copper Clustering Dynamic Followed by Dispersive X-Ray Absorption

A. Fontaine[1], **E. Dartyge**[1], **and G. Tourillon**[2]

LURE, Bât. 209 C, Université de Paris-Sud, F-91405 Orsay Cédex, France and
[1]Laboratoire de Physique des Solides, Bât. 510, F-91405 Orsay Cédex, France and
[2]Laboratoire de Photochimie Solaire, CNRS, F-94320 Thiais, France

In line with previous studies performed at LURE, works on solid solutions are still in progress. But the novelty of recent experiments on aggregates in organic conducting polymers invites us to split the paper in two parts : elastic core effect in anisotropic matrix and clustering kinetics in doped polythiophene.

1 - Elastic core effect in Zn 1 at .% Cu

The distortions of the host lattice in the immediate vicinity of the solute atom have been studied in the past at LURE in the case of aluminium matrix with Zn, Cu, Mg as solutes [1]. These experiments are in discrepancy with the predictions of the elastic continuous medium theory. In contrast, measurements in isovalent systems CuAg [2] or AuCu [3] are close to the theoretical predictions of the elastic model of FROYEN and HERRING [4].

As was pointed out by MIKKELSEN and BOYCE [5] for random solid solutions of $Ga_x In_{1-x} As$, good standards are needed in order to measure accurately the radius of the first shells around the solute atom. The Zn – 1 at .% Cu solid solutions are of interest since Vegard's law predicts anisotropic elastic core effect (parallel to $\vec{c}$ axis and $\vec{a}$ axis of the hcp structure) which can be measured separately using the polarization of the synchrotron radiation beam, and since the similarity of that peculiar solid solution with pure Zn allows

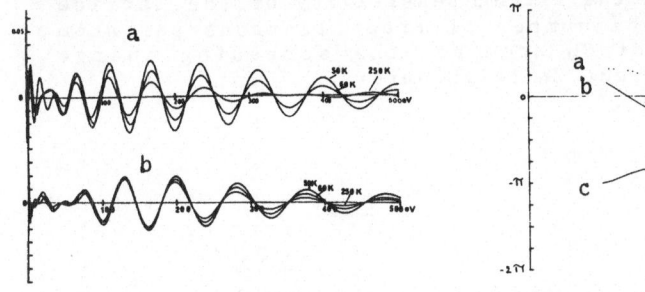

Fig. 1.1 : Filtered EXAFS spectra of pure Zn at different temperatures a) $\vec{c}//\vec{e}$ b) $\vec{a}//\vec{e}$

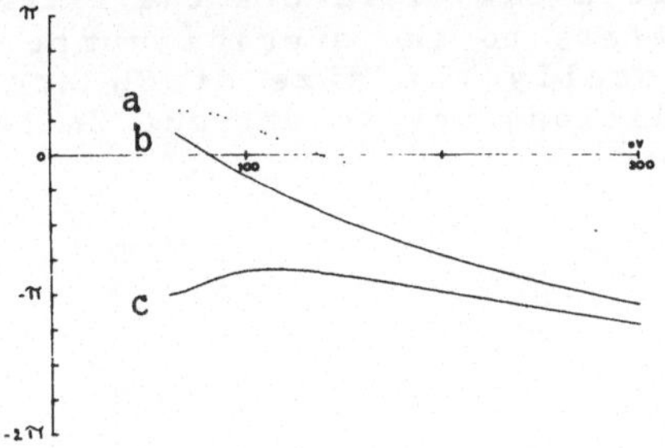

Figure 1.2 : Zn-Zn phase shifts a) theoretical (Teo and Lee) b) experimental, using 2.88 Å as Zn-Zn distance c) theoretical (Spanjaard)

one to extract quasi—experimental phase—shifts from the pure
host matrix. Following a method just proposed for AlMg solid
solutions, the experimental phase—shifts of the Zn—Zn pair has
to be slightly corrected with a computed difference to produce
the quasi—experimental Cu—Zn phase—shifts.

EXPERIMENTS

Single crystals of Zn 1 at .%_Cu and pure Zn were cut and
thinned parallel to (0001) and (01$\bar{1}$0) planes. We used the same
procedure as EISENBERGER and BROWN [6] to measure the first
neighbour distances inside and outside   the hexagonal
plane of the hcp structure (setting $\vec{a}$ or $\vec{c}$ axis parallel to
the polarization vector $\vec{e}$ of the beam). The ZnCu crystals were
water quenched from 350°C just before measurements. The EXAFS
spectra were recorded at 30 K for ZnCu crystals, at different
temperatures from ʾ250 K to 30 K in the case of pure Zn
crystals.

RESULTS

To extract the phase shifts for pure Zn one needs to avoid the
anharmonic dependance of the $\vec{e}//\vec{c}$ spectra as first shown by
EISENBERGER and BROWN [6]. That requires the cooling of the sample
below 60 K. This effect of anharmonicity is not visible for
the $\vec{a}//\vec{e}$ spectra (Fig. 1.1). Fig. 1.2 shows the Zn—Zn phase
shifts extracted from $\vec{e}//\vec{c}$ spectra. The direct comparison of
Fourier transforms (Fig. 1.3) for both Zn and ZnCu spectra
gives evidence of the reduction of the anisotropy of the Zn
matrix when Cu atoms are substituted to Zn atoms. However,
this is done by reducing the distances between nearest
neighbours both inside and outside the hexagonal plane, in
contrast from what is observed from the macroscopic effect :
the CuZn distance is 2.60 Å instead of 2.66 Å for Zn—Zn
distances within the hexagonal plane and 2.68 instead of
2.88 Å for Zn—Zn pairs of atoms outside the hexagonal plane.
From Vegard's law we expect $\frac{da}{ac}$ positive (= 0,201) and
$\frac{dc}{cc}$ = - 0.809 negative (a, c parameters of the hexagonal unit
$c\bar{e}$11, C is the concentration). The macroscopic effects
(Vegard's law) reflects the extreme sensitivity of the lattice
parameters to the average number of free electrons per atom
but locally the size of Cu atom and the screening charge
oscillations act to attract Zn neighbours.

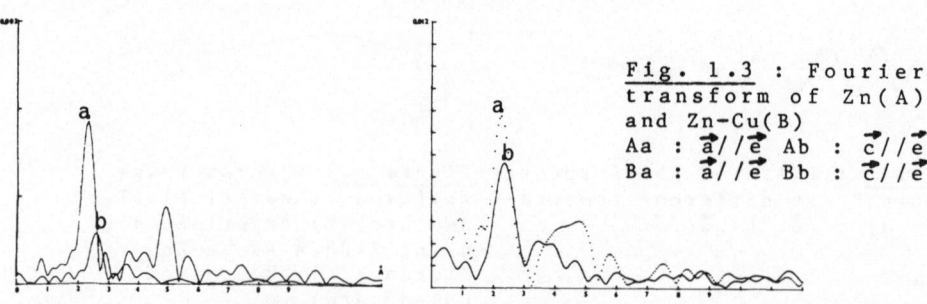

Fig. 1.3 : Fourier
transform of Zn(A)
and Zn—Cu(B)
Aa : $\vec{a}//\vec{e}$ Ab : $\vec{c}//\vec{e}$
Ba : $\vec{a}//\vec{e}$ Bb : $\vec{c}//\vec{e}$

Since the pioneering work of HEEGER and MC DIARMID on polyacetylene in 1977 considerable interest has been raised in conducting polymers [7]. The unstability of $(CH_2)_n$ against O_2 or moisture has stimulated efforts to seek other polymers able to form stronger bonds. Hence one of us (G.T.) obtained very good results with polythiophene and its derivatives, which take stability in the bridging of the carbon chain by sulfur atom [8]. Electrochemically synthetized by the oxydation of the monomer in a $CH_3CN-N(Bu)_4 SO_3 CF_3 (0.1 M)$ electrolytic medium on a Pt electrode, this polymer exhibits gain of 10 orders of magnitude in the parallel conductivity, which rises up to $1000 \, \Omega^{-1} cm^{-1}$ when p-doped. This is very similar to polyacetylene value. For aligned fibers the transverse conductivity is generally 5 times lower. In the neutral form the polymer

$$\left[(SC_4 H_2)_n \quad -\underset{S}{\bigcirc}-\underset{S}{\bigcirc}- \right]$$

behaves like a p semiconductor whose band gap is about 2 eV [9]. When fully doped one over three or two monomers units attaches the $(CF_3 SO_3)^-$ ion . When doping is achieved the diameter of the fiber increases from 200, 300 Å up to 800-1000 Å as shown by transmission electron microscopy [10]. Attempts to make crystals have not been successful yet but partial crystallization (6-8 %) has been obtained.

EXPERIMENT

We report here the first study, to our knowledge, of the direct in-situ study of the electrochemical steps leading to the inclusion of metallic particles in an organic matrix. That was done by dispersive X-ray absorption spectroscopy which uses a bent Si 111 crystal to focus the reflected beam on the sample, subsequantly detected on a 1024 photodiode array. About 10^8 photons were recorded in a single shot. Most of the spectra shown here are the sum of 8 shots generally recorded in 800 ms each, since the whole sample attenuates the transmitted beam by a hundred factor. In this first trial it was not possible to reduce the thickness of the electrochemical cell to value lower than 2 mm.

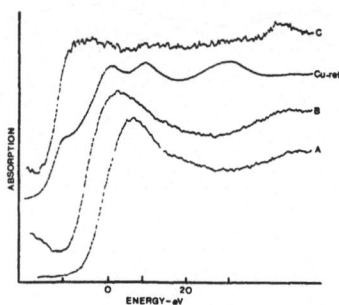

Fig. 2.1 : Near-edge spectra for Cu, aqueous $CuCl_2$ solution (A) immediate Cu^+ fixation on polymer (B), final state of reduction (C)

405

In that geometry the copper edge raises the absorption by 1. which can be compared to 4.6 for the polymer absorption cross section. That can be roughly described as an apparent 4.5 μ thick copper foil immersed within an equivalent 9 mm thick carbon layer. Thus extreme dispersion of copper is achieved.

Using this technique, we were able to follow te electrochemical aggregation of copper in poly 3 methylthiophene PMeT. The X-ray absorption in the vicinity of the Cu edge (8979.8 eV) was recorded for a 100 eV wide energy band pass. In this first attempt we used near edge features as finger prints to characterize the successive steps rather than produce quantitative analysis. Besides this first kinetic aspect of this experiment, we were taking benefit of the focussing optics to probe spatially the polymer. Since PMeT is electrochemically deposited around a Pt electrode, it is obvious that the distance of the point which is studied to the Pt electrode is an important parameter. An anterior investigation has shown that the focus spot is 350 μm wide.

RESULTS

a) The Cu foil and aqueous $CuCl_2$ -solution provides natural standards to calibrate the X-axis with energies. Cations (Cu^{2+}) yields a K-edge shifted towards higher energies by 8.5 eV (Fig. 2.1 a), compared to the Cu-metal edge which has been obtained by addition of 8 spectra each one recorded in 12 ms. Before immersion of the polymer by pouring the $Cu-Cl_2$ solution, we adjust the probing beam to be closed to the electrode so that maximum thickness of polymer is expected.

b) The immersion followed by cathodic polarization induces an immediate backward shift of the edge and associated white line (-3.5 eV) (Fig. 2.2a). That reflects copper fixation as Cu^+. That was confirmed later by red complex formation, characteristic of Cu^+ species when this as-immersed polymer was treated by bipyridine solution (Cu^{2+} species put in the same solution give, in contrast, a blue complex).

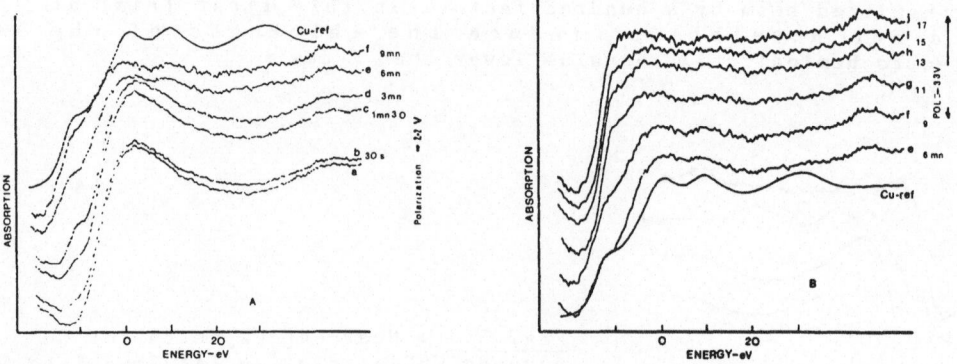

Fig. 2.2 A,B : Time-dependence X-ray absorption of Cu clustering rate as shown by edge shape

c) Holding the cathodic polarization the K-edge reveals a bump at the edge foot (Fig. 2.2b). The first hint of this bump appears after 30 sec. It is well shaped after 3' (Fig. 2.2b) but the white line is still there,which is no more true after 6 or better 9 minutes. At this stage the characteristic 9 eV splitted first peak is very similar to pure metallic features. Cu° species are formed, copper aggregates are dispersed homogeneously throughout the polymer, as checked with the scanning electron microscope.

d) Holding on the polarization achieves a total rise of the metal-like bump which ends up hidden in form of a smooth absorption jump (Fig. 2.1c, Fig. 2.2r). If the polarization is reversed (+ 1 V) Cu° species are dissolved to the initial Cu$^+$ ions : energy shift and reappearance of the white line. By applying once more cathodic polarisation, the same sequence occurred and exactly the same final state was reached. This finding still needs explanation. However,speculations upon defilling the p-states density could deal either with a dissolution of Cu aggregates under the high electric field within the polymer, or the formation of a pseudo Cu-H alloy with a partial electron transfer from Cu to proton. Such hypothesis needs extra experiments which are in progress.

e) If the same experiment is performed with a low concentrated CuCl$_2$ solution (10^{-2} M/l), large modifications are observed (Fig. 2.3).

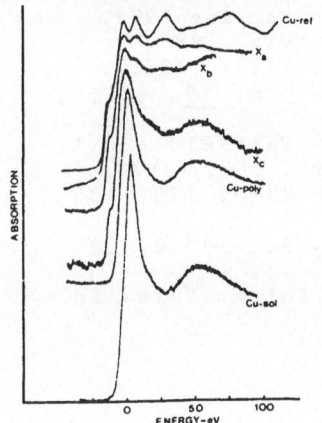

Fig. 2.3 : Spatial dependence for copper aggregates distribution Vs the position of the probing beam with respect to the Pt electrode

- No energy shift is detected when cathodic polarization is applied which shows that Cu$^+$ ions are not formed. The bipyridine solution confirms that statement since the red complex does not appear.

- After 15 minutes of polarization at - 3.3 eV, metallic clusters are detected in the polymer,but the final state previously observed is not reached.

- The Cu-cluster distribution inside the polymer is not homogeneous. Near the Pt electrode large clusters (~ μ m) are formed (Fig. 3Xa). They decrease in size and concentration when we probe progressively the outside.

- This last observation is consistent with the decrease in conductivity which leaves at the end an insulator. Hence high electric fields are located in the immediate vicinity of the Pt electrode, and drain copper ions before being reduced to the metallic state.

CONCLUSION

The two main features of dispersive X-ray absorption leads to follow kinetics at the minute time scale and to map the aggregate formation with a spatial resolution of 350 μ. It is worthwhile to point out that such studies have been performed for dilute samples. Furthermore the strong interest in that class of organic matrix to form tiny metallic clusters of high technological value provides new materials very suited for EXAFS investigations.

REFERENCES

1 - J. Mimault, A. Fontaine, P. Lagarde, D. Raoux, A. Sadoc, D. Spanjaard, J. of Physics F11, 1311, (1981).
2 - A. Craievich, E. Dartyge, A. Fontaine, D. Raoux, Proceedings EXAFS and XANES Conf. Frascatti 82, Springer Series, Chemical Physics 27, Edited A. Bianconi, L. Incoccia, S. Stipich.
3 - J. B. Boyce, J. C. Mikkelsen, SSRL User's Meeting 1983.
4 - J. C. Mikkelsen, J. B. Boyce, Phys. Rev. Lett. 1983.
5 - S. Froyen, C. Herring, J. Appl. Phys. 52, 7165, (1981).
6 - P. Eisenberger, G. S. Brown, Solid St. Com. 29, 481,4, (1979).
7 - A. G. Mac Diarmid, A. J. Heeger, in W. E. Hatfield (Ed.), "Molecular Metals", Plenum Press, New York, 161, (1979).
8 - G. Tourillon, F. Garnier, J. Electroanal Chem. 173, 135, (1982).
9 - G. Tourillon, F. Garnier, J. Electrochem. Soc. 2042, 130, (1983).
10- G. Tourillon, F. Garnier, J. Polym. Sci. Polym. Phys. Ed. 22, 33, (1984).

EXAFS Studies of Electrolyte Solutions

Donald R. Sandstrom

Department of Physics, Washington State University, Pullman, WA 99164-2814, USA

1. Direct Metal to Chloride Ion Bonding in Concentrated Transition Metal Chloride Solutions

Transition metal chlorides near saturation in aqueous solution have been the object of recent studies by X-ray diffraction, neutron diffraction, EXAFS, and other methods [1]. The local coordination environment of the metal ion in these solutions differs greatly for the different elements and valance states. For example, Fe^{3+} and Cu^{2+} show considerable first neighbor bonding to Cl^-, whereas Ni^{2+} and Co^{2+} show little or none. These differences are of technological importance in the recovery of the metals via steps involving concentrated solutions. Considerable attention has focused on Ni(II) chloride solutions, for which ordered arrangement of the cations was reported [2]. Other neutron diffraction results [3], X-ray diffraction results [4], and EXAFS results [5,6] showed full hydration of the Ni^{2+} cation, and no evidence for first neighbor cation-anion bonding.

However, subsequent studies by X-ray diffraction do show a small degree of direct bonding in that approximately equal amounts of $Ni(H_2O)_5Cl^+$ and $Ni(H_2O)_6^{2+}$ were reported for stoichiometric $NiCl_2$ solutions (3 moles/liter) [7]. This corresponds to an average coordination of each Ni^{2+} ion by approximately 0.5 Cl^-, so that Cl^- accounts for less than 10 percent of the total Ni^{2+} coordination. In non-stoichiometric $NiCl_2$ solutions (excess chloride ion), average Ni^{2+} coordination by Cl^- as high as 1.43 has been reported in X-ray diffraction studies [8].

The evident insensitivity of EXAFS to such a minority coordination species as Cl^- at the ~10 percent level in the presence of water coordination is not surprising in that both Cl and O are low-Z scatterers, with scattering amplitude functions that decay rapidly in wave vector [9]. In addition, these are highly disordered systems that require special analysis procedures for accurate coordination number determination [10]. One factor that aids in distinguishing Cl from O in EXAFS analysis is an approximately π difference in the scattering phase shift for the two scatterers [11]. In spite of this, it is likely that only the most careful analysis of high quality data, taking disorder explicitly into account, could reliably detect such a minority Cl^- coordination.

2. XAS of Solutions at the Cl K-Edge

Recent work has demonstrated that XAS can be carried out for a variety of materials on low-Z elements using the hard X-ray beam lines at SSRL [12]. These experiments have extended the useful energy range as low as the S K-edge (2472 eV) by employing fluorescence detection with the sample and detector in a He atmosphere. On unfocused wiggler beam lines, excellent resolution of ~0.5 eV was obtained.

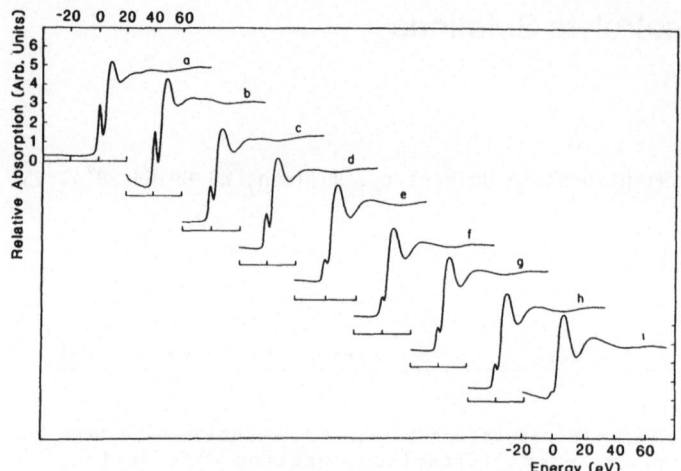

Figure 1. Cl K-edge XANES spectra for ferric chloride. a. $FeCl_3$ hydrated
salt. b. - i. $FeCl_3$ aqueous solutions in order of decreasing Fe^{3+}
concentration in range 5.75 - 0.40 moles/liter

Using the above techniques, concentrated transition metal chloride
solutions have been studied at the Cl K-edge (2822 eV) using a focused
bending magnet source of synchrotron radiation [13]. Results for ferric
chloride solutions show that both the XANES and EXAFS portions of the X-ray
absorption spectrum contain information bearing on the question of direct
cation-anion bonding. In ferric chloride solutions, such bonding is
extensive and concentration dependent [14].

In the XANES spectral region, the direct cation-anion bonding is
indicated by the presence of a strong single pre-edge peak having an
amplitude relative to the K-edge jump that is approximately proportional to
the expected degree of Fe^{3+}-Cl^- bonding, as shown in Fig. 1.

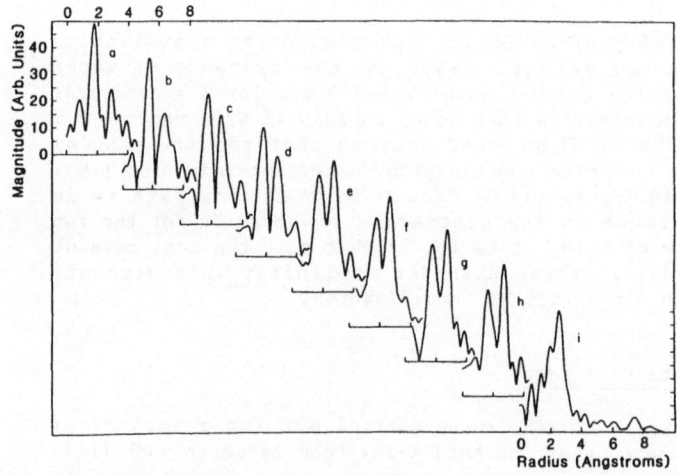

Figure 2. Fourier transforms of Cl K-edge EXAFS spectra for the same
ferric chloride hydrated salt and solutions as Fig. 1

In the EXAFS region, the bond lengths of iron-chloride vs iron-water are sufficiently different to produce relatively well resolved peaks in Fourier transforms of the EXAFS spectra. The relative heights of these peaks vary in accord with the expected degree of chloride ion vs water in the first coordination sphere, as shown in Fig. 2.

The Cl K-edge results contrast markedly with the metal K-edge results described above in that the latter show essentially no XANES information, and are dominated by the majority water coordination. In contrast, the Cl K-edge measurements take advantage of the spectral element specificity of XAS to view the bonding from the point of view of the minority constituent.

3. Cl K-Edge XANES of Transition Metal Chloride Hydrated Salts

XANES similar to that noted above for ferric chloride hydrated salt and solutions is also obtained for several other transition metals. The hydrated salts have been studied at higher resolution for several divalent [15] and trivalent [16] ions. The pre-edge peak is understood to be due to transitions to antibonding molecular orbitals derived from the metal 3d and chlorine 3p orbitals, and is thus a direct indication of direct metal-chloride bonding. For a given valency, the degree to which the pre-edge peak is separated from the main absorption edge depends on the d-electron affinity of the empty metal 3d levels. Thus, for Mn(II) there is almost no separation, and there is maximum separation (~ 5 eV) for Cu(II) [15]. Separation is enhanced for the trivalent ions, with Cr(III) and Fe(III) showing strongly separated pre-edge peaks [16].

It should be noted that corresponding pre-edge features are essentially absent in the XANES spectra for the same materials measured at the metal K-edge. This can be understood in terms of the symmetry environment of the metal ion vs that of the chloride. Dipole transitions from metal 1s to 3d in the approximately octahedral symmetry site of the metal ion are forbidden, and this is exhibited in the case of ferric chloride hydrated salt and solutions by an extremely weak single pre-edge peak. In contrast, the chloride ion is in a site of low symmetry and transitions to the metal 3d – chlorine 3p hybrids can occur, as seen in these results.

4. Cl K-Edge XANES and EXAFS Results for Nickel Chloride Solutions

Cl K-edge XAS spectra were measured for $NiCl_2$ hydrated salts and aqueous solutions having concentrations of 4.0, 2.0, and 0.4 moles/liter. The pre-edge XANES feature is clearly visible for the hydrated salt, as shown in Fig. 3. In this case, the feature is not as well separated from the main absorption edge as in the case of ferric chloride. In this structure, Ni^{2+} has two Cl^- ions in trans positions in the first coordination sphere. In contrast, the solution spectra XANES (Fig. 3) have a weak pre-edge contribution that can only be made out in difference curves, also shown in Fig. 3. If the areas under the pre-edge difference curve peaks are assumed to be proportional to the Cl^- coordination number of Ni^{2+}, with the curve for the hydrated salt corresponding to two Cl^-, then the 4.0 moles/liter curve gives a value of 0.6, in agreement with X-ray diffraction results [7].

Cl K-edge EXAFS spectra also show evidence for direct nickel-chloride bonding in these solutions. In Fig. 4, Fourier transforms show good separation of the peaks due to nickel and water coordination of chlorine, and relative peak heights show the same strong decrease in relative amount of nickel-chloride bonding in going from the hydrated salt to more dilute

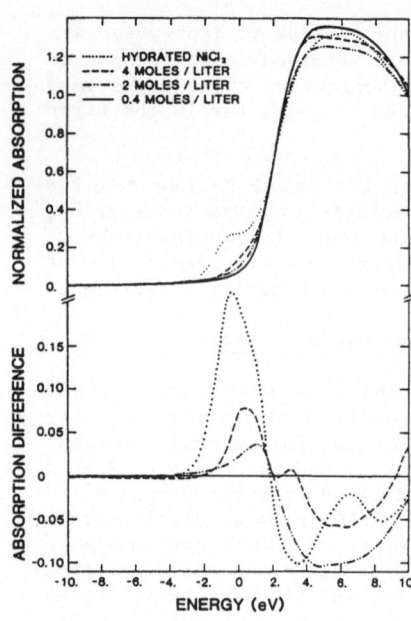

Figure 3. A. Cl K-edge XANES spectra for nickel chloride. B. Difference spectra obtained by subtracting the 0.40 moles/liter spectrum from the indicated spectra in A

Figure 4. Fourier transforms of Cl K-edge EXAFS spectra for nickel chloride solutions. a. $NiCl_2$ hydrated slat, b. 4.0 moles/liter, c. 2.0 moles/liter, d. 0.40 moles/liter

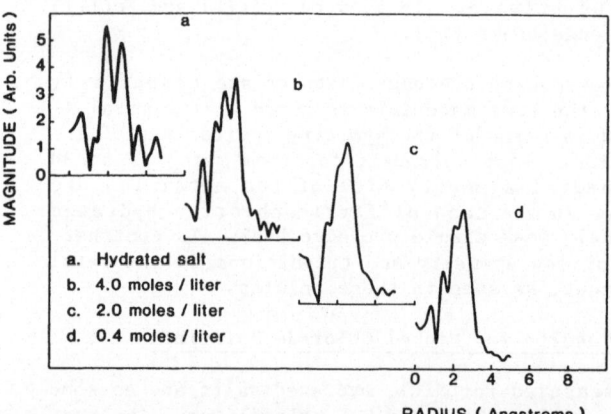

a. Hydrated salt
b. 4.0 moles / liter
c. 2.0 moles / liter
d. 0.4 moles / liter

solutions. Strong interference of short distance structure associated with background removal makes it impossible to speculate to what degree nickel-chloride bonding may persist at low concentration.

5. Summary

Cl K-edge XAS measurements overcome the limitations seen in metal K-edge studies of minority ligand bonding in transition metal chloride solutions. However, the quantitative conclusions that can be drawn from work to date is limited by the relatively poor resolution of the measurements carried out on a focused beam line. Results for other materials on unfocused wiggler beam lines [12] show much higher resolution, and further studies on these solutions will utilize this source. With attainable resolution, it should be possible to obtain resolutions for solutions comparable to those used to study the 3d transition metal hydrated salts [15, 16]. In spite of present limitations, however, these XAS measurements provide unequivocal evidence for direct nickel-chloride bonding at approximately the same level as detected by X-ray diffraction studies.

1. D. R. Sandstrom, Nuovo Cimento D, in press.

2. R. A. Howe, W. S. Howells, and J. E. Enderby, J. Phys. C. $\underline{17}$, L111 (1974).

3. A. K. Soper, G. W. Neilson, J. E. Enderby, and R. A. Howe, J. Phys. C. $\underline{10}$, 1793 (1977); G. W. Neilson and J. E. Enderby, J. Phys. C. $\underline{11}$, L625 (1978).

4. R. Caminite, G. Licheri, G. Piccaluga, and G. Pinna, Faraday Discuss. Chem Soc. $\underline{64}$, 62 (1978); R. Caminiti, G. Licheri, G. Paschina, G. Piccaluga, and G. Pinna, Z. Naturforsch. Teil A, $\underline{35}$, 1361 (1980).

5. D. R. Sandstrom, J. Chem. Phys. $\underline{71}$, 2381 (1979).

6. G. Licheri, G. Paschina, G. Piccaluga, G. Pinna, and G. Vlaic, Chem. Phys. Lett. $\underline{83}$, 384 (1981).

7. M. Magini, J. Chem. Phys. $\underline{74}$, 2523 (1981).

8. M. Magini, G. Paschina, and G. Piccaluga, J. Chem. Phys. $\underline{76}$, 1116 (1982).

9. B. K. Teo and P. A. Lee, J. Am. Chem. Soc. $\underline{101}$, 2815 (1979).

10. G. Bunker, Nucl. Instrum. Methods $\underline{207}$, 437 (1983).

11. D. R. Sandstrom, B. R. Stults, and R. B. Greegor, EXAFS Spectroscopy: Techniques and Applications, B. K. Teo and D. C. Joy, Ed., (Plenum, New York, 1981), pp. 139-57.

12. F. W. Lytle, R. B. Greegor, D. R. Sandstrom, E. C. Marques, J. Wong, C. L. Spiro, G. P. Huffman, and F. E. Huggins, Nucl. Instrum. Methods, in press.

13. D. R. Sandstrom, E. C. Marques, and R. E. Hamm, "Cl K-edge X-ray Absorption Studies of Concentrated Ferric Chloride Solutions", in Ninth SSRL User Group Meeting Proc., pp. 49-50; D. R. Sandstrom, "Study of Strong Electrolytes by X-ray Spectroscopy at the Cl K-Edge", in SSRL Activity Report for April 1, 1982 - March 31, 1983, p. VII-109.

14. M. Magini and T. Radnai, J. Chem. Phys. $\underline{71}$, 4255 (1979).

15. C. Sugiura and T. Suzuki, J. Chem. Phys. $\underline{75}$, 4357 (1981).

16. S. Muramatsu and C. Sugiura, J. Chem. Phys. $\underline{76}$, 2107 (1982).

EXAFS and X-Ray Diffraction Studies of Ni(II)- and Zn(II)-Glycinate Complexes in Aqueous Solution

K. Ozutsumi, T. Yamaguchi, and H. Ohtaki

Department of Electronic Chemistry, Tokyo Institute of Technology
Nagatsuta, Midori-ku, Yokohama 227, Japan

K. Tohji and Y. Udagawa

Institute of Molecular Science, Myodaiji, Okazaki 444, Japan

1. Introduction

EXAFS spectroscopy has been proved useful in investigating the local structure in solution. In particular, the EXAFS method is the most suitable to study the structure of dilute solutions, since information of the solvent structure is absent in the EXAFS spectra.

According to the single-electron and single-scattering theory of EXAFS [1], it is essential for structural determination to estimate the back-scattering amplitude of the absorbing atom and the phase shifts due to the absorber and scatterers. When the theoretical values calculated by Teo and Lee [2] are used, the λ and E_0 values have to be estimated reasonably to determine the reliable coordination number and the bond distance, respectively.

In the present paper we will report the EXAFS study on the structures of the bis(glycinato) complexes of Ni(II) and Zn(II) ions in aqueous solution, which could not be determined by the usual X-ray diffraction method because of their low solubilities in water. The λ and E_0 values were obtained from measurements of the primary standard samples of aqueous nitrate solutions of nickel-(II) and zinc(II), the structures of which had been determined by X-ray diffraction [3]. The same values of λ and E_0 were used in subsequent structural determination of the glycinate-complexes in solution.

2. Experimental

EXAFS measurements were performed with a laboratory X-ray absorption spectrometer using a rotating anode, which had been described elsewhere [4]. The details of X-ray diffraction measurements have been published [3]. Samples measured were aqueous solutions containing $[M(gly)_3]^-$ (1 mol dm^{-3}) or $[M(gly)_2]$ (0.1 - 0.3 mol dm^{-3}) (M = Ni(II) and Zn(II), gly = glycinate ion) as a main species as well as metal nitrate solutions (1 mol dm^{-3}). Powder samples of the bis(glycinato) complexes of both metals with known structure [5,6] and metal oxides were also measured as the structural reference.

3. Results and Discussion

3.1 Nickel(II) Complexes

In the beginning, we examined the applicability of the two structural standards to the analysis of the solution samples. From the least-squares fits, in which parameters E_0, λ, r and σ were allowed to vary with known coordination number, we obtained similar E_0 values for the both standards. However, the value of λ obtained for the metal oxides was smaller by about 1.5 than that for the aqueous solutions. This means that the analysis using the metal oxide standards will give about 50% larger coordination number than that using solution standards, though the bond length determined may be independent of the references. In subsequent calculations, the solution standards were found to be suitable for the sample solutions investigated.

414

Table 1. Parameter values obtained by the least-squares fits in the k-space. The values in parentheses are those obtained by X-ray diffraction [3,7].

	scatterer	M = Ni(II) r/Å	σ/Å	M = Zn(II) r/Å	σ/Å
$[M(OH_2)_6]^{2+}$(aq)	O	2.02 (2.04)	0.014	2.07 (2.08)	0.065
$[M(gly)_3]^-$(aq)	O or N	2.05	0.055	2.13	0.068
	O	2.01 (2.03)	0.014[a]	2.13 (2.12)	0.065[a]
	N	2.10 (2.14)	0.014[a]	2.13 (2.12)	0.065[a]
$[M(gly)_2(OH_2)_2]$(aq)	O or N	2.04	0.04	2.07	0.06
$[Ni(gly)_2(OH_2)_2]$(c)	O or N	2.04 (2.08)[b]	0.05		
$[Zn(gly)_2] \cdot H_2O$ (c)	O or N			2.03	0.07

a. Fixed. b. A mean value of 2.06 Å (2× Ni-O), 2.08 Å (2× Ni-N) and 2.10 Å (2× Ni-OH2)

Table 1 gives the results obtained by a conventional curve-fitting procedure for the Fourier filtered $k^2 \cdot \chi(k)$ values of the first peak in the radial distribution curve. The E_0 value obtained for the aqueous nickel(II) nitrate solution seemed reasonable because the Ni-OH2 bond length evaluated was compared well with that determined by the X-ray diffraction method [3]. Then, the accuracy in the structural determination by the EXAFS method was checked by analyzing the data for $[Ni(gly)_3]^-$ complex adopting the above E_0 and λ values. The least-squares calculations with a model of only one back-scatterer (O or N) gave the bond length of 2.05 Å and the σ value of 0.055 Å, the latter being about four times larger than that obtained for the Ni-OH2 bond within the $[Ni(OH_2)_6]^{2+}$ ion. The larger σ value for the Ni-O and Ni-N bonds within the $[Ni(gly)_3]^-$ complex seemed unreasonable since the temperature coefficients for the bonds estimated by the X-ray diffraction method were practically the same [3,7]. In the final calculation, therefore, both Ni-O and Ni-N interactions were taken into account with the constant σ value of 0.014 Å. The results are given in Table 1. The two different lengths of the Ni-O (2.01 Å) and Ni-N (2.10 Å) bonds were obtained, independent of the initial values inserted. The agreement between the results obtained by the EXAFS and X-ray diffraction is very satisfactory [7], and thus the structure analysis by the EXAFS method is accurate enough provided suitable standard materials are selected.

The same analytical procedure was applied to the solution and powder samples of the bis(glycinato)nickel(II) complex. The amplitude of the first peak in the Fourier transform was similar in   both samples (Fig. 1) and hence the

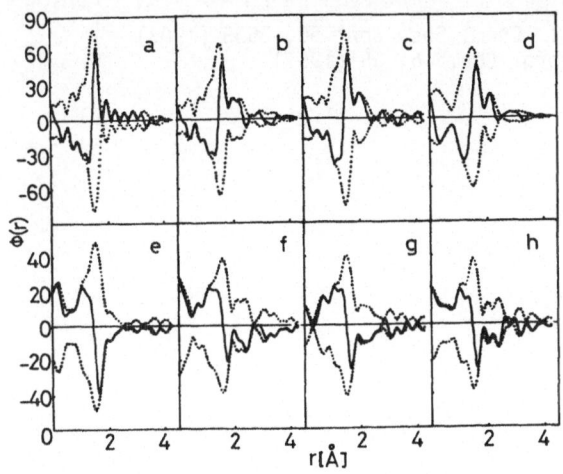

Fig. 1 The Fourier transform of the $k^2 \cdot \chi(k)$ data. The absolute values are shown by dotted lines and the values of the real part by solid lines.
(a) $[Ni(OH_2)_6]^{2+}$(aq)
(b) $[Ni(gly)_3]^-$(aq)
(c) $[Ni(gly)_2(OH_2)_2]$(aq)
(d) $[Ni(gly)_2(OH_2)_2]$(c)
(e) $[Zn(OH_2)_6]^{2+}$(aq)
(f) $[Zn(gly)_3]^-$(aq)
(g) $[Zn(gly)_2(OH_2)_2]$(aq)
(h) $[Zn(gly)_2] \cdot H_2O$(c)

curve-fits gave very similar results (Table 1). Therefore, the bis-complex in the aqueous solution was expected to have two additional water molecules to complete the octahedral configuration as in the crystalline state [5]. A mean value (2.08 Å) of the distances found in the crystal structure was slightly longer than the EXAFS result (2.04 Å for the average bond length), but the difference was almost within uncertainties of the both measurements.

3.2 Zinc(II) Complexes

The structures of the $[Zn(OH_2)_6]^{2+}$ and $[Zn(gly)_3]^-$ complexes were analyzed in the same manner as described before. The final results are given in Table 1. Both results obtained by the EXAFS and X-ray diffraction methods agreed very well. The structure of the bis(glycinato)zinc(II) complex in the solid state has been reported to be octahedral with additional two carbonyl oxgen atoms within the glycinate ions coordinated to adjacent Zn(II) ions. When the bis-complex is dissolved to water, the structure of the complex may be either octahedral with two additional water molecules or tetrahedral without coordinated water molecules. Figure 1 shows that the similar amplitude of the first peak for both the solution and the powder samples indicated the coordination number of 6 for the bis-complex in solution as in crystal. The least-squares fits of the data also confirmed this result. The mean bond distance (2.03 Å) for the powder sample seemed slightly shorter than that found in solution (2.07 Å). It has been found from the Raman spectral measurements that the Zn-N stretching frequency observed at ca. 470 cm^{-1} in the crystal was shifted to ca. 430 cm^{-1} in aqueous solution [8]. The lengthening bond distance for coordination in the solution found in the EXAFS study may correspond to the shift to the lower wavenumber of the Zn-N stretching frequency of the Raman spectra.

This work was supported by the Joint Studies Program (1983) of the Institute of Molecular Science.

References

1. D.E. Sayers, E.A. Stern and F.W. Lytle: Phys. Rev. Lett. 27, 1204 (1971)
2. B.-K. Teo and P.A. Lee: J. Am. Chem. Soc. 101, 2815 (1979)
3. H. Ohtaki, T. Yamaguchi and M. Maeda: Bull. Chem. Soc. Jpn. 49, 1482 (1976)
4. K. Tohji, Y. Udagawa, T. Kawasaki and K. Masuda: Rev. Sci. Instrum. 54, 1482 (1983)
5. H.C. Freeman and J.M. Guss: Acta Crystallogr. B24, 1133 (1968)
6. B.W. Low, F.L. Hirshfeld and F.M. Richards: J. Am. Chem. Soc. 81, 4412 (1959)
7. K. Ozutsumi and H. Ohtaki: Bull. Chem. Soc. Jpn. 56, 3635 (1983)
8. K. Krishnan and R.A. Plane: Inorg. Chem. 6, 55 (1967)

EXAFS, X-Ray, and Neutron Diffraction of Electrolyte Solutions

T. Yamaguchi
Department of Electronic Chemistry, Tokyo Institute of Technology, Nagatsuta
Midori-ku, Yokohama 227, Japan
O. Lindqvist
Department of Inorganic Chemistry, Chalmers University of Technology and
University of Gothenburg, S-412 96 Gothenburg, Sweden
J.B. Boyce
Xerox Palo Alto Research Centers, Palo Alto, CA 94304, USA
T. Claeson
Department of Physics, Chalmers University of Technology
S-412 96 Gothenburg, Sweden

1. Introduction

Since *ionic hydration* is an important phenomenon in many chemical reactions
and in biological systems, the structure of hydrated ions has been extensively
investigated so far. In particular, X-ray and neutron diffraction methods have
played a principal role for that purpose. Recently, an extended X-ray absorption fine structure (EXAFS) method has been a useful probe of the short-range
order in solutions.

For a typical four-component system M, X, O and H (e.g. MX_n in H_2O or $M(XO_n)$
in H_2O), the spacial distribution in the system can be expressed in terms of
pair correlation functions, $g(r)$, where r represents an interatomic distance.
In the present system, there are ten partial pair correlation functions classified into three kinds: those relating to ion-solvent interactions (g_{M-O},
g_{M-H}, g_{X-O}, g_{X-H}), those representing ion-ion interactions (g_{M-M}, g_{M-X}, g_{X-X}),
and those specifying solvent-solvent interactions (g_{O-O}, g_{O-H}, g_{H-H}). A
knowledge of these functions is essential for the complete understanding of
aqueous solutions. The method of X-ray diffraction gives the sum of the ten
pair correlation functions. This implies that the information on the ionic
hydration may be hidden or suppressed by the solvent structure, particularly
for species of low concentration. Although the situation is similar for a
conventional neutron scattering, an *isotopic substitution* method recently
developed has made it possible to separate a set of g_{M-O}, $g_{M-D}*$, g_{M-X}, and
g_{M-M}, or of g_{X-O}, g_{X-D}, g_{X-M} and g_{X-X} from the total correlation function [1].
The information on D atoms is obtainable only by the neutron diffraction.
However, this method has an inherent limitation in that it is applicable only to
systems with suitable isotopes. The EXAFS method can also extract similar
sets of correlation functions as the isotopic substitution method by selecting
the absorption edges. Thus, these two techniques are applicable even to dilute solutions. One of the disadvantages in the EXAFS method may be the loss
of long-range interactions in the system, but this might not be very serious
with respect to the first hydration sphere of the ions.

From the above consideration, combined EXAFS, X-ray and neutron diffraction studies may be very helpful in structure determination of solutions.
This paper will describe such structural studies on Sn(II) and Ag(I) hydration. The HOMO electrons of Sn(II), 5s in its ground state of the free ion,
usually tend to form a non-spherical charge distribution with mixing 5p electrons coming from coordinated ligands (they are *stereochemically active*).
The Sn(II) lone-pairs have been believed to be active also in solution. On
the other hand, the hydration number of Ag(I) has long been a topic of discus-

---

*Deuterium is used in neutron diffraction rather than hydrogen.

sions among coordination chemists since crystal structure determination of
hydrated Ag(I) ions is impossible because of no single crystal available.

2. Experimental

EXAFS measurements were performed on the Sn K and Ag K edges at the Stanford
Synchrotron Radiation Laboratory. An aqueous solution of tin(II) perchlorate
(3 mol dm^{-3}), and a perchlorate (3 mol dm^{-3}) and two nitrate (3 and 9 mol dm^{-3})
aqueous solutions of Ag(I) were prepared. Powder samples of SnO and Ag$_2$O were
also measured for structural standards. X-ray scattering from similar solutions was measured on a θ-θ diffractometer [2]. Neutron data of aqueous
silver(I) perchlorate solutions (3 mol dm^{-3}) were collected at the Institute
of Laue-Langevin in Grenoble [3].

3. Results and Discussion

Sn(II) hydration: The details of the results have been published elsewhere
[4]. Figure 1 shows the radial distribution curves obtained by the EXAFS
and X-ray measurements. The EXAFS spectrum showed the peak corresponding to
g_{Sn-O}, while in the X-ray case all the peaks arising from the system appeared,
e.g. at 1.5 Å (Cl-O), at 2.3 Å (Sn-O and O-O within the perchlorate ion).
The assignment of a shoulder around 2.8 Å was either to secondary Sn-O bonding often found in crystals or to the typical H_2O-H_2O interactions in the
water structure. We analyzed the EXAFS spectrum for two models; (A) one Sn-
O bond, (B) two different Sn-O bonds. The results of least-squares fits in
the r-space are given in Table 1. As seen in Table 1, no evidence for the
secondary Sn-O bonding was detected by the EXAFS analysis. This was interpreted as meaning that there was no secondary Sn-O bond, or, if any, the
corresponding pair correlation function was too broad to contribute to the
EXAFS spectrum. Thus, it was concluded that the interaction between the
active Sn(II) lone-pairs and water molecules was not significant in the solution.

Ag(I) hydration: The detailed reports of the studies appear in Refs. 2 and
5. Figure 2 shows the radial distribution curves obtained by the EXAFS,
X-ray and neutron diffraction measurements. The peak at 2.4 Å (Fig. 2b),

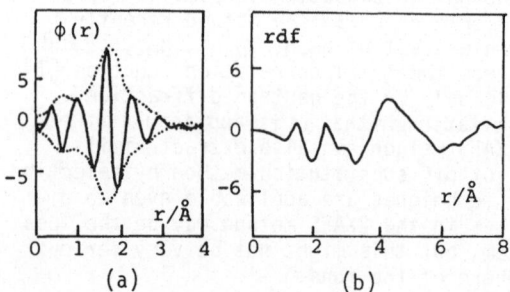

(a)   (b)

Fig. 1 The radial distribution
curves for the Sn(ClO$_4$)$_2$ solutions
obtained by (a) EXAFS and (b) X-ray
diffraction

Table 1. A comparison of Sn-O
distance r[Å] and coordination
number N

		EXAFS A	EXAFS B	X-ray
Sn-O	r	2.21	2.27	2.34
	N	3.8	3.1	3.4
Sn-O'	r	-	2.21	-
	N	-	1.1	-

Table 2. A comparison of Ag(I)-O distance
r[Å] and the hydration number N

mol/dm^3			EXAFS	X-ray	Neutron*
3	AgClO$_4$	r	2.31	2.38	2.36
		N	2.9	4	3.8
3	AgNO$_3$	r	2.36	2.42	
		N	3.9	4	
9	AgNO$_3$	r	2.34	2.43	
		N	2.6	4	

*r(Ag-D)=2.8 Å and N(Ag-D)=8

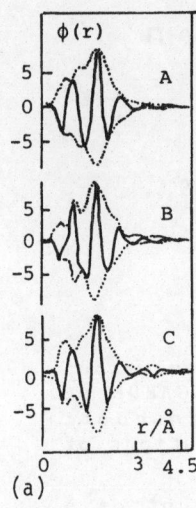

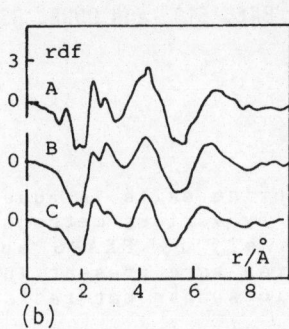

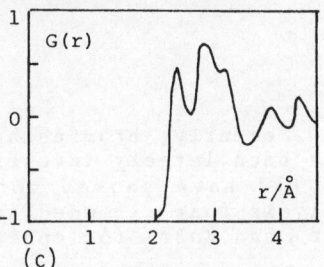

Fig. 2 The radial distribution curves for the AgClO$_4$ and AgNO$_3$ solutions obtained by (a) EXAFS, (b) X-ray and (c) neutron measurements: (A) 3 mol dm^{-3} AgClO$_4$ (B) 3 mol dm^{-3} AgNO$_3$ (C) 9 mol dm^{-3} AgNO$_3$

which was ascribed to the Ag-O interactions within hydrated Ag(I) ions, was overlapped with the 2.8 Å peak due to the typical H$_2$O-H$_2$O interactions in the bulk water. On the contrary, in the G(r) function obtained by the neutron study using isotopes Ag^{107} and Ag^{109} the contribution from the bulk water structure was absent and the peaks at 2.4 and 2.8 Å were due to the Ag-O and Ag-D interactions within the hydrated ions, respectively. The peaks were not completely resolved, however. From these standpoints, the EXAFS data were analyzed with Ag-O pairs for the solutions. The results are given in Table 2. The Ag-O distance obtained was comparable with those from the diffraction methods. The hydration number of interest was obtained as 3 - 4 (EXAFS), 4 (X-ray) and 3.8 (neutron). The results agreed well within experimental uncertainties.

From the present study, the EXAFS method has proved very useful, particularly in extracting the information of ion-solvent interactions selectively. A combined study of EXAFS, X-ray and neutron diffraction is most desirable in structure determination of electrolyte solutions.

References

1. J.E. Enderby and G.W. Neilson: Adv. Phys. 29, 323 (1980)
2. T. Yamaguchi, G. Johansson, B. Holmberg, M. Maeda and H. Ohtaki: Acta Chem. Scand. A 38, (1984), in press.
3. Unpublished works done by a group (Drs. G.W. Neilson, M. Sandström, G. Johansson and T. Yamaguchi)
4. T. Yamaguchi, O. Lindqvist, T. Claeson and J.B. Boyce: Chem. Phys. Lett. 93, 528 (1980)
5. T. Yamaguchi, O. Lindqvist, J.B. Boyce and T. Claeson: Acta Chem. Scand. A 38, (1984), in press.

EXAFS at the Cd K Edge: A Study of the Local Order in CdBr$_2$ Aqueous Solutions

A. Sadoc

Laboratoire de Physique des Solides, Bât. 510, and LURE, Bât. 209 C,
F-91405 Orsay Cédex, France

Recently, bromine and chlorine salts in aqueous solutions have been largely investigated by various methods. LAGARDE et al. [1] have showed conclusively by EXAFS spectroscopic analysis that extended structures were present in solutions of CuBr$_2$ and ZnBr$_2$ for concentrations near saturation.

The object of this paper is to give a brief account of a structural study in CdBr$_2$ aqueous solutions by EXAFS which will be discussed at length elsewhere [2]. The EXAFS spectra have been obtained at LURE (DCI, Orsay). The Cd edge has been investigated with either a Si(400) or a Si(311) monochromator and the Br edge with a Si(220) one.

Powdered samples of pure cadmium were prepared in order to test the apparatus at the Cd K edge energy (E = 26.72 keV). Cadmium has an hexagonal structure with a great c/a ratio (1.885) and two first neighbour distances (2.979 Å and 3.293 Å). For a single crystal, the EXAFS spectra obtained with synchrotron radiation should be different for polarization parallel or perpendicular to the c axis. Therefore cadmium powder was used rather than a polycrystalline foil which are generally thinned by rolling and in which a preferential ordering of the grains could have been induced.

Aqueous solutions of CdBr$_2$ were prepared for concentrations ranging from saturation (C$_S$) to C$_S$/ 50 . All data were taken at room temperature. The EXAFS was processed by using standard procedures of preedge substraction, spline removals and Fourier filtering.

1 - CADMIUM

The experimental χ (k) spectrum (fig. 1) can be fitted using 6 neighbour atoms at 2.93 Å and 6 ones at 3.26 Å with the same σ parameter for the two subshells (σ = .127 Å). These distances are in good agreement with the crystallographic data. The amplitudes and phase shifts used in this fit were calculated from TEO and LEE [3]. The E$_o$ value was taken at the first inflexion point of the edge and the Γ parameter was found equal to 0.32 Å^{-2}. It is worth noting that the Fourier transform of kχ(k) obtained for cadmium exhibit mainly one peak, as found previously by THULKE and RABE [4].

2 - AQUEOUS CdBr$_2$ SOLUTIONS

The spectra were found identical on both edges in the range of the concentrations studied. For example, normalized

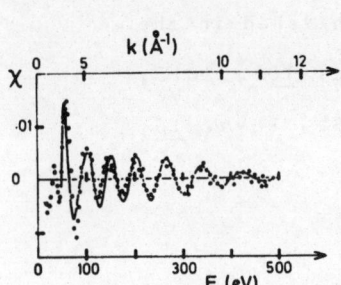

Fig.1. Experimental (dots) and simulated (solid line) X (k) for pure
cadmium

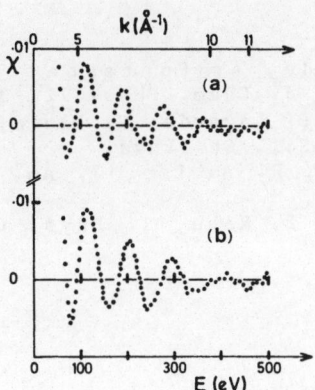

Fig.2. Experimental EXAFS above the
Cd edge for (a) the saturated, C_S,
solution of $CdBr_2$ in water, and (b)
a dilute one ($C_S/4$)

$X(k)$ EXAFS above the Cd edge are compared in fig. 2 for the
saturated solution and for a dilute one ($C_S/4$).

Fourier transformation of the EXAFS data yields a
pattern, that is qualitatively similar to a radial
distribution function. The transforms for all samples are
dominated by a peak corresponding to the Cd-Br interaction. At
distances shorter than the Cd-Br bond lengths, the transforms
of the EXAFS above the Cd edge contain features due to Cd-O
interactions. The consequence of this is evident in the
curve-fitting analysis where inclusions of Cd-O components
significantly improves the fit.

Typical fits are shown in fig. 3 : 2 Br ions at 2.55 Å
and 3 oxygen atoms at 2.18 Å are linked in average to one Cd
while 1.5 to 2 Cd ion are found at 2.58 Å around each Br ion.
Therefore these results show that a well-organized local
structure exists in the solutions of $CdBr_2$ in water, even
in dilute solutions.

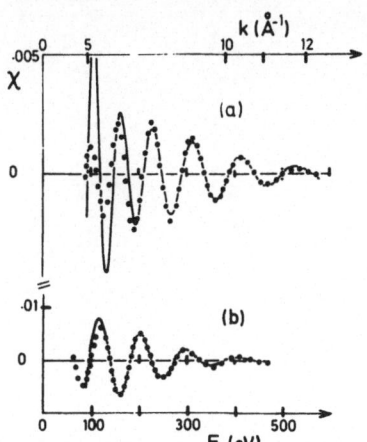

Fig.3. Fourier filtered first shell (dots)
and its fit (solid line) for $CdBr_2$ aqueous
solutions (a) on the Br edge, (b) on the Cd
one

421

REFERENCES

1- P. Lagarde, A. Fontaine, D. Raoux, A. Sadoc, P. Migliardo, J. Chem. Phys. $\underline{72}$, 3061 (1980).
2- A. Sadoc, P. Lagarde, G. Vlaic, to be published in the J. Phys. C : Sol. St. Phys.
3- B. K. Teo, P. A. Lee, J. Am. Chem. Soc. $\underline{101}$, 2815, (1979).
4- W. Thulke, P. Rabe, J. Phys. C : Sol. St. Phys. $\underline{16}$, L955, (1983).

EXAFS Studies of $Cd_2(AlCl_4)_2$: Structure of the Cd_2^{2+} Ion

A.P. Hitchcock, B. Christian, and R.J. Gillespie

Department of Chemistry, McMaster University, Hamilton
Ontario L8S 4M1, Canada

1. Introduction

Homopolyatomic ions of metals [1] are of considerable interest since they are among the simplest examples of isolated metal-metal bonding. The Hg_2^{2+} cation is well known. Recently compounds containing the Hg_3^{2+} and Hg_4^{2+} cations have been prepared and structurally characterized [2]. Compounds containing the analogous Cd_2^{2+} species are much less stable and the structure of Cd_2^{2+} has not yet been determined in any simple system. Cd_2^{2+} was first postulated to explain the properties of $Cd/CdCl_2$ melts. Raman [3] and magnetic measurements indicate that $Cd_2^{2+}(AlCl_4^-)_2$ is produced in fused mixtures of Cd, $CdCl_2$ and $AlCl_3$. Since crystals of the micaceous $Cd_2(AlCl_4)_2$ suitable for single crystal studies have not yet been obtained, we have used Cd K-shell EXAFS to investigate the structure of this species.

2. Experimental

$Cd_2(AlCl_4)_2$ was prepared by co-melting Cd, $CdCl_2$ and $AlCl_3$ in the ratio 1:1:2 [4]. The Raman spectrum was found to be very similar to that reported earlier [3]. EXAFS samples of the reactive $Cd_2(AlCl_4)_2$ species were prepared in a dry box by grinding with dried, powdered teflon and mounting in an airtight, thin-walled Kel-F holder. Cd metal, CdO and $CdCl_2 \cdot 2.5H_2O$ were used as model compounds. The metal sample was a 7μm foil. The CdO and $CdCl_2 \cdot 2.5H_2O$ samples were ground with $LiCO_3$ and sandwiched uniformly into a plastic sample holder. The Cd K absorption spectra (26.6-27.6 keV) were recorded at 77 and 300K using the C2 station of CHESS. The spectral resolution was estimated to be 10 eV from the difference between the 20 and 80% positions of the sharpest Cd K absorption edge.

3. Results and Discussion

Figure 1a shows the $k^3\chi(k)$ EXAFS isolated by a four-section spline from the spectra of CdO, Cd and $Cd_2(AlCl_4)_2$. The magnitudes of the Fourier transforms of these spectra are shown in Fig. 1b. The CdO radial distribution function (RDF) shows a peak corresponding to the 6 nearest neighbour oxygen atoms but is dominated by the second shell of 6 Cd atoms. A third shell of Cd atoms is also seen. Only a single shell of Cd atoms is detected in the 77K-EXAFS of Cd metal. This is so even though there are 6 atoms at 2.979Å and 6 at 3.293Å in the distorted hcp structure of Cd metal. It appears that even at 77K the thermal motion in the second shell direction is so large that backscattering from this shell results in a low, broad RDF peak too weak to be detected [5].

The RDF from CdK-EXAFS of $Cd_2(AlCl_4)_2$ shows a single peak at somewhat lower R than the Cd backscattering peaks in CdO (second shell) or Cd metal. Since $Cd_2(AlCl_4)_2$ is thermodynamically unstable to disproportionation to Cd

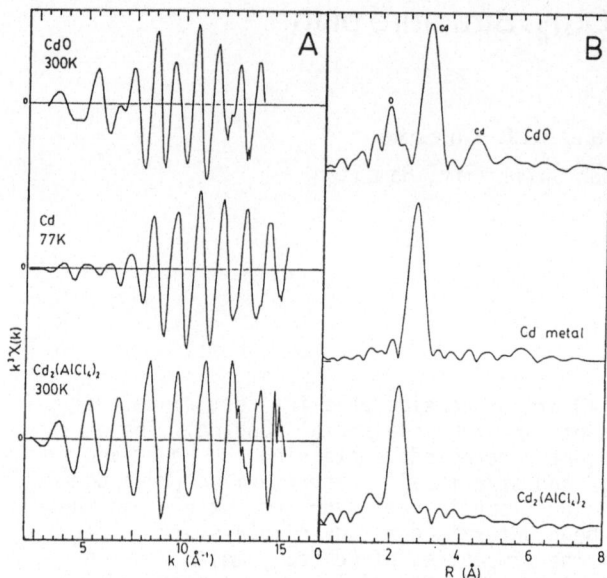

Fig. 1 A) k^3-weighted Cd K-shell EXAFS of CdO (300K), Cd metal (77K) and $Cd_2(AlCl_4)_2$ (300K). B) EXAFS radial distribution functions (uncorrected for phase shifts) of CdO, Cd metal and $Cd_2(AlCl_4)_2$ obtained from Fourier transforms of A.

and Cd^{2+} (as $CdCl_2$) [4] the amplitude function was examined to determine the chemical identity of the nearest neighbour atoms. In Fig. 2 the k^3-weighted amplitude of $Cd_2(AlCl_4)_2$ is compared to the Cd-Cd signals of Cd and CdO and the Cd-Cl signal in $CdCl_2 \cdot 2.5H_2O$. The $Cd_2(AlCl_4)_2$ nearest neighbour amplitude peaks at 11.3 A^{-1} very similar to that for Cd and CdO indicating that the nearest neighbours of Cd in $Cd_2(AlCl_4)_2$ are predominantly Cd. In addition, the position of the peak in the $Cd_2(AlCl_4)_2$ radial distribution is far too low for the Cd to be in a metallic environment. These points indicate the presence of a dominant amount of $Cd_2(AlCl_4)_2$ in our sample and are consistent with the proposed Cd_2^{2+} formulation for this species.

The Cd-Cd distance of Cd_2^{2+} derived using Cd and CdO experimental phase shifts is 2.38(3) Å (Table 1). This is somewhat shorter than the bond lengths for Hg_2^{2+} species which are typically around 2.50 Å [2]. If the M_2^{2+} species

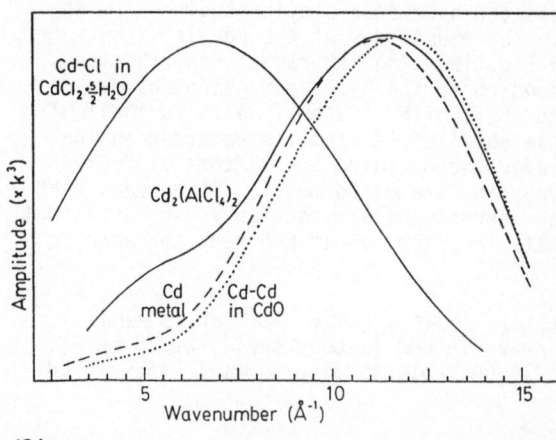

Fig. 2 k^3-weighted amplitude functions derived from the Fourier filtered single shell EXAFS of CdO (Cd-Cd distance), Cd metal, $CdCl_2 \cdot 2.5H_2O$ (Cd-Cl distance) and $Cd_2(AlCl_4)_2$

Table 1. Cd-Cd spacings, coordination numbers and rms variations of $Cd_2{}^{2+}$

T	k,R-space ranges	Model	R(Å)	N	$\sigma^2(Å^2)$
77	$\Delta k = 3.4 - 15.2$ Å^{-1}	Cd	2.39(3)	3.9	0.0084
	$\Delta R = 1.6 - 3.4$ Å	CdO	2.37(3)	1.6	0.0082
300	$\Delta k = 3.1 - 14.8$ Å^{-1}	Cd	2.40(2)	1.7	0.0080
	$\Delta R = 1.8 - 3.1$ Å	CdO	2.38(4)	0.7	0.0079
		$Cd_2(AlCl_4)_2$	2.39(1)	0.4	0.0080

1. Referenced to a 0.077 Å rms relative displacement in Cd estimated from the X-ray a axis σ [7] and a displacement correlation correction [8].

are covalently bonded their interatomic distances should reflect their covalent radii which are essentially the same for Cd and Hg (1.49 versus 1.50 Å respectively based on half the metallic bond length). In fact, the distances are more consistent with the ionic radii which are 1.14 Å for Cd^+ and 1.27 Å for Hg^+. It is possible that the shorter distance in $Cd_2{}^{2+}$ is associated with much weaker interactions with the anion. There is Raman and crystallographic evidence for relatively short Hg-anion distances in some of the Hg polyatomic cations [2] but there is no indication of any Cd-anion contact in our EXAFS results. Attempts to fit the first shell $Cd_2(AlCl_4)_2$ data to a mixture of Cd and Cl were not successful. The Cd-anion spacing is absent possibly because it is a long distance or because it is a weak interaction and thus the radial distance has a large variation associated with thermal motion.

The Cd-Cd bond length of 2.38(3)Å deduced from our EXAFS measurements is in agreement with the X-ray diffraction value of 2.35(4)Å reported for a $Cd_2{}^{2+}$ species trapped in a zeolite structure [6]. Previous work on $Hg_2{}^{2+}$ compounds shows that, although the Hg-Hg bond stretching frequency varies considerably with the nature of the anion (i.e. it ranges from 186 cm^{-1} in Hg_2F_2 to 113 cm^{-1} in HgI_2), the bond length variation is relatively small. Hg-Hg bond lengths in $Hg_2{}^{2+}$ compounds lie between 2.45 and 2.53 Å [2]. The present result is consistent with this since we find a Cd-Cd bond length for $Cd_2{}^{2+}$ in $Cd_2(AlCl_4)_2$ similar to that for $Cd_2{}^{2+}$ trapped in the zeolite cage.

Research funded by the Natural Sciences and Engineering Research Council (NSERC) of Canada. CHESS is a national facility supported by the U.S. National Science Foundation.

1. J.D. Corbett, Prog. Inorg. Chem. 21, 129 (1976).
2. B.D. Cutforth, Ph.D. Thesis, McMaster University (1975); B.D. Cutforth, R.J. Gillespie, P. Ireland, J.F. Sawyer and P.K. Ummat, Inorg. Chem. 22, 1344 (1983).
3. J.D. Corbett, Inorg. Chem. 1, 700 (1962).
4. J.D. Corbett, W.J. Burkhard and L.F. During, J.A.C.S. 83, 76 (1961).
5. The absence of anything except first-shell signal in Cd K-shell EXAFS of Cd metal has recently been noted and discussed by W. Thulke and P. Rabe, J. Phys. C. 16, L955 (1983).
6. L.B. McCusker and K. Seff, J.A.C.S. 101, 5235 (1979).
7. G.W. Brindley and P. Ridley, Proc. Phys. Soc. London 51, 73 (1939).
8. G. Beni and P.M. Platzman, Phys. Rev. B 14, 1514 (1976).

Local Structure of Pseudobinary Alloys

J.B. Boyce and J.C. Mikkelsen, Jr.

Xerox Palo Alto Research Center, 3333 Coyote Hill Road
Palo Alto, CA 94304, USA

1. Introduction

The local structural distortions around impurity atoms in alloys have, until recently, been outside the realm of direct experimental determination. Previously, they were inferred from their effects on the intensities of Bragg diffraction peaks and the associated diffuse scattering. EXAFS, on the other hand, provides a direct measure of the near-neighbor environment. Because of its spectroscopic and local nature, EXAFS is an ideal structural probe for determining the distortions of the host lattice in the vicinity of impurities in dilute binary alloys [1-5], as well as the near-neighbor environment in ternary and quaternary solid solutions [6-9].

We have measured the near-neighbor structure in a special class of ternary solid solutions, namely, the pseudobinary compounds of the form $(A_{1-x}B_x)C$. In this paper we describe the EXAFS results on two types of pseudobinary solid solutions. The first type consists of covalently bonded materials with the zincblende structure, represented by $(Ga_{1-x}In_x)As$ and $Zn(Se_{1-x}Te_x)$. The second class is the ionically bonded compounds with the rocksalt structure, $(K_{1-x}Rb_x)Br$ and $Rb(Br_{1-x}I_x)$. In each class, one alloy with cation substitution and one with anion substitution were chosen. All these alloys form random solid solutions over the entire range of x from 0 to 1, and their lattice constants accurately follow Vegard's Law; i.e., $a_0(x)$ varies linearly with x. The EXAFS was measured on the K-absorption edge of each of the constituent atoms (except for potassium) at 77K and 300K in a series of alloys with x varying from 0 to 1. The data were collected in the transmission mode and were reduced and analyzed in real space, using standard techniques [10]. The binary endpoint compounds served as structural standards. The first-neighbor and, in some cases, the second-neighbor structural information, i.e., the number and type of each neighboring atom, its near distance and the distribution (width) about this mean, were determined using least-squares fits to the data in real space.

2. Results: Covalent Materials

For (Ga,In)As, a detailed analysis of both the first and second near-neighbor distributions was performed [9]. The first-neighbor distribution is relatively simple. For the cations, it consists of four As atoms at a single, well-defined distance, r_1, from each cation for all alloy compositions, as shown in Fig. 1. From the figure, it is seen that the Ga-As and In-As distances differ markedly from one another and from the mean, or virtual, crystal distance obtained from x-ray diffraction. The latter distance is seen to accurately follow Vegard's Law.

426

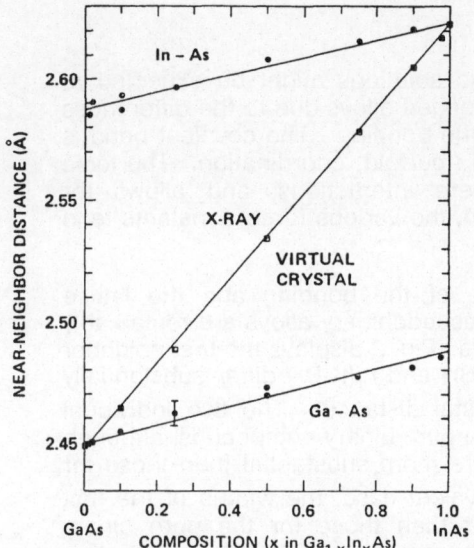

Fig. 1. Ga-As and In-As first neighbor distances as a function of alloy composition in $(Ga_{1-x}In_x)As$. The middle curve is the VCA cation-anion bond length calculated from the measured x-ray lattice constant, and is seen to accurately follow Vegard's Law. The weighted average of the first neighbor distances agrees with the VCA average.

In fact, these two individual distances are closer to the respective distances in the pure binary compounds than to the average virtual crystal distance. There is, nonetheless, a small change in r_1 with composition: $|\Delta r_1| \simeq 0.04$ Å in varying the composition over the entire range. However, this is small compared with the total charge in the virtual crystal distance, $\Delta r_{1V} \simeq 0.18$ Å.

For the anions, the first-neighbor distribution consists of two peaks, one due to Ga and the other due to In. The relative weights of the two peaks are consistent with the alloy composition, and the distances are in excellent agreement with the cation K-edge results in Fig. 1.

One other significant observation is that the widths of the first-neighbor distribution in the alloys are essentially the same as the corresponding widths in the pure binary compounds. There is no additional broadening, either static or dynamic, in the alloy, over the width in the binary compounds, to within an uncertainty of about ±0.01 A. One might expect that r_1(Ga-As), for example, would vary depending on the number of Ga atoms and the number of In atoms that are bonded to the central As atom. If this were the case, then there would be a distribution in r_1(Ga-As) due to the compositional disorder. Such a distribution, however, is not observed.

The second-neighbor distribution is more complex than that for the first neighbors. The essential results are the following: The cation-cation distribution, to a good approximation, consists of a single broadened peak at the virtual crystal distance; the anion-anion distribution, on the other hand, is bimodal, with As-Ga-As and As-In-As distances close to those in pure GaAs and InAs, respectively.

For Zn(Se,Te), which is more ionic than (Ga,In)As and which has anion substitution rather than cation substitution, the situation is almost identical to that in (Ga,In)As. Again, all the features discussed above for (Ga,In)As also apply to the results on Zn(Se,Te), down to the fact that even the distances are almost the same.

427

3. Results: Ionic Materials

The local structural distortions in ionic solid solutions might be expected to differ substantially from those in covalently bonded alloys due to the differences in the crystal structure and in the nature of the bonding. The covalent bond is strong, directional and allows for tetrahedral (fourfold) coordination. The ionic bond is central, with charged hard sphere interactions, and allows for octahedral (sixfold) coordination. In addition, the various force constants tend to decrease with increasing ionicity.

Despite these differences in the nature of the bonding and the lattice structure, the structural results on the ionic pseudobinary alloys are remarkably similar to those for their covalent counterparts. Fig. 2 displays the first neighbor distances in (K,Rb)Br. It is seen that r_1(Rb-Br) and r_1(K-Br) differ substantially from one another and from the virtual crystal distance. The two individual distances remain close to the distances in the pure binary compounds, although the variations in r_1 with alloy composition are more substantial than those for (Ga,In)As and Zn(Se,Te). But as in the covalent case, the widths of the first neighbor peaks are not substantially larger than those for the pure binary compounds used as structural standards. This argues against a large variation in r_1 at a specific alloy composition due to the local compositional disorder.

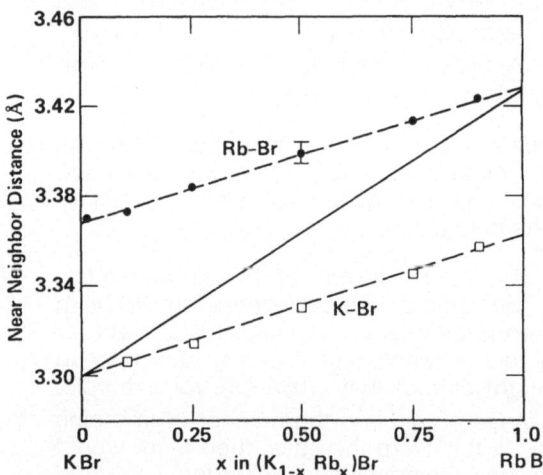

Fig. 2. K-Br and Rb-Br first neighbor distances as a function of composition in $(K_{1-x}Rb_x)Br$.

The second neighbor distribution is more complex in (K,Rb)Br than in the covalent materials, due to the higher coordination. In (Ga,In)As, for example, two As anions are bonded together via one cation only. So a bimodal distribution of As second-neighbor distances, corresponding to the two types of intervening cations, is readily understood. In (K,Rb)Br, however, two Br anions are bonded via two cations. Thus three possibilities exist in this octahedral case, rather than two as in the tetrahedral case: Br-(K,K)-Br, Br-(K,Rb)-Br, and Br-(Rb,Rb)-Br. From the data, a single Br-Br peak can indeed be ruled out. But the data were not adequate to distinguish between two Br-Br peaks or three Br-Br peaks, and reliably determine the corresponding structural parameters. The cation-cation distribution, on the other hand, was determined to be a single broad peak, as in the (Ga,In)As case. This peak also occurs at the virtual crystal second-neighbor distance.

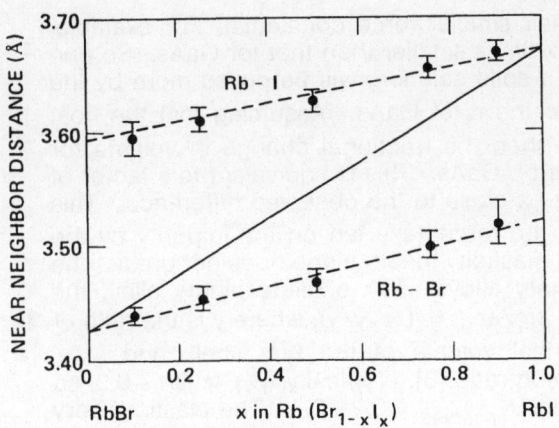

Fig. 3. Rb-Br and Rb-I first neighbor distances as a function of composition in $Rb(Br_{1-x}I_x)$.

For Rb(Br,I), which has anion rather than cation substitution, the situation is very similar to that in (K,Rb)Br. These similarities are evident from the first-neighbor distances given in Fig. 3.

4. Covalent Versus Ionic Alloys

The major difference between the first-neighbor distributions for these two classes of alloys is that the variation of r_1 with alloy composition is larger for the ionic case. This is evident in Figs. 1-3, and the pertinent parameters are listed in Table 1. For the covalent alloys, the total variation in r_1 with composition, Δr_1, is about 20% of the total variation in the virtual crystal first-neighbor distance, Δr_{1v}. In the ionic case, it is about 40 per cent. The larger variation for the ionic solid

Table 1. The total change in the first-neighbor distances, Δr_1, for composition ranging from $x = 0$ to $x = 1$ for each of the near-neighbor pairs A-C and B-C in the alloy systems $(A_{1-x}B_x)C$. These are compared with the corresponding change in the mean or average crystal first-neighbor distance, Δr_{1v}, determined by Vegard's Law. The ratio of these quantities, $R = \Delta r_1/\Delta r_{1v}$, shows that the covalently-bonded materials exhibit a smaller relative first-neighbor change with composition (~0.2) than do the ionically-bonded materials (~0.4).

Alloy system (A,B)C	$\Delta r_1(A-C)$ (A)	$\Delta r_1(B-C)$ (A)	Δr_{1v} (A)	$R(A-C)$	$R(B-C)$
(Ga,In)As	0.040	0.035	0.174	0.23	0.20
Zn(Se,Te)	0.036	0.040	0.183	0.20	0.22
(K, Rb)Br	0.062	0.059	0.127	0.49	0.46
Rb(Br, I)	0.091	0.080	0.244	0.37	0.33

429

solutions is reasonable in view of their smaller force constants. For example, the bulk modulus of RbBr is about six times smaller than that for GaAs. So one would expect that the r_1 of RbBr, in a solid solution, will be pulled more by the forces exerted on it by the host than the r_1 of GaAs. Assuming that the host pressure is the same in each case, then the fractional change in volume for RbBr will be six times larger than that of GaAs. This is equivalent to a factor of two times larger fractional change in r_1, close to the observed difference. This crude estimate ignores the fact that the forces exerted on the impurity by the different hosts will differ. In addition, elasticity theory alone does not predict the observed behavior in the pseudobinary alloys. For a dilute binary alloy, the elastic continum theory predicts $R = \Delta r_1 / \Delta r_{1v} = (1 - y/\gamma)$, where y is the ratio of the volume per atom to the spherical volume of the first shell, and $\gamma = 3(1 - \sigma)/(1 + \sigma)$, with σ being the Poisson ratio [3]. Typically, $y/\gamma \approx 0.1 - 0.2$, so that $R_{elastic} \approx 0.8 - 0.9$, compared with $R_{observed} \approx 0.2 - 0.4$. The elastic theory predicts small distortions for the host lattice (R closer to 1 than to 0), whereas we observe large distortions of the host (R closer to 0 than to 1). In any event, a detailed microscopic calculation is required to explain the observed behavior. The experiments indicate, nonetheless, that lattice relaxation differs by about a factor of two between the ionic solid solutions and the covalent alloys.

5. Summary

EXAFS has been used to determine the local structure in two classes of pseudobinary alloys: covalently-bonded solid solutions with the zincblende structure and ionically-bonded solid solutions with the rocksalt structure. In both types of materials, it is found that the first-neighbor distances remain closer to the respective distances found in the pure binary compounds than to the average, or virtual, crystal distance determined from the lattice constant of the alloy. This result implies that the virtual crystal approximation is not correct. On the other hand, there is a small but significant variation in the first-neighbor distances with composition, implying that the Pauling concept of the conservation of radii is not strictly valid in either the ionic or covalent alloys. One difference, however, between these two types of solid solutions is that a somewhat larger relative change in first-neighbor distance is observed for the ionic compounds ($\approx 40\%$) than for the covalent alloys ($\approx 20\%$), indicating that the local lattice distortions differ in the two classes of materials.

Acknowledgment

The Stanford Synchrotron Radiation Laboratory is supported by the National Science Foundation through the Division of Materials Research.

References

1. A. Fontaine, P. Lagarde, A. Naudon, D. Raoux, and D. Spanjaard, Philos. Mag. B **4**, 17 (1979).
2. T.M. Hayes, J. W. Allen, J. B. Boyce, and J. J. Hauser, Phys. Rev. B **22**, 4503 (1980).
3. D. Raoux, A. Fontaine, P. Largarde, and A. Sadoc, Phys. Rev. B **24**, 5547 (1981).
4. J. Mimault, A. Fontaine, P. Lagarde, D. Raoux, A. Sadoc, and D. Spanjaard, J. Phys. F **11**, 1311 (1981).

5. B. Lengeler and P. Eisenberger, Phys. Rev. B **21**, 4507 (1980).
6. J. B. Boyce and K. Baberschke, Solid State Commun. **39**, 781 (1981).
7. J. B. Boyce, R. M. Martin, J. W. Allen, and F. Holtzberg, in *Valence Fluctuations in Solids*, eds., L. M. Falicov, W. Hanks, and M. B. Maple (North-Holland, Amsterdam, 1981), p. 427.
8. J. Azoulay, E. A. Stern, D. Shaltiel, and A. Grayevski, Phys. Rev. B **25**, 5627 (1982).
9. J. C. Mikkelsen, Jr., and J. B. Boyce, Phys. Rev. **28**, 7130 (1983).
10. T. M. Hayes and J. B. Boyce, in *Solid State Physics*, eds., H. Ehrenreich, F. Seitz, and D. Turnbull (Academic, New York, 1982), Vol. 37, pp. 173-351.

Local Lattice Structure of Solid Solutions of Alkali Halides

Takatoshi Murata

Department of Physics, Kyoto University of Education, Fukakusa
Fushimi-ku, Kyoto 612, Japan

1. Introduction

The local lattice structure around a foreign atom or ion in a host matrix has
recently been investigated by EXAFS in various solids[1]. In most cases the
nearest neighbor distance of a foreign atom or ion is different from the value
of average distance expected from the Vegard's law, that is, a linear change
of the distance against the concentration of the guest atom or ion.

$KCl_{1-x}Br_x$ system is one of the examples and was previously investigated in
low concentration range (x lower than 0.10) of Br^- ion [2], and revealed a
local lattice dilatation around a Br^- ion in KCl matrix.

In the present work the EXAFS was measured on the K edge of Br in this alloy
in wide concentration range (x between 0.04 and 1.00), with the purpose to extend and complete the previous measurement and to observe a systematic change
in the local lattice structure from pure KBr.

2. Experimentals

X-ray absorption measurement were made with x-ray emitted from an electron
storage ring at Photon Factory in the National Laboratory for High Energy
Physics, using a Si(311) channel-cut monochromator on BL-10B EXAFS station[3].
The condition of the electron beam of the storage ring was typically 2.5GeV
and 100mA. Spectra were taken for the samples at room and at low temperature
(about 80K).

3. Results and Discussions

In Fig. 1 are shown the results of the magnitude of Fourier transform (FT) of
$k^3\chi(k)$ of the spectra at 80K in concentration x of 1.00, 0.20, and 0.04.
The positions of the first peak change only very slightly, whereas the height
of the peak increases in the lower concentration range. In contrast to the
first peak, the second peak at 4.2 Å in KBr loses contrast with decreasing
concentration of Br^- ion, and changes into band with two components.

The Fourier filtered spectra of the first shell and the results of the curve
fitting are shown in Fig. 2 for x=1.00 and x=0.04. In the fitting calculation,
phase and amplitude functions by Teo and Lee[4] were used. The number of the
first shell K^+ ion of Br^- ion was fixed to 6. Here, the letters σ and γ mean
the Debye-Waller type factor and damping parameter due to the mean free path
of photoelectrons, respectively. It is very interesting that the values of σ
and γ are rather similar in both results in contrast to a large difference in
the value of R.

The first shell distances around a Br^- ion determined from curve fitting
against the concentration x are plotted in Fig. 3. In Table 1 are listed the
parameters determined in the curve fitting. With decreasing x down to about

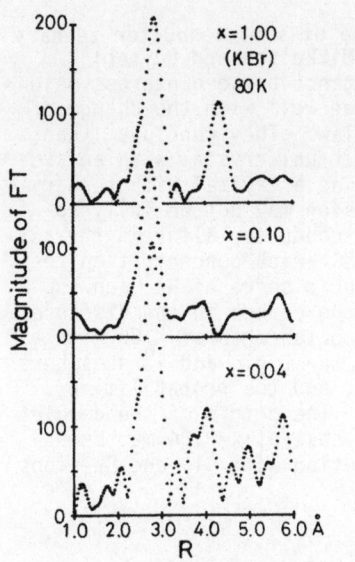

Fig. 1. Magnitude of FT of $k^3\chi(k)$ for x=1.00 (KBr), 0.20, and 0.04

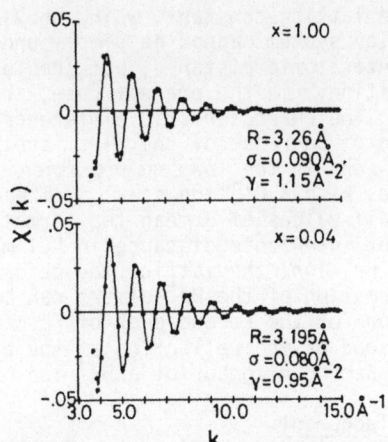

Fig. 2. Fourier Filtered inverse FT spectra (dots) and calculated spectra (line) for x=1.00 and 0.04

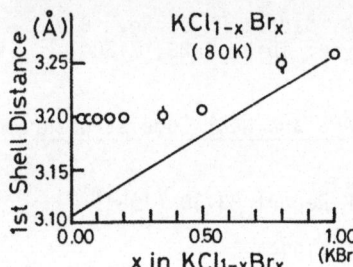

Fig. 3. First shell distance around a bromine ion against the concentration x. Straight line means Vegard's law

Table 1. Parameters determined by the curve fitting for the data at 80 K

x	R [Å]	N	σ [Å]	γ [Å^{-2}]
1.00	3.26	6	0.090	1.15
0.80	3.245	6	0.095	1.20
0.50	3.21	6	0.100	1.10
0.35	3.195	6	0.090	1.20
0.20	3.195	6	0.095	1.00
0.15	3.195	6	0.085	0.90
0.10	3.195	6	0.085	0.90
0.08	3.195	6	0.080	0.95
0.04	3.195	6	0.080	0.95

0.20, the distance decreases almost linearly, but not along the linear line of Vegard's law. Below x=0.20 the distance remains constant within experimental errors, while the value of σ tends to decrease as listed in Table 1. This means that the statistical fluctuation of the distance between Br^- and K^+ ions becomes small in the low concentration range.

The present results can be compared with the case of semi-conductor ternary alloy system, $Ga_{1-x}In_xAs$, recently investigated by Mikkelsen and Boyce[1]. They observed a small change in the interatomic distance between nearest neighbors. The distance of second nearest neighbors agree well with the change of average lattice constant, which obeys the Vegard's law. They concluded that the alloy system cannot be simply understood as a virtual crystal with an average interatomic distance, but should be considered as a crystal with ternary composition. In the present case, the above discussion may not be fully applied. The characteristic difference from the semi-conductor alloy is the saturation of decrease of the first shell distance in KCl-rich concentration region, where in the low concentration range, a hard ionic shere of Br^- ion substitutes with a Cl^- ion stably. The saturation of the change in the distance and small values of σ mean the formation of a rigid octahedron of K_6Br with a constant interionic distance in KCl matrix. Due to the rigid and large sphere of the Br^- ion, the lattice may deform considerably, and the probability of the formation of the Br^--dimers can be fairly high. The origin of the doublet structure of the second peak of FT magnitude in the case of x=0.04 may be understood as the reflection of the bimodal distribution with Cl^- and Br^- ions as the second neighbor of a Br^- ion.

Acknowledgement
The author is indebted to Drs. T. Matsushita and M. Nomura of Photon Factory for the assistance of the experiment. He also thanks   Drs. H. Maeda, M. Hida, N. Kamijo for collaboration in the measurments.

References

1 For example, D. Raoux, A. Fontaine, P. Lagarde, A. Sadoc: Phys. Rev. B 24, 5547 (1981), J. C. Mikkelsen Jr. and J. B. Boyce: ibid. B 28, 7130 (1983)

2 T. Murata, P. Lagarde, A. Fontaine, D. Raoux: EXAFS and Near Edge Structure p. 271 (Springer, Berlin Heidelberg 1983)

3 H. Oyanagi, T. Matsushita, M. Ito, H. Kuroda: KEK Report 83-30 (1984)

4 B. K. Teo and P. Lee: J. Am. Chem. Soc. 101, 2815 (1979)

An EXAFS Study of the Effect of Temperature on Short Range Ordering in the Ionic Conductor, RbBiF$_4$

C.R.A. Catlow, L.M. Moroney, and S.M. Tomlinson
Department of Chemistry, University College, 20 Gordon Street
London WC1H OAJ, United Kingdom
A.V. Chadwick
Department of Chemistry, University of Kent at Canterbury
Canterbury, Kent CT2 7NH, United Kingdom
G.N. Greaves
Science and Engineering Research Council, Daresbury Laboratory
Daresbury, Warrington WA4 4AD, United Kingdom

The fluorite phase of RbBiF$_4$ is a good F ion conductor at relatively low temperatures exhibiting a conductivity of 5×10^{-3} Ω^{-1} cm^{-1} at 100°C [1]. There are many examples of good ionic conductors possessing the fluorite structure (Fig. 1). This can be attributed to the fact that the fluorite lattice can accommodate displaced lattice anions in interstitial sites at the centre of alternate F cubes which are not occupied by cations. A neutron diffraction study of RbBiF$_4$, at room temperature, showed that, for this material, the occupation of these alternate cube-centre sites by interstitial F ions is indeed substantial [2]. It was not possible to differentiate between the different cations' environments and the data was fitted with the Rb and Bi ions statistically distributed over the available cation sites.

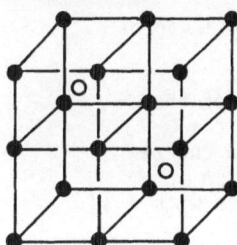

Figure 1. The fluorite structure: the filled circles depict the anions and the open circles, the cations.

One of the particular strengths of EXAFS is that one can 'tune-in' to a particular atom type and compare its local structural environment with that of another type within the same material. In the present work, we have exploited this feature to study the local ordering about Rb and Bi separately, and monitor the individual changes over the temperature range 80 K to 473 K.

In Figure 2 is shown the k^2 weighted data (solid line) obtained for the Rb K edge and Bi L(III) edge. Both series of spectra display only one clearly-resolved frequency attributed to backscattering by the nearest F neighbours. Thus no cation-cation ordering can be detected. The Rb and Bi backscattering functions are not such that this lack of structure could be attributed to cancelling effects when the profiles are summed. The crystallographic Debye-Waller factor obtained for the cations in the neutron diffraction study is large (B = 3.26 Å^2) and thus, the static disorder within the cation sublattice is expected to be substantial. The EXAFS spectra are consistent with there being a random distribution of Rb and Bi atoms over the cation sites, yielding a statistical mixture of Rb-cation and Bi-cation second nearest neighbour

435

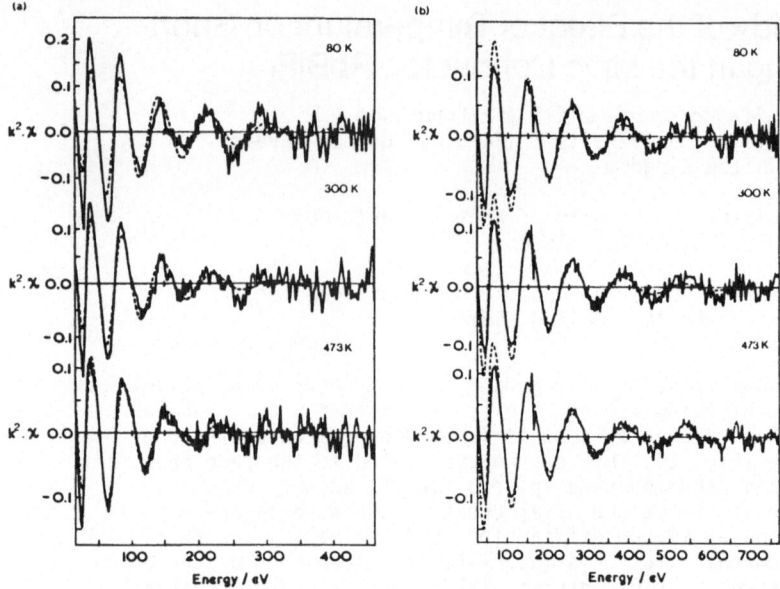

Figure 2. The k^2-weighted EXAFS data for (a) the Rb K edge and (b) the Bi(LIII) edge of RbBiF$_4$ for the temperatures 80K, 300K and 473K. The dotted line is the best fit from the parameters given in Table 1.

distances,therefore resulting in a smearing of the second shell EXAFS into the background.

There are clear differences in the temperature dependence of the EXAFS patterns of Rb and Bi. In the case of Bi, no change in the amplitude or frequency of the oscillations can be detected in the raw data. The Rb K edge EXAFS, however, displays a marked reduction in amplitude and the frequency of the oscillations also decreases with rise in temperature.

As a first approximation and to obviate the necessity of relying on fitting model compounds with the concomitant uncertainties of transferability etc., we have made the assumption that both the Rb-F and Bi-F first shell distance equals the mean cation-F distance of 2.62 Å at 80 K. On this basis we fitted the experimental data with a model of eight fluoride nearest neighbours allowing the radial distance and Debye-Waller factors to vary. This enables us to determine the relative changes in amplitude and frequency as the temperatures increases. The parameters in Table 1 yielded the best fits which are depicted by the dotted lines in Fig. 2.

Table 1

A comparison of the first shell Bi-F and Rb-F distances and their mean square variation with temperature.

Temperature[K]	R(Bi-F)[Å]	σ^2(Bi-F)[Å^2]	R(Rb-F)[Å]	σ^2(Rb-F)[Å^2]
80	2.62	0.012	2.62	0.015
300	2.61	0.012	2.57	0.021
473	2.60	0.013	2.54	0.025

These changes in the local structural environment of Rb can be explained by an increase in the number of vacancies in the F sublattice arising from F ion conduction. When a vacancy is formed in a shell of F neighbours adjacent to a Rb ion, the Rb and remaining fluorides relax towards a new configuration of minimum energy. Figure 3 shows the sort of relaxations that are envisaged. (These are the relaxations predicted to occur in response to an anion vacancy created in the isostructural UO_2 lattice [3].) This results in an asymmetric distribution of nearest neighbour Rb-F distances with a decrease in the mean Rb-F distance. Clearly, with this model, part of the increase   is necessary to accommodate the (non-symmetrical) spread in Rb-F distances and part to account for the loss of EXAFS amplitude due to the occurrence of vacancies in some of the Rb co-ordination spheres. Of course, in order to obtain meaningful Debye-Waller factors it is necessary to fit the data with an asymmetric function instead of the Gaussian one used in this inital fitting, and also, to let the co-ordination number vary.

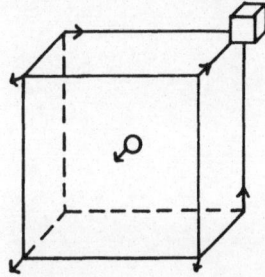

Figure 3. The predicted relaxations of cation and anions in response to a vacancy in the fluorite lattice (based on UO_2 data [3]). The cation (open circle) and anion opposite the vacancy, move in a <111> direction away from the vacancy. Three anions move along the cube edge towards the vacancy. The three remaining anions move away from the vacancy along the face-diagonal line joining vacancy and anion, and also slightly inwards towards the nearest cation.

It is interesting to note that similar changes do not occur in the EXAFS of the Bi edge. i.e. the change in Bi-F distance with increase in temperature is small (0.02 Å) and is close to the limits of accuracy with which it is possible to measure radial distances. The change in amplitude is also minimal. Given that there is a random distribution of Rb and Bi ions over the cation sites, and any vacancy in the F sublattice is in the nearest neighbour position to 4 cations, these results suggest that vacancies are preferentially stabilised in lattice sites rich in Rb nearest neighbours. This information enables us to develop a picture of likely pathways of F ions through the material. It also helps explain why the related Bi-rich solid solutions $Rb_{1-x}Bi_xF_{1+2x}$ (0.5 < x < 0.6) exhibit a lower conductivity than $RbBiF_4$ in spite of possessing more F charge carriers (needed for charge compensation of the extra Bi) [1].

Removing a F ion from a site rich in the singly-charged Rb neighbours is electrostatically more favourable than removing a F ion from a site adjacent to the more highly charged Bi ions. Furthermore, the ensuing relaxations depicted in Figure 3, stabilise the formation of a vacancy. The EXAFS data demonstrate that these factors outweigh the postulated advantage of stabilising a vacancy adjacent to Bi ions by delocalisation of the Bi lone electron pair. A further test of this model would be to attempt to synthesize Rb-rich mixed fluorite phases such as $Rb_{1-x}Bi_xF_{1+2x}$ (x<0.5) which would be predicted, on the basis of the EXAFS data, to be better ionic conductors than $RbBiF_4$.

References
[1]   S.F. Matar, J.M. Reau, C. Lucat, J. Grannec and P. Haganmuller,
        Mater. Res. Bull, 15, (1980), 1295-1301.
[2]   S.F. Matar, Ph.D. Thesis (1983) University of Bordeaux.
[3]   R. Jackson, personal communication.

Yttria Stabilized Zirconia Systems: Evolution of the Local Order Around the Yttrium and Zirconium Cations

H. Dexpert and P. Lagarde
LURE, Bât. 209 C, UPS, F-91405 Orsay, France
M.H. Tuilier
Laboratoire de Chimie Physique, Université P. et M. Curie
F-75005 Paris, France
J. Dexpert-Ghys
CNRS Eléments de Transition dans les Solides, F-92190 Meudon, France

INTRODUCTION.

The so-called "yttria stabilized zirconias (YSZ)" are compounds displaying a large number of applications as refractories and ionic conductors or diamond substitutes for example. Three crystalline forms are observed owing to the relative Y^{3+}/Zr^{4+} amount : tetragonal, cubic fluorine and cubic C- Ln_2O_3 (Ln=rare earth). Pure yttria belongs to the last type and pure zirconia is monoclinic. The substitution of tetravalent ions in ZrO_2 by tervalent ones is accompanied by the creation of vacancies in the oxygen sublattice. In view of X-rays diffraction data, the YSZ structure is disordered and usually described as a metallic network on which Y^{3+} and Zr^{4+} are statistically located, whereas oxygens and vacancies are distributed onto the anionic sublattice. A more precise insight in these phases has been already obtained using an optical fluorescent microprobe where Y^{3+} is partially substituted by a few percent of Eu^{3+}. In order to complete these results an Exafs investigation of the same phases has been undertaken and this work summarises our first conclusions. The general formulae of the YSZ phases is : $Zr_{1-x}Y_xO_{2-x/2}$ and six compositions (x close to 0,10,35,50,70 and 100%, samples F,E,D,C,B and A respectively) have been studied for the yttrium and zirconium K-edges. The data have been collected at LURE-DCI through a Si(311) double-crystal monochromator.

I X-RAY DIFFRACTION AND OPTICAL FLUORESCENCE MEASUREMENTS.

For sample F the crystalline structure is monoclinic, for sample E it is tetragonal and cubic-fluorine type for samples D and C. Sample B is diphasic : cubic-fluorine plus cubic type C-Ln_2O_3 and pure yttria (sample A) is cubic of the C-Ln_2O_3 type. The only ordered phase known in this system, $Zr_3Y_4O_{12}$, has not been detected in our samples. Europium fluorescence spectra show immediate environments of eight, then seven and six oxygens for the luminescent probe, the relative proportions of which varying continuously with x. The observed large linewidths come from a great number of slightly different point sites for europium ions and therefore indicate the lack of short range order due to oxygen relaxations towards vacancies (1).

II EXAFS DATA AND DISCUSSION.

Figures 1 and 2 are the Fourier transforms (FT) k^3 weighted of the different compositions. The Hanning window limits are 30 to 690 eV for the yttrium edge and 45 to 425 eV for the zirconium one.

-The Y203 spectrum.

A complete investigation has been tempted on Y_2O_3. Its spectrum exhibits a fine structure extended to 900 eV. The radial distribution function shows up to 5 coordination shells (Fig 1A). The filtered back FT of the first one, attributed to six oxygen neighbors,is rather well fitted using

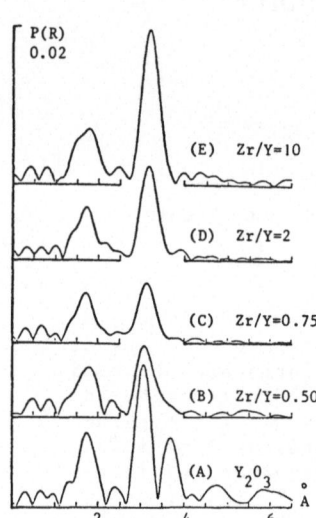

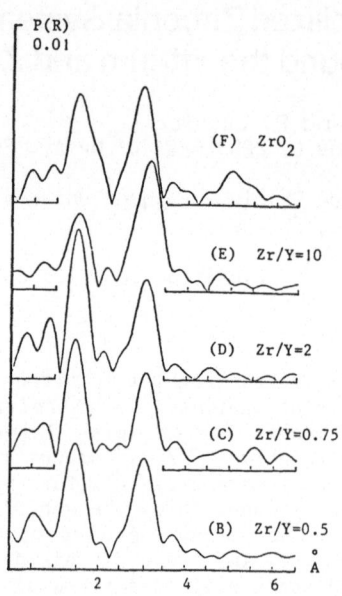

Fig.1. $Zr_{1-x}Y_xO_{2-x/2}$ Fourier transforms of the $k^3X(k)$ data at the Y
K-edge

Fig.2. $Zr_{1-x}Y_xO_{2-x/2}$ Fourier transforms of the $k^3X(k)$ data at the Zr
K-edge

amplitude and phase shifts from Teo and Lee. The averaged distance we
found, 2.285 Å, takes into account the S_6 and C_{2v} sites distribution
and agrees well with neutron diffraction results (2). The yttrium second
neighbor shell contains two subshells separated by 0.5 Å, respectively at
the average radius of 3.54 and 4.01Å. The second and third peaks shown in
Fig 1A are expected to correspond to those Y-Y distances, but the
structural oxygen-yttrium symmetry may be the source of a large focussing
effect. Therefore, up to now, we did not try a careful analysis of this
third peak, which could have at the same time both origins. Nevertheless,
its presence is a sign of the degree of order around one given cation.
Moreover, in the analysis of the YSZ, we will not distinguish, as backscatterers, between yttrium and zirconium in view of their consecutive position
in the Mendeleev's table.

      -YSZ phases at the yttrium edge (Fig 1).
As soon as zirconium is included in the yttria structure, the radial
distribution function around the cation (Fig 1B,C,D) shows a loss of order
demonstrated mainly by the disappearance of the third shell seen in pure
yttria. A decrease of the magnitude of the first and second peaks is also
clearly visible. The six to eight expected increase in oxygen number for
the Y first shell is not obvious following the 1A to 1E sequence. This lack
of amplitude variation comes from a large distribution of the Y-0 bond
lengths, characteristic of the large disorder wich seems to affect strongly
this distance, even when we are close to pure zirconia. That anionic
disorder is accompanied by an important decrease of the Y-Y intensity
around the middle of the binary system (1B to 1D). But when the zirconium
percentage becomes predominant, the yttrium network is now stabilized and
gives the intense peak seen in sample 1E (tetragonal zirconia).

      -YSZ phases at the zirconium edge (Fig 2).
From 2E to 2B the yttrium goes through 10 to 70%. As it reaches 35% there
is a definite change in the behavior of the Zr-0 bond. For the tetragonal

zirconia (E) the first metal-oxygen peak is broadened, owing to the two distances expected from the network symmetry. This first shell becomes very well organized and presents a strong increase of its amplitude. This ordering is due to a narrowing of the Zr-O distances dispersion strong enough to compensate the decrease of the coordination number from 2E to 2D. The Zr-Zr distances seem less affected during these yttrium substitutions. The Zr-O/Zr-Zr ratios changes slowly from 2D to 2B with the continuous lowering of oxygen neighbouring from eight to six.

These Exafs analyses appear to be very complementary to the previous ones. They confirm the lack of short range order around the yttrium when zirconium is incorporated in the structure; but new important points are enlighted:

    -the first shell degree of disorder (for yttrium) or order (for zirconium) is strong enough to mask the oxygen number variations.

    -the yttrium cationic network is very sensitive to the oxygen vacancies relaxations, the zirconium one remaining rather unchanged.

    -it seems that the ordering of the first shell of oxygen around the zirconium is reinforced by a small amount of yttrium while the metal-metal second shell remains stable.

Comparisons between experimental and calculated filtered FT are underway to detail these preliminary results.

1 J. Dexpert-Ghys, M. Faucher and P. Caro J. Sol. St. Chem 54 (1984)
2 M. Faucher and J. Pannetier Acta Cryst. B26 , 3209 (1980)

EXAFS Studies of Yttria Stabilized Zirconia

A.I. Goldman, E. Canova, and Y.H. Kao

Department of Physics, S.U.N.Y. Stony Brook, Stony Brook, NY 11794, USA

W.L. Roth and R. Wong

Department of Physics, S.U.N.Y. Albany, Albany, NY 12222, USA

1. Introduction

Yttria stabilized zirconia is a well known solid electrolyte which exhibits fast anion conduction at elevated temperatures. Below 1200° C pure zirconia (ZrO_2) is monoclinic, but forms a stabilized cubic phase with the fluorite structure in a solid solution with 9.4 to 24 mole percent Y_2O_3.

The high anionic conductivity is related to the creation of oxygen vacancies accompanying the substitution of Y^{3+} ions for Zr^{4+} ions on the cation sites [1]. X-ray studies [2] of a closely related compound, calcia stabilized zirconia, has indicated that the oxygen vacancies tend to associate with the stabilizing cation (Ca) as first nearest neighbors. While it is not unreasonable to assume that this is also the case for YSZ, no evidence for Y^{3+} -V_O^{2-} nearest neighbors has been offered. However, x-ray and neutron scattering studies [2,3] of YSZ have indicated that there is a net displacement of about 60 percent of the oxygen ions from their ideal positions in the fluorite structure by 0.2 to 0.3 Angstroms in ⟨100⟩ directions.

This paper describes the results of EXAFS investigations of the Metal - Oxygen (M-O) nearest neighbor distributions around both the Zr^{4+} and Y^{3+} cations.

2. Experiment and Data Analysis

Absorption spectra were taken at room temperature above both the Y and Zr K-edges of YSZ compounds ranging from 9.4 to 24 mole percent Y_2O_3. Spectra of ZrO_2 , $SrZrO_3$, $CaZrO_3$ and Y_2O_3 were also collected for use as model compounds in the analysis.

After background subtraction and normalization using standard procedures [4] the EXAFS oscillations, $k^2\chi(k)$, and the respective Fourier Transforms of ZrO_2 (Zr edge), Y_2O_3 (Y edge) and both the Zr and Y K-edges of YSZ for all compositions are shown in Fig. 1.

The M-O peak of each spectrum was isolated, backtransformed, and fit with a parameterized form of the standard EXAFS expression [4]. Initial values for the phase shift and backscattering amplitude functions were obtained from TEO and LEE [5] , and refined by fitting the M-O peaks of the model compounds using the known distances and coordination.

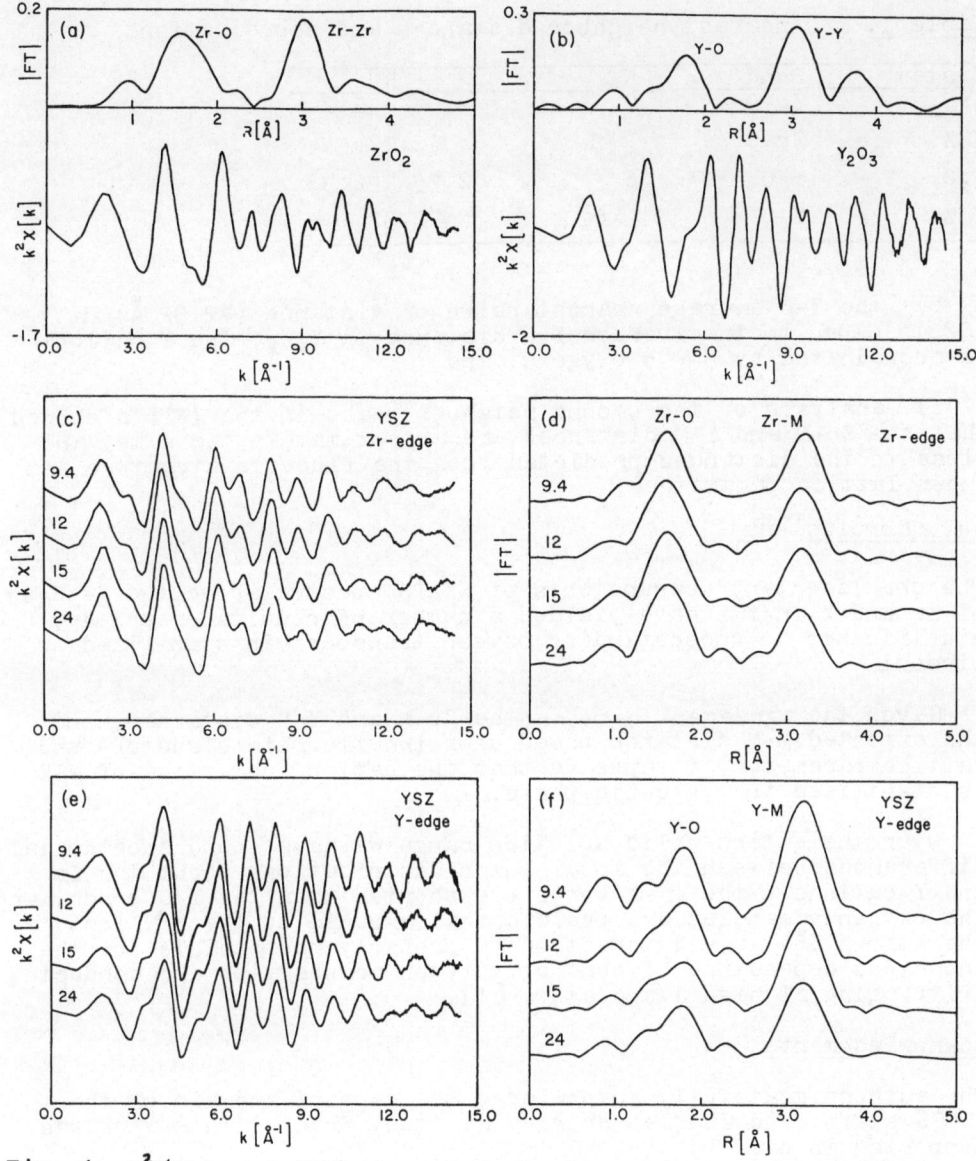

Fig. 1. $k^2\chi(k)$ and $|FT|$ for a) Zr K-edge of ZrO_2, b) Y K-edge of Y_2O_3, c) and d) Zr K-edge of YSZ, e) and f) Y K-edge of YSZ

3. Results

Table 1 summarizes the results of single shell fits to the Zr-O and Y-O nearest neighbor distributions in the YSZ compounds. It was found that:

a) The Zr-O average nearest neighbor distance (≈ 2.17 Å) in YSZ is close to the average Zr-O distance in ZrO_2. The Zr cation is coordinated by 7 - 8 oxygens,

Table 1. M-O nearest neighbor distances and coordination

System	Bond	N	R Å
ZrO_2	Zr-O	7	2.04 - 2.26
YSZ	Zr-O	7 - 8	≈ 2.17
Y_2O_3	Y-O	6	2.25 - 2.35
YSZ	Y-O	5 - 6	≈ 2.32

b)   the Y-O average nearest neighbor distance (≈2.32 Å) in
YSZ is close to the average Y-O distance in Y_2O_3. The Y cation
is coordinated by 5 - 6 oxygens, and

c)   analysis of the second neighbor peaks in the |FT|'s showed
that the Zr-M and Y-M distances are approximately the same, and
close to the distances predicted from the fluorite structure and
known lattice parameters.

4.   Conclusions

The complementary measurements of EXAFS spectra above the K-edges
of Zr and Y in YSZ have yielded a number of results of primary
significance to understanding oxygen transport in stabilized
zirconia.

Given the agreement between the Zr-M and Y-M distances, and
the expected M-M distance based upon the fluorite structure and
lattice parameter, it appears that the cation sublattice of YSZ
is stabilized in the cubic phase.

Over the entire solid solution range we have found substantial
differences between the local anion distributions about the Zr
and Y cations - the Zr-O bonds are shorter than the Y-O bonds, and
the oxygen vacancies are preferentially bound to the Y^{3+} ions.
The strong Y^{3+} -V_O^{2-} interaction is probably responsible for the
anomalous dependence of conductivity on composition and conductivity aging at high temperature [1].

Acknowledgments

The authors gratefully acknowledge the support and aid of the
CHESS staff, and Charles Wrigley at Stony Brook. This work was
supported in part by the ONR.

References

1. T. H. Etsell and S. N. Flengas, Chem. Rev. 70, 339 (1970)
2. M. Morinaga, J. B. Cohen and J. Faber, Acta Cryst A36, 520
   (1980)
3. J. Faber, M. H. Mueller and B. R. Cooper, Phys. Rev. B17,
   4884 (1978)
4. P. A. Lee, P. H. Citrin, P. Eisenberger and B. M. Kincaid,
   Rev. Mod. Phys. 53, 769 (1981)
5. B. K. Teo and P. A. Lee, J. Amer. Chem Soc. 101, 2815 (1979)

EXAFS Studies of Xenon Atoms in Matrices of Argon and of Krypton Atoms in Matrices of Neon

W. Malzfeldt, W. Niemann, P. Rabe, and R. Haensel

Institut für Experimentalphysik der Universität Kiel
D-2300 Kiel, Fed. Rep. of Germany

EXAFS measurements on solid Xe, solid Ar, Xe in Ar matrices and on Kr in Ne matrices are presented. Xe atoms are statistically distributed in a distorted Ar lattice. A strong clustering is observed for Kr atoms in Ne matrices. The Kr-Kr distances in the cluster are smaller compared to the solid.

1. Introduction

It is well known that atoms of pure rare gas solids arrange in a fcc lattice at low temperatures. On the other hand knowledge about the arrangement around rare gas impurities embedded in matrices of another rare gas species is poor. All structural informations of matrixisolated rare gases are derived from indirect methods like luminescence spectroscopy or photoelectron spectroscopy. Distance between unlike pairs of rare gas atoms calculated from potential curves differ considerably.

In this paper EXAFS measurements of various concentrations of Xe atoms in Ar and of Kr atoms in Ne are reported. The pure rare gas solids were investigated for comparison of its structural parameters(nearest neighbour distance, Debye-Waller factor and coordination number)with those of the matrices and to extract experimental scattering phases.

2. Experimental

The experiments were performed at the EXAFS II beam line of the synchrotron radiation facility HASYLAB in Hamburg. We used a Si(111) double crystal X-ray monochromator equipped with a focusing premirror(1). The rare gases were condensed onto an aluminium foil of 200 nm thickness cooled down to 5 K by a liquid helium cryostat. The mixtures of the pure gases were prepared in the gas phase in molar concentration.

The measurements were carried out at the K-edges of Ar(3202.9eV)and Kr(14330eV) and the L_3-edge of Xe(4782.2eV). The energy resolution amounted to 0.9eV at the Ar K-edge, 1.5eV at the Xe,and 8.0eV at the Kr edge.

3. Results

The fine structure $\chi(k)$ was extracted from the background in a conventional way (2). In most cases the Fourier transform F(r)was calculated using Gaussian windows. The structural information of a shell was derived from the backtransform of a distinct structure in F(r)into the k-space. Interpolated theoretical phase data(3) have been used to determine the nearest neighbour distances of the pure rare gas solids and of the Kr-Ne system. For the determination of Xe-Ar atom distances a combination of theoretical central atom phases and experimental backscattering phases have been used.

3.1. Solid Argon

The fine structure $\chi(k)$ of solid Ar is shown together with the corresponding Fourier transform in Fig.1. Due to the large Debye-Waller factor $\chi(k)$ yields,no

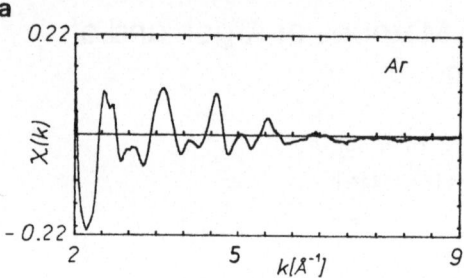
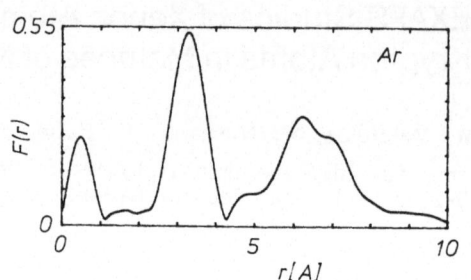

Fig.1a:K-edge absorption fine structure of solid Ar
Fig.1b:Fourier transform of the spectrum in Fig.1a

information beyond $k=9A^{-1}$.The nearest neighbour distance derived from the
EXAFS amounts to (3.76 ± 0.01)A in excellent agreement with the known value of
3.755A(4)in solid Ar.The root mean square displacement(rmsd)has been determined
to (0.141 ± 0.005)A.

3.2 Solid Xenon

According to the large interatomic spacing the peaks in the Fourier transform
of solid Xe calculated from the fine structure $\chi(k)$(Fig.2a)between the L_3 and
L_2 edge are slightly better resolved compared to the case of Ar.The nearest
neighbour distance amounts to $R=(4.355 \pm 0.010)$A which is slightly larger than
that of Horton et al. (4)of 4.335A derived from x-ray diffraction.The rmsd
amounts to (0.090 ± 0.005)A.

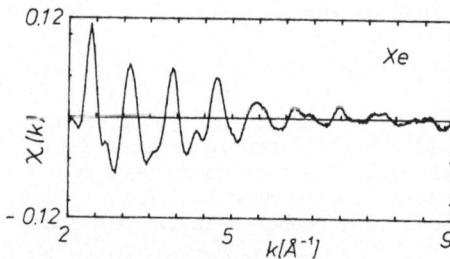
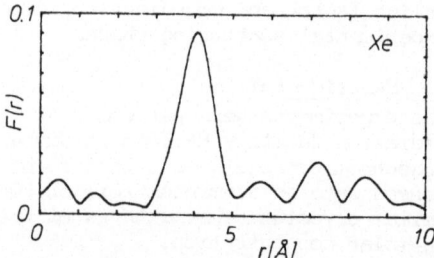

Fig.2a:L_3-edge absorption fine structure of solid Xe
Fig.2b:Fourier transform of the spectrum in Fig.2a

3.3 Xenon in Argon

The fine structure $\chi(k)$ of a sample with 3% Xe in Ar is shown in Fig.3a.The
relative large widths of structures in the Fourier transform (Fig.3b) are due
to the small range of information in (k). The nearest neighbour distance
between Xe and Ar atoms of $R=(3.99^{+0.02}_{-0.01})$A is independent on the concentration
between 1% and 10 %. This value is smaller than that derived from the Xe-Ar
molecular potentials of Parson et al.(5)(4.10A)and Aziz et al.(6)(4.067A)but
larger than that of Arora et al.(7)(3.945A). For Xe concentrations smaller
than 3 % a coordination number of $N=(12^{+3}_{-1})$atoms has been calculated.The Debye-
Waller factor is the same as for solid Ar. We conclude that the Xe atoms show
a statistical distribution and take substitutional sites in Ar matrices in
agreement with conclusions drawn from UV-absorption experiments(8). At similar
geometry was observed in the case of Kr atoms in Ar(9).

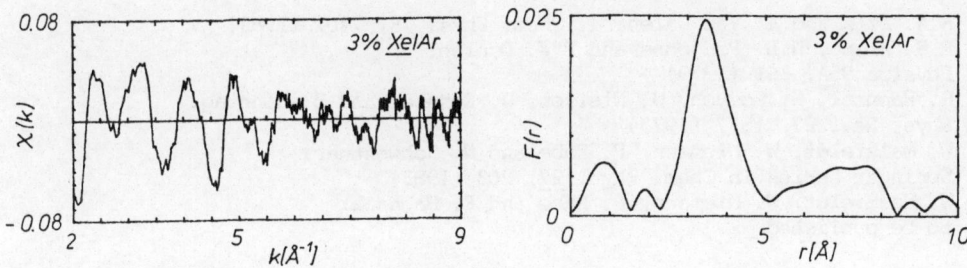

Fig.3a:Fine structure of the sytem 3% Xe/Ar at the Xe L_3-edge
Fig.3b:Fourier transform of the spectrum in Fig.3a

3.4 Krypton in Neon

Kr, Xe and Ar atoms in Ne matrices have a completely different behaviour compared to Kr or Xe atoms in Ar matrices. The system Kr/Ne is presented as an example. The fine structure of a 3% Kr/Ne sample is shown in Fig.4 together with the corresponding Fourier transform.The large structure around 1A in Fig.4b is caused by small variations of the atomic cross section of Kr(10). The peak around 3.7A changes its magnitude but keeps constant in position for all concentrations between 1% and 10%. This indicates the formation of Kr-Kr clusters.The following structural parameters have been obtained from two-shell fits of the backtransform using theoretical scattering phases and amplitudes for the Kr-Ne shell and corresponding experimental parameters for the Kr-Kr shell:$R_{Kr-Ne}=(3.485\pm0.030)$A, $N_{Kr-Ne}=(6\pm3)$atoms, $\sigma_{Kr-Ne}=(0.21\pm0.05)$A, $R_{Kr-Kr}=(3.915\pm0.030)$A, $\sigma_{Kr-Kr}=(0.13\pm0.02)$A. The coordination number for the Kr-Kr distance is 2 for the 1% sample which indicates the formation of trimers, and it is 9 for the 3 % and 10 % samples. This large coordination number of 9 points to the arrangement of Kr atoms in small clusters.It should be noted that the Kr-Kr distance in the cluster is smaller than that in solid Kr(3.97A). This tendency of the guests has also been observed for the systems Ar/Ne and Xe/Ne.

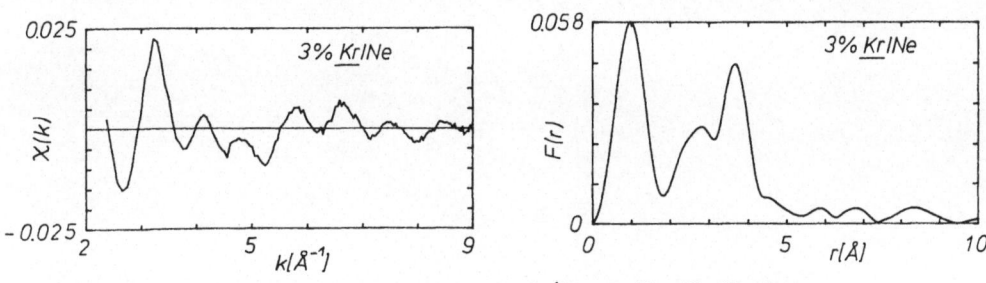

Fig.4a:Fine structure of the system 3% Kr/Ne at the Kr K-edge
Fig.4b:Fourier transform of the spectrum in Fig.4a

References
1. W. Malzfeldt, W. Niemann, R. Haensel and P. Rabe:
   Nucl. Instr. and Meth. 208, 359 (1983)
2. G. Martens, P. Rabe, N. Schwentner and A. Werner:
   Phys. Rev. B 17, 1481 (1978)
3. B.K. Teo and P.A. Lee: J. Amer. Chem. Soc. 101, 2815 (1979)
4. G.K. Horton: Am. J. Phys. 36, 93 (1968)
5. J.M. Parson, T.P. Schafer, F.P. Tully, Y.C. Wong and Y.T. Lee:
   J. Chem. Phys. 53, 3755 (1970)

6. R.A. Aziz and A. van Dalen: J. Chem. Phys. $\underline{78}$, 2402 (1983)
7. P.S. Arora, H.L. Robjohns and P.J. Dunlop:
   Physica $\underline{95A}$, 561 (1979)
8. R. Haensel, N. Kosuch, U. Nielsen, U. Rössler and B. Sonntag:
   Phys. Rev. B$\underline{7}$, 1577 (1973)
9. W. Malzfeldt, W. Niemann, P. Rabe and N. Schwentner:
   Springer Series in Chem. Phys. $\underline{27}$, 203 (1983)
10.W. Malzfeldt, W. Niemann, P. Rabe and R. Haensel:
   to be published

Soft X-Ray Absorption Measurements at the K-Edges of Sulphur and Chlorine

J. Goulon[1,2], R. Cortes[3], A. Retournard[1], A. Georges[4],
J.P. Battioni[5], R. Frety[5], and B. Moraweck[6]

1. Introduction

Regarding the importance of sulphur and chlorine in several problems of
coordination chemistry, catalysis or biochemistry, a better knowledge of
the chemical bonding of these elements is often desirable. X-ray absorption
spectroscopy is a most attractive tool for that purpose, but soft X-ray
measurements are made difficult by the high absorbance of air, Be windows
and of the sample itself below 4 keV. The presence of large amounts of harmonics has also been seen to induce dramatic distortions of the spectra [1], and is a major source of difficulty. The aim of this contribution
is to show that good quality spectra have however been recorded at LURE
(France) using the synchrotron radiation of DCI running at high energy
(1.85 GeV, 250 mA). A considerable gain of sensitivity resulted from the
adjunction, behind the monochromator, of an efficient harmonic rejector
made of two parallel SiO_2 mirrors.

2. Experimental Setup

The experiments were carried out with the EXAFS-II spectrometer [2] installed on beam line D2. Several minor modifications were made in order to
keep the sample and the front window of the ion chambers under primary
vacuum. The level of harmonics was first reduced by detuning the parallelism
of the two crystals by 90 % or more. This procedure resulted in a poor
signal/noise ratio and was abandoned after the adjunction of the two-mirror
assembly sketched in Fig.1 . These mirrors are quartz plates (SOPTEL :

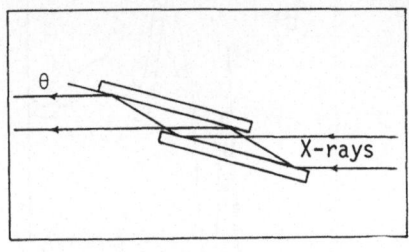

Fig. 1. Harmonic rejector design : mirror
assembly

[1]ERA 22 (CNRS), Université de Nancy I, B.P. 239,
F-54506 Vandoeuvre-lès-Nancy, France
[2]LURE (CNRS), Bât. 209C, F-91405 Orsay, France
[3]GR 4 (CNRS), Physique des Liquides, 4 Place Jussieu, F-75230 Paris, France
[4]LA 155 (CNRS), INPL, Parc de Saurupt, F-54042 Nancy, France
[5]LA 32 (CNRS), ENS, 24 rue Lhomond, F-75231 Paris, France
[6]IRC (CNRS), Avenue A. Einstein, F-69626 Villeurbanne, France

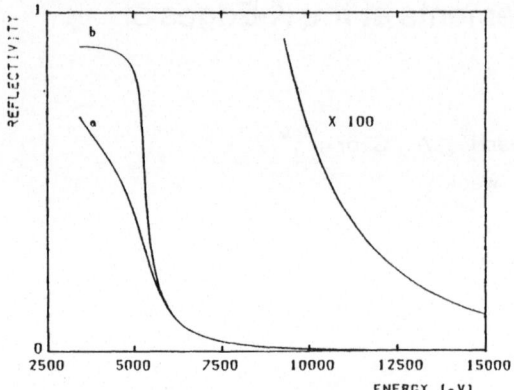

Fig. 2. Energy dependence of the reflectivity at $\theta = 0.95\ \theta_c$: a) for a gold coated mirror ($\theta_c = 15.5$ mrad at 5 keV) , b) for a quartz uncoated mirror ($\theta_c = 6.25$ mrad at 5 keV)

350x40x30 mm) with local surface distortions below $\lambda/4$. As far as we are concerned with harmonics rejection, the reflectivity profiles in energy reproduced in Fig.2 show that there is no advantage in metal coating of the reflecting planes. Total extinction of the third harmonic was achieved at 5 keV while the measured level of harmonics at E = 2.85 keV was better than 10^{-4} vs. fundamental. A reflectivity of ~ 0.85 was measured at 5 keV for each mirror. As illustrated by Fig.2a, no serious degradation of the energy resolution was observed at the titanium K-edge.

3. Sulphur and chlorine K-edge absorption spectra

XANES spectra of three metalloporphyrins [3-5] are presented in Fig.3. Of particular significance are the huge prepeaks observed,which are characteristic of the axial symmetry of the X(S,Cl)...Metal chemical bond. Strong distortion of these signatures occurred without mirror. Figure 4 also re-

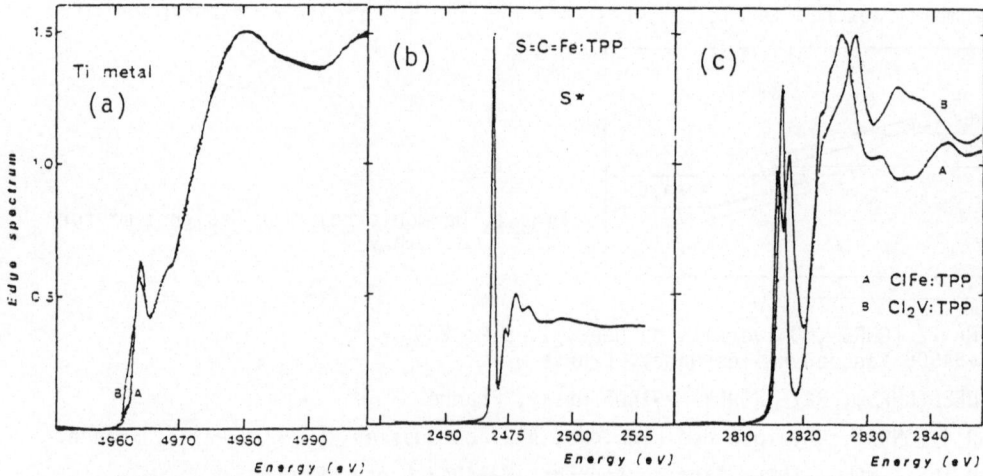

Fig. 3. Edge spectra : a) titanium metal with (A) / without (B) mirror , b) S edge spectrum of S=C=Fe:TPP (single scan $\tau = 1$ s/pt) , c) Cl edge spectra of ClFe:TPP and Cl_2V:TPP ($\tau = 1$ s/pt)

450

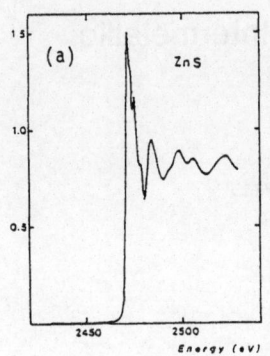 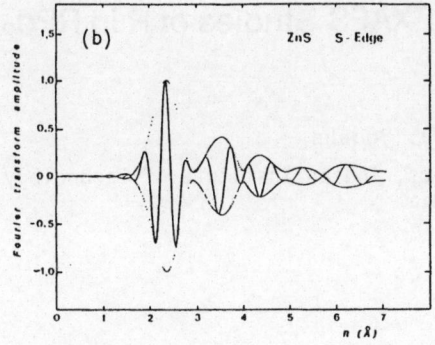

Fig. 4. S edge XANES/EXAFS spectra of ZnS

produces the XANES (Fig.4a) and the EXAFS radial distribution around S in ZnS. Phase shift/amplitude corrections relevant to the S*-Zn shell were included in the latter FT-analysis. The results are quite consistent with crystal structure data.

Acknowledgements
Thanks are due to G. VLAIC (Inst. Donegani/LURE) for technical assistance in conditioning the samples on membranes.

References

1   J. Goulon, C. Goulon-Ginet, R. Cortes, J.M. Dubois, J. Physique 43 (1982) 539.
2   J. Goulon, M. Lemonnier, R. Cortes, A. Retournard, D. Raoux, Nucl. Instrum. Meth. 208 (1983) 625.
3   J.P. Battioni, D. Mansuy, J.C. Chottard, Inorg. Chem. 19 (1980) 791.
4   P. Richard, J.L. Poncet, J.M. Barbe, R. Guilard, J. Goulon, D. Rinaldi, A. Cartier, P. Tola, JCS Dalton Trans. (1982) 1451.
5   J.L. Hoard, G.H. Cohen, M.D. Glick, J.Am.Chem.Soc. 89 (1967) 1992.

L$_3$-XANES and EXAFS Studies of R in RPd$_2$Si$_2$ Intermetallic Compounds

B. Darshan and B.D. Padalia

Department of Physics, Indian Institute of Technology, Powai, Bombay 400 076, India

1. Introduction

Rare earths (R) intermetallic compounds of composition, RM$_2$X$_2$ (R = rare earths; M = transition metal ions; X = Si or Ge) have evoked a great deal of interest in recent years. These compounds crystallize in ThCr$_2$Si$_2$ tetragonal structure [1]. Each unit cell contains two formula units. The R ion occupies the body centred position and is surrounded by Si or Ge. The transition metal ions (M) lie outside the first coordination sphere of R. The lattice parameter and the magnetic measurements on RM$_2$X$_2$ indicate that the R ions are in trivalent state. This appears to be an interesting system for the study of L$_3$-XANES and EXAFS of R. The results of such a study of RPd$_2$Si$_2$ (R = Sm, Gd and Tm) ternary intermetallic compounds are discussed here.

2. Experimental

RPd$_2$Si$_2$ samples were prepared by standard argon–arc melting technique. The single phase formation of the samples was checked by X-ray diffraction method. A conventional X-ray absorption spectroscopic (XAS) set–up consisting of a focussing X-ray spectrograph of Cauchois type and a laboratory source of continuous radiation were used in the present work [2].

3. Results and Discussion

The L$_3$ X-ray absorption near edge structure (XANES) and the extended X-ray absorption fine structure (EXAFS) of R ion in RPd$_2$Si$_2$ (R = Sm, Gd, Tm) were recorded. A sample spectrum is shown in Fig. 1. The L$_3$-XANES is dominated by a resonance absorption (white line) at the L$_3$-edge. A single white line is the characteristic feature of the L$_3$-XANES of the pure trivalent rare earths. The measured values of the position (A), half width (Δ) and height ratio (r) of the white line in RPd$_2$Si$_2$ system are given in Table 1. It is noted that Δ and r values are not affected significantly by changing the R ion. Similar observation was made earlier in the case of RM$_2$X$_2$ compounds [4]. Studies of diverse sets of rare earth intermetallics carried out by others for testing the single particle band structure calculations [3,5], support the present results.

We have recorded the EXAFS associated with the L$_3$-absorption of R in RPd$_2$Si$_2$. The energies of maxima (B,C,...) and minima (β, γ, ...) in EXAFS measured with respect to the corresponding L$_3$-edge position are included in Table 1.

A cursory look at the curves shown in Fig. 1 indicates that the gross features

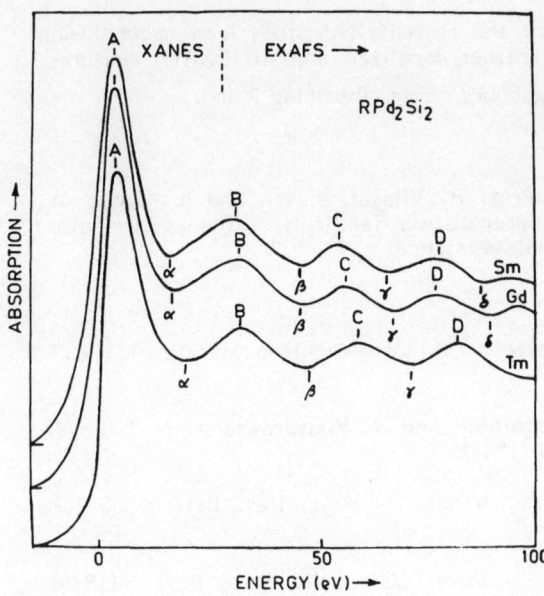

Fig. 1.   L$_3$-XANES and EXAFS of R in RPd$_2$Si$_2$ (Sm, Gd, Tm) intermetallic compounds

Table 1.   Half width (Δ) and height ratio (r) of white line, A, and EXAFS of rare earth, R, in RPd$_2$Si$_2$ intermetallic compounds

Sample	XANES						R-Si Bond lengths (Å)	
	White line		EXAFS					
R	Δ (eV)	r$^+$	B	β	C	γ	Present	Crystallographic
Sm	4.1	2.1	28.3	43.1	52.1	63.1	3.24	3.21
Gd	4.0	2.0	31.2	44.1	54.0	63.0	3.20	3.20
Tm	4.2	2.0	33.0	48.3	58.9	72.4	3.14	-

+ $r = I_A/I_\alpha$, where I_A and I_α are the intensities of white line, A,
and the first absorption minimum, α , measured with respect to pre-
threshold background [3]

of EXAFS of R (Sm, Gd and Tm) resemble   each other. This may be attributed to the fact that the nearest neighbours (Si) of R in these three samples
of RPd$_2$Si$_2$ are identical, so the back scattering amplitude is the same.   This
suggests that the major contribution to EXAFS comes from the nearest neighbours.   The R-Si distance in RPd$_2$Si$_2$ is estimated from EXAFS data using the
graphical method [6].   The chemical bonding parameter, α_1, has been first
obtained for GdPd$_2$Si$_2$ and then it is assumed as a constant for other samples.
This is a reasonable assumption because α_1 depends on the iono-covalent character of the bond , which remains nearly the same in the case of R-Si bond
in RPd$_2$Si$_2$.   The R-Si distance in RPd$_2$Si$_2$ estimated by the EXAFS method
agrees well with the available crystallographic data.

It may be stated that the results of the present XAS study lead to conclusion that (i) the d states in RPd_2Si_2 are rather localized and (ii) EXAFS features are mainly governed by the nearest neighbours of the absorbing R ion.

Acknowledgement

We express our sincere thanks to Prof. R. Vijayaraghavan and his group at TIFR, Bombay, for providing sample preparation facilities. Thanks are also due to the CSIR, New Delhi, for financial assistance.

References

1. D. Rossi, R. Marazza and R. Ferro : J. Less Common Metals <u>66</u>, P17 (1979) and references therein.

2. B. Darshan, B. D. Padalia, R. Nagarajan and R. Vijayaraghavan : J. Phys. C : Solid State Phys. <u>17</u>, L445 (1984).

3. G. Materlik, J. K. Miller and J.W. Wilkins : Phys. Rev. Lett. <u>50</u>, 267 (1983).

4. B. D. Padalia and B. Darshan : J. Phys. C : Solid State Phys. (1984) — To appear.

5. J. M. Lawrence, M. L. den Boer, R. D. Parks and J. C. Smith : Phys. Rev. B (Preprint).

6. F.W. Lytle, D. E. Sayers and E. A. Stern : Phys. Rev. B<u>11</u>, 4825 (1975).

EXAFS Study of Intermetallics of the Type RGe$_2$ (R=La, Ce, Pe, Nd, Sm, Gd, Tb, Dy, Ho, Er, and Y)

A.R. Chourasia, S.D. Deshpande, V.B. Sapre, and C. Mande
Department of Physics, Nagpur University, Nagpur-440 010, India
V.D. Chafekar
Department of Physics, Indian Institute of Technology, Kanpur-208 016, India

1. Introduction

In the present work we are reporting our study of the EXAFS associated with the K absorption discontinuity of germanium in pure germanium and in the intermetallics of the type RGe$_2$ (R = La, Ce, Pr, Nd, Sm, Gd, Tb, Dy, Ho, Er and Y), undertaken with the view of determining Ge-Ge distances in the compounds. These compounds have crystal structure of the α-ThSi$_2$ type [1,2]. They form a **framework** of covalently bonded atoms of.germanium with the rare earth metal atoms located in the voids of the germanium framework. The nearest neighbours of each germanium atom are three other germanium atoms with distances close to d$_{Ge-Ge}$ in the germanium structure (2.45 $\overset{\circ}{A}$·). Thus the only difference in the environment of Ge atoms in pure germanium and in these compounds is that,while in the former each atom is surrounded by 4 Ge neighbours, in the compounds each Ge atom is surrounded by 3 Ge neighbours.

2. Experimental

The compounds were prepared by arc melting the mixture of component metals in stoichiometric proportion in argon atmosphere. The formation of the compounds was checked by the X-ray diffraction technique.

The X-ray absorption experiments were made using a Seifert X-ray spectrometer equipped with an auto-step scanning mechanism and a scintillation counter. A sealed X-ray **tube** with Cu-target operated at 50 kV and 30 mA was used as the source of white radiation. The data processing was done using a programme developed by us for this purpose on a DEC 1090 computer in the Indian Institute of Technology, Kanpur.

3. Data Analysis

The values of the EXAFS function $\chi(k)$ were obtained from the raw curves obtained experimentally using the procedure give by SCHMÜCKLE et al [3]. The EXAFS data were then multiplied by the Hanning window function [4] in order to minimise the termination ripples, and by a weighting function, k^n, to compensate for amplitude reduction as a function of k, where n is 1, 3 or 5. The Fourier transform of the data was made in order to obtain different peaks corresponding to each coordination shell. The positions of the peaks were found to depend upon the value of the energy zero, E_o and n. Curves were drawn following the method given by STEARNS [5] between the values of the peak position for the first coordination shell and those of the energy zero for different values of n. The point of intersection of these curves gave the value of E_o.

The peak corresponding to the first shell was then inverse Fourier transformed. Because of the complex nature of the EXAFS function, it could be **easily** decomposed into the amplitude function A(k) and the phase function $\psi(k)$.

Using the concept of chemical transferability, the value of the bond distance r_u in an unknown system which is chemically similar to a model system (i.e. with known distance r_m) was obtained, following the procedure described by LEE et al [6]. from the following equation

$$\psi_m(k) - \psi_u(k) = 2k(r_m - r_u) \qquad \cdots\cdots\cdots \quad (1)$$

where $\psi_m(k)$ and $\psi_u(k)$ are the phase functions for the model and the unknown system respectively. According to (1) the plot of $[\psi_m(k) - \psi_u(k)]$ as a function of k should be a straight line line with zero intercept and slope $2(r_m - r_u)$. In actual practice, the value E_o was varied,in such a fashion as to reduce the intercept of the line to a minimum. Taking $r_m = 2.45$ Å for crystalline germanium, the Ge-Ge distances in the RGe_2 compounds were evaluated using (1). The values thus obtained for the different compounds are given in Table 1. In this table are also given the values obtained crystallographically wherever available. We observe that there is a very good agreement between our values of the Ge-Ge distances and the crystallographic values. We are reporting the Ge-Ge distances for the first time in the compounds $TbGe_2$, $HoGe_2$ and $ErGe_2$, structure data on them not being available.

456

Table   1

Data on bond length

Sample	Value of r (in Å) Observed ± 0.01Å	Crystallographic
$LaGe_2$	2.40	2.38
$CeGe_2$	2.34	2.35
$PrGe_2$	2.35	2.34
$NdGe_2$	2.32	2.33
$SmGe_2$	2.33	2.30
$GdGe_2$	2.25	2.27
$TbGe_2$	2.25	-
$DyGe_2$	2.24	2.23
$HoGe_2$	2.24	-
$ErGe_2$	2.23	-
YGe_2	2.22	2.24

Acknowledgements

We are thankful to Dr. E.E. Haller, University of California, Berkeley, U.S.A. for sending us ultrapure germanium. Thanks are also due to J.W. Allen, Technical Representative, Rare Earth Products Limited, Cheshire, U.K. for providing us the rare earth metals. Two of us (ARC & SDD) are thankful to the Council of Scientific & Industrial Research and the University Grants Commission, New Delhi respectively for financial support.

REFERENCE

1.   E.I. Gladyshevskii: J. Struct. Chem. (USSR) 5,523 (1964)

2.   K. Sekizawa: J. Phys. Soc. Japan 21, 1137 (1966)

3.   F. Schmükle, P. Lamparter and S. Steeb: Z. Naturforsch 37a, 572 (1982)

4.   G. H. Via, J.H. Sinfelt and F.W. Lytle: J. Chem. Phys. 71a 690 (1979)

5.   M.B. Stearns: Phys. Rev. B. 25, 2382 (1982)

6.   P.A. Lee, P.H. Citrin, P. Eiserberger and B.M. Kincaid: Rev. Mod. Phys. 53, 769 (1981)

Extended X-Ray Absorption Fine Structure (EXAFS) Studies of the Actinides*

S.-K. Chan and G.S. Knapp

Argonne National Laboratory, 9700 South Cass Avenue, Argonne, IL 60439, USA

The basic EXAFS parameters of the backscattering amplitude and phase shift have been computed as a function of k for the neutral atoms of Th, U, Np and Pu. It is found that an anomalous variation of the backscattering phase shift with respect to k occurs for the actinide elements at approximately $k=7A^{-1}$. However, this variation is monotonic, unlike those for Pb and a few other lighter elements. It is also found that the detailed behavior of the parameters is sensitive to the distribution of electrons in the 5f open shell. Comparison with experiments on uranium arsenide and selenide emphasizes the need for careful modeling of the 5f electronic configuration.

At the present time, there are no calculations of the EXAFS parameters beyond the element Pb (Z=82). Such calculations are of interest for two reasons. First, we have an extensive program to study actinide-containing materials. Second, TEO and LEE [1] showed that the backscattering phase shifts became almost discontinuous near $k=7A^{-1}$ for atoms with 78<Z<82, so that it is interesting to see what happens to the phase shifts and amplitudes of the higher Z elements. In this paper, exploratory results will be presented for the neutral atoms of Th, U, Np and Pu. A comparison with the experimentally determined parameters for UAs and USe will be made.

Scattering phase shifts, δ_1 for the individual partial waves of angular momentum 1 have been computed for the neutral actinide atoms by solving the one electron Schroedinger equation self-consistently in the Hartree-Fock approximation and matching the radial wave functions inside and outside a muffin-tin radius, $R_0=1.5$ times the covalent radius. The charge densities of core electrons were obtained from Herman-Skillman wave functions [2]. Wave functions of electrons in the outer shells were truncated when they exceeded R_0. Electrical neutrality ouside R_0 was maintained by adding an appropriate uniform charge density to the spherical region inside. To check for relativistic effects, the self-consistent potential obtained for Th has been compared with the result from a relativistic calculation [3]. The difference is found to be insignificant and does not influence the individual partial waves phase shifts. The total backscattering amplitude A and phase shift ψ are finally obtained from

$$f(\theta=\pi) = Ae^{i\psi} = \frac{1}{k} \sum_1 (21+1)(-1)^1 e^{i\delta_1} \sin\delta_1 \qquad (1).$$

The results for the neutral atoms Th, U, Np and Pu are presented in Figs. (1a) and (1b).

---
*Work supported by the U. S. Department of Energy.

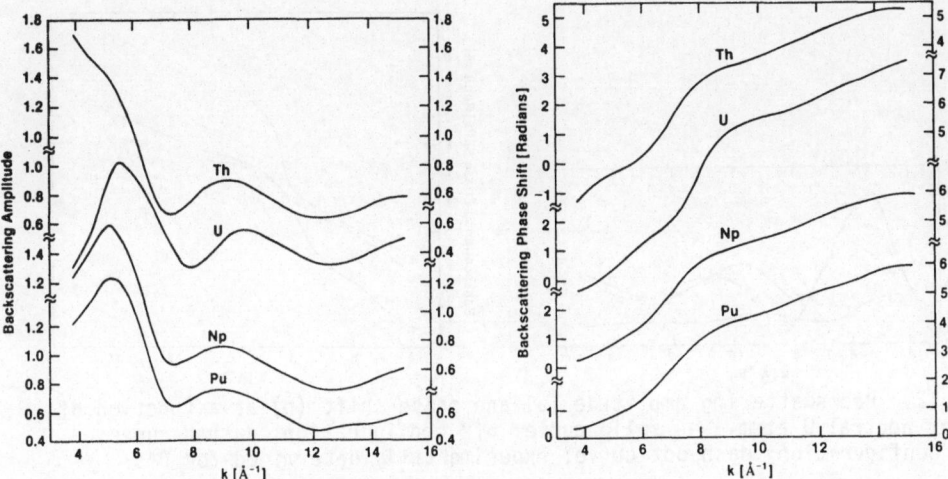

Fig. 1. Backscattering amplitude (a) and phase shift (b) as a function of k for neutral atoms of Th, U, Np and Pu

The backscattering amplitudes A as a function of k show decreasing modulations with three maxima and two minima interposed between one another within the range k=3.7 to 15.2 A^{-1}. The corresponding curves of the backscattering phase shifts ψ are more monotonic but do exhibit changes in their slopes. Of particular interest is the region around k=7 A^{-1}. There the phase shifts have the largest slopes. This behavior represents the remnant of the gradual build up and steepening of the "precipice" that occurs for elements with Z near 82, but the sign of the slope is opposite here. The reason for this change of slope is that when this near-discontinuity exceeds π, it is more natural to shift the low k phase shifts by 2π so as to minimize the anomaly. (This is correct since ψ is defined to only 2π). Since $d\psi/dk$ has the meaning of a time delay in scattering theory [4], this result suggests that when the atomic number is near Z=82, the shape and size of the potential is such that a wave packet with k=7 A^{-1} satisfies a resonance condition and is trapped for a long time between the rather steep walls of the potential.

The results of the backscattering amplitude A and phase shift ψ are sensitive to the distribution of electrons in the 5f open shell of the actinide elements. As an illustration, A and ψ for a neutral uranium atom were calculated under two different assumptions to mimic two different electronic configurations. First, all electrons in the 5f shell were assumed to hybridize with the more extended 6d and 7s electrons and became delocalized. The deficiency in charge is compensated by an appropriate charge density so that electrical neutrality is achieved outside R_0. The results are indicated by the solid curves in Figs. (2a) and (2b). Next, two of the 5f electrons were allowed to remain localized as in a free atom and the rest replaced by an appropriate uniform charge density to maintain electrical neutrality outside R_0. The corresponding backscattering amplitude and phase shift are shown by the dashed curves.

To test these model electronic configurations, we have made comparison with EXAFS measurements on UAs and USe at the arsenic and selenium absorption edges. The results are similar for both systems and we show only the parameters for UAs determined through the use of the standard

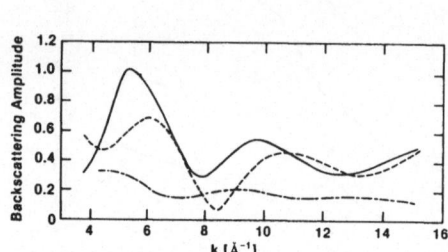 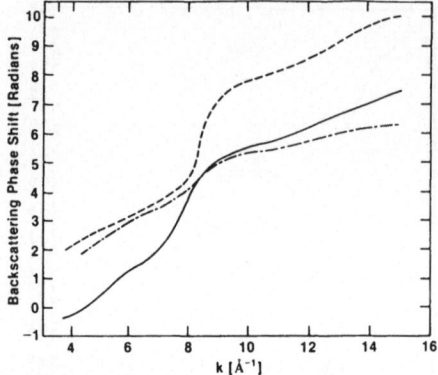

Fig. 2. Backscattering amplitude (a) and phase shift (b) as a function of
k for neutral U atom ---- solid curve: 5f⁰ configuration; dashed curve:
5f² configuration; dash-dot curve: experimentally determined for UAs

EXAFS formula [5] and the As central atom phase shift calculated by TEO
and LEE [1]. Experimental results for the Th, Np and Pu arsenides are
similar to those of UAs and will be presented later in a full paper. The
experimentally obtained A and ψ values are shown by the dash-dot curves.
Due to the limitations coming from the plane wave approximation and the
exclusion of inelastic scattering effects, we might not expect better
agreement. Clearly, neither of the two models is able to mimic the
experimentally determined parameters. What is apparent is that the
experimental results show similar structure in the amplitude (displaced in
k) and the increased slope in the phase near $k = 7\text{Å}^{-1}$ in qualitative
agreement with theory. These results emphasize the sensitivity of the
EXAFS parameters to the 5f electronic configurations and point to the need
of close coordination between experimental and theoretical efforts.

References
1. B.-K. Teo and P.A. Lee: J. Am. Chem. Soc. 101, 2815 (1979).
2. F. Herman and S. Skillman: Atomic Structure Calculations (Prentice-
   Hall, Englewood Cliffs, N. J., 1963).
3. S.-H. Chow and D.E. Ellis: to be published.
4. M.L. Goldberger and K.M. Watson: Collision Theory (Wiley, New York
   1964).
5. P.A. Lee, P.H. Citrin, P. Eisenberger and B.M. Kincaid: Rev. Mod.
   Phys. 53, 769 (1981).

XANES in SbSI, Sb_2S_3, Sb_2S_5

G. Dalba and P. Fornasini

Dipartimento di Fisica, Università di Trento, I-38050 Povo, Italy

E. Burattini

INFN, PWA group, I-00044 Frascati, Italy

1. Introduction

SbSI and Sb_2S_3, representative compounds of the $A^VB^{VI}C^{VII}$ and $A_2^VB_3^{VI}$ groups respectively, have awakened a remarkable interest for their physical and mechanical properties related to the low dimensional structure.

Both these compounds, which have a double ribbon structure parallel to the [001] crystallographic direction, are ferroelectric and semiconducting materials with high photoconductivity [1]. Moreover, the optical properties of the V_2VI_3 compounds change considerably from the crystalline to the amorphous state.

There are few calculations of the electronic band structure of these crystals owing to their very complex geometrical structure and to the lack of a sufficient amount of experimental data.

Sb_2S_5 has not yet been much studied; it is a non-stoichiometric compound with variable composition and structure.

In this work we analyse the X-ray absorption spectra at the L edges of antimony and iodine in SbSI, Sb_2S_3, Sb_2S_5, measured at high resolution with synchrotron radiation.

The interpretation of the edge structure within the first 10 eV has been carried out in terms of density of the unoccupied levels in the conduction band. By analysing the structure between 10 and 50 eV some stereochemical information has been obtained.

2. Results and discussion

In Fig.1 the structures within the first 10 eV above the Sb L_1 edges of SbSI, Sb_2S_3, Sb_2S_5 are compared with the L_1 edge of crystalline antimony. The absorption limit of each spectrum has been associated with the origin of the energy axis. The spectra of the three compounds are very similar, in particular the white peaks in Sb_2S_3 and Sb_2S_5 are almost identical; the white peak in SbSI is 17% higher and slightly narrower.

In Fig.2 the structure within the first 10 eV above the Sb L_3 edges are compared; they are characterized by two humps which are equidistant in Sb_2S_3 and in Sb_2S_5 and slightly closer in SbSI.

In a previous work the white peak at the Sb L_1 edge in SbSI has been interpreted in terms of a high density of unoccupied states of p symmetry in the bottom of the conduction band [2]. The comparative analysis of the edges L_1 and L_3 of antimony and iodine in SbSI and the available calculated DOS sug-

461

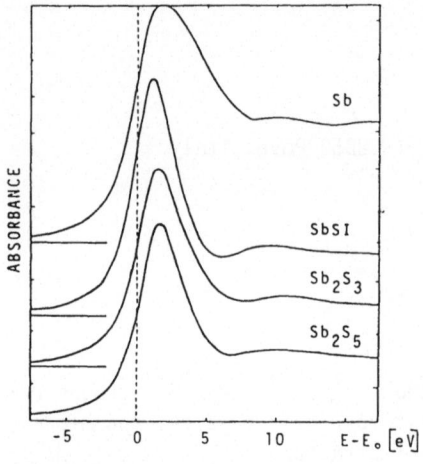

Fig.1. Edges L_1 of antimony.

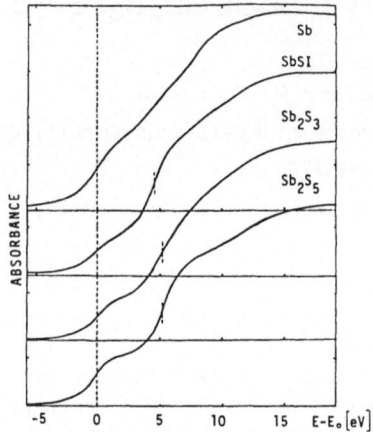

Fig.2. Edges L_3 of antimony.

gests that the low-lying group of the conduction band is formed by a mixing of states of s and p symmetry, in agreement with theoretical calculations.

Moreover the white peak at the L_1 edge of antimony in SbSI is much more pronounced than the white peak at the L_1 edge of iodine. This has been related to a partial localization of the unoccupied orbitals of p symmetry near the antimony atoms [3].

The similarity of the structures at the edges L_1 and L_3 of antimony in Sb_2S_3 and Sb_2S_5 suggests a substantial equality of the projected DOS at the bottom of the conduction band. Also for Sb_2S_3 and Sb_2S_5, like for SbSI, the bottom of the conduction band is thus formed by a mixing of s and p levels of sulphur and antimony; the p levels prevail within the first 5 eV.

The similarity of the spectra has to be related to an equal oxidation state of the antimony atoms in SbSI, Sb_2S_3, Sb_2S_5. In particular, for what concerns Sb_2S_5, the X-ray absorption measurements strengthen the analogous results obtained by Mössbauer isomeric shift [4].

A confirmation of the similarity of the antimony chemical bond for the three compounds has been obtained by analyzing the energy position of the absorption edges. It is well known that the edge shift is related to the nature of the chemical bond and changes with its ionicity.

In Table I the XAS and XPS energy shifts are compared; the shifts of the same core level measured with the same technique are equal for all the compounds.

| | XAS | | XPS [5] | |
	$2\,s$	$2\,p_{3/2}$	$3\,d_{3/2}$	$3\,d_{5/2}$
SbSI	1.4±0.3	1.3±0.3	2.0±0.3	2.0±0.3
Sb_2S_3	1.4±0.3	1.3±0.3	2.1±0.3	2.2±0.3
Sb_2S_5			1.8±0.3	1.9±0.3

Table I: Energy shifts (eV) of core levels measured by XAS and XPS.

The information given by the structure within the first 10 eV above the edge can be completed by the study of the structure between 10 and 50 eV which is sensitive to the stereochemical coordination.

The spectra obtained for SbSI, Sb_2S_3, Sb_2S_5 by calculating the relative variation with respect to the atomic absorption coefficient are shown in Fig.3. They present common features labelled A,B,C,D.

The structure between 10 and 50 eV contains information about the geometrical symmetry around the absorbing atom till the third shell. The SbSI spectrum differs from the other spectra for what concerns the amplitude and the shape of the features A and C. The difference can be due to the fact that two crystallographic sites for the Sb atoms exist in Sb_2S_3 (Sb_I and Sb_{II}) while in SbSI the Sb atom has a coordination like that of Sb_{II}. Moreover, it has to be remarked that, under the same geometrical coordination, two sulphur near neighbours of Sb_{II} in Sb_2S_3 are substituted by two iodine atoms in SbSI and that the backscattering amplitudes and phases of iodine and sulphur are considerably different.

The analysis of XANES structure strengthens the hypothesis, already put forward on the grounds of Mössbauer isomeric shift, that the short range coordination of antimony in Sb_2S_5 is nearly equal to that in Sb_2S_3. This confirms the non-existence of Sb_2S_5 as a stoichiometric compound.

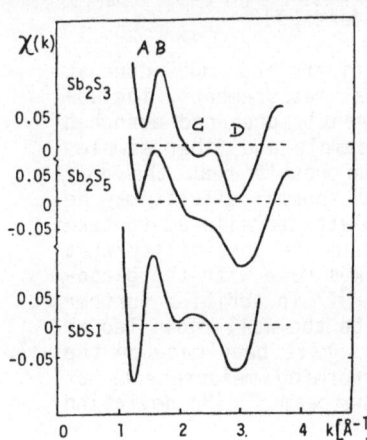

Fig. 3. Structure between 10 and 50 eV at the Sb L_1 edge. On the horizontal axis K is the wavevector of the emitted photoelectron.

References

1. W.M.Fridkin: "Ferroelectric semiconductors", Consultant Bureau (1982)
2. G. Dalba, P.Fornasini,E.Burattini: J.Phys.C:Solid St.Phys. 16,L1091 (1983)
3. E.Burattini,G.Dalba,P.Fornasini: to be published
4. G.G.Long et al.: Inorg.Nucl.Chem.Lett. 5, 21 (1969)
5. I. Ikemoto - Bull.Chem.Soc.Jap. 54, 2519 (1981)

A Comprehensive Investigation of Residual Compounds of Br in HOPG and Graphite Fibers via X-Ray Absorption

J.L. Feldman, W.T. Elam, A.C. Ehrlich, E.F. Skelton, and D.D. Dominguez
Naval Research Laboratory, Washington, DC 20375, USA
S.B. Qadri
Sachs/Freeman, Maryland 20715, USA
D.D.L. Chung
Carnegie-Mellon U., Pittsburgh, PA 15213, USA
F.W. Lytle
The Boeing Co., Seattle, WA 98214, USA

I. Introduction

The Br intercalated graphite materials have been the subject of much investigation. However, questions regarding charge transfer and Br cites
for variously prepared samples are still unresolved. For example, diffraction analyses can provide accurate information on only in-plane Br
positions, as little or no correlation appears to exist among different
intercalated layers. Furthermore, possibly multiphase intercalate layers
[1] complicate analyses of diffraction patterns. Important complementary
information may be gotten from Br K-edge x-ray absorption data using
linearly polarized x-rays as obtained from synchrotron produced radiation.
A description of the various regimes of the absorption spectrum, i.e.,
white line (WL), Br-C EXAFS and Br-Br EXAFS, has been given previously in
the case of desorbed samples of Br intercalated grafoil [2], highly
oriented pyrolitic graphite (HOPG) and graphite fibers [3].

Following our initial measurements [3] on Br-fibers and -HOPG done at
room temperature (RT), we have performed additional measurements, including RT measurements on Br_2 vapor and on a differently prepared desorbed
HOPG sample. Measurements on our previous HOPG sample and fiber samples
at 360 K and 400 K were also performed with the result that the only
temperature dependence seen is in Br-C EXAFS. The thermal effects may be
of importance because of the melting of the Br sublattice believed to take
place in Br intercalated materials [4] at 380 K and the possibility that
the effective angle, α, which the Br_2 molecular axes make with the graphite planes is an out-of-plane vibrational effect [2]. In addition further
analysis has been done, such as least square fits to the well-known Teo et
al.-Lee et al. (T-L) parameterizations, and checks have been made on the
polarization of the beam at the samples by performing measurements at
appropriate sample orientations with respect to the beam. (No deviation
from linear polarization was observable.)

II. Experimental

The Si (220) monochromating Bragg reflection was used in these measurements. We present details of our HOPG results performed at angles of 0^0
(parallel) and 45^0 (rotated) of the x-ray polarization direction with
respect to the carbon planes. Sample HOPG-1 was uniformly thick, with
a $\Delta\mu x$ of 1.4 for normal incidence; it was prepared by desorbing in air at
373 K for 1/2 hour from a fully intercalated (in liquid) state. Sample
HOPG-2 was of nonuniform thickness and was desorbed in vacuum at RT for
about 2 days from a fully intercalated (in vapor) state. Since the thickness of HOPG-2 (with $\Delta\mu x \sim 0.2$) was much smaller than that of HOPG-1 the
nonuniformity presumably had no effect on the results. The two initially

pure HOPG samples used were also of different origins. Angles between the polarization direction and the c-axis of the HOPG samples are accurate to within one degree, as verified by our results for the orientational dependence of $\Delta\mu x$ for our uniformly thick sample.

III. Results and Discussion

A few of our results are given in Figures 1 through 3 and Table 1.

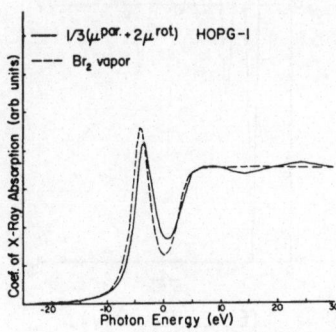

Fig.1. Comparison of white line spectra between Br-HOPG and Br_2 vapor

Fig.2. Br-Br EXAFS in Br-HOPG. The dashed curves correspond to the T-L parameterization; parameter values are in Table 1

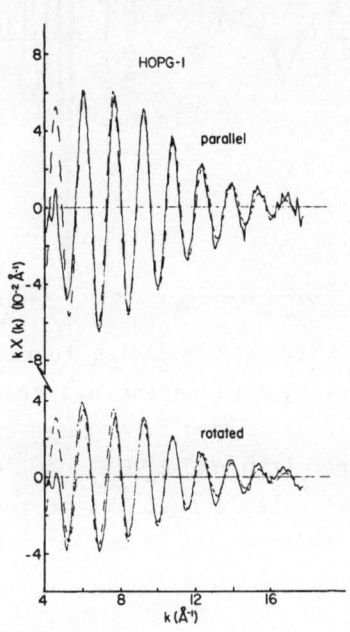

Table 1. Room temperature results of fits to Br-Br EXAFS*

	Vapor	Parallel: HOPG-1	(HOPG-2)	Rotated: HOPG-1	(HOPG-2)
$r(\text{Å})$	2.28	2.31	(2.31)	2.31	(2.31)
$\sigma(\text{Å})$	0.045	0.06	(0.06)	0.06	(0.06)
$A(\text{Å}^{-1})$	0.50	0.50	(0.41)	0.30	(0.24)
E_o-E_{WL} (ev)	9.8	3.0	(3.0)	3.0	(3.0)

* $F(k)/A$ and $\Phi(k)$ were from Teo et al.(exptl.)-Lee et al. The same $F(k)/A$ and $\Phi(k)$ were used for all fits.

Amplitude differences between HOPG-1[1] and Br_2 vapor cannot be explained in terms of differences in Debye-Waller (DW) factors; our fits to T-L (Table 1) indicate that the amplitude parameter, A, for HOPG-1 is 27% less than that for Br_2 vapor assuming coordination numbers of 1 for both materials. (We note that our results for the vapor are in excellent agreement with those of Stern et al. [5]) Furthermore the 20% amplitude decrease for HOPG-2 with respect to HOPG-1 cannot be explained by a possible difference in DW factors between HOPG-1 and HOPG-2.

---

[1]For comparison to χ (or A) of Br_2 vapor consider the spherical average, $1/3 \chi^{par.} + 2/3 \chi^{rot.}$.

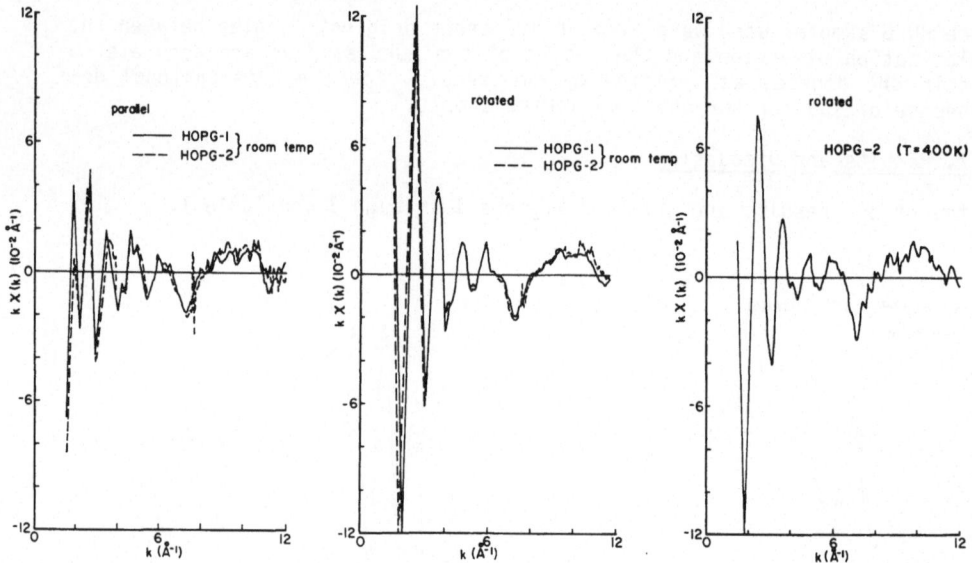

Fig. 3. $k\chi(k)$ corresponding to extracted Br-C EXAFS. $(E_o-E_{WL} = 3ev)$.
The following replacement has been made: $\chi(k) \rightarrow \chi(k) - \chi^{calc.}(k)$ where
$\chi^{calc.}$ is a calculated curve based on both a parameterization of our data
on Br_2 vapor (assuming $E_o-E_{WL} = 9.8$ ev) down to $k = 2A^{0-1}$ and on a fitted
amplitude scale factor, fit to the Br- HOPG data in the Br-Br EXAFS
region. The structure at $k \sim 7A^{0-1}$ is perhaps related to the poorness of
the fit.

We observe an agreement of the anisotropy of the WL areas and EXAFS
amplitudes for all our materials (to within ~10% for ratios of rot. to
par.). Also the spherically averaged WL areas are consistent with each
other and with the WL area for Br_2 vapor. Thus it appears that all of
the bromine is in the form of Br_2 molecules in Br-graphite and that
differences in electronic effects in these materials give rise to the
amplitude discrepancies. Perhaps further evidence that a "structural"
explanation is incorrect is that the HOPG-1 and HOPG-2 Br-C EXAFS are
quite similar. Indeed preliminary studies based on the Eeles and Turnbull
structural model [6] suggest that the data in the Br-C EXAFS region are
understandable in terms of single atom scattering Br-C EXAFS. On the
other hand it is interesting that recent x-ray diffraction data on a
desorbed sample of Br-single crystal graphite have been interpreted in
terms of a model for which 1/3 of the Br is in atomic form [1], which
would lead to a decrease in EXAFS amplitude from Br_2 vapor similar to
what we found.

The agreement which we obtained among WL areas also indicates that
no charge is transferred to Br $4p\sigma$ antibonding states as had been suggested for intercalated grafoil previously [2]. We should also mention
that differences between Br_2 vapor and Br-graphite are apparent in the
shape of the main edge with Br-graphite corresponding to a broader edge
than Br_2 vapor. This could be due to a combination of differences in
final state broadening and in density of states.

References

1. D. Ghosh and D.D.L. Chung, Synth. Metl. $\underline{7}$, 283 (1983).
2. S.M. Heald and E.A. Stern, Phys. Rev. B $\underline{17}$, 4069 (1978).
3. J.L. Feldman, E.F. Skelton, A.C. Ehrlich, D.D. Dominguez, W.T. Elam, S.B. Qadri and F.W. Lytle, Sol. State Commun., $\underline{49}$, 1023 (1984).
4. K.K. Bardhan and D.D.L. Chung in <u>Ordering in two Dimensions</u>, p. 395, S. K. Sinha ed., Elsevier North Holland (1980).
5. E.A. Stern, S.M. Heald and B. Bunker, Phys. Rev. Let. $\underline{42}$, 1372 (1979).
6. W.T. Eeles and J.A. Turnbull, Proc. of the Phys. Soc. of London A $\underline{283}$, 179 (1965).

EXAFS Study of a Decomposed Fe-Grafoil System

Yanjun Ma and Edward A. Stern

Department of Physics, FM-15, University of Washington, Seattle, WA 98195, USA

Hanan Shechter

Physics Department, Technion, Haifa, Israel

Polarization dependence of the EXAFS is a useful tool for measuring the anisotropy in bonding of oriented samples [1]. We measured the EXAFS of an Fe-Grafoil sample with x-ray polarization both parallel and perpendicular to the Grafoil surfaces to determine whether the Fe was in a two-dimensional state, with the goal of studying two-dimensional ferromagnetism.

The sample was made by decomposing $Fe(CO)_5$ loaded in Grafoil [2]. When heated above 105°C, $Fe(CO)_5$ in contact with Grafoil, which serves as a catalyst, decomposes into Fe and CO. The gas emitted was pumped away, leaving the residual Fe-Grafoil in a sealed cell. A 0.88 monolayer of $Fe(CO)_5$ was loaded into Grafoil; when decomposed, it gave less than the equivalent of 0.1 monolayer of iron on Grafoil. The sample cell was configured so that EXAFS could be measured with polarization both parallel and perpendicular to the Grafoil surface. Measurements of the K edge of Fe were made at room temperature. Because of the large surface-to-volume ratio of Grafoil the measurements could be made in the transmission mode at SSRL.

Because of the leakage caused by nonuniformity of the sample, especially in the direction in which the incident x ray is parallel to the Grafoil surface (clearly seen in the x-ray photos we took), experimental error can be as large as 10%-15%. Standard EXAFS data analysis, results of which are given in Table 1, shows that the differences between the two polarizations are within the error bars for both first and second coordination shells around Fe, indicating that the system is not two-dimensional.

Table 1. Polarization comparison. Subscripts a and b indicate polarization parallel (a) and perpendicular (b) to the surface

	First Shell	Second Shell
N_a/N_b	0.96 ± 0.1	0.96 ± 0.1
$R_a - R_b$ [Å]	0.0 ± 0.05	0.0 ± 0.05
$\sigma_a^2 - \sigma_b^2$ [Å^2]	0.0015 ± 0.002	0.002 ± 0.002

To determine the Fe environment of the sample, we compared the sample with Fe_3O_4 at 80K, Fe-glycine [$Fe_3O \cdot (glycinato)_6(H_2O)_3(ClO_4)_7$] at 80 K, and undecomposed $Fe(CO)_5$-Grafoil as standard compounds. We found for the first shell that only oxygen atoms as obtained from Fe_3O_4 and Fe-glycine standards gave reasonable results, as shown in Table 2, namely, four oxygen atoms at an Fe-O distance of 2.0 Å, while assuming carbon atoms as obtained from an $Fe(CO)_5$ standard gave about ten carbon atoms, which is unreasonable, and an unreasonable phase shift. The second-shell results using Fe-glycine are also listed in Table 2. We find about three Fe at a mean distance of 3.38 Å. It is possible that there are some carbon or oxygen mixed in the second shell, but we isolated the Fe contribution by emphasizing the high k portion of the data. We also found a larger than usual pre-edge peak which corresponds to the $1s \rightarrow 3d$ transition, indicating that the Fe environment has some lack of inversion symmetry and that the four oxygen may be tetrahedrally coordinated about the Fe.

Table 2. Fe environment. The two polarization differences are within the error bars. Disorder $\Delta\sigma^2$ is relative to that of Fe-glycine at 80 K

	N	R [Å]	$\Delta\sigma^2$ [Å^2]
First-shell oxygen	4.0 ± 0.7	2.06 ± 0.05	0.002 ± 0.002
Second-shell Fe	3.0 ± 1.0	3.38 ± 0.1	-0.001 ± 0.002

In conclusion, the EXAFS shows that the reduced Fe is not two-dimensional and is not a metallic or a standard iron oxide. Because we have less than 0.1 monolayer of Fe, it is possible that the Fe atoms cluster around the defects of Grafoil. Even if there is no clustering around defects, because of the catalytic effect of Grafoil the compound that is formed may not be a standard one. Further studies of this type with greater Fe coverage are planned.

Support for this work was provided by the National Science Foundation, grant no. DMR80-22221. The help of the staff at SSRL is gratefully acknowledged. SSRL is supported by the DOE Office of Basic Energy Sciences, and by the NIH, Biotechnology Resources Program, Division of Research Resources.

References

1. S.M. Heald and E.A. Stern: Phys. Rev. B 17, 4069 (1978)
2. R. Wang, H. Taub, H. Shechter, R. Brener, J. Suzanne, and F.Y. Hansen: Phys. Rev. B 27, 5864 (1983)

Part IX **Related Techniques and Instrumentation**

Fundamental Aspects in X-Ray Absorption in Dispersive Mode

E. Dartyge[1], A. Fontaine[1], A. Jucha, and D. Sayers[2]

LURE, Bât. 209 C, Université de Paris-Sud, F-91405 Orsay Cêdex, France and
[1]Laboratoire de Physique des Solides, Bât. 510, F-91405 Orsay Cêdex, France and
[2]Physics Department, North Carolina State University, Raleigh, NC 27650, USA

X-ray absorption in dispersive mode associates a dispersive and focussing triangle-shaped bent crystal with a position sensitive detector which is able to achieve a fast record of the energy-dependant X-ray absorption spectrum. The key-element of this method resides in the detector, which has the task to collect and locate high flux of photons ($\sim 10^{10}$, 10^{11} X-ray photons per s. at LURE). Thus a brief description of the salient features of the detection system will introduce this paper which gives in addition information about the observed energy resolution and the structure of the focus spot, which is certainly not a single point.

A – Detection system

As shown in the simplified block diagram (fig. 1) the RETICON 1024 array has been housed in a cryostat, currently cooling the photodiodes at $\sim - 100°C$ which is held constant within 0.03°C over ten minutes, within 0.1 C over a couple of hours. The system divides into five sections : derive-independant hardware, noise sensitive electronics mounted close to the back of the cryostat, signal processing electronics, μ processor (200 ns 16 bits) and fast RAM (40 ns), and computer and peripherals. Thus it has been possible to record in 16 ms the copper spectrum (Fig. 2) over a 300 eV-wide range. One has to stress that absolute measurements are achieved because the same detector is used for I_0 and I_1.

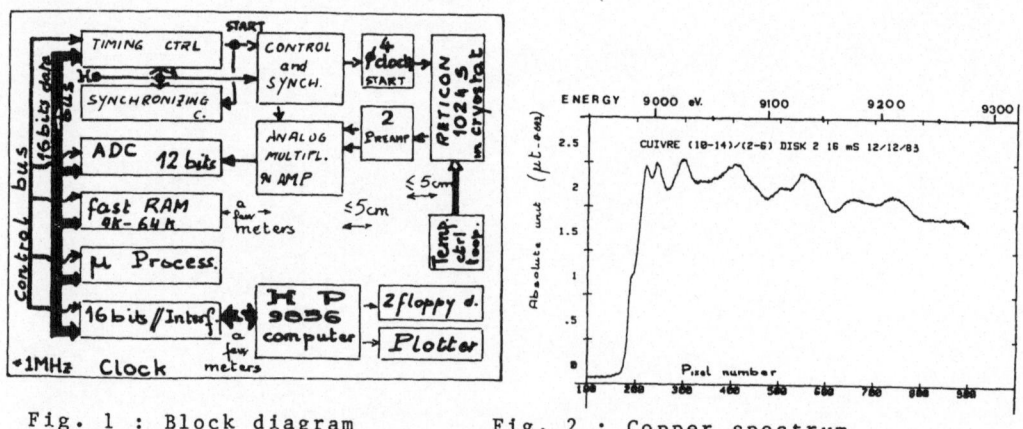

Fig. 1 : Block diagram   Fig. 2 : Copper spectrum

The ratio signal/noise is relevant to the photon statistics since the electronic readout noise is currently estimated to 2000 $\bar{e}$. This figure has to be compared to the 8.8 $10^7\bar{e}$ amount stored in one saturated pixel : assuming a quantum efficiency of 1/2 for 8 keV photons, each photon creates 1000 e-h pairs and thus 10^5 photons are needed to achieve saturation, which means 300 photons (300 000 $\bar{e}$) as a shot noise.

B - Energy resolution

It is obvious from fig. 2 that energy resolution does not compare well to the best spectrum where the bump in the rise of the edge is resolved. But it is still pretty good and even with the estimated 4 eV resolution at 9 KeV very useful near-edge data can still be collected. For the EXAFS modulations, the energy resolution is even better than is necessary. On the preedge feature (5975 eV) and the Cr^{6+} K-edge position [2] (Fig. 3) the energy-resolution affects mainly the peak height (0.65 instead of 1), while the width of the prepeak is hardly increased. Another example (Fig. 4) concerns the PtO_2 white line [3], which reaches a maximum $\mu t = 2.5$ when scanned on a well-adjusted double crystal station, and is only $\mu t = 1.87$ (t beeing chosen to get $\mu t = 1$ behind the white line). Again the energy resolution has affected the white-line shape. Furthermore the width is clearly larger (FWMH = 5.6 eV, and 7.3 eV respectively) suggesting a 6.2 eV resolution. All these features lead one to think that the curved crystal creates a tail on the lower energy side of the Darwin profile.

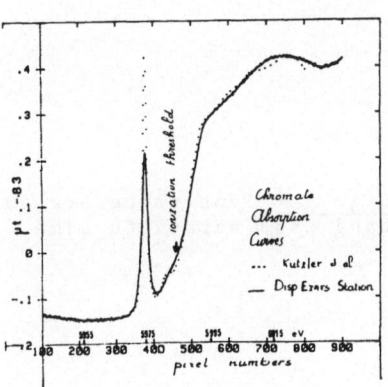

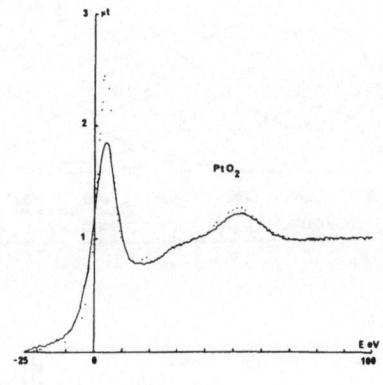

Fig. 3 : Experimental chromate absorption : block line,(Disper sive): dotted line (ref. 2)

Fig. 4 : White line of Pt 02 as recorded in dispersive mode or in a step by step scan

C - Focus structures

If the focus structure is mainly dependant on the crystal curvature, one expects a nose-shaped energy distribution vs. position. If the focus spot deals with the large source size, the nose-shape is not drawn with a line but with a wide strip, which can eventually be larger than the whole distribution

from a single point source. Thus, in that case, energies are
smeared throughout the whole focus spot. Horizontal slitting
at the focus plan in that case will result in improved energy
resolution. A rapid way to investigate the spot structure is
to make a 30 μ Pt wire-intercepting energy band out of the
nose-like distribution of energy vs. position. It is clearly
dependent on the photon-energy considered. Fig. 5 and Fig. 6
plot the absorption from the Pt wire. A zero value is thus
relevant of fully transmitted beam. Absorption equal to 1 has
been chosen to determine the width of the intercepted band.
Inserts in fig. 5 and 6 contain centered rectangles, whose
widths are the 30 μ size of the wire and heights are the
breadth of the intercepted energy band. For the S 111 crystal
set across the Pt edge (fig. 6), the low-energy tail originates
from the flat part of the triangle-shape crystal which is
illuminated, since these photon energies need Bragg angle lower
than in the copper case. Thus the spot size is about 350 μ
when working around 9000 eV with a 200 eV bandpass, while for a
similar curvature, around 11600 eV, the 400 eV bandpass are
distributed within a full mm. In this last case, because of the
assymetry between the two arms of the distribution, one may ask
if we were exactly on the focus spot and not a little bit
toward the detector.

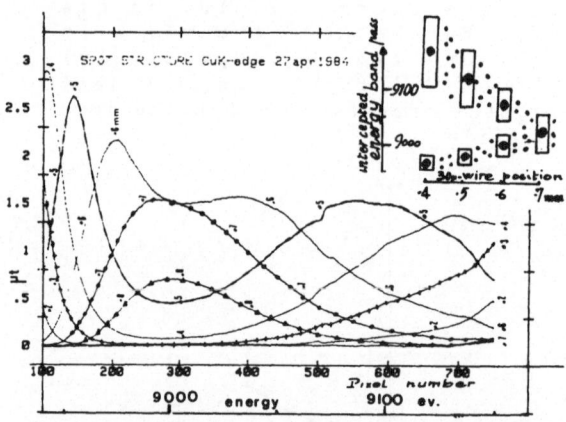

Fig. 5 : Spot structure for
band pass across Cu edge

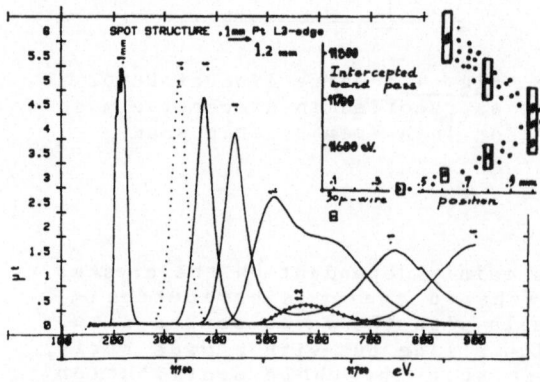

Fig. 6 : Spot structure for
band pass across Pt edge

474

REFERENCES

1 - A. Jucha, D. Bonin, E. Dartyge, A. M. Flank, A. Fontaine,
    D. Raoux, NIM, to be published (1984, July).
2 - F. Kutzler, C. R. Natoli, D. K. Misemer, S. Doniach,
    K. O. Hodgson, J. Chem. Phys. 73, 7, (1980, October).
3 - D. Sayers, D. Bazin, E. Dartyge, H. Dexpert, A. Fontaine,
    A. Jucha, P. Lagarde, This volume.

An Energy Dispersive X-Ray Absorption Spectrometer and Its Application to Stopped-Flow Experiment

T. Matsushita
Photon Factory, National Laboratory for High Energy Physics, Oho-machi
Tsukuba-gun, Ibaraki 305, Japan
H. Oyanagi
Electrotechnical Laboratory, Umezono, Sakura-mura, Niihari-gun
Ibaraki 305, Japan
S. Saigo and H. Kihara
Department of Physics, Jichi Medical School, Minamikawachi-machi
Kawachi-gun, Tochigi 329-04, Japan
U. Kaminaga
Rigaku Corporation, Matsubara-cho, Akishima, Tokyo 196, Japan

1. Introduction

With an energy dispersive X-ray absorption spectrometer, the entire spectrum of interest can be recorded simultaneously. Hence, this method is suitable for time-resolved study of structural or chemical changes in materials which are not easily repeated hundreds or thousands of times. We report a preliminary result of time-resolved (1.0 second resolution) X-ray absorption measurement with the energy dispersive spectrometer for reacting aqueous solutions of Fe $(NO_3)_3$ and $Na_2S_2O_3$ mixed by a stopped-flow method. A time-dependent energy shift of Fe-K absorption edge was observed.

2. Spectrometer

The experimental arrangement is shown in Fig. 1. This spectrometer is essentially of the same design as the one reported earlier [1,2]. A cylindrically-bent triangular-shaped silicon (111) crystal is used as a dispersing and focusing optical element. Convergent X-ray beams reflected by the crystal have one-to-one correspondence between energy and direction. By placing a sample at the focal point and by measuring the intensity distribution across the beam direction behind the focus, an X-ray absorption spectrum can be taken simultaneously. A platinum coated mirror is placed

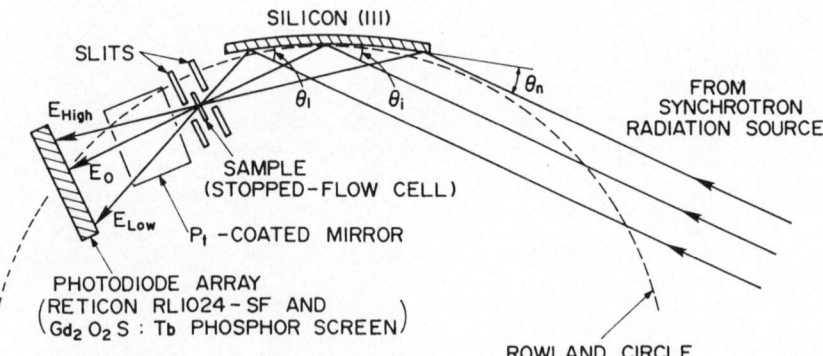

Fig.1. Geometrical arrangement of an energy dispersive X-ray absorption spectrometer

between the sample and the detector in order to suppress unwanted high energy harmonics from the dispersing crystal.

As a detector, a photodiode array (RETICON RL1024-SF) with an optical fiber face-plate is used. It has 1024 diode elements and an effective area of 2.5 mm high and 25.4 mm wide. On the surface of the face-plate, a green light emitting phosphor (Gd_2O_2S:Tb) screen was formed with a thickness of about 50 μm.

With a specially designed data aquisition system, sixteen spectra can be recorded successively with an integration time of 10 msec - 999 sec and a time interval of 1 msec - 999 sec.

For evaluating the performance of the spectrometer, an EXAFS spectrum for copper metal foil was measured in 100 msec. The result of Fourier transform showed a good agreement with the result for the spectrum measured by the conventional point-by-point method,using a channel-cut monochromator system at BL-10B at the Photon Factory.

3. Stopped-flow experiment

The stopped-flow cell is a newly developed one [3], which has a 1 mm thick sample volume sandwiched by two 50 μm thick quartz windows for X-ray transmission measurement. Flowing 0.6 M aqueous solutions of ferric nitrate ($Fe(NO_3)_3$) and sodium thiosulfate ($Na_2S_2O_3$) are mixed just before they flow into the cell. The mixed solution reaches the cell in approximately 6 milliseconds. The reaction between these two solutions is represented by 4

$$Fe^{3+} + 2S_2O_3^{2-} \longrightarrow Fe(S_2O_3)_2^- \qquad\qquad (1)$$

$$Fe(S_2O_3)_2^- + Fe^{3+} \longrightarrow 2Fe^{2+} + S_4O_6^{2-} \ . \qquad\qquad (2)$$

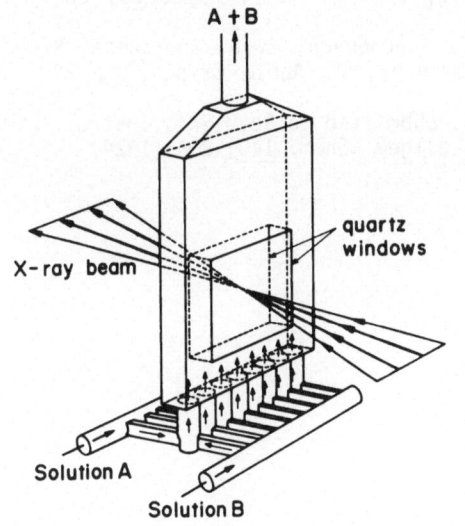

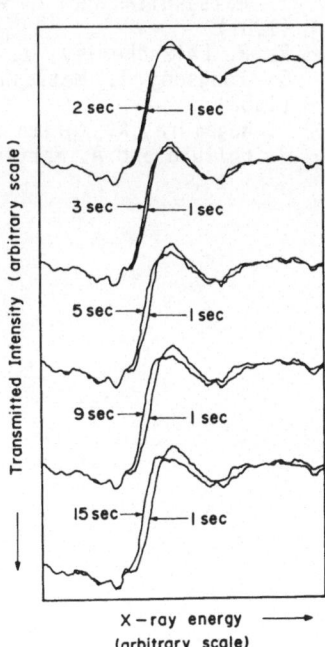

Fig.2. Stopped-flow cell

Fig.3. Time-dependent variation of the spectrum near Fe K-absorption edge after mixing two solutions of 0.6 M $Fe(NO_3)_3$ and 0.6 M $Na_2S_2O_3$

477

The first reaction takes place in a very short time, i.e. less than a few milliseconds, while the second reaction takes place in a few tens of seconds. After mixing, the color of the mixed solution suddenly turns into purple, owing to the reaction represented by (1). Then, it gradually becomes transparent again. This process is represented by (2).

The experiment was performed on Beam Line 4A at the Photon Factory. Sixteen spectra were successively taken with a time interval of one second after stopping the flow of two solutions. We observed a chemical shift of the K-absorption edge of iron due to the transition from Fe^{3+} to Fe^{2+}. Figure 3 shows spectra for 1, 2, 3, 5, 9 and 15 seconds after stopping the flow. The curve for one second is shown as a reference for others. The absorption edge is shifted by roughly 4 eV after 15 seconds. Note that raw intensity data are shown in the figure. Small dips and zig-zag shaped intensity variation are due to inhomegeneous sensitivity of photodiodes.

The time resolution can of course be improved by repeating the measurement. We actually obtained a fairly good statistic with a time resolution of 0.25 sec by repeating the measurement twenty times. With several improvements of optics and the detector, a time resolution of a few tens of milliseconds will be within reach.

Acknowledgement

The authors would like to thank Prof. K. Kohra for his continual encouragement and Dr. S. Suzuki for providing them with fine phosphor powders and for his advice in preparing thin phosphor layer. They are also thankful to Prof. Y. Sasaki and Prof. T. Iizuka for their useful suggestions. The help of Mr. H. Hirama in the experiment is acknowledged.

References

1. T. Matsushita and R. P. Phizackerley, Jpn. J. Appl. Phys. 20, 2223 (1981).
2. R. P. Phizackerley, Z. U. Rek, G. B. Stephenson, S. D. Conradson, K. O. Hodgson, T. Matsushita and H. Oyanagi, J. Appl. Cryst. 16, 220 (1983).
3. T. Nagamura, K. Kurita and H. Kihara, submitted to Rev. Sci. Instru.
4. J. Holluta and A. Martini, Z. anorg. allgem. Chem. 140, 206 (1924).

Efficiency of the Energy Dispersive Configuration for X-Ray Absorption Measurement by Total Reflection

J. Mimault[1], R. Cortes[2], E. Dartyge[3], A, Fontaine[3], A. Jucha, and
D. Sayers[4]

C.N.R.S.-LURE, Bât. 209 C, F-91405 Orsay, France

Recent progress in detection of high X-ray fluxes (1) has led to a variety of applications using the dispersive mode of X-ray absorption data collection. Here, we present the first measurement of the X-ray absorption spectrum of nickel by using total reflection in an energy-dispersed beam.

1 EXPERIMENTAL CONDITIONS

The measurements reported here include data obtained on a 300Å thick nickel layer deposited on a glass slide and a polished nickel block. All spectra were obtained in 200ms with a band width of 275eV covering the nickel K-edge. This energy range was used in order to preserve energy resolution since we were working with only a 7cm-long, triangle-shaped Si (111) crystal installed on the small angle-scattering station at LURE. The energy-dependant oscillations obtained in the total reflection geometry have been termed ReflEXAFS by MARTENS and RABE (2).

The very short data collection time compares favorably to times of more than ten minutes needed, using the step by step ReflEXAFS data collection (2,5). In fact 10^8 photons were used to produce the nickel spectrum in 200ms while the conventionnal crystals monochromator delivers 10^6 photons per second through an equal vertically slitted beam (100 μ).

The illuminated surface was set at 4.4mrad which is closed to the critical angle (6.3mrad). As pointed out by BECKER, GOLOVCHENKO, and PATEL (3), in the immediate vicinity of the dielectric surface, that condition creates standing waves with antinodes on the interface. Hence the evanescent wave, which penetrates the material past the interface, has maximum electric field in the near surface region which in any case is limited to the energy-independant value of about 25Å (4). The procedures used and the sample holder which has been built by one of us (R.C.) have been described elsewhere (5).

2 DISCUSSION

In Fig.1 is the absorption spectrum of a pure Ni foil collected in transmission mode using the dispersed beam reflected by the Ni surface as an impinging beam. Since the same detector is used with and without the Ni foil, absolute " μ t" values exhibit the edge jump expected from the 6μ thick foil. It must be mentioned also that Ni foil spectrum provides an immediate energy calibration of the pixel numbers. From the direct comparison of this spectrum with the ReflEXAFS spectrum (Fig.1) the following points can be made :

---

[1]Lab. de Métallurgie Physique, F-86022 Poittiers, France

[2]Lab. de Physique des Liquides et Electrochimie UPVI Paris, France

[3]Lab. de Physique des Solides, Bât. 510, F-91405 Orsay, France

[4]Physics Dept. North Carolina State University, Raleigh, NC 27650, USA

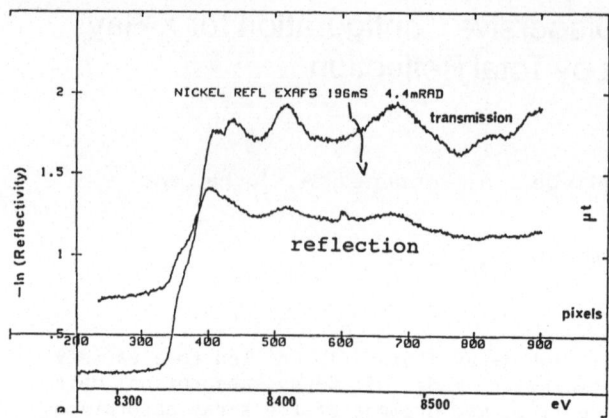

Figure 1: Comparison between Ni EXAFS spectra obtained in transmission and reflection modes

a) There is a very good agreement of the position and shape of the fine structure oscillations in the 8390-8560 eV range. This signal comes for the first 25Å past the interface (4) which are energy independant, as can be shown using Fresnel's equations. One of the features which is of importance deals with the current E^{-2} dependance of $1-\delta$, the real part of the refraction index n (n=$1-\delta-i\beta$) which is balanced by the rise in f' above the edge which follows roughly an E^2 law. This cancellation produces quasi constant values for $(1-\delta)$, in the 50-500eV band pass above the edge as experimentally shown by Snell's law measurement (6). This is no longer the case right at the edge, and far from truth below the edge where the f' decrease helps the current E^{-2} dependance to produce a rapid change of δ. An obvious result is that $\theta_c = \sqrt{2\delta}$ changes rapidly close to the edge and the reflectivity jump is not simply controlled by the absorption change.

b) For the reasons stated above the distortion of the edge shape is not an artefact. A similar feature has been observed on the Cu edge (2). This near-edge structure of the ReflEXAFS data is more complicated than the direct absorption spectrum since it is a function of δ and β ($\beta= \mu\lambda/4\pi(1-\delta)$) which are related by a Kramers-Kronig transform.

c) To convert the experimental reflectivity into an absorption value is not straightforward. In fact different successful data handlings have been used (2) (5) which utilized the normalization of the easily-extracted oscillations. But to obtain a good analysis, a good signal-to-noise is needed. This, in turn, is primarily a function of the surface roughness and the accurate achievement of the orientation of the sample to keep good parallelism between the direct and reflected beams. Under these requirements good reflectivity can be measured. From the data shown here the reflectivity below the edge is 0.47 while above the edge the average reflectivity is about 0.25 This value decreases as the glancing angle approaches θ_c.

d) For the ReflEXAFS spectrum, normalization is performed using an I_o which is indeed the direct beam collected without sample. When the sample is positioned the photodiode array is raised vertically, according to the 2θ deviation. But the exact parallelism between the direct and the reflected X-ray sheets does not exist per se. Indeed a tuning of the sagittal orientation seeks to adjust the K-edge of the reflected profile exactly on the pixel where the K-edge of the direct transmitted profile was.

480

3 FUTURE DEVELOPMENTS

The results presented here show the feasibility of doing dispersive ReflEXAFS. Even with the synchrotron sources currently available several applications of this technique seem promising. This includes studying the transformation kinetics of a metallic electrode investigated in-situ, coated with an electrolyte in such a way that the X-ray path is lower than 2mm(5). In addition, transformations and general structure of thin epilayers on semiconductors and problems in surface corrosion may be feasible with this technique. With the development of more intense sources or sources with higher brilliance it will be possible to reduce the focal spot size allowing small surfaces to be studied.

1 E. Dartyge, A. Fontaine, A. Jucha, D. Sayers: this volume
2 G. Martens, P. Rabe: Phys. Stat. Sol. a58, 415(1980)
3 J. Goulon, C. Goulon-Guinet, R. Cortes, J.M. Dubois: J. Physique 43 539(1982) and R.S. Becker, J.A.Golovchenko, J.R. Patel: P. R. L. 50, 53(1983)
4 L.G. Parratt: Phys. Rev. 95, 350(1954)
5 L. Bosio, R. Cortes, A. Defrain, M. Froment: J. of Electroanal.Chem. and Inter. Electrochem. (to be published)
6 A. Fontaine, W.K. Warburton, K. Ludwig: Submitted to Phys. Rev. B

EXAFS Investigations of Ion-Implanted Si Using Fluorescence Detection and a Grazing-Incidence X-Ray Beam

Bruce A. Bunker

Physics Department, University of Notre Dame, Notre Dame, IN 46556, USA

Steve M. Heald and John Tranquada

Brookhaven National Laboratory, Upton, NY 11973, USA

EXAFS is uniquely suited to the study of the local environment about impurities in solids. Because of technical difficulties, however, the technique has seen relatively little application to single-crystal systems. In this work, we report recent measurements made at the Cornell High-Energy Synchrotron Source (CHESS) testing the feasibility of using a grazing-incidence x ray beam, fluorescence detection, and rotation of the sample during data acquisition. The samples were three-inch Si <111> wafers implanted with Fe, a relatively fast-diffusing impurity. The implantation energies and doses were 200KeV, 10^{16} cm^{-2}; 100KeV, 5×10^{15} cm^{-2}; 40KeV, 5×10^{15} cm^{-2}.

For these implantation energies, implanted ions penetrate the material less than about 2000A. The penetration depth of 7 KeV x-rays in Si is about 200μ, so the effective volume concentration is a factor of 1000 lower than the near-surface volume concentration using typical ($\approx 45°$) x-ray incidence angles. To enhance the near-surface sensitivity, a grazing-incidence geometry was used where the x-rays impinged on the surface at a glancing angle of 5-15 mrad, slightly above the critical angle for total reflection. Because of its large acceptance and high count-rate capability, a large-area ion chamber was used to detect the Fe fluorescence x rays.

It is well known that EXAFS measurements of single crystal samples are plagued by x-ray diffraction peaks -- some of which may be much larger than the edge step itself. The solution we have chosen is to spin the sample about a vertical axis at 10 to 30 Hz -- fast compared with the data collection integration time. Figure 1 shows a schematic of the experimental configuration. This approach has proved quite successful in removal of the diffraction lines, as shown in Figure 2. This

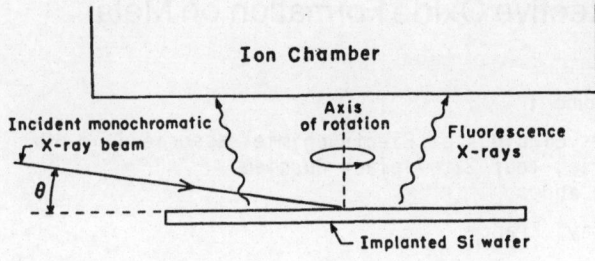

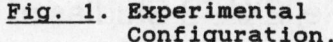
Fig. 1. **Experimental**
Configuration.

Ion Chamber

Incident monochromatic
X-ray beam

Axis
of rotation

Fluorescence
X-rays

θ

Implanted Si wafer

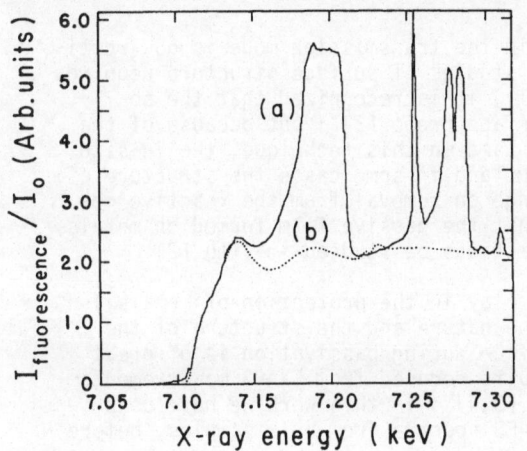

Fig. 2. Comparison of fluorescence EXAFS scan (a) without
spinning sample, and (b) spinning sample at 20Hz about a
vertical axis.

data was obtained <u>without</u> the use of x-ray filters, which would
further suppress any diffraction lines.

A striking feature of these data is the high signal-to-
background ratio -- over 3:1 without the use of x-ray filters.
Very high counting rates may also be realized by this
technique: The measured fluorescence signal corresponds to
about 3×10^6 fluorescence photons/sec. These results are
extremely promising for application of this technique to
much more dilute systems and higher-Z matrices.

ReflEXAFS Studies of Protective Oxide Formation on Metal Surfaces

L. Bosio, R. Cortes, and M. Froment

G.R. n° 4 C.N.R.S.; "Physique des Liquides et Electrochimie" associé à
l'Université Pierre et Marie Curie, Tour 22, 4 place Jussieu
F-75230 Paris, Cedex 05, France, and

L.U.R.E. Bât. 209 C, F-91405 Orsay, France

The X-ray absorption spectroscopy in the transmission mode is now routinely performed, but the application to studies of surface structure requires the use of special techniques. Actually, it is recognized that the so-called SEXAFS method is a powerful surface probe [1] ; but because of the high vacuum environment of the sample used in this technique, the in-situ measurements are definitively ruled out and in some cases the structure of the surface layers is believed to change on removal from the reactive medium. For instance, it is essential that the passive film formed on metals during chemical or electrochemical reactions be studied in-situ [2].

The importance that passive layers play in the protection of metals is well known and the determination of the nature and the structure of the oxides which are generated on the surface during passivation is of great interest. Recently, some striking results deduced from EXAFS measurements on ultrathin films have been reported [3,4] ; in this work we have used the specular reflexion to generate EXAFS spectra from bulk samples, before and after passivation process.

The method takes advantage of the fact that the index of refraction for X-rays is less than unity so that, for grazing angles lower than the critical angle Θ_c (at 7 keV, Θ_c is typically equal to about 5 mrad for a metal) the reflected intensity is high ; moreover the penetration depth in the condensed matter is low and almost only dependent on the density ρ of the surperficial zone : as an example, the angular dependence of the penetration depth Z is depicted in Fig. 1 for pure Ni and NiO (in this calculation, the oxide density has been arbitrary lowered to take into account the low density of the passivated layers [2]). By adjusting the grazing angle either slightly below the critical angle Θ_{c1} of NiO or slightly below the critical angle Θ_{c2} of Ni (Fig. 1) the reflected beam which contains the EXAFS signal will be preferentially referred to the oxide or to the substrate, respectively.

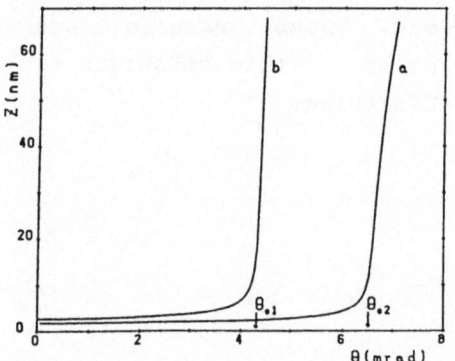

Fig. 1 : Angular dependance of the penetration depth of X-ray (E = 8263 eV)
a : for Ni
b : for porous NiO (the actual density is the half of the bulk one)

The in-situ measurements require that the spectra be taken while the sample is immersed with solution where oxidation occurs : we have developed two kinds of cells suitable for the ReflEXAFS technique provided that the edge lies about 7 keV.

i) In one (Fig. 2) the sample is settled in a flat frame which allows both the surface polishing and the alignment in the X-ray beam. The illuminated area (about 1.5 x 20 mm^2) is limited by the absorption of the radiation caused by the electrolyte in the cell ; this cell, provided with two mylar windows,contains the Pt counter-electrode for electrochemical reactions ; the reference electrode is inserted in the electrolyte circuit.

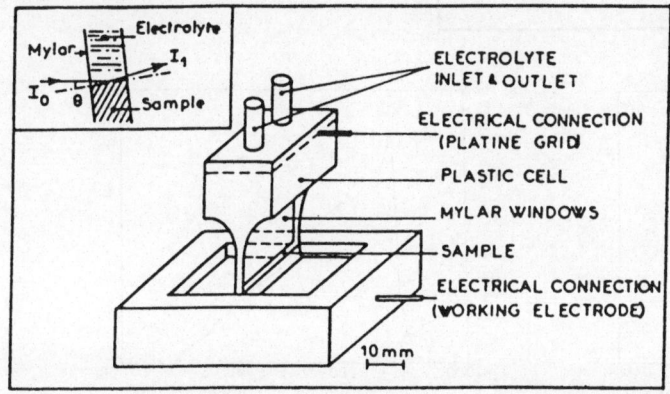

Fig. 2 : In-situ EXAFS cell for electrochemically passivated surfaces (in insert the X-ray path)

Fig. 3 shows the EXAFS modulation, $k\chi(k)$, versus the electron wave number k from a Ni-electrode before (curve b) and after (curve c) passivation in 1N sulfuric acid. For comparison, the fine structure of Ni, as determined from transmission measurement, is also reported in Fig. 3 (curve a) : the agreement with the result related to the Ni-electrode before passivation (curve b) is rather good on account of the ratio $1 : 10^{-3}$ of the irradied volumes in the two experiments. In contrast to the spectrum generated from a Ni substrate thermically oxided (curve d),the ReflEXAFS spectrum relative to the passivated electrode exhibits poor oscillations (curve c).

ii) in the other cell (Fig. 4) the sample can be brought close to the mylar window (6 μm thick) ; thus, the X-ray beam may impinge the sample at

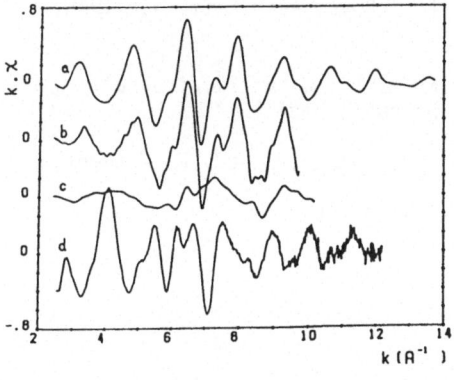

Fig. 3 : EXAFS modulation for Ni electrode before (b) and after (c) passivation in 1N sulfuric acid, at Θ = 3.4 mrad. Curves a (Ni) and d (NiO) are given for reference

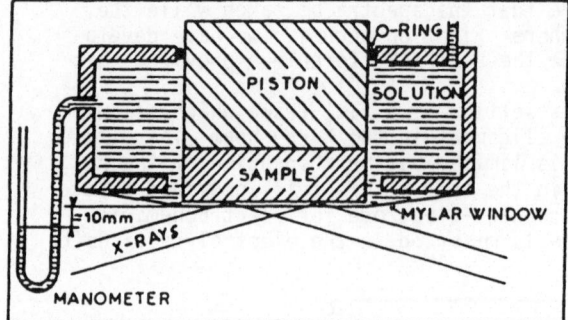

Fig. 4 : The in-situ EXAFS cell for chemically passivated surfaces

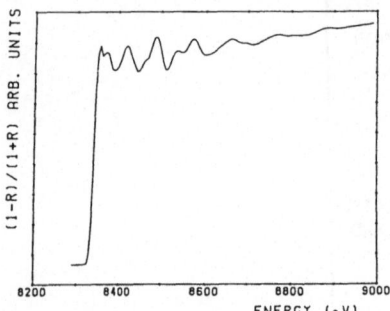

Fig. 5 : In-situ EXAFS spectrum of Ni immersed in solution

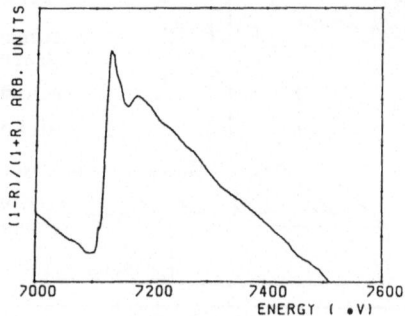

Fig. 6 : EXAFS spectrum of iron substrate passivated in concentrated nitric acid

grazing angles lower than the sample Θ_C but larger than the Θ_C of the solution on the mylar window. Fig. 5 shows the in-situ spectrum from a nickel plate immersed in solution, before passivation.

Another illustration of the capability of the ReflEXAFS technique to study the metal passivation is given by Fig. 6 which shows the spectrum generated from an iron substrate passivated in concentrated nitric acid, the most ancient example of passivation known.

REFERENCES

1   P.H. Citrin, P. Eisenberg and R.C. Hewitt : Phys. Rev. Lett. 45, 1948, (1980)

2   L. Bosio, R. Cortès, A. Defrain and M. Froment : J. of Electroanal. Chem. and Inter. Electrochem. (to be published)

3   R.W. Hoffman : Passivity of Metals and Semiconductors, ed. M. Froment (Elsevier 1983) p. 147

4   G.C. Long, J. Kruger and M. Kuriyama : ibid p. 163

Reflection Extended Energy Loss Fine Structures Above Ti $L_{2,3}$ Edge: A Comparison with EXAFS Results

M. De Crescenzi
Dipartimento di Fisica, Università dell'Aquila, I-67100 L'Aquila, Italy

G. Chiarello, and E. Colavita
Dipartimento di Fisica, Università della Calabria, Arcavacata di Rende
I-87036 Cosenza, Italy

We present a reflection energy loss investigation of the Ti $L_{2,3}$ edge.The oscillations detected above the core-edge have been analyzed following the standard EXAFS-procedure in order to obtain the radial distribution function around the excited central atom.

The Extended Energy Loss Fine Structures (EELFS) technique (1) has been used recently to perform structural studies of pure transition metals (2) and of atomic species chemisorbed on metal surfaces (3). In this view,the EELFS technique should be placed among the more conventional SEXAFS (4) and LEED spectroscopies. Moreover it could be used not only for investigating single or polycrystals,but also for local structural studies of less ordered systems such as interfaces compounds,amorphous and oxides.

This is the first EELFS investigation on $L_{2,3}$ edge (5) and the first comparison with an EXAFS study (6) carried out around the same core edge. The good agreement between spectra and results obtained with the two different probes sheds more light on the extension of the EXAFS-like interpretation and analysis procedure to the reflection energy loss features. Thus the dipole approximation on the inelastic cross section N(E),usually applied in transmission energy loss experiments (7) can be used also to interpret the EELFS spectra.

The experiment is carried out with a CMA electron analyzer in a standard UHV chamber. Fig.1 shows the extended fine structures detected above the $L_{2,3}$ edge of a clean Ti sample,in the loss range 400-900 eV. The same figure reports the EXAFS oscillations (6) above the same edge through synchrotron radia-

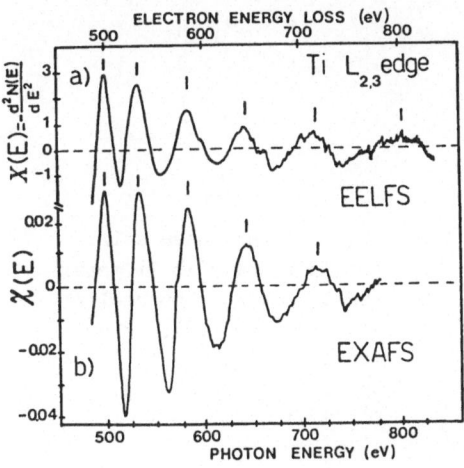

Fig.1 – a)EELFS features measured above the Ti $L_{2,3}$ edge, with a primary beam energy of E_p=1800 eV.
b) EXAFS spectrum measured by Denley et al. (6) with synchrotron radiation above the same egde.
The atomic L_1 contribution (E=566 eV) has been subtracted in both spectra.

487

tion experiments. The correspondence in energy and lineshape between the two spectra is remarkably good. Fig.2 shows the F(R) functions obtained from the two spectra using the Fourier tranform procedure. By correcting for the phase shift associated to the p-d final states,the first peak of the two F(R)-functions occurs at the same crystallographic value (within $\pm$ 0.02 Å).

Since neither the lineshape of the EELFS oscillations nor their relative intensities change when E_p is varied, the validity of the dipole approxima-tion ($q \sim 0$ Å^{-1}) seems to be confirmed:

$$N(E) \simeq \int \frac{1}{q^3} \left| \langle \psi_f | e^{i\bar{q} \cdot \bar{r}} | \psi_i \rangle \right|^2 dq \qquad \simeq \qquad \left| \langle \psi_f | \hat{\varepsilon}_q \cdot \bar{r} | \psi_i \rangle \right|^2 \qquad (1)$$

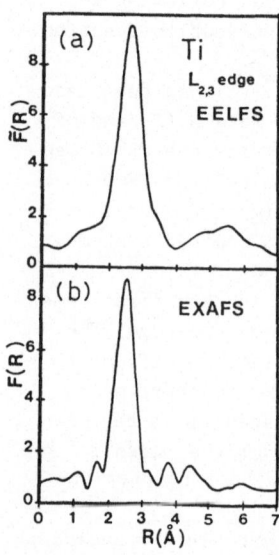

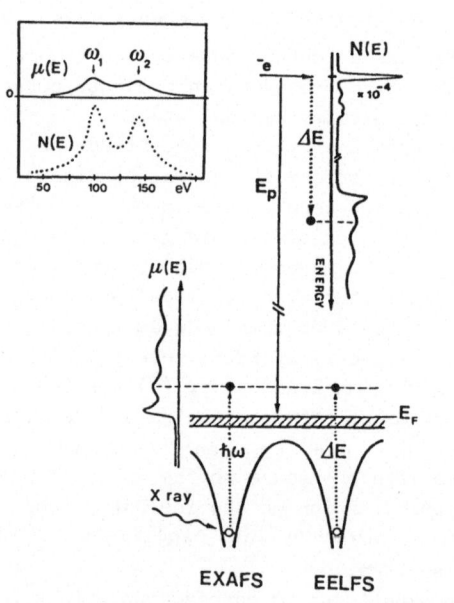

Fig.2 - a)Fourier analysis of the EELFS signal of Fig.1 a). b) EXAFS-F(R) as reported by Denley et al. (6).

Fig.3 - Schematic pictures of the exci-tation processes of EXAFS and EELFS spectroscopies.

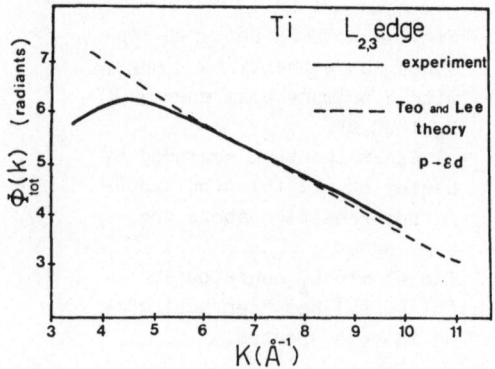

Fig.4 -Total phase shift for Ti-Ti pair in Ti metal for $L_{2,3}$ edge: theoretical (dotted line) (8), experi-mental (solid line) as ob-tained from EELFS-F(R) of Fig.2 a).

Thus the incident electron (with $E_p \gg \Delta E$) shows in the reflected inelastic cross section N(E) the same features above a threshold excited through X-rays (Fig.3). We compare in Fig.4 the EELFS phase shift derived by Fourier filtering the first shell of F(R) of Fig.2,with the Teo and Lee (8) calculations. The two curves are almost parallel in a wide k-range supporting the view that the main contribution cames from p-d transitions, as expected in the framework of the dipole slection rules.

The present EELFS result make us confident of a wider use of the technique as a structural probe,not only for clean surfaces but also for chemisorbed atomic species and interfaces at the initial stage of their formation.

References

(1)M.De Crescenzi,L.Papagno,G.Chiarello,R.Scarmozzino,E.Colavita,R.Rosei and
    S.Mobilio,Solid State Commun.40,613(1981)
(2) M.De Crescenzi,F.Antonangeli,C.Bellini and R.Rosei,Phys.Rev.Lett.
    50,1942(1983)
    R.Rosei,M.De Crescenzi,F.Sette,C.Quaresima,A.Savoia and P.Perfetti,
    Phys.Rev.B28,1161(1983)
(3) G.Chiarello,E.Colavita,M.De Crescenzi and S.Nannarone,Phys.Rev.B29,
    4878(1984)
(4) P.H.Citrin,P.Eisenberger and R.C.Hewitt,Phys.Rev.Lett.41,309(1978)
(5) M.De Crescenzi,G.Chiarello,E.Colavita and R.Memeo,Phys.Rev.B29,3730(1984)
(6) D.Denley,R.S.Williams,P.Perfetti,D.A.Shirley and J.Stöhr,
    Phys.Rev.B19,1762(1979)
(7) R.D.Leapman,L.A.Grunes and P.L.Fejes,Phys.Rev.B26,614(1982)
(8) B.K.Teo and P.A.Lee,J.Am.Chem.Soc.101,2815(1981)

X-Ray Excited Optical Luminescence (XEOL): Potentiality and Limitations for the Detection of XANES/EXAFS Excitation Spectra

J. Goulon[1,2], P. Tola[1], J.C. Brochon[2], M. Lemonnier[2], J. Dexpert-Ghys[3], and R. Guilard[4]

1. Introduction

X-Ray Excited Optical Luminescence (XEOL) has been extensively used during the past decade in analytical chemistry, especially for quantitative determinations of the fractional ppm amounts of rare earth elements contained in several host matrices [1,2]. Recent experiments have established that such a XEOL emission, when it is strong enough, does offer an alternative route for detecting EXAFS / XANES spectra [3-5]. We have already pointed out that, under favourable conditions, site selective EXAFS spectra could be obtained by this new method [5]. In the present paper, it will be shown from several examples that on the contrary, intersystem energy transfers resulting in a loss of site specificity can give rise to strong signal enhancements. Requirements for observing positive / negative edges and site selectivity will be also outlined below.

2. XEOL Spectroscopy

As a preliminary to the following discussion of optical EXAFS excitation spectra, we like to emphasize the often neglected potentiality of XEOL spectroscopy. We found that multisite rare earth systems usually exhibit different, site specific responses to u.v. vs. X-ray excitations. For instance, in c-Y_2O_3 doped with Eu^{3+} cations distributed between the two crystallographic sites of symmetry C_2 and C_{3i}, we found [6] that three lines assigned to the centrosymmetric site C_{3i} get strongly enhanced under X-ray excitation.

Our investigations were not restricted to solid inorganic matrices only. Figures 1a and 1b show the XEOL spectra of frozen solutions (liquid nitrogen temperature) of two "regular" metalloporphyrins in toluene : both fluorescence of the Q bands ($S_1 \rightarrow S_0$) and phosphorescence of the lowest triplet state ($T_1 \rightarrow S_0$) are apparent while the strong emissions below 500 nm are mostly solvent lines. We observed that for a pure polycrystalline sample of Zn:TPP (TPP = mesotetraphenylporphyrin) (Fig.2a), those solvent lines indeed disappear and the phosphorescence lines as well. Surprisingly, a

---

[1]Laboratoire de Chimie Théorique, E.R.A. 22 au C.N.R.S.,
Université de Nancy I, B.P. 239, F-54506 Vandoeuvre-lès-Nancy, France

[2]L.U.R.E., Laboratoire Propre du C.N.R.S., Université de
Paris-Sud, Bât. 209C, F-91405 Orsay, France

[3]E.R. 60210 du C.N.R.S., Place A. Briand, F-92190 Meudon Bellevue, France

[4]Laboratoire de Synthèse et d'Electrosynthèse, L.A. 33 au C.N.R.S.,
Faculté des Sciences, 6 Boulevard Gabriel, F-21100 Dijon, France

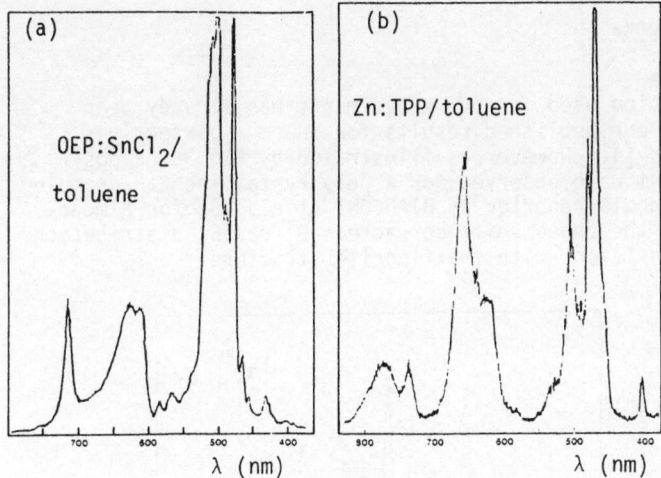

Fig. 1. XEOL spectra of frozen metalloporphyrin solutions in toluene
       a) OEP:SnCl$_2$ b) Zn:TPP

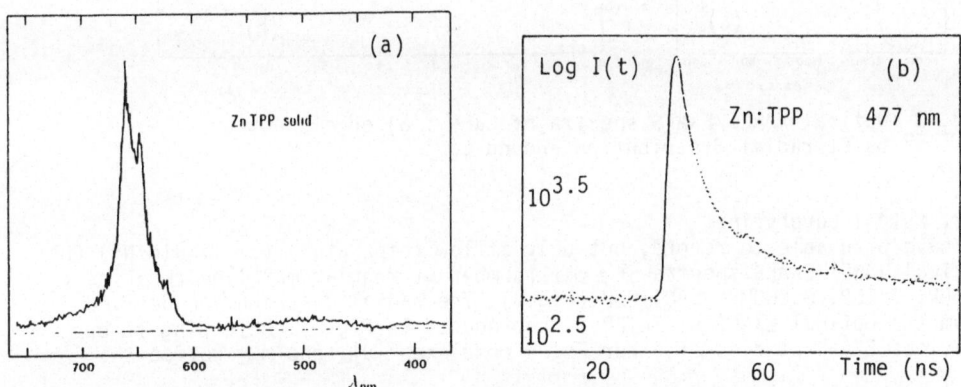

Fig. 2. a) XEOL spectrum of solid Zn:TPP at liquid nitrogen temperature
        b) Time resolved XEOL spectrum of solid Zn:TPP

weak and broad emission was however still detectable at $\sim$ 477 nm. Time resolved XEOL spectra of solid Zn:TPP recorded at a sub-nanosecond time scale
did not rule out yet definitively the possible contribution, but with a
fairly low quantum yield, of the fast decaying ($\tau_f \simeq 1$ ns) B-band ($S_2 \rightarrow S_0$)
fluorescence observed by SOLOV'EV et al. for other zinc porphyrins [7] e.g.
zinc tetrabenzoporphyrin (Zn:TBP). On the other hand, the slow multiexponen-
tial decay reproduced in Fig.2b ($\tau_c \simeq 0.75_3$ µs , $\sigma_c \simeq 0.11_2$ µs) is quite
reminiscent of the results obtained for the frozen solutions and might well
support the classical formation of a toluene solvate Zn:TPP,x toluene [8].
Synchrotron radiation does not only offer a tremendous increase in sensitivity [6], but its pulsed structure makes such time resolved XEOL studies
now practicable : our experiments were carried out at LURE, on the DCI machine running in parasitic beam time sessions with a typical pulse duration
τ_{puls}=0.96 ns and a pulse repetition rate of τ_{rep}=280 ns.

3. Optical XANES/EXAFS Spectra

3.1. ZnO and CaF$_2$ standards

The experimental configuration used for our experiments has already been detailed elsewhere [5] and our published results for ZnO perfectly agreed with previous reports [4]. However, as illustrated by Fig.3a, a positive edge and EXAFS spectrum were observed for a polycrystalline CaF$_2$ sample instead of the negative signals reported by BIANCONI et al. [3] for a monocrystalline sample. Indeed the phase shift corrected [9] radial distribution shown in Fig.3b is well consistent with the fluorite structure.

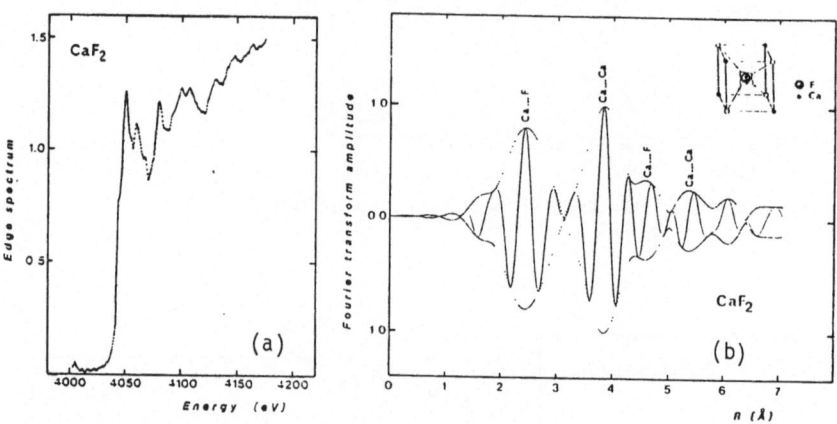

Fig. 3. Optical XANES/EXAFS spectra of CaF$_2$: a) edge spectrum
b) FT radial distribution around Ca

3.2. Metalloporphyrins

We have been able to record, but only at low temperature (T = liquid N$_2$) the optical XANES/EXAFS spectra of a small number of regular metalloporphyrins : ZnTPP, ZnTBP, GaOHOMP, SnCl$_2$OEP (Fig.4). The radial distribution derived from the optical EXAFS of ZnTPP is reproduced in Fig.5 . Referring to our previous EXAFS studies of other ZnTPP complexes [9], one may notice the

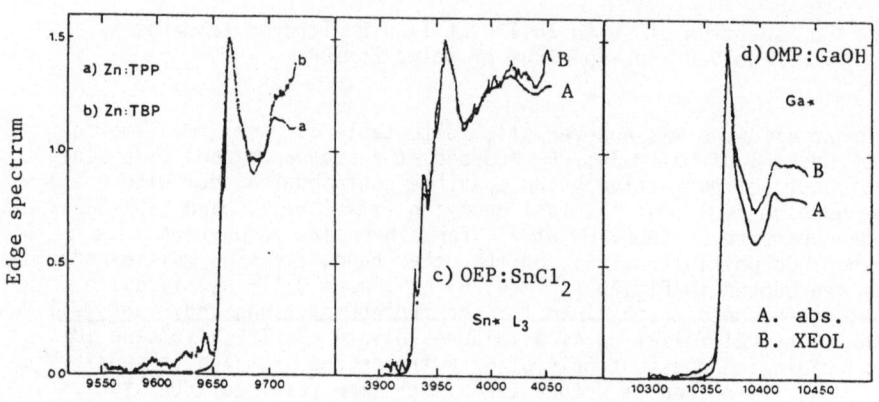

Fig. 4. Optical XANES spectra of metalloporphyrins at liquid nitrogen temperature

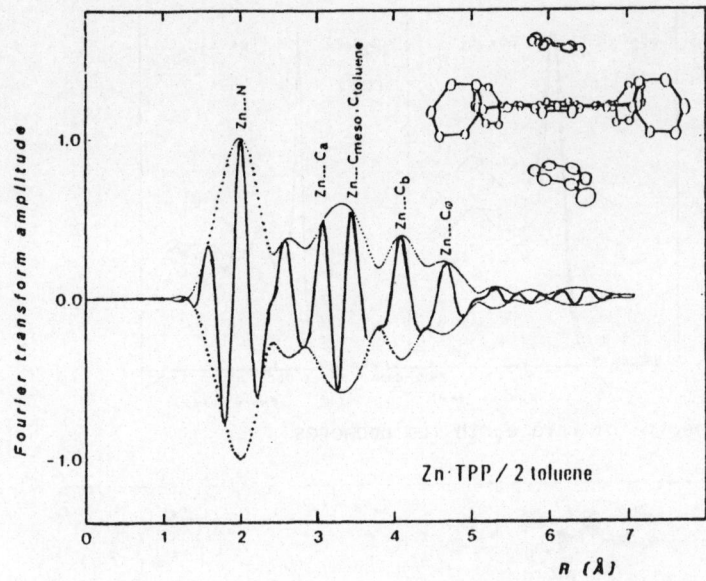

Fig. 5. Radial distribution in the ZnTPP toluene solvate from optical EXAFS data

distortion of the signature of the meso-carbons of the porphyrin in the range of 3.4 Å. One should also keep in mind that in the known bis-toluene ZnTPP solvate, the average distance $Zn...C_{toluene}$ is $\sim$ 3.34 Å [8]. Therefore this result appears well consistent with our interpretation of the above discussed time-resolved XEOL experiments.

3.3. Rare earth luminophores

The strong XEOL emission of TbP_5O_{14} made it relatively easy to detect optical XANES/EXAFS spectra of the $Tb*(L_3,L_2)$ edges (Fig.6). Of greatest interest to us was, however, the detection of the XANES/EXAFS spectra at the $La*(L_3,L_2,L_1)$ edges in $La_{0.98}Tb_{0.02}P_5O_{14}$ (Fig.7a). As confirmed by our XEOL spectra, the optical emission requires the primary excitation of the terbium activator, no direct excitation of the La sites being possible. In the present case, we observed for the first time not only intersystem energy conversion but also EXAFS information transfer without loss of coherence. Although the proximity of the $La(L_3)$ and $La(L_2)$ edges restricts dramatically the resolution, the radial distributions obtained in the conventional ab-

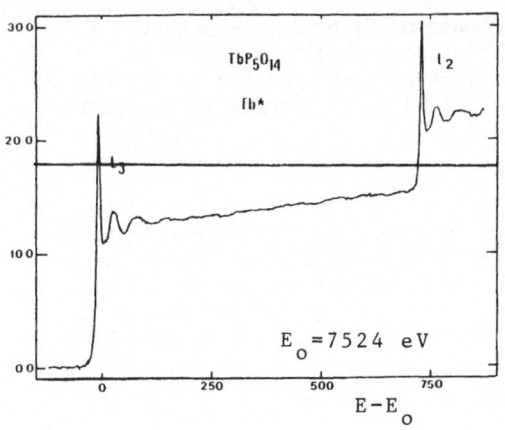

Fig.6. Optical EXAFS spectrum of TbP_5O_{14}

493

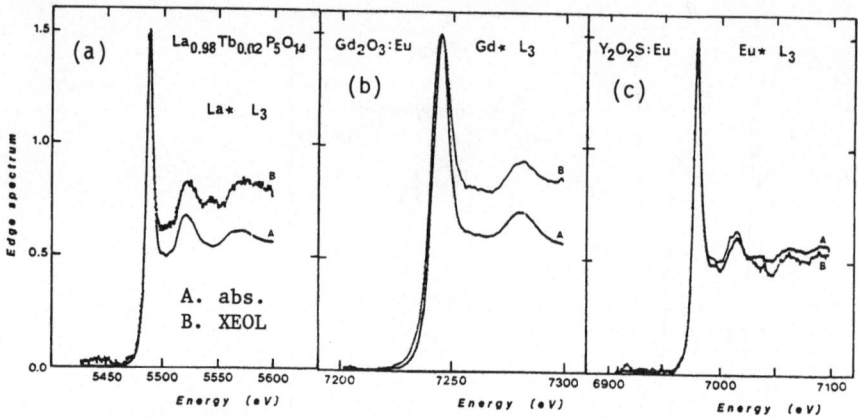

Fig. 7. Optical XANES spectra of rare earth luminophores

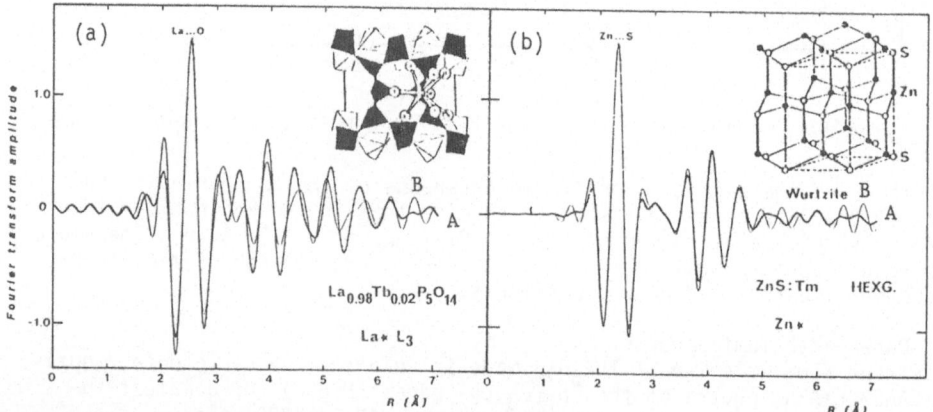

Fig. 8. Comparison of the FT spectra $Im\tilde{x}(R)$ obtained in absorption (A) and optical (B) modes : a) $La_{0.98}Tb_{0.02}P_5O_{14}$ b) ZnS:Tm

sorption mode or by optical detection look rather similar (Fig.8a). Quite in the same way, we have recorded the optical EXAFS spectra of Gd_2O_3:Eu at the $Gd*(L_3)$ edge (Fig.7b). The detection of the $Eu*(L_3,L_2,L_1)$ XANES/EXAFS spectra turned out to be quite possible in Y_2O_2S:Eu (Fig.7c) but required scan accumulations.

For ZnS matrices doped with trace amounts of Tm or Ce, we observed a spectacular enhancement of the matrix emission resulting in quite comfor- table detection conditions of the optical XANES/EXAFS spectra at the $Zn*(K)$ edge. Again, no denaturation of the EXAFS structural information is appa- rent from Fig.8b .

3.4. Site selective EXAFS in a solid dispersion
As illustrated by Fig.9, the optical EXAFS of a solid, homogeneous disper- sion of ZnO (30 mg) in ZnTPP (70 mg) was found [5] to reproduce selectively the radial distribution around Zn* in ZnO. This result is a consequence of the vanishing intensity of the XEOL emission of ZnTPP at room temperature. It gives a demonstration of the potentiality of the method for the selec- tive detection of a specific chemical species in the absence of quenching effects or intersystem conversion.

494

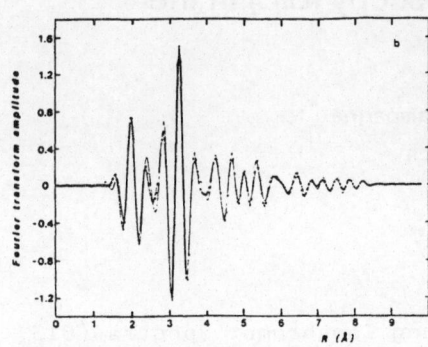

Fig. 9. FT spectra $\text{Im}\tilde{\chi}(R)$ of pure ZnO (dotted line) and of the ZnO/ZnTPP mixture (full line) both recorded in the optical mode

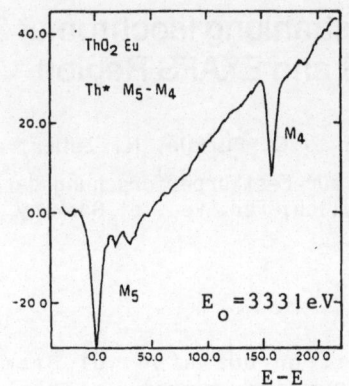

Fig. 10. Optical EXAFS spectrum of ThO_2:Eu at the Th* (M_5,M_4) edges

4. Discussion

An analytical formulation has been established in reference [5] for the XEOL emission of powdered samples in both the "reflection" (RL) or "transmission" (TL) configurations. If μ_0 refers to the edge jump of the X-ray linear absorption coefficient, thus the sign of $\partial RL/\partial\mu_0$ and $\partial TL/\partial\mu_0$ is governing the occurrence of positive/negative optical edges. We found that in the "reflection mode", positive edges only should be observed while both positive edges ($\mu x < 2$) or negative edges ($\mu x \gg 1$) are to be encountered for the transmission mode. These predictions are supported by our experimental observations : e.g. negative M_5,M_4 edges of Th* have been recorded (Fig.10) for a thick sample in transmission. For optically transparent materials (e.g. CaF_2 monocrystals) some mixing of RL and TL might occur due to reflections at the front and back interfaces. Requirements for site selectivity were also discussed elsewhere [5] using the same analytical formulation.

References

1 A.P. D'Silva, V.A. Fassel, Anal. Chem. 45 (1973) 542.
2 R. Feltin, M. Thomas, DGRST Report 72.7.0808.00221.75.01 (1975).
3 A. Bianconi, D. Jackson, K. Monahan, Phys. Rev. B17 (1978) 2021.
4 F.W. Lytle, D.R. Sandstrom, R.B. Greegor, SSRL Users Group Meeting 1977.
5 J. Goulon, P. Tola, M. Lemonnier, J. Dexpert-Ghys, Chem. Phys. 78 (1983) 347.
6 P. Tola, A. Retournard, J. Dexpert-Ghys, M. Lemonnier, M. Pagel, J. Goulon, Chem. Phys. 78 (1983) 339.
7 I.E. Zaleski, V.N. Kotlo, A.N. Sevchenko, K.N. Solov'ev, S.K. Shkirman, Dokl. Akad. Nauk SSSR 210 (1973) 312.
8 W.R. Scheidt, M.E. Kastner, K. Hatano, Inorg. Chem. 17 (1978) 706.
9 J. Goulon, C. Goulon-Ginet, F. Niedercorn, C. Selve, B. Castro, Tetrahedron 37 (1981) 3707.

Bremsstrahlung Isochromat Spectroscopy (BIS) in the XANES and EXAFS Region

W. Speier, J.C. Fuggle, R. Zeller, and M. Campagna

Institut für Festkörperforschung der KFA Jülich
D-5170 Jülich, Fed. Rep. of Germany

We report on our study of Bremsstrahlung Isochromat Spectra (BIS) and unoccupied density of states of transition and noble metals up to 400 eV above the Fermi level.

Our main aim is to establish quantitative limits on the validity of one electron approximation and related correlation-exchange correction, energy dependent potential, influence of the core hole and phase-shift at higher energies. To this end we also relate BIS with conventional X-ray absorption (XAS) measurements. Bremsstrahlung Isochromat Spectroscopy is complementary to XAS in that the selection rules of XAS do not apply, and consequently it measures the <u>total density of states.</u> It is also <u>not site-selective</u> and <u>does not involve any core-level excitation.</u>

We shall first present results of BIS measurements and DOS calculations in the region 0-10 eV above the Fermi level investigating the unoccupied d-band of transition metals and secondly density of states effects up to 90 eV. Finally we discuss Extended-BIS-Finestructure up to 400 eV above E_F.

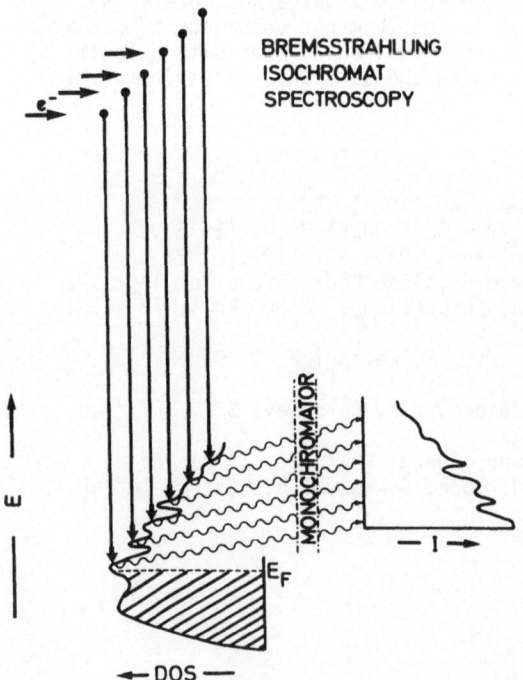

Figure 1 illustrates the principle of Bremsstrahlung Isochromat Spectroscopy (BIS). Electrons incident on a solid can relax into states above the Fermi level via the Bremsstrahlung process. The Bremsstrahlung Isochromat technique measures the intensity of the emitted light at a fixed frequency and by varying the kinetic energy of the electron beam scans the short-wavelength limit of the Bremsstrahlung. This way the direct transitions provide information on the number of states, transition probability and energy above the Fermi level [1].

BIS-Spectra of the 3d-, 4d-, and 5d metals are dominated by direct transitions and in fact over a wide energy range which is in contrast to the earlier ideas of the Bremsspectrum from a solid material, as superposition of emission after successive losses of the electrons with penetration depth [2].

Our experimental set-up, (similar to [3]), is designed for high intensity measurements under Ultra-High Vacuum conditions by using a high current electron beam [4] and a monochromator (fixed at the Al K_α line = 1486.7 eV) with a large solid angle of 0.1 Ster radians [5].

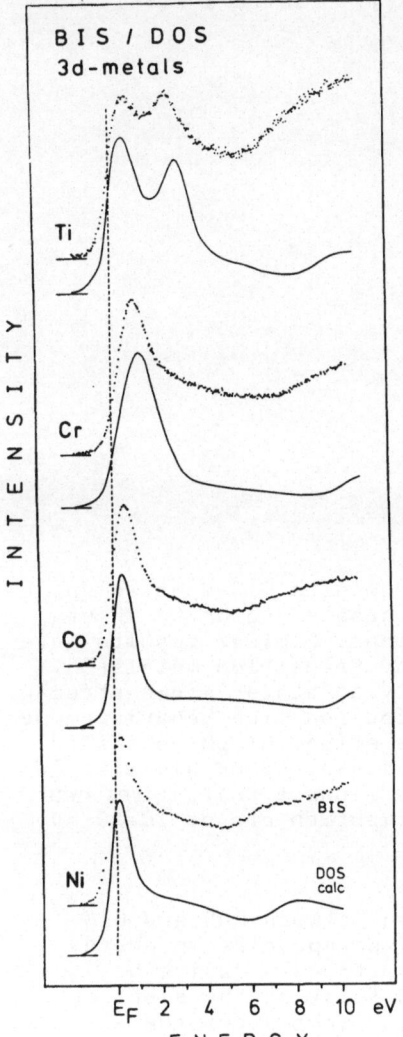

◀ **Figure 2.** The BIS-Spectra of the transition metals up to 10 eV above E_F are dominated by the unoccupied part of the d-band. We made a systematic study of BIS and density of states (DOS) calculations for the 3d and 4d metals [6], finding in fact good agreement between the spectra (...) and broadened DOS (—) as shown here for a number of 3d metals.

The overall shape of the spectra is the same for all transition metals, with the dominating unoccupied d-band just at the Fermi level above a valley due to the flat sp-contribution and a subsequent rise by higher bands.

Experimentally obtained widths for the unoccupied d-part of the bands agree with theoretical numbers and decrease monotonically with band filling, showing, however, small effects of crystal-structure and an enhanced width for the late 3d metals due to magnetic effects. The transition probability (matrix elements plus number of available states) into d-states depends on the position in the periodic table and as it shows a maximum in the middle of the metal row,it causes a stronger contribution of sp-bands for the early transition metals (see e.g. Ti) and the tails on the high-energy side of the d-peak at the end of the row (see e.g. Ni).

It should be emphasized that electron losses do not add any new structure to the spectra,but only give rise to a slowly varying and increasing background.

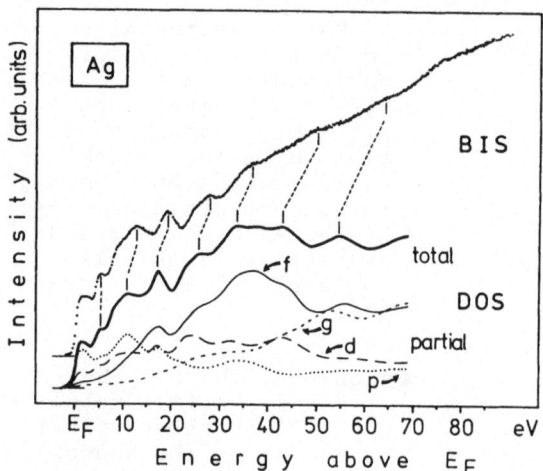

Figure 3. Density of state structures dominate BIS-spectra up to high energies as shown here for the case of Ag up to 90 eV above E_F by comparison to DOS-calculations. Similar results have been obtained for all the noble and many transition metals[7]. Even at these energies the total density of states shows structure and thus deviates from the free-electron like behaviour due to the actual crystal potential [8]. The effect of phase shifts can be observed when comparing BIS spectra of 4d-metals with 3d metals,as the different strength in the f-phase shift gives rise here in the case of Ag to a big f-contribution around 20-50 eV.

Although there is a close correlation between DOS and BIS-spectra,we find however, substantial discrepancies in energy which can amount up to 5-9 %. These deviations arise because the "electron sea" cannot respond fully,as the electron moves through the solid with high kinetic energy and the exchange and correlation contribution is reduced [9].

The discrepancy between BIS and DOS **does not vary linearly** with kinetic energy. In the case of the noble metals it is small,up to 10 eV and rises then by approximately 5-8 % and for late transition metals we find small differences up to 16 eV and a following increase of 4-7 %. For the description of these states one should use an energy dependent and **not** an energy independent potential. Also note that energy dependent potentials calculated within the free electron gas model [10] do not fully describe this behaviour.

498

Similar discrepancies between theory and experiments have also been found in X-ray absorption measurements [11]. This effect has been attributed to the influence of the core-hole potential. Since BIS does not involve any excitation from core levels, we conclude that the deviations should be really correlated to the **energy dependence of the ground-state potential,** in contrast to the suggestion by MATERLIK et al. [11]. Clearly such effects need to be taken into account for a correct, quantitative interpretation of XANES.

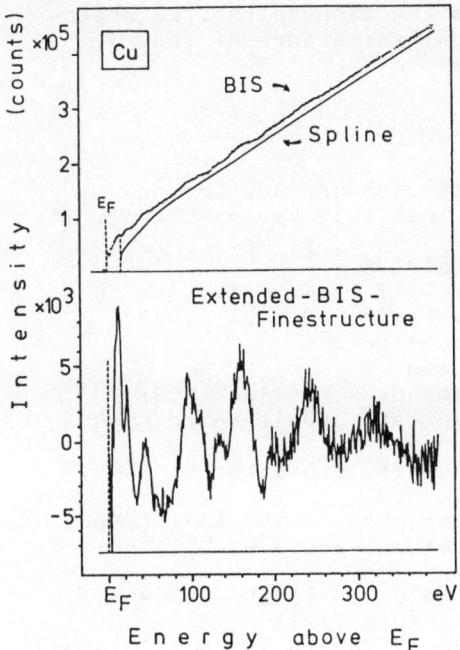

Figure 4. The BIS-Spectrum of Cu up to 400 eV above E_F (upper part) shows fine-structure of the order of 1-2 % which, by using a smoothly varying spline curve, has been extracted, as shown in the lower part of the figure. These structures can be correlated to neighbour-scattering and are similar to EXAFS by comparison with XAS-edge data of Cu [12]. The Fourier transform of the Extended-BIS-Fine structure agrees in in fact with a K-edge analysis as it gives all the scattering shells and shows only small deviations in the neighbour distance and relative intensities.

However, Extended-BIS-Finestructure should be rather thought as a variation in the "total density of states" and a full description should in principle take into account of all the l-quantum numbers. In addition, as no core level excitation is involved, **the phase shift of the absorbing atom is missing,** which, and this is a novel aspect, opens up the possibility of analysing this contribution.

The observation of the finestructures depends on the material, in a way not yet fully understood, as they are measurable in Cu and Ge [13] but **not** for Si, Ag, and Au.

It should be pointed out that the weak intensity (360 sec/channel for our Cu) and the inability of site-selection **implies quite limited technical application** for Extended-BIS-Fine-structure.

Acknowledgement

We would like to thank J.E. Müller, J. Fink, F.U. Hillebrecht, J.W. Allen, B. Lengeler, P.H. Dederichs, J. Zaanen, and G.A. Sawatzky for interesting and stimulating discussions. We gratefully acknowledge the expert technical assistance of Ing. J. Keppels.

References

1. See the review by Y. Baer in "Emission and Scattering Techniques" NATO Advanced Study Institute Series C, Vol. 73, ed. P. Day (1980)
2. D.L. Webster, Phys. Rev. 9, 220 (1917)
   B.R.A. Nijboer, Physica (Utrecht) XII 7, 461 (1946)
3. J.K. Lang and Y. Baer: Rev. Sci. Instr. 50, 221 (1979)
4. F.U. Hillebrecht et al., to be published
5. J.C. Fuggle et al., to be published
6. W. Speier, J.C. Fuggle, B. Ackermann, R. Zeller, F.U. Hillebrecht, K. Szot, and M. Campagna, submitted to Phys. Rev. B (1984) and to be published
7. W. Speier, R. Zeller, and J.C. Fuggle, submitted to Phys. Rev. Lett. (1984) and to be published
8. J.E. Müller and J.W. Wilkins: Phys. Rev. B 29, 4331 (1984)
   J.E. Müller, O. Jepsen, O.K. Andersen, and J.W. Wilkins: Phys. Rev. Lett. 40, 720 (1978)
9. B.I. Lundqvist: Phys. Stat. Sol. 32, 273 (1969); L. Hedin and B.I. Lundqvist: J. Phys. C 4, 2064 (1971)
10. P.O. Nilsson and C.G. Larsson: Phys. Rev. B 27, 6143 (1983)
11. G. Materlik, J.E. Müller, and J.W. Wilkins: Phys. Rev. Lett. 50, 267 (1983)
    L.A. Grunes: Phys. Rev. B 27, 2111 (1983)
12. W. Speier, J.C. Fuggle, J.E. Müller, J.W. Allen, and T.M. Hayes, in preparation
13. I. Sobczak, R. Goldberg, I. Pelka, and I. Auleytner in "Inner Shell and x-ray Physics of Atoms and Solids", p. 529 ed. R.J. Fabian, H. Kleinpoppen, and L.M. Watson, Plenum, New York, London (1981)

X-Ray $L\beta_{2,15}$ Emission Spectrum of Ru in $Ru(NH_3)_6Cl_3$

R.C.C. Perera
Center for X-Ray Optics, Lawrence Berkeley Laboratory, University of California
Berkeley, CA 94720, USA
J. Barth
IBM Thomas J. Watson Research Center, Yorktown Heights, NY 10598, USA
R.E. LaVilla
Quantum Metrology Group, National Bureau of Standards
Washington, DC 20234, USA
C. Nordling
Institute of Physics, University of Uppsala, Box 530
S-751 21 Uppsala, Sweden

One of the broader applications of synchrotron radiation has been to
EXAFS studies for material structure determination, i.e., for an analysis
of x-ray absorption over an extended energy region beyond a core ionization limit. Studies of the near edge structure (XANES) give a different
type of information, characteristic of the local symmetry and electronic
configuration of the absorbing atom. This type of information is reflected also in the x-ray emission spectra, in particular for transitions
involving the valence levels. Examination of the near edge absorption or
the emission spectrum does not require an instrument capable of scanning
a wide energy range with high counting statistics, as does EXAFS; the
needs are rather for good resolution and a reliable calibration of the
energy scale.

Some of the problems of near edge spectra were particularly evident in
our investigation of $Ru-L\beta_{2,15}$ emission from $Ru(NH_3)_6Cl_3$. The
$Ru-L\beta_{2,15}$ emission was measured with a laboratory Rowland circle x-ray
spectrometer with a curved quartz $(10\bar{1}0)$ crystal (radius = 22 inches) in
a fixed position appropriate to the energy range, and a position-sensi-
tive detector which can be positioned along the Rowland circle [1]. The
Ru spectrum was excited mainly by $Sn-L_\alpha$ primary radiation from a Sn
anode in a demountable x-ray tube operating at 13 kV and 120 mA. The resolution of the instrument in this region is 1.5 eV. An accurate calibration of the energy scale was conveniently obtained by measuring a reference x-ray emission line in the same instrumental configuration. In
the present case the $Pd-L_\alpha$ emission line [2] at 2838 eV was used to establish the energy scale. The energy dispersion of the instrument was
determined from the $Cl-K_\beta$ emission spectrum of CH_3Cl between 2810 eV
and 2830 eV [1] and $Pd-L\alpha_{1,2}$ and extrapolated to the energy region of
the recorded emission spectrum.

Figure 1 shows a composite of our experimental $Ru-L\beta_{2,15}$ emission
spectrum and the corresponding $Ru-L_{III}$ absorption recorded by Sham in a
recent investigation using the focussed EXAFS spectrometer at SSRL
[3,4]. In the $Ru-L_{III}$ absorption spectrum the peak labelled A was assigned to a transition to the partially occupied $t_2(4d^5)$ orbital
[3,4]. The emission spectrum corresponds to transitions involving the
same orbital and also, the $e(4d)$ and $a_1(5s)$ orbitals with relatively
higher binding energies. One would usually expect the peak A to have an
energy position at or above the high energy edge of the emission spectrum, in contrast to Fig. 1.

This observation suggested an investigation of the binding energies in
$Ru(NH_3)_6Cl_3$ to precisely place x-ray absorption and emission spec-

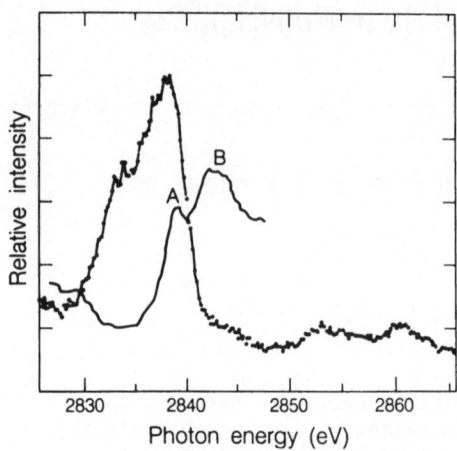

Figure 1. A composite of the measured Ru-L$\beta_{2,15}$ emission spectrum
(left) and Ru-L$_{III}$ absorption spectrum (A & B structures)
from Ru(NH$_3$)$_6$Cl$_3$ displayed on the calibrated emission
spectrum energy scale. Binding energy measurement (see text)
suggest a 2 to 3 eV shift of the Ru-L$_{III}$ absorption to a
higher energy.

tra on a common energy scale. Due to the energetic coincidence between
the Pd-L$_\alpha$ emission line and the Ru-L$_{III}$ binding energy in the metal
[5] we were able to use this energy as a fixed reference point to which
other binding energies from Ru compounds are connected by relative measurements.

We determined the Ru-L$_{III}$ binding energy for Ru(NH$_3$)$_6$Cl$_3$ by
means of an X-Ray Photoelectron Spectroscopy (XPS) measurement using
Ti-K$_\alpha$ radiation on "Kratos XSAM-800" instrument. The samples were
physically mixed with finely divided silver to provide an experimental
silver L$_{III}$ binding energy to correct for charging of the sample. We
determined that the Ru-L$_{III}$ binding energy for Ru(NH$_3$)$_6$Cl$_3$ to be
2840.4 ± 0.8 eV, which is a chemical shift of +2.4 eV relative to Ru metal. The Ru-L$_{III}$ binding energy agrees well with the fluorescence
emission spectrum (Fig. 1) in that the inflection point of the high energy edge is at the same energy position. Thus, we suggest a shift to the
Ru-L$_{III}$ absorption spectrum from Ru(NH$_3$)$_6$Cl$_3$ by 2 to 3 eV towards
higher energy [6].

Beyond the implications of this result for the particular compound
studied, we conclude with some general remarks. A conventional laboratory spectrometer would be useful as a complementary tool for the investigation of near edge absorption spectra,since it provides information
about the occupied states and reliable energy calibration with the help
of reference emission lines. This is of interest for synchrotron radiation experiments in general,where the energy calibration of monochromators is a continuing problem due to the heating of the dispersive
crystals by the absorbed x-ray power.

Acknowledgments

We would like to thank R.D. Deslattes for his encouragement and continuing interest in this study and S. McCartney for the binding energy

measurements. One of the authors (R.C.C.P.), acknowledges support by the U.S. Department of Energy (Contract No. DE-ACO3-76SF00098).

References

1. R.C.C. Perera, J. Barth, R.E. LaVilla, R.D. Deslattes, and A. Henins: (to be published).
2. J.A. Bearden: Rev. Mod. Phys. 39, 78 (1967)
3. T.K. Sham: J. Am. Chem. Soc. 105, 2269 (1983)
4. T.K. Sham and B.S. Brunschwig in "EXAFS and Near Edge Structure," eds., A. Bianconi, L. Incoccia, and S. Stipcich (Springer Verlag, Berlin 1983)
5. J. Barth, R.C.C. Perera, R.E. LaVilla, and C. Nordling: (to be published).
6. Our suggested shift appears to be within the limits of accuracy of the energy calibration of the absorption measurement which was extrapolated from the Pd-L_{III} edge at 3173 eV (T.K. Sham: private communication).

X-Ray Photoconductivity Measurements of EXAFS in Liquids: Techniques and Applications

T.K. Sham and R.A. Holroyd

Chemistry Department, Brookhaven National Laboratory, Upton, NY 11973, USA

1. Introduction

We report recent developments of a novel technique, conductivity measurements of x-ray absorption of liquids [1-3]. The aims are: (a) to develop this as a technique for liquid state x-ray studies, (b) to study the implication of different decay channels to the yield spectrum and (c) to investigate the x-ray ion yield of the pure hydrocarbon liquids [3].

2. Techniques and Applications

Conductivity measurements [1,2] of EXAFS involve the utilization of liquid cells equipped with parallel plate electrodes. Two configurations (with electrodes parallel or perpendicular to the incoming beam) have been used. The photocurrent (i) from the cell is recorded as a function of photon energy. The ratio i/I_0 (I_0 is the current in the photon beam monitor) is given by

$$i/I_0 = Y \cdot F(\text{cell, } h\nu, \text{ } I_0) \cdot A(1 - e^{-\mu(h\nu)t}) \qquad (1)$$

where Y is the ion yield per eV absorbed (Each photoelectron produces from 4 to 140 additional ionizations of solvent molecules, dependent on wavelength, solvent and electric field [3].); F is the collection efficiency of the cell — this is a function of electrode separation, length, voltage, photon energy and photon flux; A is a conversion factor; $\mu(h\nu)$ is the absorption coefficient; t is the thickness.

Conditions can be adjusted so that F = 1 and can be ignored. However, for higher photon fluxes and certain cell geometries, F ≪ 1, in which case it has been shown [2] that:

$$i/I_0 = B\mu^{1/2}t \qquad (2)$$

for a very thin cell $(1 - e^{-\mu t} \approx \mu t)$ and

$$i/I_0 = C\mu^{-1/2} \qquad (3)$$

for a thick cell (total absorption), where B and C are constants. We report below recent results obtained at the Cornell High Energy Synchrotron Source (CHESS). A 0.0364 M, 2,2,4-trimethylpentane (TMP) solution of $Re_2(CO)_{10}$ was used for the measurements. A typical current yield spectrum has a background of several picoamp and a signal of about a nanoamp when a flux of ~ 10^9 photons/sec is used. Spectra shown in

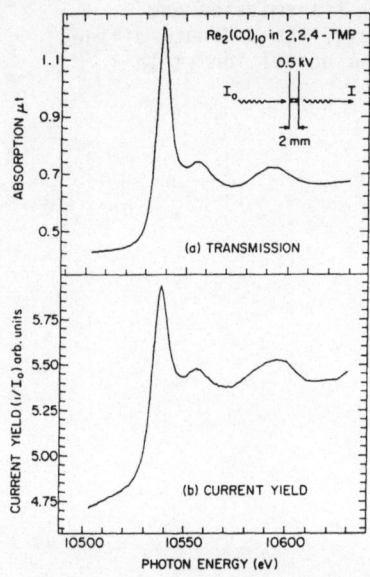

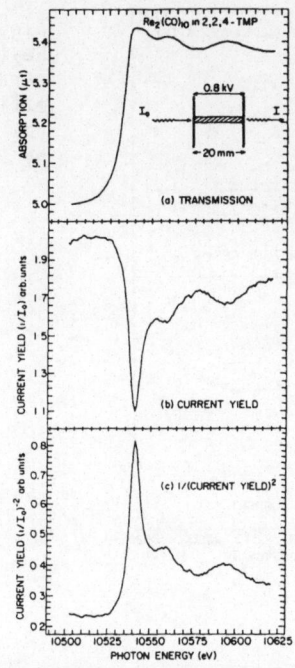

Fig. 1 Re L_{III} edge spectra of $Re_2(CO)_{10}$ in 2,2,4-TMP: (a) transmission and (b) current yield

Fig. 2 ReL_{III} edge spectra of $Re_2(CO)_{10}$ in 2,2,4-TMP in a 20 mm cell: (a) transmission, (b) current yield, and (c) $\mu \sim 1/($current yield$)^2$

Figs. 1 and 2 were obtained under inefficient ($F \ll 1$) conditions for a thin and a thick cell respectively. Several interesting features are immediately noted: First, the i/I_0 curve in Fig. 1b reproduces the same profile observed in the transmission spectrum; second, the photocurrent (Fig. 2b) drops at the edge in the thick cell configuration; and finally, according to (3), Fig. 2b can be converted to Fig. 2c which is nearly identical to the normal spectrum (Fig. 1a). Fig. 2a is the corresponding transmission spectrum which suffers from severe thickness effect [4]. The most interesting observation is seen in Fig. 3 which shows a spectrum obtained with efficiency close to unity ($F \simeq 1$). A drop in current is still observed at the Re L_{III} edge. In this case, (1) applies, the term $(1 - e^{-\mu t})$ is unity both below and above the edge. Thus $i/I_0 = YA$. The change in conductivity must be due to a change in the yield Y [5]. This can be understood in terms of the absorption coefficients μ_{HC} and μ_{Re} of the hydrocarbon and Re respectively. Below the edge, μ_{HC} and μ_{Re} are monotonic and are comparable for this solution. Above the edge, 75% of the absorption is by Re. The decay of the Re L_{III} core hole produces fewer ionization because: (a) twenty percent of the energy goes into fluorescence and most of this is lost and (b) The photoelectron and the Auger cascade electrons have a lower average energy and the yield of ionization decreases with electron energy [3]. In conclusion, the current drop at an edge is always expected in a thick sample (for both $F \ll 1$ and $F \simeq 1$). This technique would greatly facilitate measurements of liquids with low atomic numbers (organo-sulfur and organo-chlorine compounds for example).

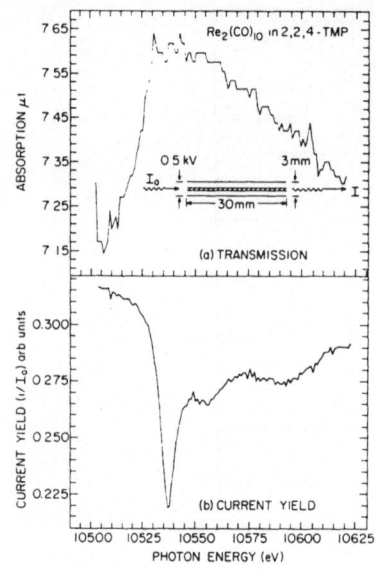

Fig. 3 Re L_{III} edge spectra of $Re_2(CO)_{10}$ in 2,2,4-TMP obtained in a parallel configuration with high efficiency: (a) transmission and (b) current yield, the ordinate divided by 336 gives the actual ion yield Y

Acknowledgement

We thank E. Ritter and K. Walther for the construction of different versions of the cells, and B. Brunschwig, J. Hastings, and S. Heald for their discussions. X-ray absorption sepctra were obtained at CHESS which is supported by the U.S. National Science Foundation. This work was carried out at Brookhaven National Laboratory under contract DE-AC02-76CH00016 with the U.S. Department of Energy and supported by its Division of Chemical Sciences, Office of Basic Energy Sciences.

References

1. T.K. Sham and S.M. Heald: J. Amer. Chem. Soc. 105, 5142 (1983).
2. T.K. Sham and R.A. Holroyd: J. Chem. Phys. 80, 1026 (1984).
3. R.A. Holroyd and T.K. Sham, to be published.
4. E.A. Stern and K. Kim: Phys. Rev. B23, 3781 (1981).
5. R.A. Holroyd and T.K. Sham: Abstracts of 32nd Mtg, Radiation Research Society, Orlando, Florida (March 1984).

Magnetic XANES

Edward Keller and Edward A. Stern

Department of Physics, FM-15, University of Washington, Seattle, WA 98195, USA

We report the results of a new experimental technique for measuring the spin-orbit contribution to x-ray absorption, called *magnetic absorption* in this paper Magnetic absorption is the x-ray version of the magneto-optic Kerr effect, used here on the ferrimagnet alloy $Gd_{18}Fe_{82}$. The size of magnetic absorption is proportional to the final spin density of states [1]. We have measured the magnetic absorption coefficient μ_m at the Gd L_3 edge ($2p_{3\ 2} \rightarrow 5d$ transition) for the alloy to quantify the extent of spin polarization of the Gd(5d) states. We find that its size relative to the normal x-ray absorption coefficient μ_n is: $\mu_m/\mu_n \lesssim 0.002$.

Bennett and Stern have shown that the magnetic absorption coefficient will be proportional to the difference in absorption for right- and left-handed circular polarization [2]. We have obtained circularly polarized x rays by two different means. Synchrotron radiation from a bending magnet is partially circularly polarized when viewed above and below the synchrotron plane. We obtained circularly polarized x rays at CHESS using this method. A silicon quarter phase plate can be constructed [3,4] to take advantage of the high degree of linear polarization of the synchrotron beam. We obtained circularly polarized x rays at SSRL using this second method. For both cases the percentage of circular polarization of the incident x rays was on the order of 5%.

The sample chosen for study was an evaporated 2.4 micron film of $Gd_{18}Fe_{82}$ on a Kapton substrate. The Kapton substrate was water cooled, and a magnetic field of 40 gauss was applied during evaporation. $Gd_{18}Fe_{82}$ is magnetically ordered at room temperature, and the Gd L_3 edge provides a convenient edge for conventional x-ray studies at a synchrotron source. These materials have evoked some interest concerning the extent of the Fe(3d)-Gd(5d) inter-action present [5,6], an effect probed by these experiments. The magnetization curve of the sample was measured on a conventional vibrating magnetometer and the magnetic moment at saturation was found to be 620 emu/cm³, consistent with earlier results cited in the literature [7]. X-ray diffraction studies showed the sample to be amorphous, as is typical for evaporated thin films of the cited composition.

The fact that the magnetic absorption will be proportional to the difference in absorption for right- and left-handed circular polarization provides a convenient way to use lock-in techniques in the experiment. The sense of polarization can be varied at a fixed frequency by alternating the magnetic field used to align the magnetic domains in the sample parallel and anti-parallel to the beam direction. The output of the detector chamber is then input to a lock-in amplifier operating at the aligning magnetic field frequency.

The results for the CHESS and SSRL measurements of magnetic absorption at the Gd L_3 white line can be summarized as follows:

507

CHESS (bending magnet): $\mu_m/\mu_n \lesssim 0.004$

SSRL (silicon phase plate): $\mu_m/\mu_n \lesssim 0.002$

These results indicate that the 5d electrons of the Gd in the alloy are only slightly polarized. We hope to quantify the spin density in the near future.

The authors wish to extend their thanks to the staffs of CHESS and SSRL for their assistance. Research was supported in part by the National Science Foundation under Grant No. DMR80-22221. The SSRL is supported by the DOE Office of Basic Energy Sciences; NSF, Division of Materials Research; and NIH, Biotechnology Resources Program, Division of Research Resources. CHESS is supported by the National Science Foundation.

References

1. L. Erskine and E.A. Stern: Phys. Rev. B 12, 5016 (1975)
2. H. Bennett and E.A. Stern: Phys. Rev. A 137, 448 (1965)
3. M. Hart: Phil. Mag. B 38, 41 (1978)
4. J. Golovchenko et al.: to be published
5. I.A. Campbell: J. Phys. F: Metal Phys. 2, L47 (1972)
6. K. Baschow: Rep. Prog. Phys. 40, 1179 (1977)
7. J. Orehotsky and K. Schroder: J. Appl. Phys. 43, 2413 (1972)

Modulated EXAFS Studies of Ferroelectrics

Olivier Petitpierre and Edward A. Stern

Dept. of Physics, FM-15, University of Washington, Seattle, WA 98195, USA

Yizhak Yacoby

Physics Department, Hebrew University, Jerusalem, Israel

Single crystals of $KTaO_3$ and $KTa_{0.92}Nb_{0.08}O_3$ were subjected to periodically varying electric fields and the corresponding modulation of the EXAFS signal was monitored through a phase-sensitive lock-in amplifier. The EXAFS signal was measured in the fluorescence mode since single crystals thin enough for a transmission measurement were too difficult to fabricate.

Measurements were made from 10 K to room temperature and at various x-ray incidence angles. During cooldown, the capacitance of the samples was monitored to determine the transition to the ferroelectric phase. The transition temperature of $KTa_{0.92}Nb_{0.08}O_3$ is 82 K, while $KTaO_3$ remains paraelectric to the lowest temperatures.

The method of electric field modulation was chosen to eliminate electric depolarization effects. The $KTa_{0.92}Nb_{0.08}O_3$ sample had an electric field applied in the (001) plane and rotated at about 10 Hz to point in the (100), ($\bar{1}$00), and (0$\bar{1}$0) directions; an electric field along the (110) direction oscillating at 10 Hz; and an electric field oscillating along the (001) direction, also at 10 Hz. The x-ray polarization was chosen to have a component along at least some of the applied electric field direction in each case. In these configurations, the modulated EXAFS was measured at the Nb and Ta edges; it is sensitive to displacements of these atoms in all possible crystal directions except the (111) direction.

The output of the phase-sensitive lock-in is, separating the ith coordination shell, $2k^2(\Delta R)^2\chi_i(k)\alpha_i$, where ΔR is the relative displacement of Ta with respect to its neighbors, $\chi_i(k)$ is the EXAFS contributed by the ith shell, and α_i is a factor depending on the electronic shell being excited (K, L, etc.) and on the relative orientations of the x-ray polarization and the displacement with respect to the bonds to the atoms of the ith shell (in this case $\alpha_1 = 1/3$). Our signal to noise ratio was approximately $1-10^4$, which sets the sensitivity of the measurement of such displacements to 0.01 Å. A simple model of a displacement of a charge $2e^-$ to produce the known saturation polarization of $p = 0.8$ C/m^2 leads to an expected displacement of 0.16 Å. Since the signal is proportional to the square of the displacement, this expected value should produce a modulated signal ~100 times as large as the noise. No EXAFS modulated signal was detectable in the measurement, indicating that the actual displacement was ≤ 0.01 Å or was in the (111) direction.

During the EXAFS measurements both the electric fields and the polarizability of the samples were monitored. The $KTa_{0.92}Nb_{0.08}O_3$ sample was driven to saturation at all temperatures, but such was not the case for the $KTaO_3$ sample at lower temperatures. The x-ray beam introduced a large photoconductivity at low temperature in the $KTaO_3$ sample which shorted out the electric field.

509

A more sophisticated calculation using the shell model of H. Bilz, as described in MIGONI et al. [1], and the values for the force constants from BILZ et al. [2] gave an expected displacement of only ~1.5 10^{-2} Å, consistent with our measurements.

Regular fluorescence EXAFS measurements were also made. For the KTa$_{0.92}$Nb$_{0.08}$O$_3$ samples, EXAFS measurements were made at five different x-ray incidence angles (40°, 45°, 50°, 55°, and 60°) and five different temperatures (30, 80, 140, 200, and 300 K), and at both the Ta and the Nb edges. For the KTaO$_3$ sample, measurements at 113, 220, and 300 K (powder sample) were made. Due to the crystalline nature of the samples, the fluorescence data presented many large peaks resulting from Bragg reflections. This introduces distortions which complicate the analysis, especially for data collected at the Nb edge because of the low Nb concentration.

We would like to thank Fred Ellis, Zhang Ke, and Yanjun Ma, who helped us prepare and run the experiments. This research was supported by grants from the National Science Foundation (DMR80-22221) and the US-Israel Binational Science Foundation. We thank the staff of SSRL for their help. The SSRL is supported by the DOE Office of Basic Energy Sciences; NSF, Division of Material Research; and NIH, Biotechnology Resources Program, Division of Research Resources.

References

1. R. Migoni, H. Bilz, and D. Bauerle: Phys. Rev. Lett. 37, 1155 (1976)
2. H. Bilz, A. Bussmann, G. Benedek, H. Buttner, and D. Strauch: Ferroelectrics 25, 343 (1979)

Analysis and Application of Multiple Diffraction Phenomena in Perfect Crystal Monochromators

Z.U. Rek, G.S. Brown, and T. Troxel

Stanford Synchrotron Radiation Laboratory, Stanford University
Stanford, CA 94305, USA

Since the introduction of EXAFS measurements using synchrotron radiation, experimentalists have been troubled by the occurrence of numerous dips in the primary intensity of the beam emerging from the monochromator system. These dips, or "glitches," cause serious complications in the Fourier transform of EXAFS spectra. The physical source of this phenomenon, originally called the "Aufhellung Effect" and interesting in its own right, has been well understood since the early twenties: the intensity of the primary beam is modified by the presence of one or more secondary reflections satisfying Bragg's law simultaneously. However, no universal way of analyzing and dealing with this effect has been proposed so far. In this note we first review and present a critical analysis of a sample of solutions of the problem and then discuss some potentially useful applications of glitches to the absolute energy calibration and precise geometric orientation of crystals.

One can classify the efforts to avoid glitches in EXAFS spectra into two basic groups:

1. Detector improvement. In this method one accepts the presence of glitches in the primary and secondary beam intensity but tries to reduce their effect on the ratio by careful monitoring of intensity. This approach requires high experimental precision, and ideal alignment in particular. Although some successes along this line have been reported, most experimental groups have problems reaching the required optimal conditions.

2. Manipulation of the geometrical parameters of the experimental setup. If the area of highest sensitivity of a given experiment is very limited in energy, say to 100 - 150 eV, one can avoid glitches in this region by careful choice of crystal orientation and azimuthal angle in the crystal plane. Over the last three years we have built up and extensively tested against experiment a software package that produces a local "map" of glitches in any given energy range and for arbitrary crystal type and orientation. Our calculations are based on two equations defining the glitch positions in energy and angle:

$$\sin\Theta = (H/H3)\sin\Theta_o = \alpha_1\cos\Theta_o\cos\phi + \alpha_2\cos\Theta_o\sin\phi \pm \alpha_3\sin\Theta_o \qquad (1)$$

$$\text{where} \quad \alpha_i = (h^*h_i + k^*k_i + l^*l_i)/H^*H_i \ , \quad H_i = \sqrt{h_i^2 + k_i^2 + l_i^2}$$

and the meaning of all geometrical parameters is illustrated in Fig. 1. The "+" sign in (1) corresponds to a multiple reflection for the primary beam and the "-" sign to a possible reflection of the scattered beam. An example of a map is given in Fig. 2. We are currently compiling an "atlas" of the most popular maps used at SSRL. Seven different crystal orientations are calculated, corresponding to three reflection planes: (111), (220) and (400). As

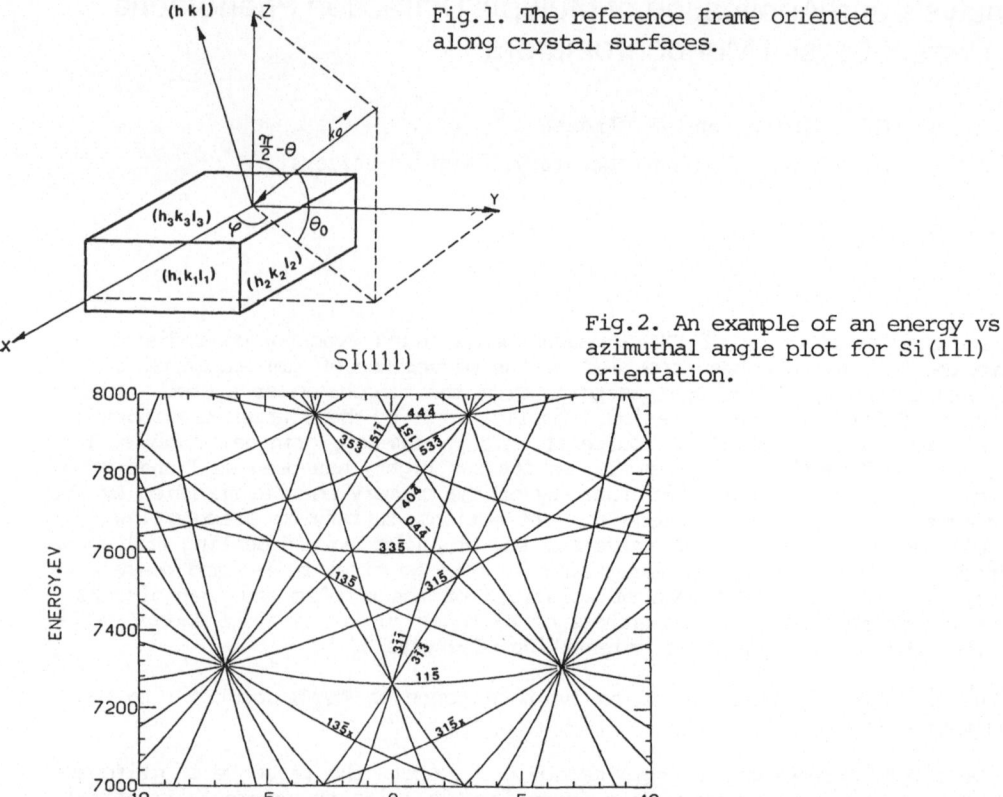

Fig.1. The reference frame oriented along crystal surfaces.

Fig.2. An example of an energy vs azimuthal angle plot for Si(111) orientation.

follows from our calculations, (400) turns out to have the largest glitch-free areas and would be the most useful for EXAFS purposes if not for the weak primary intensity (factor of 5.8 as compared to (111) and 2.4 compared to (220)) which makes it in many cases impractical. The ultimate choice of orientation is largely dictated by the requirements of experiment and depends on the type of studies: transmission or fluorescence, kinetic or static, etc.

In general, the usage of maps to "navigate" among glitches has turned out to be successful in curing some of the problems well localized in energy. However, the applicability of this method is quite limited. In particular, for most physically interesting energy regions no crystal orientation can produce a glitch-free area on the order of 1000 eV. Furthermore, rotating one of the monochromator crystals by 90 degrees, often conjectured as a good method of reducing glitches, actually causes a substantial increase in the number of glitches.

Although it is difficult to avoid the effects of multiple reflections altogether, one can pose a question: How can we use this phenomenon to our advantage? One obvious application would be to use glitches for precise absolute energy calibration in X-ray experiments. It turns out that the relative positions of some dips are highly sensitive to the orientation of the azimuthal angle (in the crystal plane) and allow for a surprisingly precise measurement of azimuthal angles of both crystals, in most cases within an error of 0.01 - 0.005°. We have developed a program to determine these angles

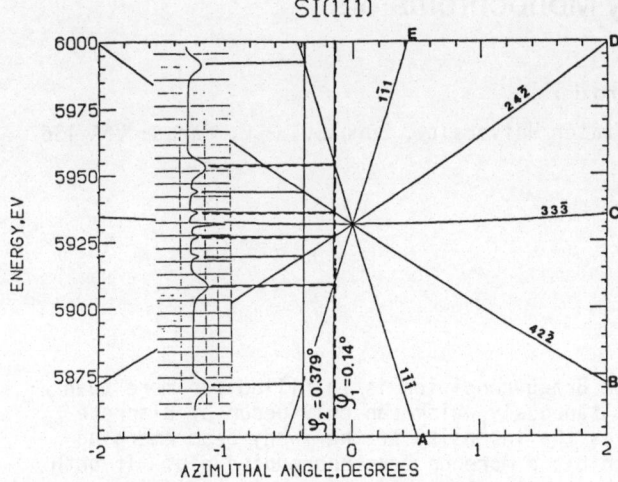

SI(111)

Fig.3. Illustration of energy calibration. On the left hand side there is the insert of original experimental curve with several experimental glitches. The energy vs azimuthal angle diagram has been initially used for glitch identification. The final results illustrate the good agreement between the theory and experiment. Obviously the steepest glitches (A,B,D,E) are the most sensitive to ϕ rotation and allow precise ϕ determination ($\pm 0.005°$). Then the "flat" glitch (C), highly insensitive to ϕ change may be used for precise energy calibration.

from the relative dip positions in primary intensity and then to predict the exact values of energies corresponding to each dip (see Fig. 3). Any useful program calibrating energy should take into account the following effects: 1) refraction correction for both primary and secondary reflection; 2) possible temperature effects; 3) precise azimuthal angle determination; 4) natural width and shape of glitch and the effect of the beam divergence.

We will briefly discuss the first three effects here, deferring a more thorough treatment to a subsequent publication. We get the following approximate modification of Eq. (1) to refraction corrections to primary and glitch reflection (Δ and Δ_o)

$$\sin\Theta_B = (H/H3)\sin\Theta_{oB} , \qquad (2)$$

$$\sin(\Theta_B + \Delta) = \alpha_1\cos(\Theta_{oB} + \Delta_o)\cos\phi + \alpha_2\cos(\Theta_{oB} + \Delta_o)\sin\phi + \alpha_3\sin(\Theta_{oB} + \Delta_o)$$

Note that despite elementary refraction corrections being always positive, the final corrections to glitch positions in energy vary in sign. In general this correction turns out to be negligible but in some cases we obtained corrections as large as 0.4 eV.

The temperature effect does not change Eq.'s (1) and (2) at all as they depend only upon the symmetry of the crystal. However, the relation between energy and angle is influenced by the value of the lattice constant,a, and the resulting shift in energy will be given by the formula

$$\Delta E/E = \Delta a/a = \lambda\Delta T$$

where λ is the linear expansion coefficient and ΔT the change of temperature. In the case of Si, and for $\Delta T = 10°C$ the fractional energy shift would be 0.0000575 corresponding to $\Delta E = 0.5$ ev around Cu edge.

Crystal Glitches of X-Ray Monochromators

K.R. Bauchspiess and E.D. Crozier

Department of Physics, Simon Fraser University, Burnaby, B.C. Canada V5A 1S6

1. Introduction

Crystal glitches arise when the Bragg condition is fulfilled for more than one set of lattice planes simultaneously, which can only occur at discrete energies [1]. At these energies the intensity of the x-ray beam emerging from the monochromator will exhibit a more or less pronounced dip. If both detectors used in an absorption experiment are linear over a sufficient dynamic range, any variation in intensity of the incident beam will cancel out upon forming the ratio I_0/I. In practice, glitches appear in absorption spectra. For example, in high pressure experiments the large non-linearity introduced by the absorption of the anvils favours glitches. This paper indicates the advantages of avoiding certain glitches by selecting the crystallographic axis about which the monochromator crystal(s) rotates. It also demonstrates that the locations of glitches can be used to calibrate the x-ray energy scale. A related method of calibration has been described by Acrivos et al. [2].

2. Glitch Locations

Let us refer to the Bragg planes that characterize the principal reflection of the monochromator with unprimed quantities, and any other set of planes that fulfills Bragg's law simultaneously with primed quantities. The basic equation is:

$$d_{hk\ell} \cdot \sin\theta_{hk\ell} = d_{h'k'\ell'} \cdot \sin\theta_{h'k'\ell'}$$

From this we can calculate the energy of a glitch as shown in [3]. The problem is completely specified by two continuous parameters ϕ and Ω (see Fig. 1) and the three discrete parameters h', k' and ℓ'. Fig. 2 shows the position of the glitches along three axes of rotation, which are axes of symmetry of the (220) crystal. The ordinate is essentially F^2 which is taken as a measure of the relative strength of the glitches. F is given by the product of the atomic form factor and geometrical structure factor.

3. Energy Calibration

Let us assume that the monochromator is miscalibrated by an amount $\Delta\theta_{hk\ell}$. If only one monochromator crystal is used then there are three parameters to be fitted to the observed glitches $\Delta\theta_{hk\ell}$, ϕ and Ω. If we have two separate monochromator crystals, there is a further value of Ω possible corresponding to the second crystal. However, if ϕ deviates only slightly from 90° it merely produces a shift of the Ω-scale. Since practically $|\phi-90°|$ is small we can therefore absorb ϕ in Ω. A linear least squares fit will yield the shift in Bragg angle. Fig. 3 shows a quartet of glitches above the Sm

514

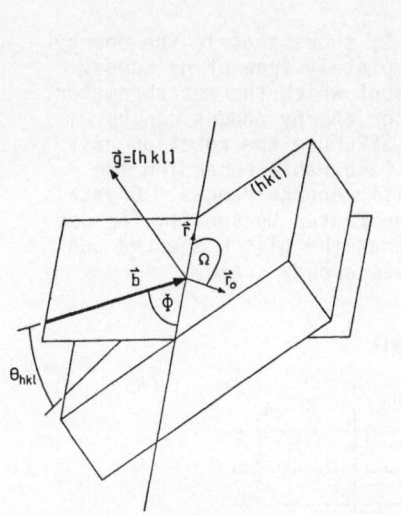

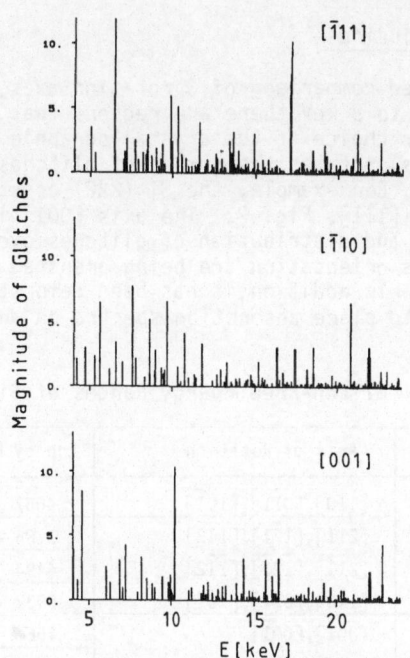

Fig.2

Fig. 1. Monochromator crystal and incident x-ray beam.

Fig. 2. The location of glitches for three different rotation axes of a Si (220) monochromator. The ordinate is $10^{-3} \sum F^2_{h'k'\ell} \sin^2\alpha$ for each glitch where α is the angle between the beam reflected off an (h'k'ℓ') plane and the polarization direction of the primary beam. Glitches are detectable when this quantity exceeds 1.5.

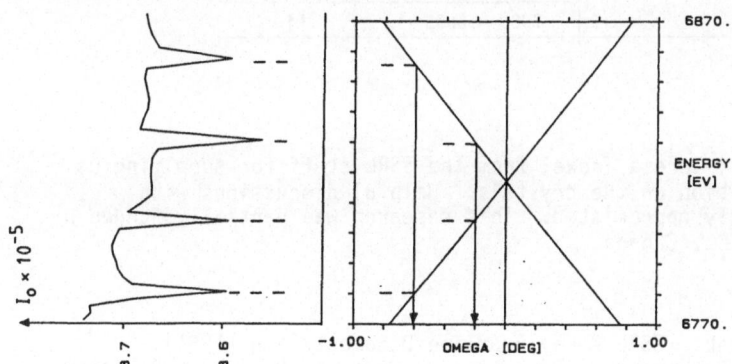

Fig. 3. Glitches in the EXAFS region of the Sm L_{III} edge.

L_{III} edge. A Si (111) monochromator with two separate crystals was used on beam line IV-1 at SSRL in obtaining the data - hence the two values for Ω. If both crystals had been aligned perfectly and if ϕ were 90° there would have been only one glitch. In this example the energy calibration is correct because the horizontal lines and the E vs. Ω curves (right-hand part of Fig. 3) intersect such that only two (and not 3 or 4) values for Ω result. Assuming $\phi = 90°$ we see that one crystal deviates by 0.2° from the optimum orientation (vertical line) and the other by 0.6°.

4. Conclusions

A detailed comparison of 3 rotation axes, Table 1, shows that in the energy range up to 8 keV there are regions that are completely free of glitches. By proper choice of the crystallographic axis about which the monochromator crystal(s) rotates, the number of glitches in other energy ranges can be reduced. For example, the Si (220) crystals at SSRL use the rotation axis $[\bar{1}11]$ or $[1\bar{1}1]$, Fig. 2. The axis [001] is more favourable regarding the strength and distribution of glitches over specific energy ranges. Crystals with this orientation are being prepared at Simon Fraser University for use at SSRL. In addition, it has been demonstrated that the glitch spectra can be used to place absorption spectra on an absolute energy scale.

Table 1: Glitch-Free Energy Ranges of Si Crystals

Crystal	Axis of Rotation	Energy Range [eV]	K-Edge
Si(111)	$[\bar{1}10],[0\bar{1}1],[10\bar{1}]$	4637 - 5931	Ti, V
	$[\bar{2}11],[1\bar{2}1],[112\bar{}]$	3786 - 4494	Ca
	$[2\bar{1}\bar{1}],[\bar{1}2\bar{1}],[\bar{1}\bar{1}2]$	4494 - 5230	Sc
	$[1\bar{1}0],[01\bar{1}],[\bar{1}01]$	5932 - 6817	Cr
Si(220)	$[001],[00\bar{1}]$	4566 - 6046	Ti, V
		7220 - 7949	
	$[\bar{1}11],[1\bar{1}1],[1\bar{1}\bar{1}],[\bar{1}1\bar{1}]$	5293 - 6457	V, Cr
Si(400)	$[001],[0\bar{1}0],[00\bar{1}],[010]$	5286 - 6457	V, Cr
		7610 - 8422	Co
	$[011],[0\bar{1}1],[0\bar{1}\bar{1}],[01\bar{1}]$	4637 - 5369	Ti
		6276 - 7264	Mn
		7909 - 8855	Ni

5. Acknowledgements

We would like to thank Teresa Troxel from the SSRL staff for supplying us with detailed information on the crystals. Helpful discussions with N. Alberding are kindly appreciated. This research was partially funded by a grant from N.S.E.R.C., Canada.

6. References

1. G. Brown and Z. Rek: SSRL X-ray Beamline Documentation (1981).
2. J.V. Acrivos, K. Hathaway, J. Reynolds, J. Code, S. Parkin, M.P. Klein, A. Thompson and D. Goodin, Rev. Sci. Instrum. 53, 575 (1982).
3. K.R. Bauchspiess and E.D. Crozier, to be published.

A Linear Spectrometer Designed for EXAFS Spectroscopy

P. Brinkgreve, T.M.J. Maas, D.C. Koningsberger*, J.B.A.D. van Zon*, M.H.C. Janssen, and A.C.M.E. van Kalmthout

Technological Development and Design Group and C.T.D.

M.P.A. Viegers

Philips Research Laboratories, P.O. Box 80000
NL-5600 JA Eindhoven, The Netherlands

A linear spectrometer, especially designed for EXAFS applications, has been built to maintain the Rowland geometry with the anode focal spot as fixed point on the Rowland circle. The performance of the mechanical construction of this spectrometer allows one to take incremental steps of 5 μm with a resetability of ± 5 μm.

INTRODUCTION

Several EXAFS spectrometers have been described in literature (1-5), which are based upon the Rowland circle configuration with the anode as fixed source point. One design uses 4 A.C. stepping motors to drive the Rowland circle mechanism,which seems to be more complicated than necessary for routine measurements. Other designs involve a much simpler drive-mechanism but lack adequate read-out or positioning control for sufficient reproducibility of the energy scale. Here we will describe a linear spectrometer suitable for EXAFS spectroscopy which has been especially designed to fulfil the following demands:

- Minimum incremental steps of 0.2 eV at 9000 eV with a resetability of ± 0.1 eV.
- Accurate and reproducible position read-out.
- Rapid and easy change of the radius of the Rowland circle, which makes the system adaptable to different, commercially available, monochromator crystals.
- Easy alignment procedure.
- Exchange of monochromator crystals without tedious alignment procedures.
- A possible load on the sample stage of at least 25 kg without any relevant influence on the positioning accuracy. To handle a wide variety of different samples and measuring conditions it must be possible to fit liquid- and in-situ cells and cryostats on the sample stage.

THE EXAFS LINEAR SPECTROMETER

The Rowland circle configuration as shown on the photograph (Fig.1) is based upon a mechanism consisting of the following elements:

1) The main slide which moves along the main guide in order to change the distance between the X-ray focal spot and the monochromator crystal.
2) The monochromator support which is attached to the main slide by means of a rotating axis.
3) The sample/detector slide which moves along a guide which is connected to the main slide by means of the same rotating axis as used for the

---

* Laboratory for Inorganic Chemistry, Eindhoven University of Technology
P.O. Box 513, NL-5600 MB Eindhoven, The Netherlands

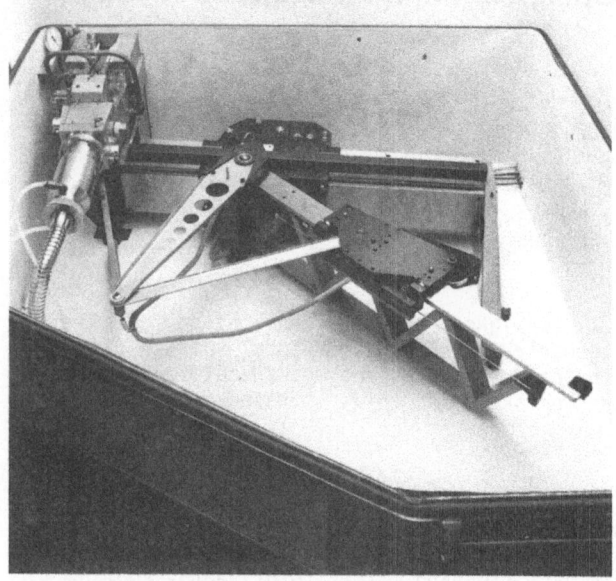

Figure 1 The linear spectrometer with Elliot type GX 21 rotating anode X-ray generator

monochromator support.
4) Three arms of equal length with one common central axis (center of the Rowland circle).
    - One arm from the anode fixed source point.
    - The monochromator arm rigidly attached to the monochromator support.
    - The sample/detector arm from the sample focal spot.

The elements described above maintain the Rowland circle configuration as long as the distances from the X-ray source point to the monochromator axis and from the monochromator axis to the sample focal spot are kept equal within the design specifications.

The angle between the monochromator crystal and the X-ray beam changes during the displacement of the main slide along the main guide. The wavelength (λ) of the photons reflected by the monochromator crystal is linearly dependent on the distance (x) between the crystal and the anode source point:

$$\lambda = \frac{d}{R} x$$

where R is the radius of the Rowland circle and d is the lattice spacing of the diffraction planes used for monochromatization.

The high positioning accuracy of the total mechanism has been achieved by paying full attention to the kinematical and statical design specifications of the total lay-out and its components. In addition all bearings and bearing points are preloaded in order to eliminate all virtual and actual clearance. Each slide is driven by statically and kinematically balanced friction wheels which are coupled to a D.C. motor. Using D.C. motors and ruler systems of high mechanical resolution, the positioning accuracy is realized by a computer electronic feedback system. With a radius R=500 mm

518

of the Rowland circle and Si(400) crystal (d=1.3576 Å) "mechanical" incremental steps of 0.2 eV can be obtained with a resetability of ± 0.1 eV
at the Cu-edge (9000 eV). The sample/detector slide can easily accomodate a
weight of 25 kg without a change in relevant positioning accuracy.

PRELIMINARY RESULTS

A simple Si(400) Johann crystal as monochromator gives for a 3 μm Pt foil
the results as presented in Fig.2.

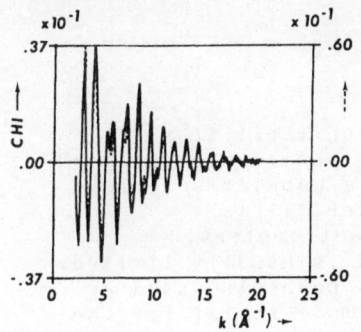 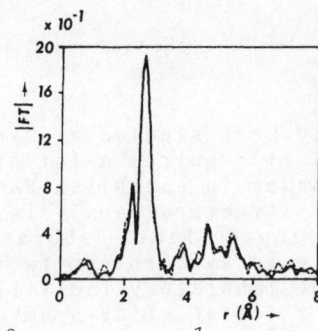

Fig.2 *EXAFS and Fourier transform (k^2, $\Delta k = 2.6\text{-}19\ Å^{-1}$) of Pt-foil (RT).*
Solid line: Eindhoven EXAFS facility, with $\Delta E \sim 14$ eV, 20 mm horizon-
tal crystal spot: Mo anode: 22 kV, 200 mA, 3×10^6 photons/sec.
Dotted line: SSRL, 3 GeV, 40-80 mA.

With a Johansson Si(111) a resolution of 11 eV is obtained at the Cu-edge
with a photonflux of 3×10^7/sec.. Conditions: 70 mm horizontal crystal
spot, Mo-anode 26 kV, 280 mA,. Test experiments with Johannson Si(311) and
Ge (311) crystals will be carried out in due course.

LITERATURE

1) P. Georgopoulos and G.S. Knapp, J. Appl. Crystal. 14 (1981) 3.
2) G.G. Cohen, D.A. Fisher, J. Colbert and N.J. Shevchik, Rev. Sci. Instrum.
   51 (1980) 273.
3) S. Khalid, R. Emrich, R. Dujari, J. Schultz and J.R. Katzer, Rev. Sci.
   Instrum. 53 (1982) 22.
4) A. Williams Rev. Sci. Instrum. 54 (1983) 193.
5) W. Thulke, R. Haensel and P. Rabe, Rev. Sci. Instrum. 54 (1983) 277.

Laboratory EXAFS Spectrometer with a SSD and a Fast Detection System

Yasuo Udagawa and Kazuyuki Tohji

Institute for Molecular Science, Myodaiji, Okazaki, Aichi 444, Japan

1. Introduction

EXAFS has mainly been studied at synchrotron facilities for the reason that it requires a lot of x-ray photons to get good S/N ratio. In order to establish EXAFS as a popular spectroscopic tool for structural analysis, however, it is necessary to develop reliable laboratory spectrometers, since the access to synchrotron facilities is usually limited. An importance of laboratory facilities has been discussed by STERN [1], and a lot of efforts have been made so far for the construction [2-5].

Compared to synchrotron radiation, x-rays from conventional sources have two disadvantages; weak intensity and sharp characteristic lines overlapping the continuous background. The former problem can partially be solved by the use of a bent crystal monochromator, which can collect over a million photons/sec with resolving power adequate for EXAFS analysis. In order to overcome problems inherent to the latter, a SSD and a fast detection electronics have been employed in this study. Combining these, performances adequate for transmission measurement have been obtained.

2. Monochromator and Data Collection System

The basic design of the monochromator and data collection system have been described previously [5]. Principally it consists of a rotating anode x-ray generator (Rigaku RU-200, 60 kV, 200 mA), a Johansson cut bent crystal, and detection electronics, as is shown in Fig.1. A SSD modified for fast response with a sacrifice of energy resolution was employed as a detector. I_0 and I are measured by changing the sample and reference by a sample positioner. Since the resolution of the SSD is enough for the complete discrimination of the undesired reflections, a high voltage can be applied to the anode, without worrying about the contamination of the harmonics. Therefore, in addition to the high efficiency of the curved monochromator, more intense flux can be utilized by the use of the SSD. It also makes it possible to use higher order reflections, if high resolution is required or when high energy region is of interest. For the cases if higher order reflection is negligible, a semi-transmitting ion chamber can be used for simultaneous monitoring of I_0, reducing the measurement time to less than a half. Because the movement of the crystal and the receiving slit is controlled mechanically and the motions required for scanning are just two translations, the alignment is very easy.

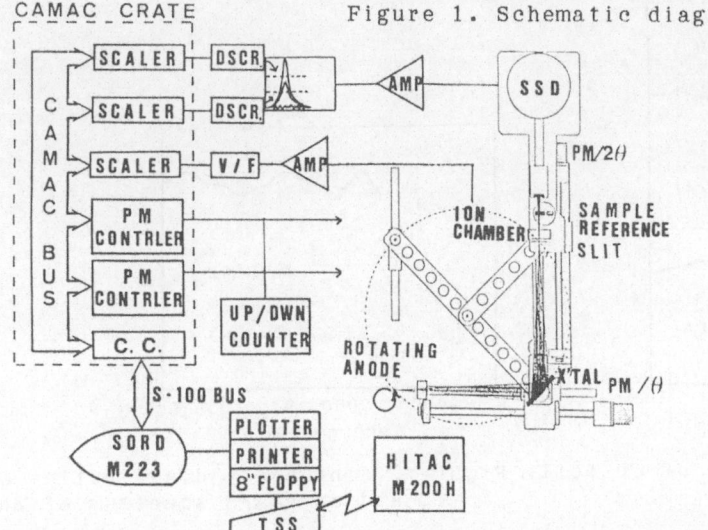

CAMAC CRATE   Figure 1. Schematic diagram of the system

3. Performances

So far LiF(220), Si(220), Ge(220), and Ge(311) crystals have
been tested. Photon numbers obtained by these crystals with
slit width of 100 μ is shown in Table 1, as well as the
resolving power as HWHM of W Lα line. Of these Ge(311) is the
best compromise between the resolution and the reflectivity
for the low energy region, and the second order reflection of
Ge(220) is used for high resolution work because it still has
a fairly high reflectivity.
     As an example, an EXAFS spectrum of ZnO is shown in Fig.2.
The ion chamber was used as I_0 detector, and an Ag target
operated at 20 kV and 100 mA was employed with Ge(311)
crystal. Also shown in Fig.2 is the variation of I_0 in the
energy region measured. Due to W Lβ lines and a tracking
error of the monochromator, the intensity distribution is far
from flat, but the effect is completely removed in the EXAFS
spectrum. In this measurement typically 5 million counts were
accumulated for 100 sec at each data point.
     Fig.3 is the near edge region of the K absorption spectrum
of Cu by the use of Ge(440) reflection in order to show the
resolving power of the monochromator. Both I_0 and I were
detected by the SSD and the intensity of I_0 amounted to about
50000 photons/sec.

Table 1. Photon numbers (fundamental/harmonics) obtained with
the monochromator by various crystals. X-ray generator is
operated at 20 kV, 10 mA.
Resolution is estaimated by HWHM of W Lα line at 8.398 keV.

	7 keV	9 keV	11 keV	Resolution
Si(220)	3200/2800	21000/1000	50000/700	5 eV
LiF(220)	2700/1600	31000/700	100000/2700	5.5
Ge(220)	8500/1500	47000/1000	75000/5000	5.5
Ge(311)	2300/80	12600/50	24000/150	4.5

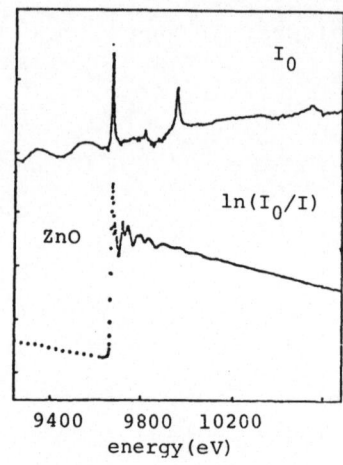
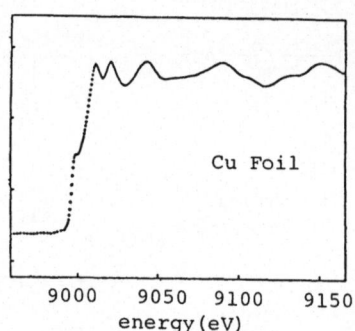

Figure 3. Near edge of Cu foil. Figure 2. Intensity distribution of I_0 and an EXAFS spectrum of ZnO

An application of the system for the study of a catalyst preparation procedure is described elsewhere in this conference.

References.
1. E.A.Stern: AIP Conference Proceedings, No.64, 1980.
2. S.Khalid, R.Emlich, R.Dujari, J.Schultz, J.R.Katzer: Rev. Sci.Instrum. 53, 22 (1982).
3. A.Williams: ibid, 54, 193 (1983).
4. W.Thulke, R.Haessel, P.Rabe: ibid, 54, 277 (1983).
5. K.Tohji, Y.Udagawa, T.Kawasaki, K.Masuda: ibid, 54,1482(1983).

Multi-Detector Fluorescence EXAFS Apparatus Applied to Very Thin Film Studies

H. Oyanagi, T. Ishiguro and H. Tanoue

Electrotechnical Laboratory, Umezono, Sakura-mura, Niihari-gun
Ibaraki 305, Japan

T. Matsushita and K. Kohra

Photon Factory, National Laboratory for High Energy Physics, Ohomachi
Tsukuba-gun, Ibaraki 305, Japan

1. Introduction

Interest in structural characterization of very thin films has rapidly
grown because of its importance in recent electronic device technology.
Fluorescence-detected x-ray absorption spectroscopy (XAS) using synchrotron
radiation as a light source can be a powerful structural tool for thin
films and interfaces in that: (1) the local structure and bonding states of
a particular atomic species are obtained over a wide range in thickness
from a few thousand Å down to even a few Å, (2) because XAS is based on the
short range order, disordered systems formed by ion beam implantation
at low substrate temperatures are studied, (3) since surface cleaning or
in-situ sample preparation in UHV are unnecessary, thin film samples prepared by various epitaxial growth techniques are characterized.

In this paper, we present the design, performance, and feasibility of
the newly built multi-detector system (MDS) for fluorescence-detected XAS
experiments. To evaluate the performance of MDS and its feasibility as a
structural tool for thin films, the ion-beam-induced structural modification of thin nickel layers on Si(100) were studied.

2. Experimental results and discussion

2.1 System configuration

For fluorescence-detected XAS experiments, the detector system should be
flexible with respect to: (1) the dynamic range in photon counting, (2) the
detector geometry to cover a wide solid angle, (3) the energy discriminat-
ing capabilities if necessary [1]-[3]. However, these requirements are
hardly attained simultaneously by any one kind of detector. Moreover,
these requirements depend on the concentration and/or matrices which tend
to vary sample by sample. Therefore, a flexible multi-detector counting
system has been designed and built to cover a wide range in fluorescence
detection rate and to provide the optimum detection scheme.

Three kinds of detector have been newly developed for the MDS: (1) plastic scintillation detectors for a system ranging from 10^{-2}-10^{-3} in atomic
fraction where the fluorescence signal-to-background (s/b) ratio $\gg 1$, (2)
NaI scintillation detector for a more dilute system down to 10^{-4} in atomic
fraction where the s/b ratio >1, (3) a high-count-rate Si(Li) detector for
the system where the s/b ratio <1. The most appropriate detectors are
chosen and arranged around a sample to achieve the best performance. Nine
plastic scintillation detectors, nine NaI scintillation detectors, and
one Si(Li) detector have been prepared. Efforts have been made to integrate the counting electronics. A multi-channel delay line amplifier for a

NaI detector and a multi-channel SCA were newly designed and built into a NIM and CAMAC module.   The counting electronics and the double crystal monochromator are controlled by ECLIPSE S/140 computer with a CAMAC system via a branch driver.

2.2 Performance

The x-ray beam line optics [4] features a constant beam-height double crystal monochromator and a bent cylindrical mirror.   The photon flux of 10^9-10^{10} photons/sec was achieved at 9 keV with an energy resolution of 2 eV. The approximate beam size at the focal point was 4.5 mm x 2 mm.   At the moment (Phase I), the experiment is limited to the energy range below 10 keV.   However, in Phase II, a sagitally bent crystal monochromator will expand the energy range to cover from 4 keV to 30 keV.   A computer software feedback is used to keep the Bragg planes of the two monochromator crystals parallel.   This "software feedback" peak search method is also used to reduce the higher harmonics components by a slight detuning. Figure 1 illustrates the top view (a) and the side view (b) of the MDS.

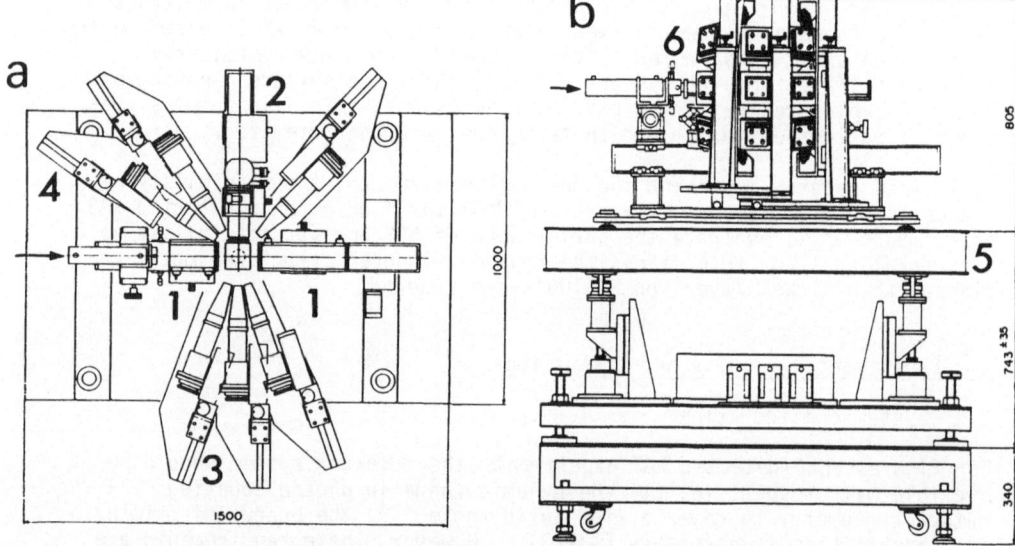

Fig. 1   Top view (a) and side view (b) of the multi-detector
         fluorescence EXAFS apparatus
1: Ion chamber, 2: Cryostat, 3: NaI scintillation detector,
4: Plastic scintillation detector, 5: Lifting table, 6: Xy slit

The fluorescence extended x-ray absorption fine structure (EXAFS) spectra of thin nickel layers (50-1000 Å in thickness) deposited on Si(100) were measured on the Ni K-edge to estimate the sensitivity.   The typical s/b ratio for 50 Å thick Ni/Si(100) was 5 with the total count rate of 10^5 photons/sec using nine NaI detectors subtending 18% of 4π steradian when the storage ring was operated at 2.5 GeV, 100 mA.   Since the s/b ratio would be ~1 for 10 Å thick Ni/Si(100), these results imply that even *a monolayer experiment is feasible* with the use of an x-ray filter and preferably, an increase of the incident beam intensity by an order of magni-

tude.   A multipole wiggler magnet matches such an experiment to enhance
the brightness by orders of magnitude.

2.3 Application to thin film studies

The formation of nickel silicide by thermal annealing usually takes three
forms: Ni_2Si, NiSi, and $NiSi_2$ which grow sequentially from thin nickel layers on Si [5].   In this study, the ion-beam-induced structural modification
of thin nickel layers on Si(100) has been investigated by fluorescence EXAFS.
Due to the absence of long range order in the modified structure by ion
beam bombardment at low substrate temperatures, structural studies using a
glancing angle x-ray diffraction have been hindered.

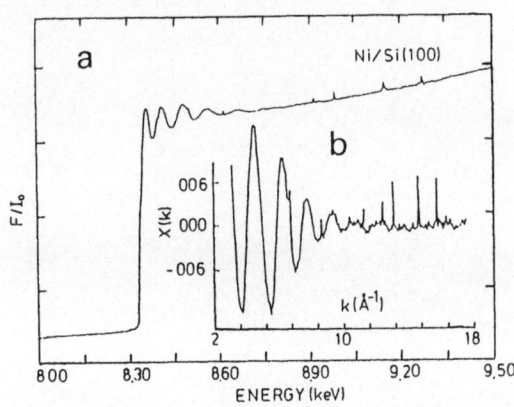

Fig. 2 Ni K-edge absorption
spectrum of ion beam
bombarded Ni/Si(100)
(a) and extracted
EXAFS oscillation (b)

Figure 2 shows the Ni K-edge absorption spectrum (a) and the extracted
EXAFS oscillations (b) for 300 Å thick Ni/Si(100) after the argon ion bombardment at 100 keV ($Ts=100°C$, $Nd=3 \times 10^{16}/cm^2$).   The most striking result
is the similarity between this spectrum with that of Si-rich nickel silicide
$NiSi_2$ [6].   From the k-dependence of the envelope function, it is readily
expected that nickel atom is surrounded by Si near neighbors.   The Fourier
transform further confirmed that Ni-Si bonds are located at 2.3 Å.   One
might be tempted to attribute this phase to amorphous Si-rich silicide, $NiSi_2$.
However, a drastic change in the short range order has been observed for
Ni/Si(100) which was ion beam bombarded at substrate temperatures above 400
°C.   This phase is crystalline and has the similar radial distribution with
that of Ni/Si(100) annealed at 400°C which is characterized by the x-ray
diffraction patterns as Ni-rich silicide, Ni_2Si.

These results imply that *amorphous* Si-rich silicide is formed by the ion
beam bombardment below 200°C while Ni-rich *crystalline* silicide grows above
400°C.   There are also possibilities that nickel atoms are knocked on to
Si lattice by the ion bombardment and take the interstitial voids.   Details
of this work will appear elsewhere.   In summary, we have shown that fluorescence EXAFS can serve as a powerful structural tool for thin films with
the thickness of a few tens of Å even with the absence of long range order.

525

References

1. J. Jaklevic, J. A. Kirby, M. P. Klein, A. S. Robertson, G. S. Brown, and P. Eisenberger: Solid State Commun. 23, 679 (1977)
2. J. B. Hastings, P. Eisenberger, B. Lengeler, and M. L. Perlman: Phys. Rev. Lett. 43, 1807 (1979)
3. R. G. Shulman, P. Eisenberger, B. K. Teo, B. M. Kincaid, and G. S. Brown: J. Mol. Biol. 124, 305 (1978)
4. Photon Factory Activity Report, 1982/83, V-8 (1984)
5. N. W. Cheung, R. J. Culbertson, L. C. Feldman, P. J. Silverman, K. W. West, and J. W. Mayer: Phys. Rev. Lett. 45, 120 (1980)
6. F. Comin, J. E. Rowe, and P. H. Citrin: Phys. Rev. Lett. 26, 2402 (1983)

Fabrication of Large-Area X-Ray Filters for Fluorescence EXAFS Measurements

Joe Wong

General Electric Corporate Research and Development, P.O. Box 8
Schenectady, NY 12301, USA

1. Introduction

X-ray transmission filters are made in a variety of ways. Ductile metals may be rolled to thin foil. Non-ductile metals may be ground and polished with abrasive, electropolished or chemically etched. Metals may be electroplated or vacuum-evaporated on thin x-ray transparent substrates. Metal or oxide powder may be slurried with an organic binder and spread on similar substrates or glass, dried and stripped off. Filter papers may be impregnated with solution, then dried. The latter two procedures involve the use of powder materials and are applicable for fabricating small-area filter (~ 1" diameter) used in conjunction with laboratory x-ray tubes. For utilizing high intensity, highly collimated synchrotron radiation sources, filters of size 3" X 3" or larger are required in the case of fluorescence EXAFS measurements, in order to obtain a large solid angle (~ $\frac{2\pi}{10}$ of collection of the fluorescent signals). This paper describes a simple, reproducible technique for fabricating uniform, large-area, pinhole and crack-free filters from materials that do not have a corresponding ductile metal.

2. Method and Procedure

Oxide powders, commonly used to fabrication x-ray filters, [1,2] are ideally suitable for the case of large-area transition metal filters used in fluorescence EXAFS measurements. The absorption length of the element in a corresponding oxide is given by $(\mu_m f \rho_{oxide})^{-1}$ where μ_m is the mass absorption coefficient of the element above an x-ray edge, f is the weight fraction of metal in the oxide and ρ_{oxide} is the density of the oxide. The weight of oxide required [3] to fabricate a filter of n absorption lengths in a dish of diameter d (in cm) is then given by

$$W_{oxide} = \frac{n \pi d^2}{4 \mu_m f} \tag{1}$$

with d = 13.8 cm and using the known values of μ_m, [4] the required weights of oxide for fabricating filters of 3 and 6 absorption-lengths thick are tabulated for the series Ti, V ......Ge in Table 1.

The fabricated steps are as follows. A pre-determined amount of oxide powder is weighed out in an analytical balance. An optimal amount of Duco[R] cement [5] is then weighed out on a 13.8 cm petri dish (tared) on a top loading balance. Acetone is added to the Duco[R] cement in the dish and stirred with a small artist brush equipped with a metal tip to form a uniform, low viscosity liquid. The weighed out oxide powder is now added to the above liquid and stirred gently to form a dispersion. The whole dish is placed over a bright light box so that uniformity of the slurry over the whole dish can visually be monitored. More acetone is added as necessary

527

to maintain a uniform dispersion. Whirling the whole slurry or ultrasonic stirring also helps to achieve a uniform dispersion. The stirring process takes about five minutes for this size dish. The dish is now carefully placed on a <u>level</u> table and the acetone is allowed to evaporate gently at room condition. No forced evaporation (as in a vented hood) should be made since this may cause cracks in the film or incomplete sedimentation of the particles. The surface of the drying dispersion is shiny when wet and dull when dry. After the film acquires a dull appearance and becomes non-deformible under fingernail pressure, an x-acto knife is used to detach the film at the meniscus point around the rim of the dish. A pair of teflon tweezers is used to peel the film radially inward from the cut edge.

Table 1 Weights of Oxide for Various Absorption-Length Thickness. (d = 13.8 cm)

Oxide	$\mu_m(cm^2/gm)$ [4]	f	Weight (gm)	
			n=3	n=6
TiO_2	712.3	.599	1.05	2.10
V_2O_5	625.9	.560	1.28	2.56
Cr_2O_3	551.3	.684	1.19	2.38
MnO_2	471.5	.632	1.51	3.02
Fe_2O_3	415.2	.699	1.55	3.10
Co_3O_4	377.7	.734	1.62	3.24
NiO	334.6	.786	1.71	3.42
CuO	288.9	.798	1.94	3.88
ZnO	265.9	.803	2.10	4.20
Ga_2O_3	215.3	.744	2.80	5.60
GeO_2	194.8	.694	3.32	6.64

The weight ratio of Duco[R] cement to oxide powder and the curing time are found to be critical in determining the ease with which the cast film can be detached from the dish and its quality. Films cured overnight or with low cement/oxide weight ratio usually stick to the dish and tear during the removal process. Optimal weights of Duco[R] cement are 10 and 15 gm for n=3 and 6 respectively.

3. <u>Concluding Remarks</u>

The advantages of this fabrication technique are quite evident. It is simple and fast. Any bad films can be re-processed by merely redissolving in acetone, so that there is no waste of filter material. Also, since high-purity metal oxides such as those of Ti, V, ..., Zn, Ga, Ge are readily available in fine particle powders (-400 mesh), it becomes cost-effective even to fabricate large-area x-ray filters from oxides of ductile metal, since rolling of large sheet of ductile foils are cost (capital investments on rollers) and time-consuming, involving a graded thickness reduction and/or annealing to remove cold work from a previous rolling. Large-area filters of the above metal oxides have been fabricated and good uniformity obtained by optical micrographic examinations.[3] In principle there is no limit to size.

528

4. References

1. B.W. Roberts and W. Parrish in "International Tables for X-ray Crysta-
     raphy", Vol. III. (1968) pp. 77-78.
2. E.P. Bertin, "Principles and Practice of X-ray Spectrometric Analysis"
     Plenum Press, 2nd Edition (1975) p. 400.
3. J. Wong, Nucl. Instrum. Methods (1984) in press.
4. W.H. McMaster, N. Nerr del Grande, J.H. Hallet, and J.H. Hubbell, "Com-
     pilation of X-ray Cross Sections" Lawrence Radiation Laboratory
     UCRL-50/74, Sec. 2, Rev. (1969).
5. A Trademark of the DuPont Company.

Index of Contributors

Achard, J.C. 199
Ainsworth, S. 258
Alberding, N. 30,151
Antonini, M. 349
Antonioli, G. 355
Asakura, K. 190,392

Baldeschwieler, J.D. 80
Balerna, A. 222
Barrault, J. 199
Barth, J. 501
Bassi, F.A. 101
Battioni, J.P. 449
Bauchspiess, K.R. 514
Baxter, D.V. 77
Bazin, D. 195,209
Beinert, H. 105
Berding, M. 33,58
Bernieri, E. 222,314
Beyersmann, D. 136
Bianconi, A. 52,164,
  167,331
Bienenstock, A.I. 280
Binsted, N. 226,297
Blackburn, N.J. 124
Boland, J.J. 80
Bosio, L. 484
Boudart, M. 187,217
Bouldin, C.E. 261,273,
  278,290
Bourdillon, A.J. 86
Bournonville, J.P. 195
Boyce, J.B. 417,426
Brinkgreve, P. 517
Britt, R.D. 130
Brochon, J.C. 490
Brooks, R.S. 258
Brown, F.C. 61
Brown, G.S. 511
Brown, Jr., G.E. 308,336

Davoli, I. 52,331
De Crescenzi, M. 23,487
Del Pennino, U. 23
Dell'Ariccia, M. 164

denBoer, M.L. 264
Deshpande, S.D. 455
Dexpert, H. 195,199,209,
  400,439
Dexpert-Ghys, J. 439,490
Dominguez, D.D. 464
Doniach, S. 33,58
Dumas, T. 291,311
Durham, P.J. 164

Ehrlich, A.C. 464
Eidsness, M.K. 83
Elam, W.T. 464
Elder, R.C. 83
Ellis, F. 287
Emili, M. 317
Evangelisti, F. 26,284
Evans, J. 226
Ewing, R.C. 343

Fagherazzi, G. 317
Farrell, N. 151
Feldman, J.L. 464
Fiorini, P. 284
Fischer-Colbrie, A. 302
Flank, A.M. 321
Foger, K. 187
Fontaine, A. 209,385,
  403,472,479
Fornasini, P. 314,461
Frank, K.H. 394
Frety, R. 449
Froment, M. 484
Fuggle, J.C. 496
Fukushima, T. 192

Gargano, A. 26
Garner, C.D. 136
Georges, A. 449
Gillespie, R.J. 423
Ginsberg, D.M. 61
Giovannelli, A. 164
Glover, B. 368
Goldman, A.I. 442
Goodin, D.B. 130

Goulon, J. 449,490
Greaves, G.N. 226,297,
  435
Greegor, R.B. 176,302,
  343,362,368,371
Grunthaner, F.J. 67
Guglielmi, M. 317
Guilard, R. 490
Guiles, R. 130
Guilleminot, A. 199
Gzowski, A. 331

Haaker, R.F. 343
Haensel, R. 445
Halaka, F.G. 80
Harris, J. 142
Hasnain, S.S. 124,136,
  139,142,145
Heald, S.M. 261,482
Hecht, M.H. 67,264
Hedman, B. 64
Henderson, C.M.B. 297
Heron, A.M. 291
Hida, M. 328,352
Hitchcock, A.P. 43,423
Hodgson, K.O. 33,58,64,
  105
Holroyd, R.A. 504
Holt, C. 139
Horsley, J.A. 46
Huang, H.W. 158
Huffman, G.P. 371
Huggins, F.E. 371
Hukins, D.W.L. 139

Ichikawa, M. 192
Ida, T. 206
Incoccia, L. 26,284,317
Ingalls, R. 374,388
Irlam, J.C. 139
Ishiguro, T. 397,523
Ito, Y. 352
Iwasawa, Y. 190

531

Springer
Proceedings in Physics

Springer Proceedings in Physics is a new series dedicated to the publication of conference proceedings. Each volume is produced on the basis of camera-ready manuscripts prepared by conference contributors. In this way, publication can be achieved very soon after the conference and costs are kept low; the quality of visual presentation is, nevertheless, very high. We believe that such a series is preferable to the method of publishing conference proceedings in journals, where the typesetting requires time and considerable expense, and results in a larger publication period. Springer Proceedings in Physics can be considered as a journal in every way: it should be cited in publications of research papers as *Springer Proc. Phys.,* follow by the respective volume number, page and year.

Volume 1

Fluctuations and Sensitivity in Non-equilibrium Systems

Proceedings of an International Conference, University of Texas, Austin, Texas, March 12–16, 1984

Editors: **W. Horsthemke, D. K. Kondepudi**

1984. 108 figures. IX, 273 pages. ISBN 3-540-13736-X

Contents: Basic Theory. – Pattern Formation and Selection. – Bistable Systems. – Response to Stochastic and Periodic Forcing. – Noise and Deterministic Chaos. – Sensitivity in Nonequilibrium Systems. – Contributed Papers and Posters. – Index of Contributors.

This volume contains the invited papers and a selection of contributed papers and posters presented at the Workshop on Fluctuations and Sensitivity in Nonequilibrium Systems, held in Austin, Texas, in March 1984. The papers deal with the subject from a macroscopic phenomenological viewpoint and address questions ob basic theory, pattern formation, bistable systems, response to stochastic and periodic forcing, noise and chaos, and sensitivity in nonlinear systems. The book contains review articles as well as papers reporting recent theoretical and experimental results. This volume will be of particular value to researchers and graduate students who wish to become acquainted with the field and to obtain an overview of its current state.

Springer-Verlag
Berlin
Heidelberg
New York
Tokyo

EXAFS
and Near Edge Structure

Proceedings of the International Conference, Frascati, Italy,
September 13–17, 1982

Editors: **A. Bianconi, L. Incoccia, S. Stipcich**

1983. 316 figures. XII, 420 pages. (Springer Series in
Chemical Physics, Volume 27). ISBN 3-540-12411-X

Contents: Introduction: Historical Perspective of EXAFS
and Near Edge Structure Spectroscopy. – Theoretical
Aspects of EXAFS and XANES. – EXAFS Data Analysis. –
XANES. – Special Crystalline Systems. – Liquids and Disor-
dered Systems. – Catalysts. – Biological Systems. – Related
Techniques. – Anomalous Scattering. – Related Techniques
– Electron Energy Loss. – Instrumentation. – Index of
Contributors.

The field of X-ray spectroscopy using synchrotron radiation
is growing so rapidly and expanding in such different
research areas that it is difficult to follow the literature. In
fact, EXAFS and XANES are becoming interdisciplinary
methods used in solid-state physics, biology, and chemistry.
This book surveys the panorama of research activity in the
field. It contains the papers presented at the International
Conference on EXAFS and Near Edge Structures held in
Frascati, Italy, in September 1982. This was the first international conference devoted to EXAFS spectroscopy (Extended X-ray Absorption Fine Structure) and its applications.
The other topic of the conference, the new XANES (X-ray
Absorption Near Edge Structure), was shown to have
advanced beyond its infancy in experiment as well as in
theory.

The applications of EXAFS concern the determination of
local structures in complex systems, and the contributions
in this area are arranged according to the different types of
material under consideration: amorphous metals, glasses,
solutions, biological systems, catalysts, and special crystals
such as mixed valence systems, ionic conductors. For each
kind of system the application of EXAFS gives unique information, but EXAFS data analysis is faced with special problems. A chapter is devoted to general problems on EXAFS
data analysis and another to developments of instrumentation for X-ray absorption using synchrotron radiation and
laboratory EXAFS. There is also a chapter on related techniques for local structure studies such as the new X-ray
anomalous scattering using tunable synchrotron radiation
and electron loss spectroscopies.

Springer-Verlag
Berlin
Heidelberg
New York
Tokyo